Civil Engineer's Handbook of Professional Practice

Civil Engineer's Handbook of Professional Practice

Karen Lee Hansen
and Kent E. Zenobia

WILEY
John Wiley & Sons, Inc.

Copyright © 2011 by John Wiley & Sons, Inc. All rights reserved

Published by John Wiley & Sons, Inc., Hoboken, New Jersey
Published simultaneously in Canada

For general information about our other products and services, please contact our Customer Care
Department within the United States at (800) 762-2974, outside the United States at (317) 572-3993
or fax (317) 572-4002.

Wiley also publishes its books in a variety of electronic formats. Some content that appears in print may
not be available in electronic books. For more information about Wiley products, visit our web site at
www.wiley.com.

Library of Congress Cataloging-in-Publication Data:

Hansen, Karen Lee.
Civil engineer's handbook of professional practice/Karen Lee Hansen and Kent E. Zenobia.
    p. cm.
  Includes index.
  ISBN 978-0-470-43841-1 (cloth), ISBN 978-0-470-90161-8 (ebk.); ISBN 978-0-470-90162-5 (ebk.);
  ISBN 978-0-470-90164-9 (ebk.); ISBN 978-0-470-95004-3 (ebk.); ISBN 978-0-470-95164-4 (ebk.);
  ISBN 978-0-470-95186-6 (ebk.)
  1. Civil engineering–Handbooks, manuals, etc.   I. Zenobia, Kent E.   II. Title.
TA151.H295 2011
624.023–dc22

                                                                                            2010031086

Printed in the United States of America

V10017893_031020

Contents

Chapter **3**   Ethics   63

Chapter   **6**   What Engineers Deliver   157

Chapter   **7**   Executing a Professional Commission—Project Management   183

Chapter 8   Permitting 227

Chapter 9   The Client Relationship and Business Development 247

Chapter 10   Leadership   265

Chapter 11   Legal Aspects of Professional Practice   281

Chapter **12**   Managing the Civil Engineering Enterprise 341

Chapter 15   Globalization   399

Chapter 16   Sustainability   439

Chapter **17**   Emerging Technologies 471

Preface

The American Society of Civil Engineers (ASCE) has made a concerted effort to work with ABET (formerly named the Accreditation Board for Engineering and Technology) in order to assure that civil engineering education anticipates and responds to the profession's evolving needs. The ASCE has formed several task forces over the last decade not only to address these needs in the present but also to foresee significant trends.

The ASCE has incorporated these findings in multiple reports and policy statements, including: Policy 465—Academic Prerequisites for Licensure and Professional Practice; the vision articulated by the Summit on the Future of Civil Engineering— 2025; and the *Civil Engineering Body of Knowledge for the 21st Century* (*BOK1-2004* and *BOK2-2008*). Policy 465 supports the concept of the master's degree or equivalent as a prerequisite for licensure and the practice of civil engineering at the professional level. The attendees of the Summit on the Future of Civil Engineering—2025 articulated a vision that sees civil engineers as being entrusted by society to be leaders in creating a sustainable world and enhancing the global quality of life. (More information is available at: www.asce.org/raisethebar).

Each of the *BOK2*'s 24 outcomes could command its own textbook. The goal of this book is to provide an easily understood and readily usable resource for civil engineering educators, students, and professional practitioners that develops overall understanding and points readers to additional resources for further study. The book distills 15 of the *BOK2*'s outcomes (six technical outcomes and all nine professional outcomes), as well as other relevant issues.

The *Civil Engineer's Handbook of Professional Practice* targets both academia and industry. The book can be used as a textbook for Professional Practice, Senior Project, Infrastructure Engineering, and Engineering Project Management courses.

It is appropriate for upper division and graduate level students in the major. Additionally, the book is a helpful reference for practicing civil engineers.

The information contained in the 191-page *BOK2* provides a vision for a civil engineering body of knowledge. The *Civil Engineer's Handbook of Professional Practice* builds on that vision by providing illuminating techniques, quotes, case examples, problems and information to assist the reader in addressing the many challenges facing civil engineers in the real world. This book:

- Focuses on the business and management aspects of a civil engineer's job, providing students and practitioners with sound business management principles
- Addresses contemporary issues, such as permitting, globalization, sustainability, and emerging technologies
- Offers proven methods for balancing speed-quality-price with contracting and legal issues in a client-oriented profession
- Includes guidance on juggling career goals, life outside work, compensation, and growth

Additionally, the authors and publisher have established a website:

www.wiley.com/go/cehandbook

Wiley and the Authors wish to support this book and to enable communication between the readers and authors and offer this website address as a convenient mechanism to do so.

Acknowledgments

This book was born through our involvement with the students of the Department of Civil Engineering at California State University, Sacramento (CSUS) and a desire to help them become highly functioning, competent, ethical, and successful Civil Engineers. We have been guided by the vision established by the American Society of Civil Engineers (ASCE) in the *Bodies of Knowledge 1 (2004)* and *2 (2008)* and other ASCE policy statements. We would like to acknowledge both our students and the many professional Civil Engineers, both past and present, who have inspired us.

We have relied heavily on the insights and professional experience of our many expert contributing authors and technical reviewers and are most grateful for their participation. To engage with these professionals, who are part of an engineering community that is dedicated to continuous improvement, mentoring, public health and safety, was a pleasure. The contents of this book truly reflect a national and international flavor and represent the diversity of our fellow engineers in academia, public service, and the private sector. These dedicated professionals are acknowledged and listed with their credentials in the following pages.

The authors also would like to thank our colleagues in the CSUS Department of Civil Engineering for their assistance with this project and for helping to provide an environment that is both stimulating and nurturing. Specifically, we wish to thank Dr. Ramzi Mahmood, Department Chair, for his support. Keith Bisharat is thanked for great leadership and insight into the initial mystery of book publishing. Keith was able to show us the true end product, his book titled *Construction Graphics,* and often made himself available for consulting and coaching. Dr. Ed Dammel is acknowledged for his leadership and contributions from the Civil Engineering (CE) Senior Project class, which are samples of actual engineering problems prepared by graduating CE students under the tutelage of volunteer professional Civil Engineers. We also are grateful for additional guidance and encouragement provided by Dr. Cyrus Aryani and Dr. John Johnston.

On a personal level, Karen Hansen would like to thank all of those who have assisted in this book-writing-publishing odyssey. Several good friends and relatives have provided warmth as well as homes away from home. I am forever indebted to Martha Padilla-Borrego, Susan Padilla-Riney, and Maxine Padilla-Selby and to my aunts and uncles, Gordon and Peggy Winlow and Blanche and Herbert Jensen, for their hospitality. These friends and family used all of their considerable collective creative powers to help me keep on track. My parents, Barbara Lee Winlow and Robert W. Hansen, have given me the curiosity and drive required to see this project through to completion. How fortunate I have been to have these people in my life!

There are many others, who have offered intellectual counterpoints, good humor, and strong shoulders. Among these are: Sandra Benedet, my cousin Kristie Denzer, Jan Escamilla and Steve Sheridan, Carole Hyde, John and Lana Kacsmaryk, Marion Lee, Irene McNay, Jane Millar, Marie-Lorraine Muller, Ronald Speake, Noel (Bill) Stewart, and Dr. Jorge Vanegas. Thank you all!

Kent Zenobia wishes to thank several people that helped immensely with the production of this book. I would like to thank my wife, Ellen, for her love, support and patience during the past three or so years it has taken me to collect and produce this work. She demonstrated great patience and understanding throughout the process. She helped with subject matter presentation, editing and actual manuscript preparation. I am so fortunate to have her as a partner in life and love. I would like to thank my two children, Taylor and Jack for their love, support and patience waiting for their playmate (Dad).

I am treated to another dimension of engineering by my fellow colleagues at CSUS. Working as an adjunct professor at California State University, Sacramento provides me with another family of colleagues for which I am truly grateful.

Producing this handbook has been stimulating, numbing, satisfying, frustrating, and always challenging. Each author wishes to thank the other for their patience, grace under pressure, and insights we anticipate our readers will find constructive. Together, we hope our multi-dimensional views from academic, public service and industry perspectives enhance readers' professional practice of Civil Engineering.

Finally, we thank John Wiley and Sons, Inc. for their efforts producing this handbook. We whole-heartedly thank Jim Harper, Editor, who helped initiate this project; Daniel Magers, Senior Editorial Assistant; Kerstin Nasdeo, Production Manager; Nancy Cintron, Senior Production Editor; and Robert L. Argentieri, Executive Editor for their patience, craftsmanship, and experience in the actual publication of this work.

Karen Lee Hansen and Kent Zenobia
March 2011

Contributing Authors

Keith A. Bisharat, MS, is a professor in the Construction Management Program at California State University in Sacramento. He is also a licensed general contractor with more than 25 years of experience in construction as a sole proprietor, partner, forensic construction consultant, developer, building designer, project manager, superintendent, project engineer, carpenter, and laborer. He is author of *Construction Graphics: A Practical Guide to Interpreting Working Drawings*, a book that shows how construction graphics "translate" into construction methods and practices.

Dr. Tim Brady has been researching innovation and innovation management since 1980. He is a Principal Research Fellow at the Center for Research and Innovation Management (CENTRIM), at the University of Brighton, United Kingdom. He joined CENTRIM in 1994 to work on a study of the management of innovation within complex product systems (CoPS) and later became Deputy Director of the Economic and Social Research Council (ESRC)-funded CoPS Innovation Centre. His current research interests include learning and capability development in project-based business, and the emergence of integrated solutions. He was a member of the Engineering and Physical Sciences Research Council (EPSRC) network: Rethinking Project Management, and organized the eighth International Network on Organizing by Projects (IRNOP) research conference, which took place in Brighton in September 2007. He previously worked at the Science Policy Research Unit (SPRU), University of Sussex, and at the University of Bath. Dr. Brady's Ph.D. dissertation examined business software 'make-or-buy' decisions.

Jody Bussey has worked for architects, general contractors and construction management firms since 2000. She graduated magna cum laude from California State University, Sacramento with a BS in Construction Management and a minor in

Business Administration. Her involvement on a LEED Gold high rise construction project introduced her to sustainable design and construction. Jody recently joined PMA Consultants, acting as a senior engineer assisting with construction management services on the San Francisco Water System Improvement Program. She is currently working on multiple pipeline, water treatment facility, and crossover valve facility projects totaling $300M. The projects include the $85M Tesla UV Water Treatment Plant, a LEED-certified facility that will be the third largest in the country and the largest in California. These projects are part of a $4B overall program utilizing state of the art construction management software and award winning best practices procedures.

E.J. Koford is a biologist and project manager with 20 years of experience preparing environmental permitting documents, wildlife and fisheries investigations, threatened and endangered species surveys, EIS/EIRs, water quality evaluations, and environmental regulatory compliance with requirements of CEC, FERC, SMARA, CERCLA, RCRA, NEPA, and CEQA. He has performed field surveys in 18 states and countries. Mr. Koford has an M.S. in Ecology from the University of California at Davis, an A.B. in Zoology from the University of California at Berkeley, and is a Certified Wildlife Biologist of the Wildlife Society.

Dr. Iain A. MacLeod, a Chartered Engineer and Fellow of both the Institute of Civil Engineers (ICE) and Institution of Structural Engineers (IStructE), is Professor Emeritus in the Department of Civil Engineering, Strathclyde University. He has worked as a design engineer and consultant in the United States and Canada and in design research with the Portland Cement Association in the United States. He was Professor of Structural Engineering at the University of Strathclyde in Glasgow for 23 years and Professor and Head of Department at Paisley University. He is a former Lecturer at the University of Glasgow. His research work has spanned a range of topics in the design of buildings, including the analysis of tall buildings, the use of information technology (IT) in design and studies in design process. He is author of *Modern Structural Analysis: Modelling Process and Guidance*, published by Thomas Telford Ltd., a book that redresses the imbalance in risk between computer models based around generally determinate calculation outputs and possibly non-determinate understandings of the actual modeling process.

Dr. Jane E. Millar, principal of Jane Millar & Associates in Brighton, United Kingdom, consults in Policy Research. She has been a Senior Research Fellow at the Migration Research Unit (MRU), University College London; at the Institute for Public Policy Research in London; and at the Policy Research Unit (SPRU), University of Sussex. She holds a Ph.D. in Cognitive and Computing Sciences from the University of Sussex and has managed a wide range of projects in both industry and academia.

Brian S. Neale, a Chartered Engineer and Fellow of both the Institution of Civil Engineers (ICE) and Institution of Structural Engineers (IStructE) and member of the Council of Management of the Institute of Demolition Engineers (IDE) in the United Kingdom, is an independent consultant and Secretary of the UK based Hazards Forum. He formerly worked for the Health and Safety Executive and other professional Civil Engineering organizations. He chaired the drafting of BS6187:2000 Code of Practice for Demolition standard and its 2010 revision. As a European Committee for Standardization (CEN) convenor, Mr. Neale oversaw the drafting of one of the Structural Eurocodes related to the topic of demolition. He was editor of the 2009 Thomas Telford Ltd. book, *Forensic Engineering: From Failure to Understanding*, and chaired the Organizing Committees of all four International Conferences on Forensic Engineering organized by the Institution of Civil Engineers and supported by the American Society of Civil Engineers (ASCE). His published papers include an international dimension and his consultancy includes a training element.

Greg Oslund, P.E. has more than 22 years of experience in the planning, approval, design, management and oversight of transportation projects. He has spent his entire career developing a comprehensive understanding of the project development phases required for these projects including project initiation, planning, programming, project approval and environmental design (PA&ED), design (PS&E), utility coordination, permitting, R/W acquisition and engineering support during construction. He has served as project engineer, project manager and/or principal in charge for more than 25 large transportation projects. In addition, Mr. Oslund has more than 15 years business development experience involving major transportation project pursuits as the prime consultant. He has served as client service manager, pursuit manager and regional business development manager responsible for setting and implementing the business develop and marketing strategy for a large engineering and construction firm.

George T. Qualley, P.E., is a licensed professional engineer with 40 years of civil engineering design, construction, operation, and maintenance experience for the State of California. He served for 13 years as Flood Management Division Chief for the California Department of Water Resources, responsible for a staff of over 300, carrying out an integrated statewide flood management program including flood and water supply forecasting; flood emergency operations; assuring adequate maintenance and repair of existing flood control projects; promoting effective management of unprotected floodplains to discourage unwise and damageable development; and collaborating with federal, state, and local partners in developing new multi-objective projects in areas of critical need that integrate structural and nonstructural approaches to flood risk reduction. Mr. Qualley holds a Bachelor of Science Degree from North Dakota State University.

Tony Quintrall, P.E. is a geotechnical project engineer with HDR Engineering, Inc. in Folsom, CA. At HDR he has been involved in numerous geotechnical investigations and design and construction activities for levees and small dams throughout Northern California. He has been involved with all aspects of the design process, from preliminary investigations and analysis to construction management, functioning as a technical specialist performing analysis as well as providing oversight and quality control.

Dr. Matthew Salveson, P.E. is a licensed civil engineer and has been working in the transportation engineering field since 1991. His project experience includes the planning and design of various transportation facilities in California, including bridges, freeways, local roads, and interchanges. He has also managed the construction, retrofit and repair of numerous bridges. Dr. Salveson received his Bachelor of Science, Master of Science, and Doctor of Philosophy in Civil Engineering from the University of California, Davis. He is currently an Assistant Professor of Civil Engineering at California State University, Sacramento.

Michael A. Turco, P.E., BCEE is a licensed professional engineer and certified project manager, with 40 years of engineering, design, and management experience in and for the oil, chemical, hazardous waste management and environmental consulting industries. He is board certified by the American Academy of Environmental Engineers in hazardous waste management and holds a BS in Chemical Engineering, an MS in Environmental Engineering and an MBA, all from Drexel University.

Scott D. Woodland, P.E., M. ASCE is a licensed professional engineer in the State of California. With experience in design and construction, operations and maintenance and planning for the California Department of Water Resources he is an 18 year veteran of California's on-going struggles to deliver water and protect the State's citizens from floods. He currently is helping with the implementation of the California FloodSAFE and Integrated Regional Water Management Programs. Scott has a BS in Civil Engineering from the University of California, Davis. Scott contributed to portions of this book related to executing a professional commission, engineer's role in project development, and professional engagement.

Phil Welker, PMP is a chemical/environmental engineer with nearly 20 years of experience managing complex large-scale toxic and hazardous waste remediation projects for both the private and public sector, particularly the federal government. He is a certified project management professional (PMP), and is an Associate at GeoEngineers, Inc., where he monitors and assists project managers with their daily project oversight activities. Phil has a BS in Chemical Engineering from Trinity University, Texas. Phil contributed to portions of this book related to executing a professional commission, products that engineers deliver, and professional engagement.

Contributing Editors

Dr. Cyrus Aryani, P.E., G.E. is professor of geotechnical engineering and graduate program coordinator in the Department of Civil Engineering at California State University, Sacramento. Prior to joining the university, he worked as a consulting geotechnical engineer in southern California where he planned and supervised subsurface exploration programs, conducted feasibility studies for site selection and development, analyzed slope stability and designed landslide stabilization plans, and incorporated geosynthetic materials on a wide variety of projects, including: commercial/industrial tracts, residential development, bridges, road embankments, airports, oil storage and landfill facilities, earth dams and water storage reservoirs, utility tunnels, and distressed structures. He is the author of several publications and professional reports including a three volume text book, *Applied Soil Mechanics and Foundation Engineering*, California State University, Sacramento 2008, 2009, and 2010.

Dr. Sandra M. Benedet holds a Ph.D. in Spanish from Stanford University and a BA from California State University, San Francisco. Dr. Benedet currently is a Professor at DePaul University in Chicago and has taught at Stanford University, Roosevelt University, Northwestern University, and the University of Iowa. She has instructed a wide range of courses, including language, composition, and literature, as well as a course on urban literature that examines the way in which the Latin American city has been imagined in the 20th century. She has worked extensively on questions of modernity as they relate to the avant-garde. Her work has appeared in "La palabra y el hombre: Revista de la Universidad Veracruzana," and "Contratiempo," a Chicago-based publication.

Phil Brozek, P.E., is a Professional Engineer in the State of California and has more than 30 years of professional experience in contract management, construction management, and project management on large US Army Corps of Engineers projects. Phil is currently a partner in Brozek & Associates providing project leadership for natural resource conservation projects.

Dr. Janis E. Hulla, D.A.B.T., has worked with the U.S. Army Corps of Engineers since 2002. She provides environmental health and toxicological expertise to the Corps, Army and Department of Defense. She identifies and frames national issues at the intersection of policy, science, and field practice to resolve both longstanding and emerging issues. She serves as an advisor to, and project manager for, the Physical Sciences and Life Sciences Divisions of the Army Research Office located in Research Triangle Park, NC. Prior to moving to Sacramento, Dr. Hulla was a senior fellow at the National Institute of Environmental Health Sciences, RTP, NC. A former faculty member of the University of North Dakota and North Dakota State Toxicologist, Dr. Hulla earned her B.S. in Microbiology and M.S. in Biochemistry from Montana State University. Her Ph.D. was earned in Pharmacology from the University of Washington School of Medicine. Dr. Hulla is certified as a Diplomate of the American Board of Toxicology (ABT) and currently serves on its Board of Directors.

Dr. John Johnston, P.E. is professor of environmental engineering in the Department of Civil Engineering at California State, Sacramento (CSUS) and Technical Advisor in the CSUS Office of Water Programs where he has guided stormwater research for all Caltrans projects. He served as Senior Environmental Engineer, Camp Dresser and McKee, Inc., in Boston, MA, managing EPA-sponsored technology evaluation of in-vessel composting systems for municipal sludge, and a study of sludge dewatering system options for the City of Fall River, MA. Dr. Johnston also was a Civil Engineer with U.S. Army Corps of Engineers, Sacramento District, where he designed water and wastewater systems, roads, and facilities at Corps reservoirs in California.

Thomas J. Kelleher, Jr. is an attorney and Senior Partner with Smith, Currie, & Hancock LLP, a nationally recognized firm that practices in the areas of construction law, government contracts, and environmental law. He graduated cum laude from Harvard University and graduated from the University of Virginia School of Law. He served in the U.S. Army from 1968 through 1973 including positions as the Assistant Chief and Instructor in the Procurement Law Division at the U.S. Army Judge Advocate General's School, Charlottesville, Virginia. Mr. Kelleher has extensive government and construction contract experience on the spectrum of issues involving bidding, changes, differing site conditions, delays, and terminations. He has represented clients on hospital projects, airport facilities, research laboratories, convention facilities, prisons, federal and state courthouse and office complexes, and resort hotels and has practiced before the various federal government boards of contract appeals, as well as federal and state courts. In addition, he has represented clients in mediations, as well as arbitration proceedings. Mr. Kelleher is co-editor of *Common Sense Construction Law: A Practical Guide for the Construction Professional.*

Dr. Debra Larson, P.E. is Associate Dean of the College of Engineering, Forestry and Natural Sciences at Northern Arizona University (NAU). She joined in 1994 as an Associate Professor after completing a Ph.D. in Civil Engineering from Arizona State University and working in industry as a civil and structural engineer for ten years. Her research interests have included alternative building materials and techniques, value-added wood products, low-rise structures, and engineering pedagogy. Dr. Larson has designed and managed numerous American Society of Civil Engineers (ASCE)-sponsored Excellence in Civil Engineering Education (ExCEEd) Teaching Workshops for civil engineering educators and participated actively as a member of the ASCE's Body of Knowledge (BOK) Educational Fulfillment Committee. She also has lead ABET, Inc.—formerly Accreditation Board for Engineering and Technology—specialized evaluation teams in reviewing academic institutions and programs to ensure that they are meeting established standards of educational quality.

Todd Kamisky, P.E., G.E. is a licensed civil and geotechnical engineer, and has been working in the geotechnical engineering field since 1994. His project experience includes all geotechnical aspects of residential subdivisions, detention basins, bridges, communication towers, schools and commercial/industrial developments. Mr. Kamisky received his Bachelor of Science degree in Civil Engineering from California State University, Chico and a Master of Science degree in Civil Engineering with emphasis in Geotechnical Engineering, from University of California, Davis.

Bridget Crenshaw Mabunga is an Adjunct Professor of English in the Los Rios Community College District and a Writer/Editor. She also volunteers as an Assistant Editor for Narrative Magazine. She holds a BA in English (cum laude) from California State University, Chico and an MA in English (emphasis Creative Writing) from California State University, Sacramento.

Janet Riser, MBA, CFM, CRPC obtained her undergraduate degree from the University of Pittsburg, and an MBA from Drexel University before entering the financial investment community as a financial advisor for over 25 years with Merrill Lynch and now with Janney, Montgomery, Scott LLC as a First Vice-President. Janet earned her Chartered Retirement Planning Counselor designation from the College of Financial Planning in 2007 and in 2009 received Five Star Wealth Manager Award in the Delaware Valley. Janet specializes in the financial planning process, helping her clients deal with life cycle and market transitions. One of Janet's greatest pleasures in her work is the long-term relationships working with and growing extended families through multiple generations. Janet contributed to portions of this book related to the client relationship, communication, and professional engagement.

List of Abbreviations

A

AA	Affirmative action
AAA	American Arbitration Association
ABET, Inc.	Accreditation Board for Science and Technology (formerly)
ACEC	American Council of Engineering Companies
ACI	American Concrete Institute
ACLC	Administrative civil liability complaint
ADA	Americans with Disabilities Act
ADR	Alternative dispute resolution
AAP	Affirmative action program
A / E	Architect / engineer
AEA	Atomic Energy Act
AEC	Architectural / engineering / construction
AGC	Associated General Contractors
AIA	American Institute of Architects
APN	Assessor's parcel number
ASCE	American Society of Civil Engineers
ASTM	American Society for Testing and Materials (formerly)

B

BCEE	Board Certified Environmental Engineer
BIM	Building Information modeling
BOK1	*Civil Engineering Body of Knowledge for the 21st Century* (ASCE, 2004)
BOK2	*Civil Engineering Body of Knowledge for the 21st Century* (ASCE, 2008)
BPR	Business process reengineering

C

CAD	Computer-aided design
CAM	Computer-aided manufacturing
CBD	*Commerce Business Daily*
CBS	Cost breakdown structure
CEQA	California Environmental Quality Act
CERCLA	Comprehensive Environmental Response, Compensation and Liability Act
CM	Construction manager or management
CPI	Cost performance index
CPM	Critical path method
CSA	County Service Area
CVRWQCB	Central Valley Regional Water Quality Control Board
CWA	Clean Water Act

D

DA	Design assist
DB	Design build
DBB	Design-bid-build
DL	Design (team) leader
DPM	Design performance measure
DRB	Dispute review board

E

EEOC	Equal Employment Opportunity Commission
EIR	Environmental impact report
EJCDC	Engineers Joint Contract Development Committee
EO	Presidential Executive Order
EO	Equal opportunity
EPA	U.S. Environmental Protection Agency
EPCRA	Emergency Planning and Community Right-to-Know Act
EEOC	Equal Employment Opportunity Commission
ESA	Endangered Species Act

F

FAR	Federal acquisition regulation
FIATECH	Fully Integrated and Automated Technology (formerly)
FIFRA	Federal Insecticide, Fungicide, and Rodenticide Act
FS	Feasibility study

G

GC	General contractor
GINA	Genetic Information Nondiscrimination Act
GMP	Guaranteed maximum price

| GPS | Global positioning systems |
| GIS | Geographic information systems |

I

ICE	Institution of Civil Engineers (UK)
IPCC	Intergovernmental Panel on Climate Change
IPD	Integrated project delivery
ISO	International Organization for Standardization
IT	Information technology

L

LCCA	Lifecycle cost analysis
LEED	Leadership in Energy and Environmental Design
LLC	Limited liability company
LOE	Level of effort

M

MEP	Mechanical, electrical, plumbing (engineers)
MP	Multiple prime
MSA	Master services agreement

N

NEPA	National Environmental Policy Act
NBIMS	National BIM Standards
NOA	Naturally occurring asbestos
NPDES	National Pollutant Discharge Elimination System

O

OBS	Organizational breakdown structure
OFCCP	Office of Federal Contract Compliance Programs
OSHA	Occupational Safety and Health

P

PERT	Performance evaluation review technique
PM	Project manager
PMI	Project Management Institute
PMP	Project management plan
PPA	Pollution Prevention Act
PS&E	Plans, specifications, and (cost) estimates

Q

| QBS | Qualifications-based selection |

R

RCRA	Resource Conservation and Recovery Act
RF	Radio frequency
RFI	Request for information
RFP	Request for proposal
RFQ	Request for qualifications
RP	Responsible party
RTC	Response to comment
R / W	Right of way

S

SARA	Superfund Amendments and Reauthorization Act
SDWA	Safe Drinking Water Act
SMD	Sewer maintenance district
SOQ	Statement of qualifications
SOW	Statement, or scope, of work
SPCC	Spill prevention, containment, and contingency
SPI	Schedule performance index

T

TBD	To be determined
TSCA	Toxic Substances Control Act
TQM	Total Quality Management

V

VR	Virtual reality

W

WBS	Work breakdown structure
WWT	Wastewater treatment
WWTP	Wastewater treatment plant

Introduction

Ethics

History

Professional Engagement

What Engineers Deliver

Engineer's Role in Project
Development

Permitting

Leadership   Managing

Emerging Technology

Having
a Life

Sustainability

Executing a Professional
Commission

Client Relationship

Legal Aspects

Globalization

Communicating

1 2 3 4 5 6 7 8 9 10 11 12 13 14 15 16 17 A B C D E F

Chapter **1**

Introduction

Big Idea

"Entrusted by society to create a sustainable world and enhance the global quality of life, Civil Engineers serve, competently, collaboratively, and ethically as: master planners, designers, constructors; stewards of the natural environment and its resources; innovators and integrators; managers of risk and uncertainty; and leaders in discussions and decisions shaping public environmental and infrastructure policy."

—*ASCE Body of Knowledge 2*

Key Topics Covered

- The Need for Accreditation
- American Society of Civil Engineers (ASCE)
- 21st Century Engineer
- Goal of This Book
- Reader's Guide

Related Chapters in This Book

- Chapters 2 through 17 and Appendices A, B, C, D, E, F

(Continued)

Related to *ASCE Body of Knowledge 2* Outcomes

ASCE BOK2 outcomes covered in this chapter

Foundational
1. Mathematics
2. Natural sciences
3. Humanities
4. Social sciences

Technical
5. Materials science
6. Mechanics
7. Experiments
●8. Problem recognition and solving
●9. Design
●10. Sustainability
●11. Contemporary issues & historical perspectives
●12. Risk and uncertainty
●13. Project management
●14. Breadth in civil engineering areas
●15. Technical specialization

Professional
●16. Communication
●17. Public policy
●18. Business and public administration
●19. Globalization
●20. Leadership
●21. Teamwork
●22. Attitudes
●23. Lifelong learning
●24. Professional and ethical responsibility

BACKGROUND

The *Civil Engineer's Handbook of Professional Practice* is a professional practice guide for civil engineers. The first decade of the 21st century has afforded many opportunities to reflect on the role civil engineers will play in coming years. The global economy and world banking system, national security, climate change, dwindling natural resources, technological advances, and societal changes have provided sufficient food for thought. In retrospect, the 2001 American Society of Civil Engineers (ASCE) report, titled *Engineering the Future of Civil Engineering,* which acknowledged that civil engineering must respond proactively to increasingly complex challenges related to public health, safety, and welfare, appears prophetic.

As a university program, civil engineering has been growing in the 21st century. Enrollment in most universities across the nation continues to increase, partially due to shrinking opportunities in other technical fields as a result of outsourcing. Civil engineers work very closely with government agencies and on projects requiring significant local knowledge, making outsourcing of their work difficult. According to the U.S. Bureau of Labor Statistics:

> **Civil engineers** are expected to experience 24 percent employment growth during the projections decade [2008–2018], faster than the average for all occupations. Spurred by general population growth and the related need to improve the Nation's infrastructure, more civil engineers will be needed to design and construct or expand transportation, water supply, and pollution control systems and buildings and building complexes. They also will be needed to repair or replace existing roads, bridges, and other public structures.

For several years the country's infrastructure has been given a grade of "D" on the ASCE's infrastructure report card; in 2009 the ASCE estimated that a $2.2 trillion investment was needed over the next five years to rectify this problem. Significant public and private funding sources have been established to address this challenge and, as a result, the demand for well-educated and competent civil engineers should continue.

> *"Infrastructure is a multitrillion-dollar marketplace with enormous need for private investment."*
>
> Source: Henry Kravis in the New York Times, 5/16/08

THE NEED FOR ACCREDITATION

ASCE has made a concerted effort to work with ABET, Inc., formerly named the Accreditation Board for Engineering and Technology, to assure that civil engineering education anticipates and responds to the profession's evolving needs. ASCE has formed several task forces not only to address these needs in the present but also to foresee significant trends.

ABET, Inc. accredits civil engineering programs within U.S. universities and plays a significant role in determining the development of the profession. University Departments of Civil Engineering undergo extensive, periodic reviews by ABET in order to maintain their accreditation.

ABET, Inc. was established more than 75 years ago as the Engineers' Council for Professional Development (ECPD). A survey of multiple engineering societies revealed the need for quality control, and in 1932, seven societies founded ECPD. These societies included: the American Society of Civil Engineers (ASCE); the American Society of Mining and Metallurgical Engineers (now the American Institute of Mining, Metallurgical, and Petroleum Engineers); the American Society of Mechanical Engineers (ASME); the American Institute of Electrical Engineers (now IEEE); the Society for Promotion of Engineering Education (now the American Society for Engineering Education—ASEE); the American Institute of Chemical Engineers (AIChE); and the National Council of State Boards of Engineering Examiners (now NCEES). By 2009, ABET accredited approximately 2,700 programs at more than 550 universities and colleges nationwide.

ABET OUTCOMES

Following a long period of development, in 1997, ABET adopted Engineering Criteria 2000 (EC2000), which took a completely new approach to engineering education. By defining *outcomes* of engineering education, EC2000 focused on what is learned rather than what is taught. ABET has identified 11 outcomes of civil engineering education:

1. *Mathematics, science, and engineering*—an ability to apply knowledge of mathematics, science, and engineering

2. *Experiments*—an ability to design and conduct experiments, as well as analyze and interpret data

3. *Design*—an ability to design a system, component, or process to meet desired needs

4. *Multidisciplinary teams*—an ability to function on multidisciplinary teams

5. *Engineering problems*—an ability to identify, formulate, and solve engineering problems

6. *Professional and ethical responsibility*—an understanding of professional and ethical responsibility

7. *Communication*—an ability to communicate effectively

8. *Impact of engineering*—the broad education necessary to understand the impact of engineering solutions in a global and societal context

9. *Lifelong learning*—a recognition of the need for, and an ability to engage in, lifelong learning

10. *Contemporary issues*—a knowledge of contemporary issues
11. *Engineering tools*—an ability to understand techniques, skills, and modern engineering tools necessary for engineering practice

AMERICAN SOCIETY OF CIVIL ENGINEERS

Meanwhile, the American Society of Civil Engineers has made a concerted effort to work with ABET to assure that civil engineering education anticipates and responds to the profession's evolving needs.

The ASCE has formed several task forces not only to address these needs in the present but also to foresee significant trends. Policy 465 expresses the vision articulated by the Summit on the Future of Civil Engineering–2025 held in 2006. The attendees of the Summit saw civil engineers as being entrusted by society to be leaders in creating a sustainable world and enhancing the global quality of life. As depicted in Figure 1.1, Policy 465 supports the concept of the master's degree or equivalent as a prerequisite for licensure and the practice of civil engineering at the professional level.

The 2001 ASCE report *Engineering the Future of Civil Engineering,* mentioned above, concluded that for civil engineers to maintain leadership in the infrastructure and environmental arena, an implementation master plan was needed; and the basis of this master plan is a document called the *Body of Knowledge.* The *Body of Knowledge 1*

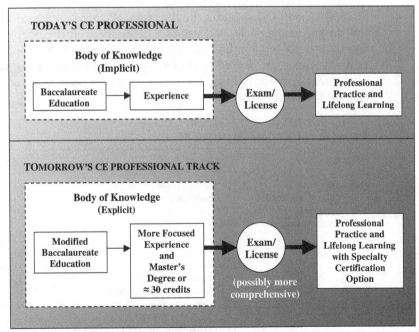

Figure 1.1 ASCE's vision of preparation for a career in civil engineering
(Adapted from ASCE Policy Statement 465)

(BOK1), published in 2004, defines categories of knowledge and recommends 15 outcomes that collectively prescribe a "substantially greater depth and breadth of knowledge, skills, and attitudes required of an individual aspiring to the practice of civil engineering at the professional level (licensure) in the 21st Century." The first 11 outcomes are those identified by ABET, but the BOK1 included four additional outcomes that broaden and deepen these ABET outcomes. The new outcomes are:

12. *Specialization*—an ability to apply knowledge in a specialized area related to civil engineering
13. *Management*—an understanding of the elements of project management, construction, and asset management
14. *Policy and administration*—an understanding of business and public policy and administration fundamentals
15. *Leadership*—an understanding of the role of the leader and leadership principles and attitudes

The BOK1 also emphasized the importance of attitude: "knowledge and skill, while necessary, are not sufficient to be a fully functioning civil engineer." (Note: ABET has incorporated outcomes 13, 14, and 15 into its Criterion 9 for civil engineering programs.)

ASCE published the second edition of BOK1, the *Body of Knowledge 2* (BOK2), in 2008. The BOK2 also uses the "outcomes" approach developed by ABET to define the knowledge, skills, and attitudes necessary to enter civil engineering practice at the professional level in the 21st century. The BOK2 further adopts Bloom's Taxonomy to indicate the desired level of achievement for each outcome. The BOK2's 24 outcomes are organized into three categories: foundational, technical, and professional. (See Table 1.1.)

Table 1.1   BOK2 Outcomes (2008)

Foundational	Technical	Professional
1) mathematics	5) materials science	16) communication
2) natural sciences	6) mechanics	17) public policy
3) humanities	7) experiments	18) business and public
4) social sciences	8) problem recognition	administration
	and solving	19) globalization
	9) design	20) leadership
	10) sustainability	21) teamwork
	11) contemporary issues/	22) attitudes
	historical perspectives	23) lifelong learning
	12) risk and uncertainties	24) professional and
	13) project management	ethical responsibility
	14) breadth in civil engineering areas	
	15) technical specialization	

ASCE Has Developed a Global Vision of the Profession:

Entrusted by society to create a sustainable world and enhance the global quality of life, Civil Engineers serve, competently, collaboratively, and ethically as master:

- Planners, designers, constructors, and operators of society's economic and social engine, the built environment
- Stewards of the natural environment and its resources
- Innovators and integrators of ideas and technology across the public, private, and academic sectors
- Managers of risk and uncertainty caused by natural events, accidents, and other threats
- Leaders in discussions and decisions shaping public environmental and infrastructure policy

—*Civil Engineering Body of Knowledge for the 21st Century* (BOK2).

The first and second editions of the *Civil Engineering Body of Knowledge for the 21st Century* stress the need for change in the way civil engineers practice their profession and in the way they are educated. Though not strictly prescriptive, BOK1 and BOK2 offer guidance to academia in helping to educate future engineers. Summary findings are highlighted below.

Key issues facing engineering education

BOK1 identifies the chief issues facing civil engineering as:

- Escalated complex risks and challenges to public safety, health, and welfare
- Vulnerability to human-made hazards and disasters (such as terrorism)
- Globalization
- Four-year bachelor's degree inadequacy in providing formal academic preparation for the practice of civil engineering at the professional level

BOK2 adds further concerns:

- Sustainability
- Emerging technology

Teaching/learning modes

BOK1 identifies four teaching/learning modes:

- Undergraduate study typically leading to a BSCE
- Graduate study or equivalent

- Co-curricular and extracurricular activities
- Post-B.S. engineering experience prior to licensure

BOK1 also concludes that distance learning increasingly will improve accessibility to high-quality formal education.

Faculty member characteristics

BOK1 identifies characteristics of the model full- or part-time civil engineering faculty member:

- *Scholars* having and maintaining expertise in the subjects they teach
- *Teachers* who effectively engage students in the learning process
- *Professionals* with practical experience, preferably with professional engineering licenses
- *Positive role models* for the profession

Table 1.2 depicts the relationships among the ABET, BOK1, and BOK2 outcomes.

What Is the Role of Engineers in Society and How Is that Role Changing?

- By 2020, we aspire to a public that will understand and appreciate the profound impact of the influence of the engineering profession on sociocultural systems, the full spectrum of career opportunities accessible through an engineering education, and the value of an engineering education to engineers working successfully in non-engineering jobs.

- We aspire to a public that will recognize the union of professionalism, technical knowledge, social and historical awareness, and traditions that serve to make engineers competent to address the world's complex and changing challenges.

- We aspire to engineers who will remain well grounded in the basics of mathematics and science, and who will expand their vision of design through solid grounding in the humanities, social sciences, and economics. Emphasis on the creative process will allow more effective leadership in the development and application of the next-generation technologies to problems of the future.

—National Academy of Engineering, The Engineer of 2020.

Table 1.2 From ABET to BOK2 Outcomes
(Adapted from Table H-1. From ABET program criteria to BOK2 outcomes. *Civil Engineering Body of Knowledge for the 21st Century*, February 2008, p. 101.)

ABET Outcomes	BOK1 Outcomes (2004)	BOK2 Outcomes (2008)
a. Mathematics, science, and engineering	1) Mathematics, science, and engineering	1) Mathematics 2) Natural Sciences 5) Materials Science 6) Mechanics
b. Experiments	2) Experiments	7) Experiments
c. Design	3) Design	9) Design 10) Sustainability
	3) Design	12) Risk and uncertainties
d. Multidisciplinary teams	4) Multidisciplinary teams	21) Teamwork
e. Engineering problems	5) Engineering problems	8) Problem recognition and solving
f. Professional and ethical responsibility	6) Professional and ethical responsibility	24) Professional and ethical responsibility
g. Communication	7) Communication	16) Communication
h. Impact of engineering	8) Impact of engineering	11) Contemporary issues/ historical perspectives
i. Lifelong learning	9) Lifelong learning	23) Lifelong learning
j. Contemporary issues	10) Contemporary issues	11) Contemporary issues/ historical perspectives 19) Globalization
k. Engineering tools	12) Engineering tools	8) Problem recognition and solving
	13) Specialized area related to civil engineering	15) Technical specialization
Program Criteria for Civil and Similarly Named Engineering Programs	14) Project management, construction, and asset management	13) Project management
	15) Business and public policy	17) Public policy 18) Business and public administration
Program Criteria for Civil and Similarly Named Engineering Programs	16) Leadership	20) Leadership 22) Attitudes
ABET Criterion for General Education	ABET Criterion for General Education	3) Humanities 4) Social sciences
Program Criteria for Civil and Similarly Named Engineering Programs	Program Criteria for Civil and Similarly Named Engineering Programs	14) Breadth in civil engineering areas

Table 1.3 Entry into the Practice of Civil Engineering at the Professional Level Requires Fulfilling 24 Outcomes to the Appropriate Levels of Achievement

Outcome Number and Title	Level of Achievement					
	1 Knowledge	2 Comprehension	3 Application	4 Analysis	5 Synthesis	6 Evaluation
Foundational						
1. Mathematics	B	B	B			
2. Natural sciences	B	B	B			
3. Humanities	B	B	B			
4. Social sciences	B	B	B			
Technical						
5. Materials science	B	B	B			
6. Mechanics	B	B	B	B		
7. Experiments	B	B	B	B	M/30	
8. Problem recognition and solving	B	B	B	M/30		
9. Design	B	B	B	B	B	E
10. Sustainability	B	B	B	E		
11. Contemp. issues & hisL perspectives	B	B	B	E		
12. Risk and uncertainty	B	B	B	E		
13. Project management	B	B	B	E		
14. Breadth in civil engineering areas	B	B	B	B		
15. Technical specialization	B	M/30	M/30	M/30	M/30	E
Professional						
16. Communication	B	B	B	B	E	
17. Public policy	B	B	E			
18. Business and public administration	B	B	E			
19. Globalization	B	B	B	E		
20. Leadership	B	B	B	E		
21. Teamwork	B	B	B	E		
22. Attitudes	B	B	E			
23. Lifelong learning	B	B	B	E	E	
24. Professional and ethical responsibility	B	B	B	B	E	E

Key:

B	Portion of the BOK fulfilled through the bachelor's degree
M/30	Portion of the BOK fulfilled through the master's degree or equivalent (approximately 30 semester credits of acceptable graduate-level or upper-level undergraduate courses in a specialized technical area and/or professional practice area related to civil engineering)
E	Portion of the BOK fulfilled through the prelicensure experience

The BOK2 not only defined outcomes but also identified what level of proficiency should be achieved for each outcome through the use of Blooms' taxonomy. Table 1.3 depicts the BOK2's 24 outcomes with the level of proficiency expected for each outcome.

21ST CENTURY ENGINEER

Aspiring civil engineers face challenges posed by the unique attributes and characteristics of facilities and civil infrastructure systems, as well as the complexities of the current processes and the diverse set of resources required for both their delivery and their use.

BOK2 gives some structure to what is a large educational challenge. These new outcomes and approaches have raised the bar substantially for civil engineering educators. Twentieth-century civil engineering education focused on learning about engineering mechanics; doing calculations; writing essays and lab reports; acquiring knowledge; and working with determinant processes. Twenty-first-century civil engineering practice requires innovative thinking and relies heavily on tacit knowledge—understanding, judgment, associativity, and intuition. (MacLeod, 2009) (See Figure 1.2.)

GOAL OF THIS BOOK

Given these complexities, the question is: How can the new BOK outcomes be achieved? Clearly, each of the BOK2's 24 outcomes could command its own textbook. The goal of this book is to provide an easily understood and readily usable resource for civil engineering educators, students, and professional practitioners that develops overall understanding and points readers to additional resources for further study. The book distills 15 of the BOK2's outcomes (six technical outcomes and all nine professional outcomes) as well as other relevant issues.

The *Civil Engineer's Handbook of Professional Practice* targets both academia and industry. The book can be used as a textbook for Professional Practice, Senior Project, Infrastructure Engineering, and Engineering Project Management courses. It is intended for junior, senior, and graduate level students in the major. As the issues addressed in the 2008 BOK2 are disseminated and better understood by educators, all Civil Engineering Departments will need to offer a course on Practice Management, if they do not do so already.

Additionally, the book is a helpful reference for practicing civil engineers. The information imbedded in the 191-page BOK2 provides a vision for a civil engineering body of knowledge. The *Civil Engineer's Handbook of Professional Practice* builds on that vision.

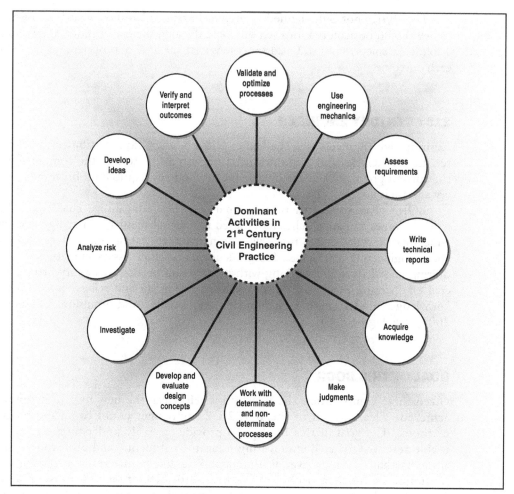

Figure 1.2 Dominant activities in 21st century practice
(Source: Dr. Iain A. MacLeod, Department of Civil Engineering, Strathcyle University, Glasgow, Scotland)

READERS' GUIDE

Of the 24 outcomes discussed in BOK2, this book addresses the following:

8. Problem Recognition and Solving	17. Public Policy
9. Design	18. Business and Public Administration
10. Sustainability	19. Globalization
11. Contemporary Issues/Historical Perspectives	20. Leadership
12. Risk and Uncertainties	21. Teamwork
13. Project Management	22. Attitudes
16. Communication	23. Lifelong Learning
	24. Professional and Ethical Responsibility

The *Civil Engineer's Handbook of Professional Practice* offers additional relevant information such as: the design professional's role in the project development process; the legal infrastructure in the United States; the fundamental contents of contracts; the origin of conflicts; the various roles that the civil engineer plays in construction projects; how the legal world views construction disputes; the basic economics of civil engineering practice; and emerging technologies relevant to civil engineering. Each chapter concludes with references for further reading and or study.

The book presents information in three levels of increasing detail through the use of graphics (photographs, illustrations, line drawings, graphs, text boxes, and cartoons) and text. These illustrations form one level of information, the commentary included in text boxes forms another, and the third is the actual text. The first page of each chapter outlines the key concepts presented and contains a unique graphic that helps to orient the reader.

The chapters of the *Civil Engineer's Handbook of Professional Practice* can be read in the order that best suits the reader. Following is a brief summary of the chapters and appendices:

- Chapter 1—Introduction
    This chapter addresses the overall issues outlined in the ASCE's Body of Knowledge, first and second editions (BOK1 and BOK2) and the need for a new approach to civil engineering.
- Chapter 2—Background and History of the Profession
    This chapter covers BOK2 Outcome 11 – Historical Perspectives and gives an overview of the Architectural/Engineering/Construction (AEC) industry.
- Chapter 3—Ethics
    This chapter covers BOK2 Outcome 24 – Professional and Ethical Responsibility.
- Chapter 4—Professional Engagement
    This chapter covers BOK2 Outcome 8 – Problem Recognition and Solving.
- Chapter 5—The Engineer's Role in Project Development
    This chapter covers BOK2 Outcome 9 – Design.
- Chapter 6—What Engineers Deliver
    This chapter covers BOK2 Outcome 8 – Problem Recognition and Solving and Outcome 9 – Design.
- Chapter 7—Executing a Professional Commission
    This chapter covers BOK2 Outcome 13 – Project Management.
- Chapter 8—Permitting
    This chapter covers BOK2 Outcome 17 – Public Policy.
- Chapter 9—The Client Relationship
    This chapter covers BOK2 Outcome 18 – Public Administration.
- Chapter 10—Leadership

This chapter covers BOK2 Outcome 20–Leadership and Outcome 21–Teamwork.

- Chapter 11—Legal Aspects of Professional Practice
  This chapter covers BOK2 Outcome 12–Risks and Uncertainties as well as the additional legal aspects.
- Chapter 12—Managing the Civil Engineering Enterprise
  This chapter covers BOK2 Outcome 18–Business Administration.
- Chapter 13—Communicating as a Professional
  This chapter covers BOK2 Outcome 16–Communication.
- Chapter 14—Having a Life
  This chapter covers BOK2 Outcome 22–Attitudes and Outcome 23–Lifelong Learning.
- Chapter 15—Globalization
  This appendix covers BOK2 Outcome 19–Globalization.
- Chapter 16—Sustainability
  This appendix covers BOK2 Outcome 10–Sustainability.
- Chapter 17—Emerging Technologies
  This appendix covers a primary concern identified in BOK2.

SUMMARY

The demands of society and the related high standards required by both ABET and ASCE present civil engineers and civil engineering educators with numerous challenges. The authors hope that the *Civil Engineer's Handbook of Professional Practice* will provide both aspiring and practicing civil engineers, as well as civil engineering educators, with useful information that assists them in meeting the needs of society and achieving their own personal goals.

REFERENCES/FURTHER READING

American Society of Civil Engineers. (2008). *Civil Engineering Body of Knowledge for the 21st Century*, 2d edition. ASCE report, Reston, VA.

American Society of Civil Engineers. (2004). *Civil Engineering Body of Knowledge for the 21st Century*, 1st edition. ASCE report, Reston, VA.

American Society of Civil Engineers. (2006). *Policy 465*. ASCE report, Reston, VA.

Galloway, Patricia D. (2008). *21st Century Engineer: A Proposal for Education Reform*. American Society of Civil Engineers. Reston, VA.

National Academy of Engineering (2004). *The Engineer of 2020: Visions of Engineering in the New Century.* National Academies Press, Washington, D.C. ISBN-10: 0-309-09162-4.

Rockefeller Foundation's 2050 Forum. (2008). *Rebuilding and Renewing: 21st Century Infrastructure Agenda*, May 9, 2008, Washington, D.C.

www.abet.org/history.html.

www.infrastructurereportcard.org/ (accessed November 7, 2009).

www.bls.gov/oco/ocos027.htm#outlook (accessed November 7, 2009).

Introduction
1
Ethics
2
3
Professional Engagement
History
4
5
What Engineers Deliver
Engineer's Role in Project
Development
6
Permitting
7
8
Executing a Professional
Commission
9
Leadership
10
Managing
11
Client Relationship
12
13
Having
a Life
14
15
Emerging Technology
16
17
A
B
C
D
E
F
Sustainability
Globalization
Communicating
Legal Aspects

Chapter **2**

Background and History of the Profession

Big Idea

" . . . lessons learned from the behavior and especially the failure of even ancient designs are no less relevant today . . . good design practice of engineers in centuries past can serve as models for the most sophisticated designs of the modern age."

—Henry Petroski

Key Topics Covered

- Background
- Civil Engineering as a Profession
- Civil Engineering's Historical Inheritance
- The Ancient Engineers
- Engineering in Medieval Times
- Engineering in the Renaissance and the Age of Enlightenment
- The Industrial Revolution
- Modern Civil Engineering
- Civil Engineering Education
- Civil Engineering Careers
- Summary

Related Chapters in This Book

- Chapter 1: Introduction
- Chapter 3: Ethics
- Chapter 8: Permitting
- Chapter 11: Legal Aspects of the Profession

(*Continued*)

Related to *ASCE Body of Knowledge 2* Outcomes

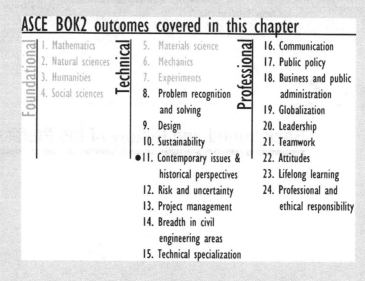

ASCE BOK2 outcomes covered in this chapter

Foundational
1. Mathematics
2. Natural sciences
3. Humanities
4. Social sciences

Technical
5. Materials science
6. Mechanics
7. Experiments
8. Problem recognition and solving
9. Design
10. Sustainability
● 11. Contemporary issues & historical perspectives
12. Risk and uncertainty
13. Project management
14. Breadth in civil engineering areas
15. Technical specialization

Professional
16. Communication
17. Public policy
18. Business and public administration
19. Globalization
20. Leadership
21. Teamwork
22. Attitudes
23. Lifelong learning
24. Professional and ethical responsibility

BACKGROUND

Chapter 2 examines civil engineering as a profession and the significant contributions civil engineers have made to civilization. The chapter explores civil engineering's historical inheritance, provides examples of outstanding projects—from ancient to modern times—and profiles several legendary civil engineers. The chapter also gives background on various career specializations, as well as typical educational and licensure requirements for achievement of professional status.

CIVIL ENGINEERING AS A PROFESSION

Until modern times there was no clear distinction between civil engineering and architecture, and the term *engineer* or *architect* referred to the same person. In the western world, the origins of civil engineering as a profession can be found in the years immediately preceding and including the Industrial Revolution, the late 18th and early 19th centuries. The scientific discoveries of the Age of Enlightenment and the new commercial needs of the Industrial Revolution converged to create an ideal environment for innovation. During this period, certain *military* engineers began to work on nonmilitary, or *civil*, projects. The term *civil engineer* was adopted to emphasize this difference. In response to the growth of these new civil projects, the British Institution of Civil Engineers (ICE) was chartered in 1818 and the American Society of Civil Engineers (ASCE) was founded in 1852. Other professional civil engineering organizations followed: Institution of Civil Engineers India (ICEI) in 1860; Spanish Asociación de Ingenieros de Caminos, Canales, y Puertos (AICCP) in 1903; South African Institution of Civil Engineers (SAICE) in 1903; Japan Society of Civil Engineers (JSCE) in 1914; and Chinese Institute of Civil Engineering (CICE) in 1936, among others.

These organization, as well as those in other countries, helped to formalize civil engineering as a profession. The geotechnical engineer and author, John Philip Bachner, lists five characteristics of a profession. These are:

- Systematic body of theory
- Authority
- Community sanction
- Ethical codes
- A culture

These characteristics help define today's professional civil engineer, who must be adequately prepared with a *systematic body of theory* that incorporates a spirit of rationality. This theory is based on mathematics and natural sciences, such as physics and chemistry. Like other professions, for instance, law and medicine, civil engineers are granted *authority* based on their extensive education and are afforded *community sanction*, in the form of licensure or registration. Civil engineers are held to well-documented *ethical*

Attributes of a Profession

1. Systematic body of theory
   - Skills flow from an internally consistent system
   - Spirit of rationality; expansion of theory
2. Authority
   - Extensive education in systematic theory highlights the layperson's comparative ignorance
   - Functional specificity
3. Community sanction
   - State-sponsored boards
   - License or registration
4. Ethical codes
   - Ethical
     - professional
   - Client-professional
     - impulse to perform maximally
   - Colleague to colleague
   - Cooperative
     - egalitarian
     - supportive
5. A culture
   - Social values
   - Services valuable to the community
   - Various modes of ''appropriate'' behavior
     - sounding like a professional
     - saying ''no'' gracefully
     - making presentations and conducting meetings
   - Symbols
     - argot, jargon
     - insignia, emblems
     - history and folklore

Adapted from John Bachner, *Practice Management for Design Professionals: A Practical Guide to Avoiding Liability and Enhancing Profitability.*

codes and are expected to be their clients' trusted advisors. Civil engineers also have *a culture* of their own that involves providing valuable services to society, behaving appropriately, and sharing a rich history and folklore.

Knowledge of civil engineering history and culture helps civil engineers communicate the importance of their profession to the world. Noted engineering historian, Henry Petroski, posits that engineering history is both history and engineering. Additionally, familiarity with civil engineering history can assist with the practice of the profession. "The lessons of the past are not only brimming with caveats about what mistakes should not be repeated but also are full of models of good engineering judgment." [Petroski, p. x]

CIVIL ENGINEERING'S HISTORICAL INHERITANCE

Much of the material in this section is derived from L. Sprague de Camp's seminal work, *The Ancient Engineers*. To begin learning about civil engineers' rich historical inheritance, we have to turn the clock back 6,000 years to the dawn of civilization. Mr. de Camp observes:

> The first engineers were irrigators, architects, and military engineers. The same man was usually expected to be an expert at all three kinds of work. This was still the case thousands of years later, in the Renaissance, when Leonardo, Michelangelo, and Dürer were not only all-around engineers but outstanding artists as well. Specialization within the engineering profession has developed only in the last two or three centuries. [p. 9]

After 4000 B.C., when humans began to abandon the nomadic way of life, the need for water, permanent shelter, religious monuments and burial sites, and fortification emerged. Early river valley civilizations, such as those around the Tigris and Euphrates (Mesopotamia), Nile (Egypt), Indus (India), and Hwang-ho (China),

Studying the Past Yields Valuable Lessons

" . . . any lessons learned from the behavior and especially the failure of even ancient designs are no less relevant today, and the good design practice of engineers in centuries past can serve as models for the most sophisticated designs of the modern age. Indeed, ignoring wholesale the lessons and practices of the past threatens the continuity of engineering and design judgment that appears to be among the surest safeguards against recurrent failures."

—Henry Petroski, *Design Paradigms: Case Histories of Error and Judgment in Engineering*, p.143.

Figure 2.1 Seven wonders of the ancient world

From left to right, top to bottom: Great Pyramid of Giza, Hanging Gardens of Babylon, Temple of Artemis, Statue of Zeus at Olympia, Mausoleum of Maussollos, Colossus of Rhodes, and the Pharos Lighthouse of Alexandria.

(Wikipedia Commons)

required canal systems to irrigate surrounding land so that farmers could raise sufficient food to support the population. Kings or rulers desired houses larger than huts of stone, clay, or reed; and priests wanted homes for the gods at least as grand. To protect the growing wealth of these early settlements, walls and moats needed to be constructed. These were the challenges that occupied the first engineers.

THE ANCIENT ENGINEERS

Some early writing on stone and brick in Mesopotamia and Egypt has survived, but other written accounts of ancient engineering in those areas have perished. The same can be said about the documentation of the ancient engineering feats of the Persians, Indians, and Chinese. Because of the limited number of written accounts, relatively more is known about ancient Greek and Roman engineering. Around 100 B.C., several Greek writers created lists of the seven most magnificent engineering feats of which they were aware. Shown in Figure 2.1, the typical list included:

1. Great Pyramid at Giza, Egypt
2. Hanging Gardens of Babylon, Mesopotamia
3. Statue of Zeus at Olympia, Greece
4. Temple of Artemis at Ephesus, modern Turkey
5. Tomb of King Mausolos of Karia at Halikarnassos, Greece
6. Colossus of Rhodes, Mediterranean
7. Pharos Lighthouse of Alexandria, Egypt

Of the ancient wonders included on these lists of, the Pyramids of Egypt (circa 2700–1600 B.C.) alone survive in a recognizable form today. (See Figure 2.2 for more information on the Egyptian Pyramids.) The Greek writers could list only the wonders they had heard of, so the Great Wall of China, the dam at Ma'rib, which furnished water to a valley in southwest Arabia for about 1,000 years, the Buddhist stûpas of Sri Lanka, enormous domed structures over religious relics, and other feats of civil engineering are missing from the Greeks' lists.

Though civilization in Mesopotamia, "the land between the rivers" in Greek, may have begun several hundred years before Egypt's, little remains of its monumental architecture. Mesopotamia comprised most of the area that is modern-day Iraq. In ancient Babylon, this land was predominately desolate and barren except where water from the Tigris and Euphrates rivers provided irrigation. According to de Camp:

> In southern Mesopotamia, at the beginning of recorded history [5,000 to 6,000 years ago], the Sumerians—a people of unknown origins—built the city walls and temples and dug the canals that comprised the world's first engineering works. Here, for over two thousand years, little city-states bickered and fought over water rights. [p. 47]

Figure 2.2   Origins of the Egyptian Pyramids

Around 5,000 years ago, the typical Egyptian king or noble was buried in a rectangular, mud brick structure made of inward sloping walls and set over an underground chamber. Although unfired mud brick is a poor building material, these very early Egyptians learned that if they tapered the walls of their tombs inward from bottom to top, the walls did not crumble as quickly as if constructed vertically. Eventually these mud brick tombs gave way to stone structures; and although the reason for the sloping walls no longer existed, the stone tombs maintained the same form.

Around 2,700 B.C., the first engineer/architect known to us by name—Imhotep—emerged. Imhotep was a genius who was known as a builder, physician, statesman, writer, and overall sage. Together with his ruler, he embarked on improving the traditional tomb so that raiders would be less successful at entering. He and the king started with a stone tomb that was square rather than rectangular, 200' on a side and 26' high. Changes were made several times and the tomb was enlarged by adding stone to the sides. Before the second enlargement was finished, the king decided to build another level on top of the first. He changed his mind several times more, and the tomb that Imhotep finally built was comprised of six stages of decreasing size, or levels, over a burial chamber—the first step pyramid.

Succeeding generations of Egyptian kings also built step pyramids and eventually filled in the steps, creating true straight-sided pyramids. The largest of these was built around 2,500 B.C. near Giza, a town on the west bank of the Nile River, just upstream from Cairo. King Khufu, or Cheops as the Greeks called him, built his pyramid 756 feet square and about 480 high. The Great Pyramid is made of approximately 2,300,000 blocks of limestone, each weighing an average of two and one half tons, and is faced with a higher quality limestone. Most of the stone for the pyramids was cut from local stone outcrops. Early scholars theorized that the stone was dragged on sleds to the building sites. Based on existing evidence—tomb paintings, ancient tools found in modern times, tool marks on stones, and quarries with blocks partially detached—rollers under sleds were not used. Rather, workers may have poured liquid, possibly milk, on the soil in front of the sled to improve slipping. More recently researchers have found through experimentation that large stones fixed inside ingenious wood crates can be rolled by several workers, not the many required to pull sleds. Finer limestone used for exterior facing and granite used to line chambers had to be moved down the Nile on barges and then lugged to sites. To position these stones, agricultural workers conscripted during off-seasons, not slaves, used elaborate levers and ramps. As a pyramid rose, workers built a large earthen mound surrounding the structure. After one course was laid, the mound and accompanying ramp were raised to a new level. When the pyramid was complete, workers had to haul away this vast amount of soil; and masons standing on the ramp removed any irregularities in the facing stone. The last Egyptian pyramids were built approximately 1,600 B.C., possibly because of the prohibitive cost of construction or liberalization of religious doctrines. But Egyptian engineers learned much about quarrying, shaping, and moving heavy stones, and this knowledge became part of the world's collective technological wisdom.

—Adapted from L. Sprague de Camp, *The Ancient Engineers.* pp. 20–36.

Unfortunately for those interested in history, the Mesopotamian plain contained no stone suitable for building; and the only timber available had to be brought down the Tigris River from the Assyrian hills. Kiln-dried or burnt brick was expensive because of the lack of fuel (wood) for kilns. Consequently, the predominant building material was sun-dried mud brick, which was strong when dry but crumbled when wet. Mesopotamian temples and palaces were faced with kiln-dried brick but interiors were sun-dried. Consequently, when cracks developed and were left untended, sharp winter rains penetrated the mud brick within and the buildings eventually disintegrated. The upper part of Mesopotamian public structures has disappeared but their rubble has protected the foundations beneath.

One marvelous Mesopotamian invention was paving. On processional ways that were regular features of cities, Mesopotamian engineers lay flat bricks in a bed of mortar made from lime, sand, and asphalt. Sandstone flagstones were placed on top of the bricks. Special rules governed these sacred streets; parking the odd chariot or other vehicle along such a road could result in impalement. Eventually pavement was applied to major thoroughfares and then to heavily traveled roads outside main cities.

Mesopotamia is also home of the earliest recoded stone bridge over a river. Previous bridges had been made of tree trunks, reeds, or inflated goatskins. This first stone bridge over the Euphrates River was 380 feet long and rested on seven piers constructed of fired brick, stone, and timber. Due to shifts in the river channel over centuries, the sizes of the bridge and its large piers, 28 feet by 65 feet in plan, are known because the piers have been excavated in modern times. Most large, ancient bridges were constructed in a similar way—the piers took up half the width of the river.

Contemporary with the construction of the Egyptian Pyramids at Giza and the urbanization of Mesopotamia was the construction of Stonehenge. Stonehenge, located in what is now modern Britain, is a magnificent feat of ancient engineering and organization of human labor. Its real function and meaning are not yet clear, but the scale of effort and command of physical principles necessary to build it can be admired today. (See Figure 2.3 for more information on Stonehenge.)

Figure 2.3   Stonehenge

Based on radio carbon dating, Stonehenge was constructed more than 5,000 years ago. Its ancient British builders were working on it at the same time the Egyptians were constructing the Great Pyramids at Giza. But what was it and what purpose did this collection of giant stones fulfill? Part of a collection of remarkable stone circles in northwestern Europe, Stonehenge attracts a wide array of the curious interested in diverse topics like archeology, astronomy, meteorology, sacred geography, geomancy, and shamanism. In truth, knowledge from all of these fields is necessary to begin to explain something created by people who lived in the Neolithic (Stone) Age. Stonehenge was the centerpiece of a culture that flourished on the Salisbury plain in what is now western England. The builders moved Sarsen Stones, the tallest uprights and lintels weighing 50 tons each, from Fyfield Down over 20 miles away. The smaller blue stones came from the Preseli Mountains, Wales, over 150 miles. Evidentially, this was a very special site! The Sarsen Stones are a type of sandstone harder than granite. These stones were dressed with mauls weighing up to 65 pounds. Five trilithons [two uprights spanned by one lintel] originally stood in the center of the circle at a height of 17 to 25 feet. An approximately 100 foot diameter Sarsen Circle surrounded the trilithons. The bluestones formed an inner circle approximately 75 feet in diameter, as well as a horseshoe around the trilithons. Early researchers strove to understand the geometry of the stones, and the position of the stones is linked to the rising and setting positions of the Sun, Moon, and stars. One stone has been called the *Heel Stone* for centuries; some thought this because there is an indentation on the stone resembling a heel print. However, in the Old Welsh language *ffriw yr haul* is phonetically very similar and means, "appearance of the sun." Students of Stonehenge do not always agree on what the *ghost in the machine* is. What is known is that people very much like ourselves prepared, surveyed, and marked-out the site; transported the megaliths [very large stones]; and erected them. Adapted from: Robin Heath, *Stonehenge: Ancient Temple of Britain*

Persian Engineers

Around 550 B.C., Cyrus the Great founded the Persian Empire, modern Iran, which ruled the Near and Middle East for more than 200 years. Persian kings did not rule from a single capital but maintained four capitals among which they moved. Darius, his son Xerxes, and his grandson Artaxerxes labored for decades in the 5th century B.C. to create a magnificent royal center at what the Greeks named Persepolis, "Persian City," one of the four capitals. In 331 B.C. Alexander the Great burned Persepolis and it was abandoned. Alexander's action strangely conserved more to be appreciated in our times because the other three capitals continued as great cities, and what existed has been demolished or built upon.

The Persians spread ideas about building, such as a system of irrigation, far and wide. One of their innovations, a ghanat, was adopted widely. As shown in Figure 2.4, a ghanat is a sloping tunnel that conveys water from an underground source in a range of hills to a dry plain below. Less water is lost to evaporation than in an open-air aqueduct. To construct a ghanat, a line of vertical shafts is dug. The bottoms of these shafts are connected by the continuous tunnel. To allow workers to maintain the tunnel or to draw water, other shafts are dug at an angle from the surface to the tunnel. The water is distributed into irrigation channels when the tunnel reaches its destination. Ghanats were a family effort that could take several generations to complete. The ghanat system is still in use in the Middle East and North Africa today.

The Persians also accomplished major feats in military bridge-building. In 480 B.C. in a campaign involving the famous battle at Thermopylae that pitted the Persian King Xerxes and his forces against Leonidas and his 300 Spartans, Persian engineers constructed a pontoon bridge over the Hellespont, a narrow straight dividing Europe and Asia Minor. The price of failure for many early engineers was quite

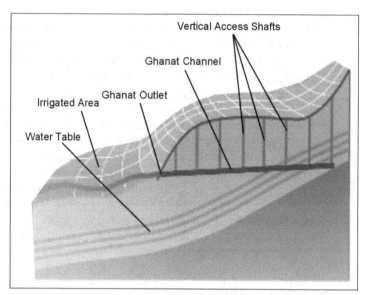

Figure 2.4 The Persian ghanat: Groundwater distribution

high—the first bridge was torn apart in a storm and the engineers were beheaded. The new engineers built a bigger bridge with larger factors of safety. The new bridge was constructed of 674 galleys anchored in a double row connected by two enormous flax cables and four cables of papyrus. Planks were laid at right angles over the cables, brush was piled on the planks, and finally soil was piled on the brush. Xerxes's army of perhaps 150,000 soldiers and an equal number of noncombatant camp followers marched over the bridge. (Early historians may have exaggerated this figure; nonetheless the number of troops was enormous.)

Other ancient civilizations around the Mediterranean—the Phoenician, Carthaginian, Lydian, Hebrew, Assyrian, Minoan—accomplished major feats of engineering. Over 3,000 years ago on the island of Crete, seafaring kings built exquisite unfortified palaces with stone walls and post-and-lintel colonnades. Ceramic drain pipes carried water away from baths and toilets, creating some of the first sanitary sewer systems (see Figure 2.5). The palaces were destroyed by earthquakes and were rebuilt more splendidly. The Minoan civilization collapsed around 400 B.C. and these magnificent buildings with their advanced sewer systems fell into disrepair.

Greek Engineers

Around 1000 B.C., when Kings David and Solomon ruled in Israel, Aryan invaders began to attack the Aegean coastline and Mediterranean islands. Three to four centuries later, the aggressors and locals had mingled to form a new people, the Greeks. The Greeks were influenced by other people—the Egyptians, Babylonians, and Phoenicians (seafaring people whose home was the coast of current-day Lebanon). But the

Figure 2.5   Sewers: Knossos, Crete, 3000–100 B.C.
(Wikipedia Commons)

Greeks began doing something different—they were starting to connect engineering and pure science, freed for the first time from the supervision of priests, who until then had controlled intellectual life. One of the earliest examples of a scientific approach to physical and mathematical problems applicable to civil engineering is the work of Archimedes in the 3d century B.C. To this day Archimedes's Principle continues to inform our understanding of buoyancy and offers practical solutions, such as Archimedes's screw, which can be used in irrigation to transfer water from a lower source (lake, creek, river) to a higher level (ditch, canal).

In the 5th century B.C. during Greece's Golden Age, the leader of Athens—Pericles—commissioned leading artists, architects, and engineers to cover the Acropolis with temples, shrines, and statues. The Acropolis is a huge rock outcropping whose top was reached through a winding, processional path. At the top of this path, worshippers entered the site through a monumental gateway, the Propylaia. The Propylaia is notable for the wrought iron bars used to reinforce marble ceiling beams, among the first known use of metal structural members in a building. One of the temples constructed on the Acropolis was the Parthenon, a temple to the goddess Athena. The Greek temple style spread all over the Mediterranean and lasted for centuries. It was revived in Renaissance Europe and again in the 19th and early 20th centuries. The style often has been used in the design of art museums, banks, churches, and memorials.

In the 4th century B.C., Alexander of Macedonia—Alexander the Great—subdued all of Greece and later conquered much of the Middle East, including Egypt, Mesopotamia, and Persia. This was a period of intellectual fervor, travel and tourism, scholarship and research, invention, and intermarriage of people and cultures, not unlike our own time. Aristotle and Plato lived then. In Egypt, Alexander ordered the

Ancient Structural Systems and Success Factors

In the ancient world, building styles depended on locally available materials: clay, stone, and wood. Buildings of antiquity utilized one or a combination of four devices to support roofs or upper stories:

1. Corbel—an "arch" that requires no falsework or shoring. Stones are layered in courses from two sides, overhanging each previous course until the two sides meet in the middle.
2. Post and lintel—a system of vertical columns crossed by horizontal beams.
3. Arch and vault
4. Truss

Mesopotamia had lots of clay but no stone or wood and, thus, preferred the corbel or arch and vault construction. Egypt had stone and clay, while Greece and China had stone, clay, and wood; these civilizations favored post-and-lintel construction. Europe had abundant sources of wood and consequently developed the truss.

Underpinning the success of ancient engineers were three factors:

1. Intensive and careful use of existing principles and tools, such as the water level and astronomical observation
2. Unlimited labor and the power to organize and command it
3. A different perspective of time

Perhaps the most important is the last factor—the Ancients seemed to have infinite patience.

Adapted from L. Sprague de Camp, *The Ancient Engineers*. pp. 26–31.

construction of the city of Alexandria. Alexander's successor created a splendid harbor and erected a magnificent lighthouse on a rocky islet called Pharos. The Pharos Lighthouse of Alexandria is one of the seven ancient Wonders of the World. A later ruler created an enormous library; and while it endured, the Library of Alexandria was the intellectual capital of the Mediterranean world.

Roman Engineers

The founding of ancient Rome is traced to the 8th century B.C., and the fall of the Roman Empire dates to the 5th century. Roman engineering, like that of other defining empires, relied on intensive application of existing principles and tools, cheap labor, and time. Rome possessed plenty of raw materials in the form of clay for brick, stone, and timber; and because of its rapid expansion, it also had an abundance of slave labor. Romans devoted more resources to constructing useful public works than their predecessors and developed civil structures throughout their empire, including aqueducts, harbors, bathhouses, markets, bridges, dams, and roads. Some scholars argue that Rome contributed little to pure science; but Roman genius had more to do with pragmatism. Remarkable Roman statesmen, soldiers, administrators, and jurists built on others' scientific findings and artistic creations.

Principally, Roman engineering is civil engineering. Romans themselves developed new building methods, which continued from the early years of the Empire's expansion in the 4th century B.C. for nearly 800 years. By the 1st century B.C., Vitruvius—author of the only surviving book on engineering and architecture from classical antiquity—wanted his ideal architect (engineer) to be a scholar, a skillful draftsperson, a mathematician, a student of philosophy, familiar with historical studies, acquainted with music, not ignorant of medicine, knowledgeable of jurisconsults' responses, and familiar with astronomy—a point of view that resonates with the ASCE's 2008 Body of Knowledge 2.

> Although Vitruvius gives sound reasons for having all these skills—for instance, a knowledge of music is useful in tuning catapults by striking the tension skeins—the difficulty is the same as in all professions in all ages. Vitruvius' requirements are a counsel of perfection, because nobody lives long enough to learn everything that might be useful to him. [de Camp, p. 174]

Perhaps this is not dissimilar to the situation in which the 21st century civil engineer finds himself or herself.

Vitruvius's vision was grand, however, the scale of invention and innovation was small in ancient times. When a city was sacked, some creations could be destroyed and then they vanished for centuries. Innovations were created, lost, and recreated. For example, a palace in Beycesultan, southwestern Anatolia (Turkey), built in 1200 B.C. and excavated in 1954, had indirect central heating. Ducts beneath the floor suggested a central heating plant. Then no evidence of this innovation appeared for a thousand years. By the 1st century B.C., Roman engineers had harnessed the power of geothermal springs for use in public baths. The concept also was applied to baths in houses, which were equipped with under-floor ceramic ducts through which air heated by fire passed. Eventually, Romans applied this system to whole buildings.

About 300 B.C., Romans discovered concrete. They found that when sandy volcanic ash was mixed with lime mortar, a cement formed and dried rock-hard, even under water. Concrete was created by mixing this cement with sand and gravel. After centuries of exposure, some examples of Roman concrete are harder than many natural rocks. At first, Romans only used concrete in limited applications—like a superior mortar. Then it began to replace the brick and stonework it was helping to bond in walls and fences.

The Romans not only invented new construction materials and ways of combining new and old materials, they also created new architectural forms. They excelled in building secular rather than religious edifices. Romans improved the arch and vault, making public buildings adaptable due to large, clear spans and giving these buildings a feeling of spaciousness. They developed methods of erecting huge, well-constructed buildings in a fraction of the time and for far less expense than other ancient engineers. They also were able to adapt circular dome ceilings to square or rectangular buildings. This was accomplished through the use of *pendentives*, triangular sections of masonry leaning in from the rectilinear base to connect with the circular base of the dome. Later in Byzantium, pendentives were used extensively. Their temples, however, followed the Greek style of post-and-lintel structures. Yet a Roman innovation was to make columns solid rather than a series of "drums" erected one on top of the other to form pieced columns, as the Greeks had done.

Romans also were master road builders. Only since the advent of the car have road-building standards returned to anything close to Roman engineering criteria. Today the Appian Way, constructed in the 4th century B.C., still exists southeast of Rome. Roman engineers designed their roads to require little maintenance and to last a minimum of 80 to 100 years; obviously, some lasted longer. Primarily, Roman roads were intended to enable an army to move swiftly. During most of the Roman Empire, the army was comprised of heavy infantry. So Roman engineers were more concerned with providing a firm footing for marching soldiers than hoofed animals. Important roads were paved, and most secondary and provincial roads were graveled. An unpaved strip might be included on either side of a paved road. Romans preferred to build roads as straight as land contours allowed and to go over hills rather than around, even if this meant grades of 20 percent.

Roman surveyors laid out routes using simple instruments, such as water levels and plumb-bobs, to establish horizontal lines. To determine right angles, they used a pair of boards nailed to make a right-angled cross, which they mounted on a post. Then they leveled and sighted along the cross pieces. A Roman paved road has been likened to a massive wall lying on its side. They started with a trench several feet deep; and if the soil was not firm at that depth, they drove piles. The rest of the road construction depended on the importance of the route and the availability of local materials. A major, fully paved road in Italy might be made of five layers totaling 4 feet thick and 6 to 20 feet wide. First was a layer of sand, mortar, or both. Then was a layer of small squared stones set in cement or mortar. Next was a layer of gravel set in clay or concrete, followed by a layer of rolled concrete made with sand aggregate. On top of everything were large blocks of hard rock, dressed on their upper surfaces, set in concrete.

The Romans also excelled in bringing water to their cities (see Figure 2.6). They were not the first to build aqueducts. The Mesopotamians, Greeks, and Phoenicians all had constructed them, but the Romans built more and bigger aqueducts. Rows of arches remaining from Roman aqueducts can be seen today in Italy, France, Spain, North Africa, Greece, and other locations in Asia Minor. Parts of several, those around Rome, Segovia (Spain), and Athens still function. Why did the Romans build

Figure 2.6   Roman aqueduct: Pont du Gard, Nîmes, France
(Wikipedia Commons)

so many? Most large cities are built on rivers; but even without the knowledge of bacteria, the ancients knew that spring water was better than river water. Additionally, Romans had inherited Greek ideals of civilized living that called for public fountains, baths, and gardens, all of which required water.

Roman engineers built aqueducts on a simple pattern consisting of a series of small round arches bearing on tall piers of stone or brick. Above was a water channel of concrete covered with an arched or gabled roof. When the aqueduct had to cross an exceptionally deep gorge, two or three rows of arches were erected on top of each other. Sometimes two or three channels shared the same arcade. Because the water flowing in open channels was moved by gravity all the way from the source to the point of distribution, the channels needed a fairly consistent downgrade of two to three feet per mile. The famous Pont du Gard in Nîmes, France, has three superimposed arcades. While arcades were the most conspicuous part, the vast majority of the Roman aqueduct system was in conduits and tunnels.

A Roman Civil Engineer's Field Report

Nonius Datus on the Difficulties of Building a Tunnel for an Aqueduct Saldae, Algeria, 152 A.D.

I found everybody sad and despondent. They had given up all hopes that the opposite sections of the tunnel would meet, because each section had already been

(*Continued*)

excavated beyond the middle of the mountain. As always happens in these cases, the fault was attributed to me, the engineer, as though I had not taken all precautions to ensure the success of the work. What could I have done better? For I began by surveying and taking the levels of the mountain, I drew plans and sections of the whole work, which plans I handed over to Petronius Celer, the Governor of Mauretania; and to take extra precaution, I summoned the contractor and his workmen and began the excavation in their presence with the help of two gangs of experienced veterans, namely, a detachment of marine infantry and a detachment of alpine troops. What more could I have done? After four years' absence, expecting every day to hear good tidings of water at Saldae, I arrive; the contractor and his assistants had made blunder upon blunder. In each section of tunnel they had diverged from the straight line, each towards right, and had I waited a little longer before coming, Saldae would have possessed two tunnels instead of one.

—Ivor B. Hart, The Great Engineers. p. 24.

Indian Engineers

Further to the east of the Mediterranean, kings kept up roads and irrigation systems, especially the great canal network in Babylonia. Iranians built many bridges, including one constructed in the 8th century near Susa that had abutments made of iron slag and lead, using these materials as concrete. East of Iran lies India, but much less is known about the ancient history of India. When the Persian King Darius conquered the Punjâb in approximately 515 B.C., Persian construction techniques were introduced into India; but the Indians continued to prefer to build in wood rather than stone. Darius's arrival corresponded with the life of Buddha, and Buddhism eventually brought changes to Indian building. Previously, Indian religious structures had been unpretentious, wooden religious shines. With the advent of Buddhism came monasteries, stupâs, and rock temples. These new structures required different building materials, and wood gave way to brick and later stone.

Indian monasteries either consisted of numerous cells built around a compound or were stepped in tiers up mountainsides. Stupâs housed relics of Buddhist saints. As can be seen in Figure 2.7, a stupâ's main feature was a domed structure over an actual relic. Also, each of the four sides of the base had a symbolic gateway. Some stupâs, such as those in Sri Lanka, were very grand, their domes having 300-foot diameters and the entire monuments' height reaching 250 feet. In the 2d century B.C., the foundations of one such stupâ were constructed of layers of stone, clay, and iron, all compacted by elephants wearing leather boots. Rock temples were built into the sides of rocky hills, similar to the one built by Rameses II at Abu Simbel, Egypt. Over many centuries, Hinduism and Buddhism vied for dominance and Hindus also constructed many magnificent temples carved from rock

Figure 2.7   Buddhist Stupâ, Polonnaruwa, Sri Lanka
(Wikipedia Commons)

hillsides. Other Hindu temples were large compounds accessed through monumental gateways.

Indians used post-and-lintel construction with domes and arches. Rather than mortar, Indian builders preferred to use iron dowels to join large stones. In fact, ancient Indians knew the secret of good steel. In Roman times, Indian steel was exported widely. Then and later, Indian steel found its way to Damascus, where it was made into sword blades that became famous for their strength and durability. The Iron Pillar of Dehli, dating approximately 415 A.D., was a 24-foot shaft that bore a manbird statue, the steed of the Hindu god Vishnu. The figure has been lost, but the column remains. Around the same time, Indians were making suspension bridges supported with iron chains.

Chinese Engineers

Due to great barriers such as jungles, mountains, deserts, forests, and seas, China remained largely cut off from the activities of the ancient Mesopotamians, Egyptians, Greeks, and Romans to the west. Based on archeological remains, a civilization similar to that of ancient Sumer existed in northwest China around 2000 B.C., located near a narrow band of passable territory that became a trade route. Another civilization arose approximately 500 years later in Hunan, also connected to a trade route. Once established, these trade routes had more or less continuous traffic. However, traders usually carried goods from point to point, perhaps over a few hundred miles.

Traders traveled several days from home and then returned. The route acted as a kind if filter, through which goods passed more easily than ideas. Therefore, Chinese engineers were largely cut off from western influences.

China had many building materials at its disposal and was not limited to clay bricks as were the ancient inhabitants of Mesopotamia. Stone foundations with wood and occasional brick superstructures topped by clay tile roofs were usual. The Chinese also knew about the barrel vault and used wood post-and-lintel construction; but they did not use the truss. By the 4th century B.C., the Chinese had discovered how to make cast iron. By the 10th century A.D. continuing to the 15th century A.D., the Chinese constructed pagodas, memorial towers adjacent to temples that were derived from the stupâ form, entirely of cast iron. They also supported suspension bridges with cables made of bamboo fiber. Through communication with India via Buddhist monks, bridges suspended from iron chains began to appear in China in the 8th century A.D.

Ancient China was not as large as it is today; most development took place in the north-central part of modern China. There were periods of relative unity and also of great division, with lesser rulers opposing the dynasty in power and imposing a sort of local feudal power. For centuries, nomadic peoples—Huns, Avars, Turks, Uighurs, Tartars, Mongols, and Uzbeks—invaded and conquered parts of China. Finally, around 220 B.C. the king of the Tsin people conquered all other contending states and founded the first centralized, autocratic rule. This emperor, Ch'in Shih Huang Ti, undertook the largest single engineering work of the ancient period—the Great Wall of China—in order to hold back the invaders. If the ancient Greeks, who compiled lists of the Seven Wonders of the World, had known of this magnificent creation, they certainly would have included the Great Wall.

As the crow flies, the Great Wall is 1,400 miles long; taking into consideration curves, branches, and loops, it stretches over 2,000 miles (3,200 kilometers). Under the direction of General Meng T'ien, construction of the wall began by first laying out farms along the route to supply workers with food. Ancient engineers had to think a lot like military generals planning for food and supplies to care for the large labor force needed to accomplish such enormous tasks. Construction varied in different sections depending on the existence of previously constructed wall sections and availability of local building materials. In most places the wall was 30 feet high, 25 feet wide at the base, and 15 feet wide at the crest. The paved road that capped the structure had a 6-foot crenellated parapet (wall with a zigzag top) on the invader side and 3-foot crenellated parapet on the homeland side. There were watchtowers at regular intervals, averaging 35 feet square and 45 feet high. In most sections, the core of the wall was rammed earth or rubble faced with cut stone, fieldstone, or brick and mortar (see Figure 2.8). The masonry was excellent—the emperor ruled that any worker who left a crack into which a nail could fit should be beheaded instantly. Quality problems are generally handled with a little more relaxed attitude today. An inexperienced autocratic manager might go as far as firing (beheading) a noncompliant engineer, but a more experienced manager would likely counsel the engineer about the need for improvement and quality control.

Figure 2.8   Great Wall of China at Jinshanling
(Wikipedia Commons)

African Engineers

Bordering the Mediterranean, the Roman Empire extended across the top of the African continent, but the Roman Empire's prosperity did not extend beyond its boundaries. The sociopolitical arrangement of mainland Africa contrasted sharply with that of the Romans. Hundreds of self-reliant groups lived in the arid sub-Saharan zone, tropical savannahs, coastal forests, and in quiet river basins.

According to Spiro Kostof, noted architectural historian:

> They were tied by the same basic verities: "a house, a family and the respect of old age." They built few religious structures. Material permanence was not a concern. On the contrary, built forms were something that responded to the changing circumstances of daily life and the domestic family cycle; they could be adapted, extended, replaced, or moved . . . . The permanence was in the land and its spirits, the generational patterns of self-preservation and reverence. (pp. 219–220)

Because of the wide variation in climate and landscape, there was no pan-African style of building. Construction methods included *banco*, a wet-clay process similar to coil pottery, and other readily available building materials—stones, wood, grass, animal skins. Groupings of homesteads and villages reflected social structures and functions. Each family unit had a grinding house and granary, stable, and beer store; these buildings were grouped and linked by straight or enclosing walls.

However, in the south of modern-day Zimbabwe on a high plateau between the Zambezi and Limpopo rivers, lie the Great Zimbabwe ruins. The Great Zimbabwe ruins are what is left of a once thriving religious and civic center. Starting in the 11th century for approximately 300 years, the ancestors of today's Shona people, who

Figure 2.9   Great Zimbabwe
(Wikipedia Commons)

currently inhabit that area, enjoyed a thriving cattle-based economy and traded luxury goods with outsiders—beads and pottery fragments have been found from China, Persia, and other Middle Eastern countries. The Great Zimbabwe site offered abundant resources: reliable rainfall, freedom from tsetse flies and malaria, ample woodlands for timber and fuel, arable soil for crops, open grasslands for grazing cattle, minerals, and gold.

From north to south, the Great Zimbabwe site is approximately one half mile (800 meters). It can be divided into three distinct zones: structures atop a kopje (ridge of granite); a Great Enclosure on a flat granite shelf across a valley from the kopje; and a number of smaller structures on the shallow slopes of the valley. The Great Enclosure, or Elliptical Enclosure, is the largest ancient structure in southern Africa. The builders used stone blocks occurring naturally and made others by heating and then quenching granite. They also made use of puddled clayey soil mixed with daga (a fine aggregate), which set hard in the sun. This material functioned as wall plaster as well as floor covering (see Figure 2.9).

American Engineers

Although war and trade connected Europe, Asia, and Africa, the American continents stood alone from the rest of the ancient world. Of course, the earliest Americans came from Asia across what is now the Bering Straight approximately 30,000 years ago. Once these inhabitants had settled in North, Central, and South America, they adopted a wide variety of building materials and systems. The earliest Americans built stone domes braced with whale bone, igloos made from snow and ice, gabled cedar houses, and partially sunken pit houses, for protection against the cold and wind. In the Southwest of the United States there is evidence of flood irrigation systems dating from several centuries B.C.

By the 5th century B.C., in central Mexico and the Gulf Coast the social order involved a ruling class with priests who were experts on the calendar and weather.

The priests interceded with the divine on behalf of the agriculturally based population. Large monumental structures began to be constructed that magnified the importance of religion and the state. On the island of La Venta, located in the mangrove wetlands in the Mexican state of Tabasco, a rounded pyramid was constructed, perhaps, earlier than the 5th century B.C. Apparently, La Venta was a civic and religious center where a small number of priests and possibly a related labor force were housed.

La Venta foreshadowed the developments at Teotihuacán, a magnificent religious center and premier market town located 25 miles (40 kilometers) northeast of modern-day Mexico City. One can visit Teotihuacán today and experience its principle characteristics, which were developed between 100 B.C. and 200 A.D. Teotihuacán's axis, now called the "Avenue of the Dead," is approximately 3 miles (5 kilometers) long. At the north end of this axis is the Pyramid of the Moon and along the east side are the Temple of Quetzalcóatl, with its famed feathered serpent heads, and the towering Pyramid of the Sun. Originally, there were hundreds of smaller platforms along this axis. As with many ancient sites, the building at Teotihuacán seeks to capture the nature of the cosmic order.

As shown in Figure 2.10, both the Pyramid of the Sun and the Pyramid of the Moon are terraced. The Pyramid of the Sun was made of horizontal layers of clay faced with unshaped stones. The newer Pyramid of the Moon was built with a core of tufa (volcanic stone) piers and rubble, which filled the shafts between the piers. Angled walls buttressed the core and determined the slope of the main terraces.

To the south, in the Yucatán Peninsula of Mexico, Guatemala, San Salvador, and Honduras, the Maya culture reached its zenith between 600 A.D. and 900 A.D. Maya land is exemplified by temples—stepped stone pyramids with temples set atop,

Figure 2.10 Teotihuacán, Mexico D.F.
(Wikipedia Commons)

ball courts, and clusters of one-story buildings sometimes referred to as "palaces," but whose actual function is not clear. Occasional burials have been found inside these temple pyramids, but their primary purpose was different from Old Kingdom Egyptian pyramids. The temple on top of Mayan pyramids contained a two-room sanctuary where human sacrifices were made. The temple used post-and-lintel construction topped with corbel vaults. Steel hard sapodilla wood formed the lintels and the walls were made of rubble, lime mortar, and a casing of cut stone. Several million Mayans built ritual centers at Palenque, Chichén Itzá, Uxmal, Tikal, and Copan, and many lesser sites.

Further to the south and many centuries later, the Incas flourished in an immerse territory including all of modern Peru, Ecuador, and northern Chile. With their capital in Cuzco, Peru, the Incas had been active in the Andean highlands since 1200 A.D. When their toughest adversaries—the Chimu—submitted, the Incas found themselves the uncontested masters of an enormous domain. The Incas and the Romans had much in common.

At the apex of their imperial power, the Incas ruled a superbly organized and well-administered domain, which was connected by a vast network of roads. All roads started in Cuzco and spread out to the four quarters into which the empire was divided, not on the basis of compass points but on the land's topography. The 3,100-mile (5,000-kilometer) royal road cut through the Andes. Draft animals and the wheel were not known, so the roads were not paved. Incan roads accommodated people and llamas by tunneling through spurs, becoming stairs at sharp ridges, and providing stone-lined causeways in swamps. Suspension bridges across valleys were made of enormous plant-fiber cables anchored by stone towers. At regular intervals, there were posts for runners carrying official messages and resthouses for bureaucrats and merchants. Masonry walls of perfectly matched polygonal blocks characterized Inca building, and examples can be seen in Cuzco today. See Figure 2.11 for more information about Incan civil engineering.

Back in the Valley of Mexico, the Aztec federation was the strong power. Contemporaries of the Inca, the Aztecs had a dazzling metropolis the size of London. Tenochtitlán is the site of modern-day Mexico City. The Aztecs were latecomers from the north of Mexico who emerged as a cohesive group in the area around 1200 A.D. They established Tenochtitlán as their capital in about 1325 A.D. on an island in Lake Texcoco, a salt lake. By 1450 A.D. they occupied a position of primacy in central Mexico. By filling large containers of wickerwork (chinampas) with mud, they reclaimed wetlands and turned them into arable land. Three causeways that doubled as levees lead from terra firma to the central plaza of Tenochtitlán, where twin temple pyramids replaced earlier shrines in the 1480s A.D. A fourth causeway stopped at the eastern bank of the island. Fresh water was brought to the city from Chapultepec and Coyoacán via aqueducts.

Figure 2.11   Machu Picchu: An Inca Engineering Marvel

Machu Picchu, the most well-known Inca archeological site, was home to a permanent population of 300 residents; but the inhabitants grew to 1,000 when the
Inca emperor was in residence. Situated near the headwaters of the Amazon
River in the Peruvian Andes Mountains, Machu Picchu was inhabited primarily
between 1450—1540 AD, prior to the arrival of the Spanish Conquistadores. Due
to its effectively engineered foundation and drainage systems, the royal retreat
survived—abandoned in a South American rainforest—for over 400 years. Hiram
Bingham, a professor from Yale University, 'discovered' Machu Picchu in 1911.
The city that he and other 20th century scientists investigated was nearly in the
same condition as it had been four centuries earlier.

Macho Picchu's engineered drainage system and foundations are the secret
of its longevity. Without good foundations and drainage, many of its buildings
would have crumbled and its agricultural terraces would have been un-
recognizable due to high levels of rainfall, sheer slopes, slide-prone soils, and sub-
sidence. The engineers and builders of Machu Picchu gave serious consideration
to both surface and subsurface water drainage. Extensive excavations have
shown a deep subdrainage system under the agricultural terraces, as well as urban and agricultural drainage channels integrated with stairways, walkways, and
temple interiors. Additionally, there are over 100 strategically placed drain outlets in numerous stone building and retaining walls. Like many other early engineers, the Inca builders constructed Machu Picchu to last an eternity, perhaps
the original version of sustainable engineering.

Adapted from: Wright, K.R., A.V. Zegarra, and W.L. Lorah (1999). "Ancient Machu Picchu
Drainage Engineering," *Journal of Irrigation and Drainage Engineering.* November/
December 1999.

Of course, most of us know the story of Cortés and the Spanish invasion that occurred in 1519:

> Bernal Díaz del Castillo, who was there, left a vivid description of what they saw. He speaks of the communities along the shoreline, of boats on the lake bringing foodstuffs and carrying out merchandise, of terrace houses, of the aqueduct coming in from Chapultepec. . . . Of the market of Tlatelolco, he writes: "There were among us soldiers who had been to many parts of the world, to Constantinople, to the whole of Italy and to Rome, and they said they had never seen a market so well organized and orderly, so large, so full of people." [Kostof, p. 438]

No wonder the Spaniards found Tenochtitlán so impressive. At the time of the Spanish invasion of the Americas, Europe was just awakening from the medieval era and the Renaissance was preparing to make its debut.

The Power of Ancient Building

"To chart a place on earth—that is the supreme effort of the built environment in antiquity . . . .To mediate between cosmos and polity, to give shape to fear and exorcise it, to affect a reconciliation of knowledge and the unknowable   that was the charge of ancient architecture."

Spiro Kostof, *A History of Architecture: Settings and Rituals*, 2d edition. pp. 240–241.

ENGINEERING IN MEDIEVAL TIMES

The term "medieval" literally means "between ages" and is used to describe the time in Western Europe between the end of the Roman era and the beginning of the Renaissance in the 15th century. Of course, the people living then had no concept that they were between anything—except perhaps a rock and a hard spot.

Much has been said of the fall of the Roman Empire, usually dated 476. While civilization continued in the eastern Mediterranean, Iran, Iraq, India, and China as before, the fall of the Western Roman Empire was no small event. Due to the lack of a strong central government, Roman roads, aqueducts, and harbors fell into ruin over a vast area. In the West, communities demolished Roman buildings to make fortifications and dismantled roads and bridges to slow down marauding Goths, Germans, and Vikings. Literacy almost vanished, science became superstition, and engineering deteriorated to rule-of-thumb craftsmanship.

In the Byzantine Empire, which was an extension of the Roman Empire in Asia Minor, certain trends continued that had started when the Roman Empire was united; but these had to do more with governance and religion than Roman engineering. In the late 11th century when the Crusaders arrived in Byzantium, the capital of which was Constantinople or modern-day Istanbul, they had little in common with their allies. The Byzantines had also spent centuries battling invaders.

Intellectual activity was the domain of churchmen, some of whom did scientific work; the general attitude toward science, however, was one of indifference or hostility.

Meanwhile, in the 7th century a religious revolution led by Muhammad ibn-Abdallah took place in Arabia. Within one century Islam had spread from Spain to Turkestan. These invaders were quicker to master the arts of those whom they conquered than the Germanic people who overran the Western Roman Empire. Starting in approximately 750 for a century and a half, the caliphs (rulers) in Baghdad employed scholars to translate Western wisdom into Arabic. For the two previous centuries the Persians had done the same at Jundishapur, translating Greek and Sanskrit into their language. Thus, the Middle East became the intellectual center of the Mediterranean-facing world.

In terms of building, the Arabs continued using the system of fortifications, walls with battlements and towers, developed by the Romans and Byzantines. The mosque was a distinctly Muslim style of building that used domes and arches. Another uniquely Muslim development was the minaret—a tall, slender tower from which the public are called to prayer. The need for and interest in irrigation and canal building continued.

In Europe during what was once called the Dark Ages, between the 6th and 10th centuries, engineering and architecture stopped being recognized as professions. Design and construction were carried out by artisans, such as stone masons and carpenters, rising to the role of master builder. Knowledge was retained in guilds and advances in technology came slowly. For many centuries in Western Europe, construction in stone became rare while wood and plaster were common, resulting in the half-timbered medieval building style. Churches were constructed in the Romanesque style. These were rather plain, massive stone buildings with small windows and many round arches.

The 12th and 13th centuries were a period when conflicts between the major monotheistic religions, Christianity and Islam, and schisms within them were an everyday reality. There were frenzied outbreaks of religious hysteria and fanaticism, including massacres of "heretics." European feudal lords fought incessantly. However, engineering began to regain some of the ground lost after the fall of Rome. Scholars pondered the nature of motion, force, and gravity; and Medieval builders made advances in structural forms. In addition to the semi-circular arch of the Romans, the Islamic pointed arch was introduced. Another advance was the use of the truss to support roofs. Unfortunately, no one could analyze these structures so Medieval roof trusses had unnecessary members that contributed to visual clutter but nothing to the trusses' load-carrying capacity.

The most significant engineering achievement of the time, however, was the development of the Gothic cathedral. The word "Gothic" meant *barbarous* to the Italians (due to the name of one of the early invading ethnic groups, the Goths), but the style spread over most of Europe. As shown in Figure 2.12, Gothic cathedrals were characterized by soaring vaulted interiors and large stained-glass windows. In anticipation of modern skyscrapers, the structure of the Gothic cathedral was a skeleton, represented by piers and flying buttresses. The walls were used to keep out the

Figure 2.12 Chartres Cathedral, France: Gothic masterpiece (Wikipedia Commons)

weather, not as structural support. Vaults were developed that enabled clear spaces of over 100 feet high. Lacking scientific principles, Medieval builders relied on trial and error. The roof of Beauvais Cathedral with a ceiling of 154 feet, the tallest of all Gothic cathedrals, collapsed twice. These massive undertakings could take several generations to complete.

The other noteworthy building type of this period was the fortified castle. Feudal warfare encouraged castle building. Until the advent of gunpowder, these edifices were so successfully engineered that they could withstand sieges for months and often were captured only through treachery. One of the best preserved European style castles, Kerak des Chevaliers, was built in modern-day Syria for the Knights Hospitallers of St. John in the 12th century A.D. Ironically, the finest Medieval Muslim palace remaining today is the Alhambra, in Granada, Spain.

Medieval times also saw advances in the use of water wheels. The ancients had used water wheels for raising water and for milling grains. The notebook of a 13th-century craftsman shows a water-powered sawmill. In the later Middle Ages, water power also was applied to the bellows of smelting furnaces, to trip hammers for crushing ore or bark in tanneries, and to grinding and polishing armor and other metal wares.

Improvements also were made in canal building. Canals enabled people and goods easier movement than did the existing rutted, unpaved roads; and the development of the lock changed everything. The origins of the canal lock are uncertain, but this innovation dates to the late 14th century in The Netherlands or Italy. In the 1450s the engineer Bertola da Novate put forward-looking ideas about locks into practice:

The dukes and republics of North Italy kept Bertola, the ablest canal builder of his time, busy all his life digging canals for them. Sometimes they quarreled over who should have priority on his services. His only trouble was that his workmen sometimes could not understand his advanced concepts.

[de Sprague, p. 381]

ENGINEERING IN THE RENAISSANCE AND THE AGE OF ENLIGHTENMENT

The term "Renaissance," which means *rebirth*, applies to Western Europe in the 15th through 16th centuries. In a narrow sense, the name refers to the revival of learning that took place in that period. Fashionable people had at least a veneer of scholarship. Study of classical antiquity, the writing and architecture of Greece and Rome, became vogue. However, many other sweeping changes also were taking place: the Reformation, world exploration, the downfall of the old astronomy that put Earth at the center of the universe, and the creation of the first patent systems for encouraging innovation.

Engineering again grew to be respected, and engineers became famous and, sometimes, well paid. They were no longer anonymous craftsmen; they promoted themselves and were not shy about arguing with employers or rivals. One of the earliest engineers of the Renaissance was the Florentine Filippo Brunelleschi. He mastered perspective drawing and competed for and won the commission to build the famous dome on Florence's cathedral, Santa María del Fiore (shown in Figure 2.13), among other accomplishments. Brunelleschi first competed for the award in 1407, received

Figure 2.13 Florence Cathedral: Brunelleschi's Renaissance dome
(Wikipedia Commons)

the order to build in 1419, and finished the task in 1436. The entire cathedral is 351 feet high, and the dome is 105 feet high (approximately ten stories) and 143 feet in diameter. The City of Florence also gave Brunelleschi the first known patent, for a canal boat fitted with cranes capable of moving heavy cargo. Like others of the same period—Leonardo da Vinci, Michelangelo Buonarroti, Andrea Palladio—Brunelleschi had to serve as both a civil and military engineer.

Most early Renaissance engineers achieved fame through word of mouth. Later in the 15th century, the printing press helped to disseminate engineering knowledge. An Italian engineer/architect/painter/philosopher/musician/poet, Leon Battista Alberti, wrote a book in Latin on rules of thumb for the proportions of structures, such as bridges. This work originally was published in 1452 and circulated in manuscript form among Alberti's friends. Later, however, it was translated into Italian, French, Spanish, and English. In 1472, Roberto Valturio published a book that surveyed the state of military engineering. In the 1580s, Palladio, who had perfected the bridge truss, wrote about that subject and others in *I quattro libri dell' architectura* (*The Four Books of Architecture*).

Through the Spanish Inquisition (starting in the 15th century and lasting several hundred years) and Counter-Reformation (16th to mid-17th centuries), a dim light shone on science, as demonstrated by the threats of torture to which Galileo Galilei was subjected for proposing that the Earth did indeed rotate around the sun, rather than the other way around. Italy did not fare well during this period because in addition to changing religious views, armies from France, Spain, and the Holy Roman Empire (centered in Vienna, Austria), kept waging war there. But technical progress did continue elsewhere. About the 16th century people began to write about "modern discoveries"; and by the 18th century, engineering schools appeared in France.

This was the Age of Enlightenment (18th century) and many unforeseen changes were taking place. In an Enlightened Europe there was a strong appetite for attack on the church. The church began to lose power to nations; the Jesuits were expelled from Portugal, France, Spain, and Naples. As in the early Renaissance when Henry VIII of England seized monastic property, many of the Jesuits' holdings became "available." Access to this property, social unrest, capitalism, and the notion that existing structures could be replaced with more up-to-date ones helped to establish land as a liquid, negotiable commodity.

THE INDUSTRIAL REVOLUTION

At the close of the 18th century, the first stirrings of the Industrial Revolution were beginning to be felt. In England, earlier than in the rest of Western Europe, the transition from an agrarian, handcraft-based economy to a machine-dominated economy was underway. The trend had earlier roots, but mechanized labor, inanimate power—particularly steam—and inexpensive raw materials accelerated dramatic changes. Workers were moving away from home-based (cottage) industry and shops to mills and factories. In England the countryside was under assault as scores of towns emerged around country plants making anything from cast iron to cotton cloth. In the country, industry could flourish away from the influence of guilds and government regulations.

Up until the late 18th century, military engineers had undertaken the construction of public infrastructure in support of expanding industry. However, in 1768, an Englishman named John Smeaton is credited with being the first person to call himself a civil engineer. By describing himself as a 'civil engineer,' Smeaton identified a new and distinct profession that encompassed all nonmilitary engineering. Smeaton's work was backed by thorough research, and he became a member of the prestigious Royal Academy of Engineering. In 1771, he founded the Society of Civil Engineers (now known as the Smeatonian Society). His objective was to bring together engineers, entrepreneurs, and lawyers to promote the building of large public works, such as canals (and later railways). These new professionals also recognized that they needed to obtain parliamentary approval necessary to execute their schemes.

The Industrial Revolution brought with it new materials and methods for producing and using them. Cast and wrought iron are good examples. As early as 1780, cast iron columns began to be substituted for wood posts supporting the roofs of cotton mills in England. Bricks and timber (lumber) were produced using industrial methods and glass began to replace oiled paper as window coverings. Structural innovations accompanied these developments enabling spectacular early applications in bridges and railroad tracks.

Iron Bridge, designed by Thomas Farnolls Pritchard, is an outstanding monument to both civil engineering and the Industrial Revolution. In 1779, Iron Bridge, the world's first cast iron bridge, opened for traffic over the River Severn in Coalbrookdale, Shropshire, England. The bridge was cast in the local foundries by a man named Abraham Darby III. His grandfather, Abraham Darby, was the first to use less-expensive iron, rather than brass, to cast strong thin pots for the poor. Under his son and grandson, the Coalbrookdale foundry flourished. In 1777, Abraham Darby III began erecting 378 tons of cast iron to build the bridge, which spans 100 feet (30 meters) (see Figure 2.14).

The development of mills and factories in the countryside attracted workers by tens of thousands. Because good roads and rail systems did not yet exist, canals connecting locks, wharves, boatyards, limekilns, and warehouses were constructed at a frantic pace. The first public railroad opened in 1825. The race was on to shrink distance and speed up time.

The use of iron and glass continued to shake up traditional construction methods. According to Kostof, "Not since the Roman invention of concrete had a building technology so radicalized architecture." [p. 595] Actually, Kostof continues to say that architects were not so thrilled about the appearance of cast iron and tended to conceal or decorate it in the sort of public buildings they specialized in designing. Its characteristics, however, were impossible not to appreciate. Iron was less expensive than stone and possessed exciting mechanical properties: It withstood fire better than wood and it could be prefabricated, shipped to the site, and assembled with relative ease.

Figure 2.14 Iron bridge over the River Severn, Coalbrookdale, Shropshire, England: First use of cast iron in a bridge
(Wikipedia Commons)

As early as 1813 an iron and glass dome was built over a granary in Paris. Fifteen years later iron and glass roofs were used to span commercial arcades and shopping streets for Parisian pedestrians. England's most innovative uses of iron were railroad stations and bridges. Civil engineers embraced these new materials and created magnificent, awe-inspiring new structural forms.

For a time, the Scot Thomas Telford, first president of the Institution of Civil Engineers (ICE) in the United Kingdom, lived near Iron Bridge; he must have been fascinated by what he saw. He later used cast iron in many innovative bridge designs, including a chain suspension bridge over the Menai Straight in Wales (see Figure 2.15). French immigrant to the United Kingdom, Marc Brunnel, and his son, Isambard Kingdom Brunnel, also pushed the limits of civil engineering design and construction with projects such as the first tunnel under the River Thames for the new underground rail system in London. Isambard Kingdom Brunnel went on to design railroads, bridges (see Figure 2.16), train stations, and a ship—he also owned the Great Western Railroad. Brunnel's design for Paddington Station in London (1849–1854) resulted in a flexible covered space without columns. New railways were regarded as sources of future prosperity for provincial cities and towns, and the public took intense interest in Brunnel's daring schemes.

As the Industrial Revolution rolled along, many social changes were taking place. One significant development was the rise of the professions. New societal needs, commerce, educational opportunities, and exciting developments in technology converged. Institutions and societies were created to lend credibility, codify conduct, and provide a place where meetings of minds could occur. The following years are important in the development of civil engineering and architecture as professions:

- Institution of Civil Engineers (ICE) —launched 1818
- Royal Institute of British Architects (RIBA) —launched 1834
- American Society of Civil Engineers (ASCE) —launched 1852
- American Institute of Architects (AIA) —launched 1857

In the United States, other civil engineers were designing and building canals, railroads, municipal water systems, and bridges. The Croton Aqueduct (Figure 2.17) was a 41-mile (66-kilometer) water distribution system constructed for New York City between 1837 and 1842. It brought water from the Croton River into reservoirs in Manhattan. During the 1830s, New York City desperately needed a fresh water supply to combat both disease and fire. After numerous proposals and a plan abandoned after two years, construction began in 1837 under the expertise of John Bloomfield Jervis.

The field of civil engineering grew with the times. A German immigrant to the United States, John Roebling, designed the first suspension bridge using steel cables—the Brooklyn Bridge. Planning for the bridge began in 1867 and construction was completed in 1883. The Brooklyn Bridge stretches 5,989 feet (1,825 meters) over

Figure 2.15   Thomas Telford: Menai chain suspension bridge, Wales; Craigellachie cast iron bridge, Scotland; elevation and view of construction
(Wikipedia Commons)

ASCE's Profile

The American Society of Civil Engineers, a professional organization representing more than 146,000 civil engineers, celebrated its 150th anniversary in 2002. When the 12 founders gathered at the Croton Aqueduct on November 5, 1852, and agreed to incorporate the American Society of Civil Engineers and Architects, one can only wonder if they dreamed the profound significance and long-lasting impact ASCE would have on the overall development of society. They laid a foundation for what proves to be one of the most prominent engineering societies in the world.

Today ASCE is a worldwide leader for excellence in civil engineering. With a mission to advance professional knowledge and improve the practice of civil engineering, ASCE is a focal point for the development and transfer of research results, and technical policy and managerial information. Through strategic emphasis in key areas, including infrastructure renewal and development, policy leadership and professional development, ASCE delivers the highest quality publications, programs, and services to its worldwide membership, demonstrating a daily commitment to sustaining the profession.

As civil engineering enters a new millennium, the American Society of Civil Engineers not only reflects on the profession's rich heritage, but equipped with this knowledge, ASCE continues to develop flexible, forward-thinking plans for the future of the society and the civil engineering profession.

—ASCE website, www.asce.org

Figure 2.16   Isambard Kingdom Brunnel; Clifton suspension bridge, Bristol, England (Wikipedia Commons)

**Figure 2.17   Croton Aqueduct: Clean water for New York City
(Wikipedia Commons)**

the East River and connects the New York City boroughs of Manhattan and Brooklyn (see Figure 2.18). At the time of its completion, it was the longest suspension bridge in the world.

MODERN CIVIL ENGINEERING

Civil engineering has continued to evolve. The 20th century saw increasing specialization and advancements in theoretical understanding, materials and methods, and technologies. (See Figure 2.19.)

Just as the Greeks compiled a list of The Wonders of the Ancient World, the American Society of Civil Engineers has compiled a list of wonders of the modern world, which are summarized and pictured in Figure 2.20.

Other innovative projects continue to excite the imagination. The Millau Viaduct (shown in Figure 2.21), a large cable-stayed road-bridge spanning the valley of the River Tarn in southern France, was completed in 2004. Designed by structural engineer Michel Virlogeux and British architect Norman Foster, it is the tallest vehicular bridge in the world. One mast's summit is 1,125 feet (343 meters), only 125 feet (38 meters) shorter than the Empire State Building. The bridge won the 2006 International Association for Bridge and Structural Engineering (IABSE) Outstanding Structure Award.

Taipei 101, completed in 2005 in Taipei, Taiwan, was the world's tallest building until being surpassed by Burj Khalifa. Designed by C.Y. Lee & Partners and constructed by Samsung Engineering & Construction, Taipei 101 incorporates many innovations necessary to build skyscrapers in earthquake and high wind zones (see Figure 2.22).

Figure 2.18 John August Roebling; Niagara suspension bridge; Cincinnati suspension bridge; Brooklyn Bridge
(Wikipedia Commons)

Figure 2.19 Founders of modern geotechnical engineering

From left to right:

Karl Terzaghi: Published his theory of consolidation in 1923, which taken together with his earlier theories on earth pressures and piping, established modern soil mechanics.

Arthur Casagrande: As a professor at Harvard University, continued and expanded upon Terzaghi's work through experimentation and development of soil testing techniques.

Ralph Peck: Taught at the University of Illinois, conducted research and an active international consulting practice, coauthored *Soil Mechanics in Engineering Practice* with Terzaghi in 1948; emphasized judgment in engineering practice.
(Wikipedia Commons)

(Wikipedia Commons)

	Modern Wonder	Started	Finished	Location
1	Channel Tunnel	1987	1994	Strait of Dover, between the United Kingdom and France
2	CN Tower • Tallest freestanding structure in the world 1976–2007	1973	1976	Toronto, Ontario, Canada
3	Empire State Building • Tallest structure in the world 1931–1967 • First building with 100+ stories	1930	1931	New York, NY, U.S.
4	Golden Gate Bridge	1933	1937	Golden Gate Strait, north of San Francisco, California, U.S.
5	Panama Canal	1880	1914	Isthmus of Panama, Panama
6	Delta Works/Zuiderzee Works	1950	1997	The Netherlands
7	Itaipu Dam	1970	1984	Paraná River, between Brazil and Paraguay

Source: American Society of Civil Engineers website: www.asce.org
Figure 2.20   ASCE's seven wonders of the modern world

Figure 2.21 Millau cable-stayed road bridge, France: tallest vehicular bridge in the world (Wikipedia Commons)

The building is 101 stories above ground (1,670 feet, 509 meters) and five stories underground. A steel-tuned mass damper (TMD) weighing 662 metric tons and consisting of 41 layered steel plates welded together to form a 5.5-meter diameter sphere is suspended from the 92d and 88th floors. The TMD acts like a giant pendulum to counteract the building's movement, reducing sway by 30 to 40 percent.

Figure 2.22 Taipei 101, Taiwan: incorporates many innovations necessary to build skyscrapers in earthquake and high wind zones (Wikipedia Commons)

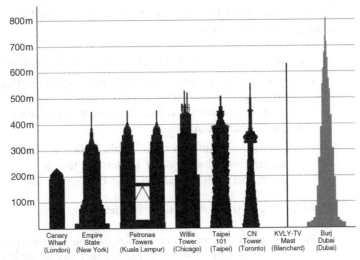

Figure 2.23 The world's tallest buildings
(Wikipedia Commons)

Burj Khalifa, formerly called Burj Dubai, has held the record for the world's tallest building at 2,717 feet (828 meters) since 2010. A collection of the world's tallest man-made structures is depicted in Figure 2.23.

CIVIL ENGINEERING EDUCATION

Increases in the civil engineering body of knowledge have resulted in a formalized approach to civil engineering education. The École Polytechnique was founded in Paris in 1794, and the Bauakademie was started in Berlin in 1799, but no such schools existed in Great Britain or the United States until several decades later. The University of Glasgow, Scotland, was the first university school of engineering in the United Kingdom to establish a chair in civil engineering. The first degree in Civil Engineering in the United States was awarded by Rensselaer Polytechnic Institute, New York, in 1835.

Today's civil engineering is linked to advances in understanding of physics, mathematics, and the social and political forces of its time. Civil engineers typically earn a Bachelor of Science (B.S.) degree with a major in civil engineering, though some universities award a Bachelor of Engineering. Students usually pursue their studies for four or five years. Typical civil engineering programs initially cover most, if not all, of the subdisciplines of civil engineering. Students then choose to specialize in one or more subdisciplines toward the end of their degrees.

As discussed in Chapter 1, ideally the degree should include units covering topics in three major categories:

- *Foundational*—mathematics, natural sciences, humanities, and social sciences
- *Technical*—materials science, mechanics, experiments, problem recognition and solving, design, sustainability, contemporary issues/historical perspectives, risk and uncertainties, project management, breadth in civil engineering areas, and technical specialization
- *Professional*—communication, public policy, business and public administration, globalization, leadership, teamwork, attitudes, lifelong learning, and professional and ethical responsibility

According to the ASCE's Body of Knowledge 2:

Engineering does not occur in a vacuum, and engineers must be able both to explain the impact of historical and contemporary issues on engineering and to explain the impact of engineering on the world. [p. 130–131]

In most countries, a Bachelor's degree in civil engineering represents the first step toward professional registration or licensure, and the degree program itself is accredited by a professional body, such as ABET. After completing an accredited degree program, the civil engineer must satisfy a range of requirements (including work experience and exam requirements) before becoming registered or licensed.

The National Council of Examiners for Engineering and Surveying (NCEES) administers the civil engineering professional engineer (Civil PE) exam. After passing the EIT (Engineer in Training) exam, the prospective engineer is tested with a:

- **Breadth exam** (morning session): This exam contains questions from all five areas of civil engineering: Construction, Geotechnical, Structural, Transportation, and Water Resources and Environmental
- **Depth exams** (afternoon session): These exams focus more closely on a single area of practice in civil engineering. Examinees must choose one of the following areas: Construction, Geotechnical, Structural, Transportation, and Water Resources and Environmental. [NCEES]

Once licensed, the civil engineer is designated the title of *Professional Engineer* (in the United States, Canada, and South Africa), *Chartered Engineer* (in most British Commonwealth countries), *Chartered Professional Engineer* (in Australia and New Zealand), *European Engineer* (in much of the European Union), and *Professional Engineer* in many Asia countries. There are international engineering agreements between relevant professional bodies that are designed to allow engineers to practice across international borders:

Civil Engineering Associations

American Society of Civil Engineers

Canadian Society for Civil Engineering

Chi Epsilon, Civil Engineering honor society

Earthquake Engineering Research Institute

Engineers Australia

Institution of Civil Engineers (UK)

Institute of Structural Engineers (UK)

Institute of Transportation Engineers

Royal Academy of Engineering (UK)

Transportation Research Board

The Institution of Civil Engineering Surveyors

A complete list of international societies with which ASCE has cooperative agreements is available at http://content.asce.org/international/AOC.html.

The advantages of registration or licensure vary depending upon location. For example, in the United States and Canada most licensing organizations use something like the following quote: "only a licensed engineer may prepare, sign and seal, and submit engineering plans and drawings to a public authority for approval, or seal engineering work for public and private clients." This requirement is enforced by state and provincial legislation. In other countries, no such legislation exists. Most professional associations of civil engineers, such as the American Society of Civil Engineers, the British Institution of Civil Engineers (ICE), and the British Institute of Structural Engineers (ISE) maintain a code of ethics by which members are expected to abide or risk expulsion. In this way, these organizations play an important role in maintaining ethical standards for the profession. (See Chapter 3, Ethics, of this text.)

Even in countries where licensure has little or no legal bearing on work, engineers are subject to contract law. In cases where an engineer's work fails he or she may be subject to the tort of negligence and, in extreme cases, the charge of criminal negligence. An engineer's work must also comply with numerous other rules and regulations, such as building codes and legislation pertaining to environmental law. (In this book, see Chapter 8, Permitting and Chapter 11, Legal Aspects of Professional Practice.)

CIVIL ENGINEERING CAREERS

There is no one typical career path for civil engineers. Most engineering graduates start with entry-level positions, and as they prove their competence, they gain more and more significant tasks. In some fields and firms, entry-level engineers are put to work primarily monitoring construction in the field, serving as the "eyes and ears" of more senior design engineers. In other areas, entry-level engineers perform routine tasks of analysis or design and interpretation. Senior engineers can execute complex analysis or design work. They also can work in project management of design projects, or management of other engineers, or specialized consulting. Civil engineers are in high demand at financial institutions and management consultancies because of their analytical skills. They can find many career opportunities in high technology for the same reason.

Areas of civil engineering specialization have changed over time due to society's needs and the complexities of projects and technologies. Currently, the ASCE incorporates the following Institutes:

- Architectural Engineering (AEI)
- Coasts, Oceans, Ports, and Rivers (COPRI)
- Construction (CI)
- Engineering Mechanics (EMI)
- Environmental and Water Resources (EWRI)
- Geo (G-I)
- Structural Engineering (SEI)
- Transportation & Development (T&DI)

The activities and responsibilities of civil engineers working in these various areas are included in Table 2.1.

Table 2.1   Civil Engineering Areas of Concentration

Area	Activities and Responsibilities
General Civil	• Focuses on the overall interface of projects with their environments • Applies the principles of geotechnical engineering, structural engineering, environmental engineering, transportation engineering, and construction engineering to residential, commercial, industrial, and public works projects of all sizes and levels of construction • Works closely with surveyors and specialized civil engineers • Designs grading plans, drainage, pavement, water supply, sewer service, electric and communications supply, and land divisions • Visits project sites, develops community consensus, and prepares construction plans and specifications
Coastal	• Helps manage coastal areas • Defends against flooding and erosion • Designs ports • Also works to reclaim land
Construction	• Plans and executes the designs from transportation, site development, hydraulic, environmental, structural and geotechnical engineers • Writes and/or reviews contracts • Evaluates logistical operations • Controls prices of necessary materials, operations, and equipment

Area	Activities and Responsibilities
Environmental	• Deals with the treatment of chemical, biological, and/or thermal waste, purification of water and air, and the remediation of contaminated sites • Works with pollution reduction, green engineering, and industrial ecology • Reports information on the environmental consequences of proposed actions and the assessment of effects of proposed actions for the purpose of assisting society and policy-makers in the decision-making process, i.e., writes environmental impact reports (EIRs)
Geotechnical	• Deals with complex nature of rock and soil, subsurface investigation and testing, foundations and earth structures (dams, levees, engineered fills, etc.) • Depends on knowledge from the fields of geology, material science and testing, mechanics, and hydraulics to design foundations, retaining structures, land fills and similar structures • Can specialize further to use biology and chemistry to devise ways of disposing of hazardous materials and groundwater contamination (called geoenvironmental engineering) • Contrasts with the relatively well-defined material properties of steel and concrete used in other areas of civil engineering
Land Surveying (considered a distinct profession in the United States, Canada, the United Kingdom, and most Commonwealth countries)	• Establishes the boundaries of a parcel of land using its legal description and subdivision plans • Lays out the routes of railways, tramway tracks, highways, roads, pipelines, and streets as well as positions other infrastructures, such as harbors, before construction • Employs surveying equipment, such as levels and theodolites, for accurate measurement of angular deviation, horizontal, vertical, and slope distances • Makes use of electronic distance measurement (EDM), total stations, global position system (GPS) surveying, and laser scanning with computerization, have supplemented (and to a large extent supplanted) the traditional optical instruments
Municipal or Urban Engineering	• Involves specifying, designing, constructing, and maintaining municipal infrastructure, such as streets, sidewalks, water supply networks, sewers, street lighting, municipal solid waste management and disposal, storage depots for various bulk materials used for maintenance and public works (salt, sand, etc.), public parks, and bicycle paths • Includes the civil portion (conduits and access chambers) of the local distribution networks of electrical and telecommunications services • Focuses on the coordination of infrastructure networks and services, as they are often built and managed by the same municipal authority
Structural	• Analyses and designs the structures of buildings, bridges, towers, overpasses, tunnels, offshore structures like oil and gas fields in the sea, and other structures • Identifies the loads which act upon a structure and the forces and stresses that arise within that structure due to those loads
	• Considers strength, stiffness, and stability of the structure when it is subjected to its own self weight, other dead loads, live loads, including furniture, wind, seismic, crowd or vehicle loads, or transitory, such as temporary construction loads • Also takes into account aesthetics, cost, constructability, safety, and sustainability wind engineering and earthquake engineering • Can specialize further (wind and earthquake engineering)

(Continued)

Table 2.2 (*Continued*)

Area	Activities and Responsibilities
Transportation	• Deals with moving people and goods efficiently, safely, and in a manner conducive to a vital community • Plans this movement using queuing theory, Intelligent Transportation Systems (ITS), and infrastructure management • Designs, constructs, and maintains transportation infrastructure, including streets, canals, highways, rail systems, airports, ports, and mass transit • Investigates and specifies paving materials • Involves transportation design, transportation planning, traffic engineering, some aspects of municipal/urban engineering
Water Resources	• Combines hydrology, environmental science, meteorology, geology, conservation, and resource management in the collection and management of water as a natural resource • Relates to the prediction and management of both the quality and the quantity of water in underground resources (aquifers) and above ground resources (lakes, rivers, and streams) • Analyzes and models very small to very large areas to predict the amount and content of water as it flows into, through, or out of a facility such as pipelines, water distribution systems, drainage facilities (including bridges, dams, channels, culverts, levees, storm sewers), and canals

SUMMARY

In the broad sense, civil engineering has been a necessary component of life since the beginning of human existence. Between 4000 and 2000 B.C. in Mesopotamia and ancient Egypt, humans started to abandon a nomadic existence, requiring increasingly complex structures such as fortifications, temples, and residences. During the medieval period, craftsmen such as masons and carpenters, carried out most building. Knowledge was retained in guilds that were not compelled to make changes to their well-guarded knowledge.

As time marched on, new materials, methods, and societal demands evolved. The need for specialization and regulation also grew. Professional organizations were launched and laws were created requiring the licensing of professional civil engineers. As a "formal" profession, civil engineering dates from the Industrial Revolution. Over time, the profession has developed several subdisciplines including coastal engineering, construction engineering, environmental engineering, geotechnical engineering, municipal or urban engineering, structural engineering, surveying, transportation engineering, and water resources engineering.

The history of civil engineering is linked to developments in mathematics and science and other fields. Civil engineers now have specialized educations involving diverse topics that enable them to recognize and solve problems. In addition to mathematics and science, practicing civil engineers need to know about design; sustainability and other contemporary issues; risks and uncertainties; project management; communication; public policy; business; leadership; teamwork; and professional and ethical responsibility. Although the average citizen may not recognize the role of civil engineers in society, civil engineers continue to shape the quality of our lives. As the ASCE puts it:

Civil engineers build the world's infrastructure. In doing so, they shape the history of nations around the world.

ASCE History and Heritage website:

http://content.asce.org/history/index.html

"We shape our buildings, and afterwards our buildings shape us."
—*Winston Churchill, address in the House of Parliament, London, October 28, 1943*

REFERENCES

American Society of Civil Engineers. (2008). *Civil Engineering Body of Knowledge for the 21st Century*, 2d edition. Prepared by the Body of Knowledge Committee on the Academic Prerequisite for Professional Practice. ASCE, Reston, VA. ISBN-13: 978-0-7844-0965-7

Bachner, John Philip. (1991). *Practice Management for Design Professionals: A Practical Guide to Avoiding Liability and Enhancing Profitability.* John Wiley & Sons, New York. ISBN 0-471-52205-8

"civil engineering." Encyclopædia Britannica. 2009. Encyclopædia Britannica Online. Nov. 5, 2009, www.britannica.com/EBchecked/topic/119227/civil-engineering.

de Camp, L. Sprague. (1993). *The Ancient Engineers.* Ballantine Books, New York. (First published 1960.) ISBN 0-345-48287-5

Hart, Ivor B. (1928). *The Great Engineers.* Methuen & Co., London.

James, William, Professor of Water Resources Engineering, "A historical perspective on the development of urban water systems," University of Guelph, Guelph, Ontario, Canada.

Kostof, Spiro. (1995). *A History of Architecture: Settings and Rituals*, 2d edition . Oxford University Press, New York. ISBN-13 978-0-19-508378

Landel, J.G. (2000). *Engineering in the Ancient World*, rev. ed. (Berkeley and Los Angeles, 2000).

Petrowski, Henry. (1994). *Design Paradigms: Case Histories of Error and Judgment in Engineering.* Cambridge University Press, Cambridge. ISBN 0-521- 46108-1 (hardcover), ISBN 0-521- 46649-0 (paperback).

Scarborough, Vernon L. (2003). *The Flow of Power—Ancient Water Systems and Landscapes. A School of American Research Resident Scholar Book*, SAR Press. Santa Fe, New Mexico. ISBN 1-930618-32-8

Vali-Khodjeini, Ali. (1995). Human impacts on groundwater resources in Iran. Man's Influence on Freshwater Ecosystems and Water Use (Proceedings of a Boulder Symposium, July 1995. IAHS Publication No. 230.

William, N. Morgan. (2008). *Earth Architecture: From Ancient to Modern*. ISBN-13 978-0-8130- 3207-8. ISBN-10: 0-813032075

Wright, K.R., A.V. Zegarra, and W.L. Lorah (1999). "Ancient Machu Picchu Drainage Engineering," *Journal of Irrigation and Drainage Engineering*. November/December 1999.

Introduction
Ethics
1 2 3 4 5 6 7 8 9 10 11 12 13 14 15 16 17 A B C D E F
History
Professional Engagement
What Engineers Deliver
Emerging Technology
Engineer's Role in Project
Development
Permitting
Leadership
Managing
Having
a Life
Executing a Professional
Commission
Client Relationship
Legal Aspects
Communicating
Globalization
Sustainability

Chapter **3**

Ethics

Big Idea

"Engineers shall act in such a manner as to uphold and enhance the honor, integrity, and dignity of the engineering profession and shall act with zero-tolerance for bribery, fraud, and corruption."

—Cannon 6—ASCE Code of Ethics

Key Topics Covered

- Introduction
- Defining The Engineer's Ethical Code
- The American Council of Engineering Companies (ACEC) Ethical Conduct Guidelines
- The American Society of Civil Engineers (ASCE) Code of Ethics
- The National Society of Professional Engineers (NSPE) Code of Ethics
- The International Federation of Consulting Engineers (FIDIC)
- Important and Relevant Policy Statement By ASCE and NSPE
- Case Studies
- Summary

Related Chapters in This Book

- Ethics are related to every chapter in this book
- Related to *ASCE Body of Knowledge 2* Outcomes

(Continued)

ASCE BOK2 outcomes covered in this chapter

Foundational
1. Mathematics
2. Natural sciences
3. Humanities
4. Social sciences

Technical
5. Materials science
6. Mechanics
7. Experiments
8. Problem recognition and solving
9. Design
10. Sustainability
11. Contemporary issues & historical perspectives
12. Risk and uncertainty
13. Project management
14. Breadth in civil engineering areas
15. Technical specialization

Professional
16. Communication
17. Public policy
18. Business and public administration
19. Globalization
20. Leadership
21. Teamwork
22. Attitudes
23. Lifelong learning
●24. Professional and ethical responsibility

INTRODUCTION

Like several other chapters in this book, the challenge with writing a chapter on ethics is that there are many volumes of references for this subject. The challenge then becomes how to sort through thousands of pages of text to produce reference material for practical use by the engineer. It's logical then, to begin with the definition of ethics.

Ethics is referred to as moral philosophy, and recommends concepts of right and wrong behavior for professionals practicing within a profession. Contemporary ethical theories can be divided into three general subject areas: metaethics, normative ethics, and applied ethics. Metaethics investigates the origin of ethical principles, their meaning and source within society. There is some question about whether metaethics involve more than expressions of our individual emotions. Metaethical subjects include issues focusing on universal truths such as the reality and the will of God and logical reasoning in ethical judgments. Normative ethics include practical issues like moral standards society sets to regulate right and wrong conduct. Sometimes this causes global or international conflict when one nation's standards appear to be violated by another nation's practice. A contemporary example of this could be human rights, an appropriate age to marry, or how males treat females in a particular society. It could also involve good habits, ethical or moral duties like caring for our family elders, or the consequences our behavior has on others like smoking in public places. Finally, applied ethics involve examining debatable, controversial issues, such as the death penalty, environmental concerns, same sex marriage, and animal rights, among others. For the purpose of ethics related to professional engineering, the focus is on *normative ethics* where the practicing professionals set standards to regulate right and wrong conduct.

DEFINING THE ENGINEER'S ETHICAL CODE

Before defining the code it's important to discuss why it even exists. An interesting study from the Illinois Institute of Technology—Center for the Study of Ethics in the Professions notes:

> The adoption of a code is significant for the professionalization of an occupational
> group, because it is one of the external hallmarks testifying to the claim that the
> group recognizes an obligation to society that transcends mere economic self-interest
> (p. 138).

Michael Davis makes a strong positive case for professional codes of ethics. Davis argues that codes of ethics should be understood as conventions between professionals. Davis writes,

> The code is to protect each professional from certain pressures (for example, the
> pressure to cut corners to save money) by making it reasonably likely . . . that most
> other members of the profession will not take advantage of her good conduct. A code

protects members of a profession from certain consequences of competition. A code is a solution to a coordination problem. (p. 154)

Davis goes on to suggest that having a code of ethics allows an engineer to object to pressure to produce substandard work not merely as an ordinary moral agent, but as a professional. Engineers (or doctors, clergy, and other professionals) can say "As a professional, I cannot ethically put business concerns ahead of professional ethics."

Davis gives four reasons why professionals should support their professions code:

First . . . supporting it will help protect them and those they care about from being injured by what other engineers do. Second, supporting the code will also help assure each engineer a working environment in which it will be easier than it would otherwise be to resist pressure to do much that the engineers would rather not do. Third, engineers should support their professions code because supporting it helps make their profession a practice of which they need not feel . . . embarrassment, shame, or guilt. And fourth, one has an obligation of fairness to do his part . . . in generating these benefits for all engineers. (p. 166)

Harris and colleagues summarize Stephen Unger's analysis of the possible functions of a code of ethics:

First, it can serve as a collective recognition by members of a profession of its responsibilities. Second, it can help create an environment in which ethical behavior is the norm. Third, it can serve as a guide or reminder in specific situations . . . Fourth, the process of developing and modifying a code of ethics can be valuable for a profession. Fifth, a code can serve as an educational tool, providing a focal point for discussion in classes and professional meetings. Finally, a code can indicate to others that the profession is seriously concerned with responsible, professional conduct. (p. 35)

To better understand the actual description of the language, reference, and detail in an engineer's code of ethics we can visit the code as accepted by several engineering organizations including:

- The American Society of Civil Engineers (ASCE)
- The National Society of Professional Engineers (NSPE)
- The American Council of Engineering Companies (ACEC)
- The International Federation of Consulting Engineers (FIDIC)

Remember, as *engineers* we regulate the ethical standards for the profession and if they appear to need revising there's an amendment process in place. The codes can be revised within the organization and agreed upon as professionals within our profession. Ethical standards should be *usable and live* codes accepted and followed by all engineers. Adherence to ethical standards makes choosing the right road for civil engineers obvious. (See Figure 3.1.)

Figure 3.1   Choosing the right road

For a clear and succinct definition we can begin with the ACEC. (The ASCE, NSPE, and FIDIC codes will be presented later on.)

Code of Ethics Definitions

- *Principles* refer to a fundamental and comprehensive doctrine (morality) regarding behavior and conduct.
- *Canons* are broad principles of conduct.
- *Standards* are more specific goals toward which individuals should aspire in professional performance and behavior.
- *Rules of Conduct* are mandatory; violation of a Rule usually is grounds for disciplinary action. Rules can implement more than one Canon or Standard.

THE AMERICAN COUNCIL OF ENGINEERING COMPANIES ETHICAL CONDUCT GUIDELINES

The ACEC Guidelines

Preamble
Consulting engineering is an important and learned profession. The members of the profession recognize that their work has a direct and vital impact on the quality of life for all people. Accordingly, the services provided by consulting engineers require

honesty, impartiality, fairness and equity and must be dedicated to the protection of public health, safety and welfare. In the practice of their profession, consulting engineers must perform under a standard of professional behavior which requires adherence to the highest principles of ethical conduct on behalf of the public, clients, employees and the profession.

I. Fundamental Canons
Consulting engineers, in the fulfillment of their professional duties, shall:

1. Hold paramount the safety, health and welfare of the public in the performance of their professional duties.
2. Perform services only in areas of their competence.
3. Issue public statements only in an objective and truthful manner.
4. Act in professional matters for each client as faithful agents or trustees.
5. Avoid improper solicitation of professional assignments.

II. Rules of Practice
1. Consulting engineers shall hold paramount the safety, health and welfare of the public in the performance of their professional duties.

   a. Consulting engineers shall at all times recognize that their primary obligation is to protect the safety, health, property and welfare of the public. If their professional judgment is overruled under circumstances where the safety, health, property or welfare of the public are endangered, they shall notify their client and such other authority as may be appropriate.

   b. Consulting engineers shall approve only engineering work which, to the best of their knowledge and belief, is safe for public health, property and welfare and in conformity with accepted standards.

   c. Consulting engineers shall not reveal facts, data or information obtained in a professional capacity without the prior consent of the client except as authorized or required by law or these Guidelines.

   d. Consulting engineers shall not permit the use of their name or firm nor associate in business ventures with any person or firm which they have reason to believe is engaging in fraudulent or dishonest business or professional practices.

   e. Consulting engineers having knowledge of any alleged violation of these Guidelines shall cooperate with the proper authorities in furnishing such information or assistance as may be required.

2. Consulting engineers shall perform services only in the areas of their competence.

   a. Consulting engineers shall undertake assignments only when qualified by education or experience in the specific technical fields involved.

b. Consulting engineers shall not affix their signatures to any plans or documents dealing with subject matter in which they lack competence nor to any plan or document not prepared under their direction and control.

c. Consulting engineers may accept an assignment outside of their fields of competence to the extent that their services are restricted to those phases of the project in which they are qualified and to the extent that they are satisfied that all other phases of such project will be performed by registered or otherwise qualified associates, consultants or employees, in which case they may then sign the documents for the total project.

3. Consulting engineers shall issue public statements only in an objective and truthful manner.

a. Consulting engineers shall be objective and truthful in professional reports, statements or testimony. They shall include all relevant and pertinent information in such reports, statements or testimony.

b. Consulting engineers may express publicly a professional opinion on technical subjects only when that opinion is founded upon adequate knowledge of the facts and competence in the subject matter.

c. Consulting engineers shall issue no statements, criticisms, or arguments on technical matters which are inspired or paid for by interested parties, unless they have prefaced their comments by explicitly identifying the interested parties on whose behalf they are speaking and by revealing the existence of any interest they may have in the matters.

4. Consulting engineers shall act in professional matters for each client as faithful agents or trustees.

a. Consulting engineers shall disclose all known or potential conflicts of interest to their clients by promptly informing them of any business association, interest or other circumstances which could influence or appear to influence their judgment of the quality of their services.

b. Consulting engineers shall not accept compensation, financial or otherwise, from more than one party for services on the same project, or for services pertaining to the same project, unless the circumstances are fully disclosed to, and agreed to, by all interested parties.

c. Consulting engineers in public service as members of a governmental body or department shall not participate in decisions with respect to professional services solicited or provided by them or their organizations in private engineering practices.

d. Consulting engineers shall not solicit or accept a professional contract from a governmental body on which a principal or officer of their organization serves as a member.

5. Consulting engineers shall avoid improper solicitation of professional assignments.

    a. Consulting engineers shall not falsify or permit misrepresentation of their, or their associates', academic or professional qualifications. They shall not misrepresent or exaggerate their degree of responsibility in or for the subject matter of prior assignments. Brochures or other presentations incident to the solicitation of assignments shall not misrepresent pertinent facts concerning employees, associates, joint ventures or past accomplishments with the intent and purpose of enhancing their qualifications and their work.

    b. Consulting engineers shall not offer, give, solicit or receive, either directly or indirectly, any political contribution in an amount intended to influence the award of a contract by public authority, or which may be reasonably construed by the public of having the effect or intent to influence the award of the contract. They shall not offer any gift or other valuable consideration in order to secure work. They shall not pay a commission, percentage or brokerage fee in order to secure work except to a bona fide employee or bona fide established commercial or marketing agencies retained by them.

THE AMERICAN SOCIETY OF CIVIL ENGINEERS CODE OF ETHICS

The ASCE Code of Ethics[1,2]

Fundamental Principles[3]

Engineers uphold and advance the integrity, honor and dignity of the engineering profession by:

    a. using their knowledge and skill for the enhancement of human welfare and the environment;

    b. being honest and impartial and serving with fidelity the public, their employers and clients;

    c. striving to increase the competence and prestige of the engineering profession; and

    d. supporting the professional and technical societies of their disciplines.

---

[1] www.acec.org/, Adopted October 1980

[2] The Society's Code of Ethics was adopted on September 2, 1914 and was most recently amended on July 23, 2006. Pursuant to the Society's Bylaws, it is the duty of every Society member to report promptly to the Committee on Professional Conduct any observed violation of the Code of Ethics.

[3] In April 1975, the ASCE Board of Direction adopted the fundamental principles of the Code of Ethics of Engineers as accepted by the Accreditation Board for Engineering and Technology, Inc. (ABET).

Fundamental Canons

a. Engineers shall hold paramount the safety, health and welfare of the public and shall strive to comply with the principles of sustainable development[4] in the performance of their professional duties.

b. Engineers shall perform services only in areas of their competence.

c. Engineers shall issue public statements only in an objective and truthful manner.

d. Engineers shall act in professional matters for each employer or client as faithful agents or trustees, and shall avoid conflicts of interest.

e. Engineers shall build their professional reputation on the merit of their services and shall not compete unfairly with others.

f. Engineers shall act in such a manner as to uphold and enhance the honor, integrity, and dignity of the engineering profession and shall act with zero-tolerance for bribery, fraud, and corruption.

g. Engineers shall continue their professional development throughout their careers, and shall provide opportunities for the professional development of those engineers under their supervision.

Guidelines to Practice Under the Fundamental Canons of Ethics

Canon 1

Engineers shall hold paramount the safety, health and welfare of the public and shall strive to comply with the principles of sustainable development in the performance of their professional duties.

a. Engineers shall approve or seal only those design documents, reviewed or prepared by them, which are determined to be safe for public health and welfare in conformity with accepted engineering standards.

b. Engineers whose professional judgment is overruled under circumstances where the safety, health and welfare of the public are endangered, or the principles of sustainable development ignored, shall inform their clients or employers of the possible consequences.

c. Engineers who have knowledge or reason to believe that another person or firm may be in violation of any of the provisions of Canon 1 shall present such

---

[4] In November 1996, the ASCE Board of Direction adopted the following definition of Sustainable Development: "Sustainable Development is the challenge of meeting human needs for natural resources, industrial products, energy, food, transportation, shelter, and effective waste management while conserving and protecting environmental quality and the natural resource base essential for future development."

information to the proper authority in writing and shall cooperate with the proper authority in furnishing such further information or assistance as may be required.

d. Engineers should seek opportunities to be of constructive service in civic affairs and work for the advancement of the safety, health and well-being of their communities, and the protection of the environment through the practice of sustainable development.

e. Engineers should be committed to improving the environment by adherence to the principles of sustainable development so as to enhance the quality of life of the general public.

Canon 2
Engineers shall perform services only in areas of their competence.

a. Engineers shall undertake to perform engineering assignments only when qualified by education or experience in the technical field of engineering involved.

b. Engineers may accept an assignment requiring education or experience outside of their own fields of competence, provided their services are restricted to those phases of the project in which they are qualified. All other phases of such project shall be performed by qualified associates, consultants, or employees.

c. Engineers shall not affix their signatures or seals to any engineering plan or document dealing with subject matter in which they lack competence by virtue of education or experience or to any such plan or document not reviewed or prepared under their supervisory control.

Canon 3
Engineers shall issue public statements only in an objective and truthful manner.

a. Engineers should endeavor to extend the public knowledge of engineering and sustainable development, and shall not participate in the dissemination of untrue, unfair or exaggerated statements regarding engineering.

b. Engineers shall be objective and truthful in professional reports, statements, or testimony. They shall include all relevant and pertinent information in such reports, statements, or testimony.

c. Engineers, when serving as expert witnesses, shall express an engineering opinion only when it is founded upon adequate knowledge of the facts, upon a background of technical competence, and upon honest conviction.

d. Engineers shall issue no statements, criticisms, or arguments on engineering matters which are inspired or paid for by interested parties, unless they indicate on whose behalf the statements are made.

e. Engineers shall be dignified and modest in explaining their work and merit, and will avoid any act tending to promote their own interests at the expense of the integrity, honor and dignity of the profession.

Canon 4
Engineers shall act in professional matters for each employer or client as faithful agents or trustees, and shall avoid conflicts of interest.

a. Engineers shall avoid all known or potential conflicts of interest with their employers or clients and shall promptly inform their employers or clients of any business association, interests, or circumstances which could influence their judgment or the quality of their services.

b. Engineers shall not accept compensation from more than one party for services on the same project, or for services pertaining to the same project, unless the circumstances are fully disclosed to and agreed to, by all interested parties.

c. Engineers shall not solicit or accept gratuities, directly or indirectly, from contractors, their agents, or other parties dealing with their clients or employers in connection with work for which they are responsible.

d. Engineers in public service as members, advisors, or employees of a governmental body or department shall not participate in considerations or actions with respect to services solicited or provided by them or their organization in private or public engineering practice.

e. Engineers shall advise their employers or clients when, as a result of their studies, they believe a project will not be successful.

f. Engineers shall not use confidential information coming to them in the course of their assignments as a means of making personal profit if such action is adverse to the interests of their clients, employers or the public.

g. Engineers shall not accept professional employment outside of their regular work or interest without the knowledge of their employers.

Canon 5
Engineers shall build their professional reputation on the merit of their services and shall not compete unfairly with others.

a. Engineers shall not give, solicit or receive either directly or indirectly, any political contribution, gratuity, or unlawful consideration in order to secure work, exclusive of securing salaried positions through employment agencies.

b. Engineers should negotiate contracts for professional services fairly and on the basis of demonstrated competence and qualifications for the type of professional service required.

c. Engineers may request, propose or accept professional commissions on a contingent basis only under circumstances in which their professional judgments would not be compromised.

d. Engineers shall not falsify or permit misrepresentation of their academic or professional qualifications or experience.

e. Engineers shall give proper credit for engineering work to those to whom credit is due, and shall recognize the proprietary interests of others. Whenever possible, they shall name the person or persons who may be responsible for designs, inventions, writings or other accomplishments.

f. Engineers may advertise professional services in a way that does not contain misleading language or is in any other manner derogatory to the dignity of the profession. Examples of permissible advertising are as follows:

- Professional cards in recognized, dignified publications, and listings in rosters or directories published by responsible organizations, provided that the cards or listings are consistent in size and content and are in a section of the publication regularly devoted to such professional cards.

- Brochures which factually describe experience, facilities, personnel and capacity to render service, providing they are not misleading with respect to the engineer's participation in projects described.

- Display advertising in recognized dignified business and professional publications, providing it is factual and is not misleading with respect to the engineer's extent of participation in projects described.

- A statement of the engineers' names or the name of the firm and statement of the type of service posted on projects for which they render services.

- Preparation or authorization of descriptive articles for the lay or technical press, which are factual and dignified. Such articles shall not imply anything more than direct participation in the project described.

  Permission by engineers for their names to be used in commercial advertisements, such as may be published by contractors, material suppliers, etc., only by means of a modest, dignified notation acknowledging the engineers' participation in the project described. Such permission shall not include public endorsement of proprietary products.

g. Engineers shall not maliciously or falsely, directly or indirectly, injure the professional reputation, prospects, practice or employment of another engineer or indiscriminately criticize another's work.

h. Engineers shall not use equipment, supplies, laboratory or office facilities of their employers to carry on outside private practice without the consent of their employers.

Canon 6

Engineers shall act in such a manner as to uphold and enhance the honor, integrity, and dignity of the engineering profession and shall act with zero-tolerance for bribery, fraud, and corruption.

a. Engineers shall not knowingly engage in business or professional practices of a fraudulent, dishonest or unethical nature.

b. Engineers shall be scrupulously honest in their control and spending of monies, and promote effective use of resources through open, honest and impartial service with fidelity to the public, employers, associates and clients.

c. Engineers shall act with zero-tolerance for bribery, fraud, and corruption in all engineering or construction activities in which they are engaged.

d. Engineers should be especially vigilant to maintain appropriate ethical behavior where payments of gratuities or bribes are institutionalized practices.

e. Engineers should strive for transparency in the procurement and execution of projects. Transparency includes disclosure of names, addresses, purposes, and fees or commissions paid for all agents facilitating projects.

f. Engineers should encourage the use of certifications specifying zero-tolerance for bribery, fraud, and corruption in all contracts.

Canon 7

Engineers shall continue their professional development throughout their careers, and shall provide opportunities for the professional development of those engineers under their supervision.

a. Engineers should keep current in their specialty fields by engaging in professional practice, participating in continuing education courses, reading in the technical literature, and attending professional meetings and seminars.

b. Engineers should encourage their engineering employees to become registered at the earliest possible date.

c. Engineers should encourage engineering employees to attend and present papers at professional and technical society meetings.

d. Engineers shall uphold the principle of mutually satisfying relationships between employers and employees with respect to terms of employment including professional grade descriptions, salary ranges, and fringe benefits.[5]

When the ACEC Code of Ethics and the ASCE Code of Ethics are compared, the ACEC code appears concise and direct while the ASCE code includes a greater amount of detail included in the "Guidelines to Practice Under the Fundamental Canons of Ethics" section above. Both codes are excellent examples and this detail helps crystallize potential questions or concerns that practicing engineers may have about the code of ethics adopted by the organization.

---

[5] www.asce.org/

NATIONAL SOCIETY OF PROFESSIONAL ENGINEERS CODE OF ETHICS

The NSPE Code of Ethics for Engineers

Preamble

Engineering is an important and learned profession. As members of this profession, engineers are expected to exhibit the highest standards of honesty and integrity. Engineering has a direct and vital impact on the quality of life for all people. Accordingly, the services provided by engineers require honesty, impartiality, fairness, and equity, and must be dedicated to the protection of the public health, safety, and welfare. Engineers must perform under a standard of professional behavior that requires adherence to the highest principles of ethical conduct.

I. Fundamental Canons

Engineers, in the fulfillment of their professional duties, shall:

1. Hold paramount the safety, health, and welfare of the public.
2. Perform services only in areas of their competence.
3. Issue public statements only in an objective and truthful manner.
4. Act for each employer or client as faithful agents or trustees.
5. Avoid deceptive acts.
6. Conduct themselves honorably, responsibly, ethically, and lawfully so as to enhance the honor, reputation, and usefulness of the profession.

II. Rules of Practice

1. Engineers shall hold paramount the safety, health, and welfare of the public.

   a. If engineers' judgment is overruled under circumstances that endanger life or property, they shall notify their employer or client and such other authority as may be appropriate.

   b. Engineers shall approve only those engineering documents that are in conformity with applicable standards.

   c. Engineers shall not reveal facts, data, or information without the prior consent of the client or employer except as authorized or required by law or this Code.

   d. Engineers shall not permit the use of their name or associate in business ventures with any person or firm that they believe is engaged in fraudulent or dishonest enterprise.

   e. Engineers shall not aid or abet the unlawful practice of engineering by a person or firm.

   f. Engineers having knowledge of any alleged violation of this Code shall report thereon to appropriate professional bodies and, when relevant, also to public authorities, and cooperate with the proper authorities in furnishing such information or assistance as may be required.

2. Engineers shall perform services only in the areas of their competence.

   a. Engineers shall undertake assignments only when qualified by education or experience in the specific technical fields involved.

   b. Engineers shall not affix their signatures to any plans or documents dealing with subject matter in which they lack competence, nor to any plan or document not prepared under their direction and control.

   c. Engineers may accept assignments and assume responsibility for coordination of an entire project and sign and seal the engineering documents for the entire project, provided that each technical segment is signed and sealed only by the qualified engineers who prepared the segment.

3. Engineers shall issue public statements only in an objective and truthful manner.

   a. Engineers shall be objective and truthful in professional reports, statements, or testimony. They shall include all relevant and pertinent information in such reports, statements, or testimony, which should bear the date indicating when it was current.

   b. Engineers may express publicly technical opinions that are founded upon knowledge of the facts and competence in the subject matter.

   c. Engineers shall issue no statements, criticisms, or arguments on technical matters that are inspired or paid for by interested parties, unless they have prefaced their comments by explicitly identifying the interested parties on whose behalf they are speaking, and by revealing the existence of any interest the engineers may have in the matters.

4. Engineers shall act for each employer or client as faithful agents or trustees.

   a. Engineers shall disclose all known or potential conflicts of interest that could influence or appear to influence their judgment or the quality of their services.

   b. Engineers shall not accept compensation, financial or otherwise, from more than one party for services on the same project, or for services pertaining to the same project, unless the circumstances are fully disclosed and agreed to by all interested parties.

   c. Engineers shall not solicit or accept financial or other valuable consideration, directly or indirectly, from outside agents in connection with the work for which they are responsible.

   d. Engineers in public service as members, advisors, or employees of a governmental or quasi-governmental body or department shall not participate in decisions with respect to services solicited or provided by them or their organizations in private or public engineering practice.

   e. Engineers shall not solicit or accept a contract from a governmental body on which a principal or officer of their organization serves as a member.

5. Engineers shall avoid deceptive acts.

   a. Engineers shall not falsify their qualifications or permit misrepresentation of their or their associates' qualifications. They shall not misrepresent or exaggerate their responsibility in or for the subject matter of prior assignments. Brochures or other presentations incident to the solicitation of employment shall not misrepresent pertinent facts concerning employers, employees, associates, joint venturers, or past accomplishments.

   b. Engineers shall not offer, give, solicit, or receive, either directly or indirectly, any contribution to influence the award of a contract by public authority, or which may be reasonably construed by the public as having the effect or intent of influencing the awarding of a contract. They shall not offer any gift or other valuable consideration in order to secure work. They shall not pay a commission, percentage, or brokerage fee in order to secure work, except to a bona fide employee or bona fide established commercial or marketing agencies retained by them.

III. Professional Obligations

1. Engineers shall be guided in all their relations by the highest standards of honesty and integrity.

   a. Engineers shall acknowledge their errors and shall not distort or alter the facts.

   b. Engineers shall advise their clients or employers when they believe a project will not be successful.

   c. Engineers shall not accept outside employment to the detriment of their regular work or interest. Before accepting any outside engineering employment, they will notify their employers.

   d. Engineers shall not attempt to attract an engineer from another employer by false or misleading pretenses.

   e. Engineers shall not promote their own interest at the expense of the dignity and integrity of the profession.

2. Engineers shall at all times strive to serve the public interest.

   a. Engineers are encouraged to participate in civic affairs; career guidance for youths; and work for the advancement of the safety, health, and well-being of their community.

   b. Engineers shall not complete, sign, or seal plans and/or specifications that are not in conformity with applicable engineering standards. If the client or employer insists on such unprofessional conduct, they shall notify the proper authorities and withdraw from further service on the project.

   c. Engineers are encouraged to extend public knowledge and appreciation of engineering and its achievements.

   d. Engineers are encouraged to adhere to the principles of sustainable development[6] in order to protect the environment for future generations.

3. Engineers shall avoid all conduct or practice that deceives the public.

   a. Engineers shall avoid the use of statements containing a material misrepresentation of fact or omitting a material fact.

   b. Consistent with the foregoing, engineers may advertise for recruitment of personnel.

   c. Consistent with the foregoing, engineers may prepare articles for the lay or technical press, but such articles shall not imply credit to the author for work performed by others.

4. Engineers shall not disclose, without consent, confidential information concerning the business affairs or technical processes of any present or former client or employer, or public body on which they serve.

   a. Engineers shall not, without the consent of all interested parties, promote or arrange for new employment or practice in connection with a specific project for which the engineer has gained particular and specialized knowledge.

   b. Engineers shall not, without the consent of all interested parties, participate in or represent an adversary interest in connection with a specific project or proceeding in which the engineer has gained particular specialized knowledge on behalf of a former client or employer.

5. Engineers shall not be influenced in their professional duties by conflicting interests.

   a. Engineers shall not accept financial or other considerations, including free engineering designs, from material or equipment suppliers for specifying their product.

   b. Engineers shall not accept commissions or allowances, directly or indirectly, from contractors or other parties dealing with clients or employers of the engineer in connection with work for which the engineer is responsible.

6. Engineers shall not attempt to obtain employment or advancement or professional engagements by untruthfully criticizing other engineers, or by other improper or questionable methods.

   a. Engineers shall not request, propose, or accept a commission on a contingent basis under circumstances in which their judgment may be compromised.

---

[6] "Sustainable development" is the challenge of meeting human needs for natural resources, industrial products, energy, food, transportation, shelter, and effective waste management while conserving and protecting environmental quality and the natural resource base essential for future development.

b. Engineers in salaried positions shall accept part-time engineering work only to the extent consistent with policies of the employer and in accordance with ethical considerations.

c. Engineers shall not, without consent, use equipment, supplies, laboratory, or office facilities of an employer to carry on outside private practice.

7. Engineers shall not attempt to injure, maliciously or falsely, directly or indirectly, the professional reputation, prospects, practice, or employment of other engineers. Engineers who believe others are guilty of unethical or illegal practice shall present such information to the proper authority for action.

a. Engineers in private practice shall not review the work of another engineer for the same client, except with the knowledge of such engineer, or unless the connection of such engineer with the work has been terminated.

b. Engineers in governmental, industrial, or educational employ are entitled to review and evaluate the work of other engineers when so required by their employment duties.

c. Engineers in sales or industrial employ are entitled to make engineering comparisons of represented products with products of other suppliers.

8. Engineers shall accept personal responsibility for their professional activities, provided, however, that engineers may seek indemnification for services arising out of their practice for other than gross negligence, where the engineer's interests cannot otherwise be protected.

a. Engineers shall conform with state registration laws in the practice of engineering.

b. Engineers shall not use association with a nonengineer, a corporation, or partnership as a "cloak" for unethical acts.

9. Engineers shall give credit for engineering work to those to whom credit is due, and will recognize the proprietary interests of others.

a. Engineers shall, whenever possible, name the person or persons who may be individually responsible for designs, inventions, writings, or other accomplishments.

b. Engineers using designs supplied by a client recognize that the designs remain the property of the client and may not be duplicated by the engineer for others without express permission.

c. Engineers, before undertaking work for others in connection with which the engineer may make improvements, plans, designs, inventions, or other records that may justify copyrights or patents, should enter into a positive agreement regarding ownership.

d. Engineers' designs, data, records, and notes referring exclusively to an employer's work are the employer's property. The employer should indemnify the engineer for use of the information for any purpose other than the original purpose.

e. Engineers shall continue their professional development throughout their careers and should keep current in their specialty fields by engaging in professional practice, participating in continuing education courses, reading in the technical literature, and attending professional meetings and seminars.[7]

THE INTERNATIONAL FEDERATION OF CONSULTING ENGINEERS

The International Federation of Consulting Engineers (Fédération Internationale Des Ingénieurs-Conseils) recognizes that the work of the consulting engineering industry is critical to the achievement of sustainable development of society and the environment. To be fully effective not only must engineers constantly improve their knowledge and skills, but also society must respect the integrity and trust the judgment of members of the profession and remunerate them fairly. All member associations of FIDIC subscribe to and believe that the following principles are fundamental to the behavior of their members if society is to have that necessary confidence in its advisors. The FIDIC Code of Ethics follows.

FIDIC Code of Ethics

Responsibility to society and the consulting industry, the consulting engineer shall:

- Accept the responsibility of the consulting industry to society.
- Seek solutions that are compatible with the principles of sustainable development.
- At all times uphold the dignity, standing and reputation of the consulting industry.

Competence

The consulting engineer shall:

- Maintain knowledge and skills at levels consistent with development in technology, legislation and management, and apply due skill, care and diligence in the services rendered to the client.
- Perform services only when competent to perform them.

(Continued)

---

[7] National Society of Professional Engineers, www.nspe.org/index.html

Integrity

The consulting engineer shall:

- Act at all times in the legitimate interest of the client and provide all services with integrity and faithfulness.

Impartiality

The consulting engineer shall:

- Be impartial in the provision of professional advice, judgment or decision.
- Inform the client of any potential conflict of interest that might arise in the performance of services to the client.
- Not accept remuneration which prejudices independent judgment.

Fairness to Others

The consulting engineer shall:

- Promote the concept of "Quality-Based Selection" (QBS).
- Neither carelessly nor intentionally do anything to injure the reputation or business of others.
- Neither directly nor indirectly attempt to take the place of another consulting engineer, already appointed for a specific work.
- Not take over the work of another consulting engineer before notifying the consulting engineer in question, and without being advised in writing by the client of the termination of the prior appointment for that work.
- In the event of being asked to review the work of another, behave in accordance with appropriate conduct and courtesy.

Corruption

The consulting engineer shall:

- Neither offer nor accept remuneration of any kind which in perception or in effect either a) seeks to influence the process of selection or compensation of consulting engineers and/or their clients or b) seeks to affect the consulting engineer's impartial judgment.
- Co-operate fully with any legitimately constituted investigative body which makes inquiry into the administration of any contract for services or construction.

It is comforting to see that globally, engineers seem to embrace a very similar set of ethics governing the profession.

Reference:

International Federation of Consulting Engineers
FIDIC, Box 311 - CH-1215, Geneva 15, Switzerland
SKYPE fidic.secretariat, Tl +41-22-799 49 00, Fx +41-22-799 49 01
www.fidic.org

IMPORTANT AND RELEVANT POLICY STATEMENTS BY ASCE AND NSPE

Referenced below are several of ASCE's very interesting and relevant "policy statements" of which engineers should be aware. These policies are shown with the policy statement title, adoption date, the actual written adopted policy and the issue. The policies regard:

- Continued education requirements for annual "ethics training," as stated in Policy Statement 376
- Engineer's judgment and adherence to the ASCE Code of Ethics, as stated in Resolution 502
- Use of the term "engineer" as stated in Policy 433

ASCE Policy Statement 376—Continuing Education in Ethics Training

Approved by the National Engineering Practice Policy Committee on March 8, 2007

Approved by the Policy Review Committee on March 9, 2007

Adopted by the Board of Direction on April 24, 2007

Policy

The American Society of Civil Engineers (ASCE) encourages all state boards of engineering licensure to institute a minimum professional development requirement consisting of at least one (1) hour per year on professional ethics for professional licensure which would be reciprocal with other states. The one hour per year should be based upon the fundamental canons of professional conduct and other appropriate administrative rules or regulations, and designed to demonstrate a working knowledge of professional ethics.

Issue

Professional ethics is the cornerstone of engineering practice. Adherence to a Code of Ethics encourages engineers to practice in areas in which they are competent and that 'they will hold the safety, health and welfare of the public as their highest duty.

The majority of complaints referred to state boards of licensure for investigation and possible penalty action involve ethics and, often, a lack of understanding of the Fundamental Canons of Professional Conduct.

Using a Code of Ethics

Codes of ethics are created in response to actual or anticipated ethical conflicts. Considered in a vacuum, many codes of ethics would be difficult to comprehend or interpret. It is only in the context of real life and real ethical ambiguity that the codes take on any meaning.

Codes of ethics and case studies need each other. Without guiding principles, case studies are difficult to evaluate and analyze; without context, codes of ethics are incomprehensible. The best way to use these codes is to apply them to a variety of situations and see what results. It is from the back and forth evaluation of the codes and the cases that thoughtful moral judgements can best arise.

ASCE Resolution 502—Professional Ethics and Conflict of Interest

Approved by the Engineering Practice Policy Committee on March 26, 2009

Approved by the Policy Review Committee on March 27, 2009

Adopted by the Board of Direction on July 25, 2009

First Approved in 2003

Policy

The American Society of Civil Engineers (ASCE) believes that:

- The engineer's judgment and adherence to the ASCE Code of Ethics must be above reproach and beyond the influence of competing interests. Even the appearance of a conflict of interest is to be avoided.

- The ability to exercise the independent judgment required of engineers to protect the public health, safety, welfare and environment should not be compromised in any way by the rules of any organization to which the engineer belongs.

- Laws, regulations, conditions of employment and collective bargaining agreements must permit engineers to maintain their independence and avoid potential conflicts of interest to protect the public health, safety, welfare, and environment.

- Engineers should not be subject to disciplinary or demeaning actions for holding the public interest above all others.

Issue

Engineering is a learned profession that has a direct impact on the environment and the safety, health and welfare of the public. Accordingly, the services provided require high standards of honesty, integrity and fairness.

ASCE's Code of Ethics recognizes the unique employment aspects of the engineer, regardless of the employer, public or private. Employment conditions for engineers must support their duty to hold paramount the health, safety, welfare and environment of the public in their engagements. To fulfill their duty, engineers must apply responsibly their independent judgment in design and construction matters. This duty to the public super-cedes any actual or perceived obligations engineers have to the owners of their projects, their employers, or any organizations to which they belong.

Rationale

Engineers must adhere to ASCE's Code of Ethics and operate under the jurisdiction of state licensure laws and are subject to discipline for violation of these laws. Engineers are also subject to discipline from the professional societies of the engineering profession for violation of the public trust. These laws and standards include the responsibility for properly preparing design documents or performing field observation and testing to document construction.

An engineer relies on a variety of resources, including non-professional personnel, in rendering professional engineering services. An engineer must oversee the performance of those resources for public health, safety, welfare and the environment.

Since ASCE is composed of individual members, the Society is concerned about matters that affect its members and will voice its concerns relative to the employment conditions of its professional members while simultaneously striving to protect the health, safety, welfare and the environment of the public it serves.

ASCE Policy Statement 433—Use of the Term "Engineer"

Approved by the National Engineering Practice Policy Committee on March 11, 2010

Approved by the Policy Review Committee on March 23, 2010 Adopted by the Board of Direction on July 10, 2010

Policy

The American Society of Civil Engineers (ASCE) believes that the following standards are the only basis on which any title or designation should include the term "engineer."

- Graduation from an accredited engineering program with a degree in engineering;
- Registration as a professional engineer or engineer-in-training under a state engineering registration law; or,
- An official ruling designating an individual or a group in an engineering capacity as meeting the definition of "Professional Engineer" (P.E.) under the Taft-Hartley Act or the Fair Labor Standards Act.

Only persons in one of these categories should be designated by the title "engineer" or "professional engineer." This policy shall not be construed to prohibit using the word "engineering" as a modifier in titles such as "engineering assistant", "engineering aide" and "engineering technologist" where the title clearly implies that the duties of the position are not those of [a] professional engineer.

ASCE further encourages registered professionals to always use their P.E. title on all professional correspondence and communication where permissible and appropriate.

Issue

Improper use of the term "engineer" is confusing and misleading to the public. Employers and employees sometimes misuse the term in titles and resumes. This misuse of the title by groups and people who are usually knowledgeable tends to diminish the value of the title which should be applied to people qualified professionally by accepted standards of education, law and engineering practice.

Rationale

There is a need within ASCE as well as within government and other organizations with practicing professional engineers to provide employee titles and/or classifications that properly identify the individual's level of responsibility or expertise within that organization. A title such as "designer" is not proper for a graduate engineer with several years of experience; "associate engineer" or similar title as used by ASCE in designating professional grades is more appropriate and strongly encouraged.

NSPE Position on Potential Incidents of the Unlicensed Practice

NSPE has issued guidance to NSPE State Societies ("State Societies") and to NSPE members on reporting potential incidents of the unlicensed practice, or offer to practice, of engineering.

The practice of engineering by unlicensed practitioners potentially places the public health, safety and welfare at risk. For this reason, it is of interest to State Societies to encourage members to report potential unlicensed practice, or offers to practice, to State Licensing Boards. Note that efforts to prevent unlicensed practice are intended solely to protect the public health, safety, and welfare, and are not intended to improperly restrict lawful activities or practices.

In many jurisdictions, a Professional Engineer has an ethical and legal obligation to report unlicensed practice to the State Licensing Board.

A recent National Council of Examiners for Engineering and Surveying (NCEES) survey of State Professional Engineering Licensing Boards ("State Licensing Boards") requested information from each Board on the categories of violations indicated for cases opened during a two-year period. Responses from 43 boards were received, with information on the reported violations for 3,369 disciplinary cases. The frequency of categories of violations reported in that survey is as follows, beginning with the most frequent:

- Incompetence/negligence
- Unlicensed practice/offer

- Ethics/professional conduct/misconduct
- Fraud, deceit, misrepresentation
- Sealing of work not prepared under the direct supervision and control of the licensee

Some case studies of actual violations posted on the State of California Board for Professional Engineers and Land Surveyors and NSPE websites follow.

CASE STUDIES

Case studies provide valuable insight and enlightenment to civil engineers regarding licensing issues and ethics. The California Business and Professions Code is presented below in a text box for reference. Each state has a similar code by which the Licensing Board may receive and investigate complaints against registered professional engineers. Ethical violations of the code may also be investigated and acted upon by these state boards.

California Business and Professions Code 6775*

The board may receive and investigate complaints against registered professional engineers, and make findings thereon.

By a majority vote, the board may reprove, suspend for a period not to exceed two years, or revoke the certificate of any professional engineer registered under this chapter:

a. Who has been convicted of a crime substantially related to the qualifications, functions and duties of a registered professional engineer, in which case the certified record of conviction shall be conclusive evidence thereof.

b. Who has been found guilty by the board of any deceit, misrepresentation, or fraud in his or her practice.

c. Who has been found guilty by the board of negligence or incompetence in his or her practice.

d. Who has been found guilty by the board of any breach or violation of a contract to provide professional engineering services.

e. Who has been found guilty of any fraud or deceit in obtaining his or her certificate.

(*Continued*)

f. Who aids or abets any person in the violation of any provision of this chapter.

g. Who in the course of the practice of professional engineering has been found guilty by the board of having violated a rule or regulation of unprofessional conduct adopted by the board.

h. Who violates any provision of this chapter.

*Note: 6775.1 states that the board may receive and investigate complaints against engineers-in-training and make findings thereon.

Several case studies are presented for reference below and show how many common issues are interpreted and acted upon with details on some fines to the licensee.

Citations Issued to Board Licensees

Citations are issued to licensed engineers and land surveyors when the severity of a violation may not warrant suspension or revocation of the licensee's right to practice. When a fine is levied with a citation, payment of the fine represents satisfactory resolution of the matter. Summaries of the citations, including each licensee's name and license number, remain on the website for five years after the citation is final, unless further action is taken against the licensee. All citations issued by the Board are matters of public record.

Case 1—Expired License

Mr. W.

Civil Engineer C 3xxxx

Citation 5L

Final: October 27, 2002

Action: Order of Abatement; $7,500 fine

An investigation revealed that Mr. W., whose Civil Engineer License, C 3xxxx, expired on September 30, 1998, violated Business and Professions Code Sections 6733 and 6737(a) and (e) by performing civil/geotechnical engineering on several projects in California during the period his license was expired. The citation ordered Mr. W. to cease and desist providing civil engineering services in California until such time as his delinquent license is renewed and reinstated and to pay administrative fines to the Board in an amount totaling $7,500.00. The administrative fines have been paid. In accordance with Section 125.9(d) of the Business and Professions Code, payment of an administrative fine does not constitute admission of any violation(s) charged but represents a satisfactory resolution of the matter.

Case 2—False, Misleading, and Deceitful Information

Mr. D.

Civil Engineer C 1xxxx

Citation 52-L

Final: March 26, 2005

Action: Order of Abatement; $250 fine

A citation was issued to Mr. D. on May 15, 2000, alleging that Mr. D. had provided false, misleading, and deceitful information on reference forms for an applicant for licensing as a civil engineer. The references were dated July 20, 1999, and October 12, 1999. An investigation, including a review by at least one licensee of the Board competent in civil engineering, determined Mr. D. had signed and sealed reference forms which contained incorrect information. At an informal conference following service of the citation, which was reissued February 25, 2005, Mr. D. admitted that he failed to adequately check the dates of employment on the reference forms he signed and that he had taken the applicant's word that the information on the forms was correct. The Board ordered Mr. D. to cease and desist from violating Business and Professions Code Section 6775(f) and to pay an administrative fine of $250. The administrative fine has been paid in full. In accordance with Section 125.9(d) of the Business and Professions Code, payment of an administrative fine does not constitute admission of any violations charged but represents a satisfactory resolution of the matter.

Case 3—Failing to Sign and Stamp a Feasibility Study Report

Ms. R.

Civil Engineer C 2xxxx

Citation 56-L

Final: July 22, 2005

Action: Order of Abatement; $250 fine

Investigation determined that Ms. R. violated Business and Professions Code Section 6735 by failing to sign and stamp a feasibility study report that was released to her client, who then released it to the public during a public meeting. When questioned about who was responsible, Ms. R. and her colleagues signed and stamped the report but failed to include the date of signing and stamping. Ms. R. was ordered to obey all laws by properly signing and stamping all final engineering reports and include both the date of expiration of her license and the date reports are signed and stamped. Additionally, she was ordered to pay an administrative fine of $250. The fine has been paid. In accordance with Section 125.9(d) of the Business and Professions Code, payment of an administrative fine does not constitute admission of any violations charged but represents a satisfactory resolution of the matter.

Case 4—Providing Structural Engineering Services Without a Contract

Mr. M.

Civil Engineer C 4xxxx

Citation 59-L

Final: October 7, 2003

Action: Order of Abatement; $500 fine

The Board found that Mr. M. violated Business and Professions Code Section 6749 by providing structural engineering services for a room addition to a residence without entering into a written contract. Mr. M. stated he was hired by an unlicensed designer to provide services on the project but was paid directly by the homeowner. The unlicensed designer is not legally authorized to provide civil engineering services to his clients unless he has a partner who is a licensed civil engineer or is part of a business that is owned or co-owned by a licensed civil engineer. The homeowner stated the project was never completed; however, Mr. M. provided the client with a refund of all of the fees paid to him concerning the project. The Board ordered Mr. M. to enter into written contracts as required by Section 6749 when providing civil engineering services and to pay an administrative fine of $500. The administrative fine has been paid. In accordance with Section 125.9 (d) of the Business and Professions Code, payment of an administrative fine does not constitute admission of any violation(s) charged but represents a satisfactory resolution of the matter.

NSPE Ethics Case Study

Case 5—Sustainabale Development

NSPE Board of Ethical Review, Case No. 0X-X, 4/8/08 - FINAL

Sustainable Development – Threatened Species Facts:

Engineer A is a principal in an environmental engineering firm and is requested by a developer client to prepare an analysis of a piece of property adjacent to a wetlands area for potential development as a residential condominium. During the firm's analysis, one of the engineering firm's biologists reports to Engineer A that in his opinion, the condominium project could threaten a bird species that inhabits the adjacent protected wetlands area. The bird species in not an "endangered species," but it is considered a "threatened species" by federal and state environmental regulators.

In subsequent discussions with the developer client, Engineer A verbally mentions the concern, but Engineer A does not include the information in a written report that will be submitted to a public authority that is considering the developer's proposal.

Question
Was it ethical for Engineer A not to include the information about the threat to the bird species in a written report that will be submitted to a public authority that is considering the developer's proposal?

References
Section I.3.—NSPE Code of Ethics: Engineers, in the fulfillment of their professional duties, shall issue public statements only in an objective and truthful manner.

Section I.5.—NSPE Code of Ethics: Engineers, in the fulfillment of their professional duties, shall avoid deceptive acts.

Section II.3.a.—NSPE Code of Ethics: Engineers shall be objective and truthful in professional reports, statements, or testimony. They shall include all relevant and pertinent information in such reports, statements, or testimony, which should bear the date indicating when it was current.

Section III.2.d.—NSPE Code of Ethics: Engineers are encouraged to adhere to the principles of sustainable development in order to protect the environment for future generations.

Section III.4.—NSPE Code of Ethics: Engineers shall not disclose, without consent, confidential information concerning the business affairs or technical processes of any present or former client or employer, or public body on which they serve.

Discussion
In January 2006, the NSPE Board of Directors approved a change to the NSPE Code of Ethics to add Section III.2.d. to the NSPE Code. The new section stated that "engineers shall strive to adhere to the principles of sustainable development in order to protect the environment for future generations." A footnote (Footnote 1) was also included at the end of the NSPE Code of Ethics. The footnote further clarified and defined the term "sustainable development." It stated that "sustainable development" is "the challenge of meeting human needs for natural resources, industrial products, energy, food, transportation, shelter, and effective waste management while conserving and protecting environmental quality and the natural resources base essential for future development." Thereafter, in July 2007, the NSPE House of Delegates voted to modify the language in NSPE Code Section III.2.d. to state that "engineers are encouraged to adhere to the principles of sustainable development in order to protect the environment for future generations. With this added language and further clarification, the NSPE Board of Ethical Review will review this language as a matter of first impression and in the context of other language in the NSPE Code and earlier NSPE Board of Ethical Review Opinions.

Not unlike earlier NSPE Board of Ethical Review cases of this type, the facts in this case present a situation that often raises very difficult issues for engineers in dealing with clients. Engineering practice sometimes places the engineer in the position where the interests of a client and the interests of the public are in open and serious conflict.

As this Board has noted on several occasions, engineers play an essential role in society by taking steps and actions to see that products, systems, facilities, structures, and the land surrounding them are reasonably safe. Sometimes engineers are placed in situations where they must balance the extent of their obligations to their employer or client with their obligations to protect the public health and safety. NSPE Code Section III.2.d. places some additional responsibilities on engineers for the protection of environment.

At the same time, as noted in ABC Case No. ZZ-Y, there are various rationales for the nondisclosure language contained in NSPE Code Section III.4. Engineers, in the performance of their professional services, act as "agents" or "trustees" to their clients. They and the members of their firms are privy to a great deal of information and background concerning the business affairs of their client. The disclosure of confidential information could be quite detrimental to the interests of their client and, therefore, engineers as agents or trustees are expected to maintain the confidential nature of the information revealed to them in the course of rendering their professional services.

SUMMARY

Ethics is referred to as moral philosophy, and recommends concepts of right and wrong behavior for professionals practicing within a profession. It's important for engineers to be aware of potential ethical issues and acceptable responses under the code of ethics. The engineer's license to practice depends upon it.

Organizations with Useful Information about the Professional Practice of Engineering

- American Council of Engineering Companies (ACEC)—www.acec.org/
- American Institute of Architects (AIA)—www.aia.org/index.htm
- American Society of Civil Engineers (ASCE)—www.asce.org/
- Association of General Contractors (AGC)—www.agc.org/
- Design Build Institute of America (DBIA)—www.dbia.org/
- National Society of Professional Engineers (NSPE)—www.nspe.org/index.html
- National Council of Examiners for Engineering and Surveying (NCEES)—www.ncees.org/
- International Federation of Consulting Engineers (FIDIC)—www.fidic.org/

REFERENCES

American Institute of Architects, website: www.aia.org/index.htm

American Society of Civil Engineers, website: www.asce.org/

American Council of Engineering Companies website: www.acec.org

DAVIS, Michael. "Thinking Like an Engineer: The Place of a Code of Ethics in the Practice of a Profession." *Philosophy and Public Affairs* 20.2 (1991) 150–167.

Harris, Charles E., Jr., Michael S. Pritchard, and Michael J. Rabins. *Engineering Ethics: Concepts and Cases*. Belmont, CA: Wadsworth Publishing, 1995.

Illinois Institute of Technology—Center for the Study of Ethics (IIT CSEP). http://ethics.iit.edu

International Federation of Consulting Engineers, website: www.fidic.org

National Society of Professional Engineers, website: www.nspe.org/index.html

Introduction
1
Ethics
2
Professional Engagement
3
History
4
What Engineers Deliver
5
Engineer's Role in Project
Development
6
Permitting
7
Leadership   Managing
8
Executing a Professional
Commission
9
Client Relationship
10   11   12   13
Legal Aspects
Communicating
14
Having
a Life
15
Globalization
16
Sustainability
17
Emerging Technology
A   B   C   D   E   F

Chapter **4**

Professional Engagement

Big Idea

Understanding how Civil Engineers obtain work is essential for those in private practice as well as their Clients. Mastering the process of Professional Engagement enhances the civil engineer's probability of success.

"Chance favors only the prepared mind."

—Louis Pasteur

Key Topics Covered

- Qualifications-Based Selection—The Federal Government Process
- Fee-Based Selection
- The Contract
- Budgeting
- Enhancing the Engineering Firm's Probability for a Successful Professional Engagement
- Summary
- A Sample RFP Is Presented
- A Sample Proposal Is Presented
- A Sample Feasibility Study Report Is Presented

Related Chapters in This Book

(*Continued*)

Related to *ASCE Body of Knowledge 2* Outcomes

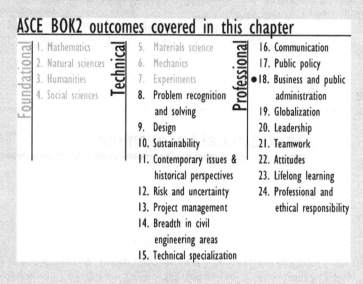

INTRODUCTION

In this reference book "professional engagement" is defined as:

To secure professional services; to hire.

One of the immediate issues recognized by someone requiring engineering services is the difficulty in defining and communicating the specific needs and/or tasks required to arrive at a solution. While potential clients may realize they have a problem to be solved, most do not understand engineering or what it takes to define and communicate their problem to the engineer.

Engineers recognize this situation and refer to it as the need to create a scope, or statement, of work (SOW). One of the challenges for engineers is that sometimes learning and understanding the client's operation and/or objectives takes quite a bit of time, and there may be several ways to address the problem with a variety of capital and/or expense scenarios.

> The situation may be analogous to a patient telling a medical doctor that they have a headache and then asking, "How will you fix it and (by the way) how much will it cost?" The doctor usually has no idea without at least examining the patient—it might be as simple as prescribing a pain reliever or it might require surgery.

In the course of assessing the client's needs, the engineer may spend quite a bit of their own (or their company's) time and expenses to provide a detailed SOW. The SOW needs to include a cost estimate, project schedule, and tabulated labor categories, which may include subcontractors, so that the client may choose an appropriate path forward. Occasionally the engineer may find that the client is surprised at the depth of their own needs or the costs associated with resolving them. The client may then provide the engineer's SOW and proposal package to a competing engineer for an alternate approach or lower cost. The original engineer may then be left "holding the bag" with respect to sunk costs for the initial services and scoping effort.

To increase the chances of preparing a winning proposal and securing the professional engagement, engineering firms often create and follow a business development process as illustrated in Figure 4.1.

This process involves the civil engineering firm identifying potential leads early in the project initiation phases and following these projects through to the RFP and proposal phases. Savvy firms work to position themselves strategically with exceptional project experiences, talented staff, and other differentiating factors that can give them distinct advantages to win these projects.

Professional engineering services are probably best provided when the client knows and trusts the engineer. We always (or should always) act in the best interest

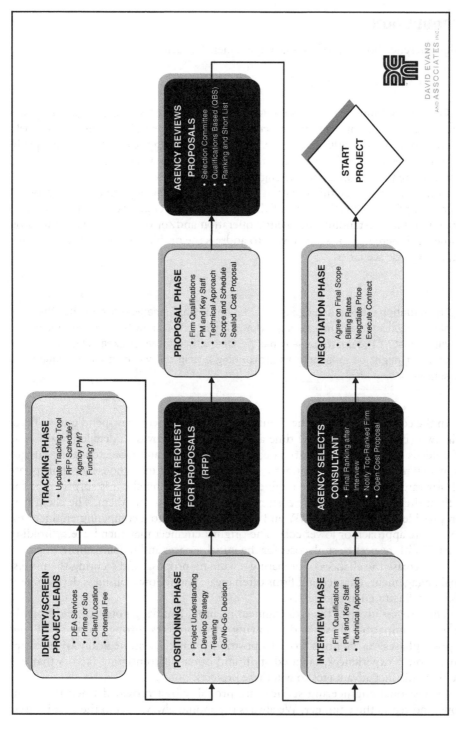

Figure 4.1 Business development process

of the client. This is why large clients (companies, facilities, or organization) have their own engineering staff. These staff engineers know the mission and requirements of their own employer and can act in the latter's best interest. Sometimes, however, the engineering staff may become overwhelmed or may require specialized help; in these cases, the staff engineers may require assistance just as companies that doesn't have engineering staffs might. The engineering staff may be in a good position to prepare a SOW for outside engineering support. Alternately, a firm could establish a support contract with an engineering firm to provide these services as the need arises.

We've now envisioned a scenario where a need for engineering services arises and the client is in a position to request these services. But, the key question is "how" does the client request these services and specifically "what" services do they request? This situation is not unique to private industry; federal, state, local governments, as well as nonprofit organizations and members of the general public, can experience the same situation. One way to arrange for engineering services is for the client to prepare a request for qualifications (RFQ) that may or may not include a specific SOW The client then solicits responses and selects an engineer based upon their general qualifications. This arrangement is referred to as qualification-based selection, or QBS. The benefit of QBS is that the client selects an engineer because of specific experience and capabilities relevant to the client's need.

The only part of the equation that may be missing is the working relationship and the "trust factor" between the client and the engineer. However, professional engineers are bound by ethics, business law, and contracts so the trust factor is not usually a problem, although it's human nature to hire or work with someone you know rather than a perfect stranger. (More on this subject is discussed in Chapter 3, Ethics; Chapter 5, The Engineer's Role in Project Development; Chapter 9, The Client Relationship and Business Development; and Chapter 13, Communicating as a Professional Engineer.)

QUALIFICATIONS-BASED SELECTION—THE FEDERAL GOVERNMENT PROCESS

Development of a Short List

Following the evaluation of the statements of qualifications, the board prepares a report that recommends the firms to be included on the short list. Short-listed firms are those that the evaluation board has chosen to interview. The evaluation board's report generally rank at least three of the firms for the purpose of discussing the project with them in another meeting referred to as a "short-list interview." In the event that only two firms submitted qualifications, the client may elect to only evaluate these two firms, but may retain the option of re-advertising with a re-written SOW. Evaluation boards are not limited in the number of firms that they can select for these "interviews"; it is left to the discretion of each board.

Qualifications Based Selection—Summary

An excellent description of the Federal Government Process has been presented by American Council of Engineering Companies (ACEC) and may be found at the following address: http://www.acec.org/advocacy/committees/brooks.cfm) The Brooks Act (Public Law 92-582), also known as Qualifications-Based Selection (QBS), enacted on October 18, 1972, establishes the procurement process by which architects and engineers (A/Es) are selected for design contracts with federal design and construction agencies. The qualifications-based selection process requires that contracts for A/Es be negotiated on the basis of demonstrated competence and qualification for the type of professional services required at a fair and reasonable price. Under QBS procurement procedures, price quotations are generally not a consideration in the selection process. (American Council of Engineering Companies website, www.ACEC.org) While this is a federal process, most states and local county and city governments use similar processes.

This QBS process, as established by the Brooks Act, has long been enthusiastically supported by professional A/E societies.

There are seven basic steps involved in pursuing federal design work under QBS:

1. Public solicitation for architectural and engineering services
2. Submission of an annual statement of qualifications and supplemental statements of ability to design specific projects for which public announcements were made
3. Evaluation of both the annual and project-specific statements
4. Development of a short list of at least three submitting firms in order to conduct interviews with them
5. Interviews with the firms
6. Ranking of at least three of the most qualified firms
7. Negotiation with the top-ranked firm

A brief explanation of each of these steps follows, along with a description of what is involved in each. The user is reminded that while QBS procedures are mandated by law, agencies may modify the procedures slightly, within the confines of the Act and the Federal Acquisition Regulation.

Public Announcement

QBS calls for public announcement of opportunities for design contracts. The government fulfills this obligation by publicizing opportunities in the *Commerce Business Daily*. The *Commerce Business Daily*, or "CBD" as it is known, is published

Monday through Friday by the U.S. Department of Commerce. The CBD lists proposed government procurements, subcontracting leads, and contract awards. A proposed procurement action appears in the CBD only once.

All intended procurement actions of some moderate fee (typically $25,000 or more), whether for military or civilian agencies, are published in the CBD. This publication also identifies contracts that have been awarded, if the contract amount exceeds $25,000 for civilian agencies and $100,000 for the Department of Defense. The CBD does not list procurements that are:

- Classified for reasons of national security
- For perishable items
- For certain utility services
- Required within 15 days
- Placed under existing contracts
- For personal professional services
- Made only from foreign sources
- Not to be given advance publicity, as determined by the Small Business Administration

These notices in the CBD give the location and scope of a project and may also contain such information as:

- Estimated construction contract award range
- Project schedule and the date and time limit for receiving replies
- Categories of evaluation criteria and weight factors
- Any requirements for submitting supplemental information

Opportunities for A/E services are usually listed under the "R" section. However, design opportunities can be included in other sections, such as those for design/build services listed under "Y," Construction of Structures and Facilities.

Statement of Qualifications

A/E firms with an interest in being considered for design services contracts must submit the required statements of qualifications (SOQ) to each agency with which the A/E wants to contract. The Standard Form 254 (SF 254), Architect-Engineer and Related Services Questionnaire, may be filed each year with the appropriate field office of each agency. This form can also be updated and resubmitted at any time. A completed form furnishes the federal agency with general

(Continued)

information on the size, capabilities, personnel, and past experience of an interested firm.

The advantage to the annual filing of SF 254 is that many federal agencies keep and review this file for prospective design firms if they have a small project that will not be advertised. However, an A/E firm can also submit this form as part of the response to a project-specific advertisement.

When a project is advertised in the CBD, the agency does not usually directly notify firms that previously have filed a SF 254. The project advertisements, or notices, that appear in the CBD are tailored to each specific project and invite interested firms to submit the SF 254 and the Standard Form 255 (SF 255), Architect-Engineer and Related Services Questionnaire for Specific Project, along with any supplemental data requested in the announcement.

Following the review of the notices in the CBD, if an A/E firm wants to be considered for a specific listed project, then it must submit SF 255. This form is submitted in response to a specific solicitation and, when completed, contains the data relative to the specific project. Firms that have a current SF 254 on file with the listed procurement office are not required to resubmit that form; however, they must submit a SF 255, to be considered for each separate project. Instructions on how to complete Standard Forms 254 and 255, which include substantial guidance on what information to add, are contained in the forms. For example, the instructions in SF 254 stress that additional data, brochures, photos, and related material, should not accompany this form unless specifically required. On the other hand, the instructions for SF 255 state that, when appropriate, respondents may supplement this proposal with graphic material and photographs that best demonstrate design capabilities of the proposer for the specific project.

Evaluation of Statements

This is a multistep process that begins with qualifying and ranking the firms based on their submittals. The evaluation/selection process is performed by an architectural/engineering evaluation board composed of members who, collectively, have experience in architecture, engineering, construction, and government and related acquisition matters. The members of the board are usually appointed from among the professional employees of the agency or other agencies. In some situations, private practitioners and/or members of the public sit on a these board if authorized by agency procedures. Of course, when private practitioners sit on an evaluation board, they or their firms are not eligible for award of a design contract.

The evaluation board reviews the statements of qualifications (Standard Forms 254 and 255). Evaluations are done in accordance with the criteria cited in the CBD notice. For example, some of the criteria in the CBD notice may include the following: professional qualifications and experience of the firm with design of a specific type of project (for example, dams, levees, roadways, pipelines, and so forth); experience and professional qualifications of the firm's staff to be assigned to the project; location of the main office of the proposing firm and its consultants or subcontractors; overall performance record of the firm; and analysis of the firm's current workload.

—American Council of Engineering Companies (ACEC), www.acec.org

Interviews/Discussions with Firms

The interviews usually involve discussions on project concepts and the relative utility of alternative methods of furnishing the required services. Before the interview, some agencies send detailed selection criteria and other information about the project to the firms recommended for further consideration. Although conceptual alternatives may be presented, under the system established by QBS, the architect-engineer designer does not produce any design product in competing for the project.

Usually these interviews are held at the agency's office. Occasionally, and in special circumstances, phone interviews are conducted. The interviews are brief, usually lasting only 30 to 60 minutes. The interview is the best chance for the engineer to display the firm's knowledge, unique strengths, project experience, and project team to the decision-makers. For more information on this topic, the reader should also review Chapter 6, What Engineers Deliver; Chapter 7, Executing a Professional Commission; Chapter 8, Permitting; Chapter 9, The Client Relationship and Business Development; and Chapter 13, Communicating as a Professional Engineer.

Ranking of the Top Three Firms

Following the interviews, the board's report is presented to the agency head or a person who is designated to act on behalf of the agency head. The report lists, in order of preference, at least three firms that are considered to be the most highly qualified to perform the services. This is considered to be the final selection of the competing firms. If the firm listed as the most preferred is not the firm that was recommended as the most highly qualified by the evaluation board, the head of the agency will provide a written explanation for the reason for the preference. The head of the agency, or that person's designate, may not add names of other firms to the final report. The report reviews the recommendations of the evaluation board and, from that, the agency head makes the final selection. Samples of federal government architect/engineer evaluation forms employed by the Veteran's Administration are illustrated in Figure 4.2, Example Short List Form; Figure 4.3, Example A/E Interview Scoresheet; and Figure 4.4, Example A/E Performance Form. A detailed evaluation of these example forms will enlighten the competing A/E firms on the criteria for independent judgments and allow potential time for the firm to prepare their presentation materials in concert with the reviewing materials.

SHORT-LIST CRITERIA UTILIZING THE SF 330 FORM

Department of Veterans Affairs – Architect/Engineer Evaluation Board

1. Specialized experience and technical competence of the firm (including a joint venture or association) with the type of services required
   Assignable point rang .... (0 to 40)

2. Specific experience and qualifications of personnel proposed for assignment to the project and record of working together as a team
   Assignable point range .... (0 to 40)

3. Professional capacity of the firm in the designated geographic area of the project to perform work (including any specialized services) within the time limitations. Unusually large existing workload that may limit A/E's capacity to perform project work expeditiously
   Assignable point range .... (0 to 20)

4. Past record of performance on contracts with the Department of Veterans Affairs. This factor may be used to adjust scoring for any unusual circumstances that may be considered to deter adequate performance by an A/E. (Firms with no previous VA experience receive a +5 rating)
   Assignable point range .... (−20 to 20)

5. Geographic location and facilities of the working office(s) which would provide the professional services and familiarity with the area in which the project is located
   Assignable point range .... (0 to 20)

6. Demonstrate success in prescribing the use of recovered materials and achieving waste reduction and energy efficiency in facility design
   Assignable point range .... (0 to 20)

7. Inclusion of small business consultant(s) (1 point), and/or minority-owned consultant(s) (1 point), and/or women-owned consultant(s) (1 points), and/or veteran owned consultant(s) (1 point), and/or disadvantage veteran owned consultant(s) (1 point), and/or HUBZone consultant(s) (1 point)
   Assignable point range .... (0 to 6)

SCORING KEY						
SCORING FACTORS	RANGE	POOR	MARGINAL	ACCEPTABLE	VERY GOOD	OUTSTANDING
1 and 2	0–40	0	5–10	15–25	30–35	40
3, 5, 6	0–20	0	5	10	15	20
4	(−20)–(+20)	(−20)	(−10)	0	10	20
7	0–6	0	1–2	3–4	5	6

Figure 4.2   Example short list form

A/E INTERVIEW SCORESHEET

Project Title:			**A/E APPLICANTS**								

Project Location:	Raw Score Key	
	0.9 to 1.0	Excellent
Project #:	0.7 to 0.8	Very Good
	0.4 to 0.6	Acceptable
Date:	0.2 to 0.3	Marginal
	0.0 to 0.1	Poor

FACTORS	WEIGHT	RAW SCORE	WEIGHTED SCORE	RAW SCORE	WEIGHTED SCORE	RAW SCORE	WEIGHTED SCORE	RAW SCORE	WEIGHTED SCORE	RAW SCORE	WEIGHTED SCORE
I - TEAM PROPOSED FOR THIS PROJECT											
Background of the personnel											
1. Project Manager											
2. Other key personnel											
3, Consultants											
II - PROPOSED MANAGEMENT PLAN											
Team organization											
1. Design Phase											
2. Construction Phase											
III- PREVIOUS EXPERIENCE OF PROPOSED TEAM											
Project Experience											
IV - LOCATION AND FACILITIES OF WORKING OFFICES											
A. Prime firm											
B. Consultants											
V - PROPOSED DESIGN APPROACH FOR THIS PROJECT											
A. Proposed design philosophy											
B. Anticipated problems and potential solutions											
VI - PROJECT CONTROL											
A. Techniques planned to control the schedule and costs											
B. Personnel responsible for schedule and cost control											
VII - ESTIMATING EFFECTIVENESS											
Ten most recently bid projects											
VIII - SUSTAINABLE DESIGN											
Team design philosophy and method of implementing											
IX- MISCELLANEOUS EXPERIENCE & CAPABILITIES											
A. Interior Design											
B. CADD & Other Computer Applications											
C. Value Engineering & Life Cycle Cost Analyses											
D. Environmental & Historic Preservation Considerations											
E. Energy Conservation & New Energy Resources											
F. CPM & Fast Track Construction											
X - AWARDS											
A. Awards received for design excellence											
XI - INSURANCE AND LITIGATION											
A. Type and amount of liability insurance carried											
B. Litigation involvement over the last 5 years & its outcome											
TOTALS											

Remarks: ....

....

....

Signature of Chairman	Signature of Member
Signature of Member	Signature of Member
Signature of Member	Signature of Member

08-INT 2/12/98

Medical - General

Figure 4.3   Example A/E interview scoresheet

A/E PERFORMANCE

Project #:
Project Title:
Location:

A/E Name:
Architect:
Interior Designer:
Structural Engineer:
HVAC Engineer:
Plumbing Engineer:
Civil Engineer:
Fire Protection Engr:
Electrical Engineer:
Landscape Architect:
Estimator:

Stage of Service: ☐ Schematic Design   ☐ Design Development   ☐ Contract Documents

Performance

Legend:	Rating Symbol	Score	Discipline	Not Applicable	Accuracy	Completeness	Cooperation	Coordination	Management	Meeting Schedule	Personnel Ability	Work Quality	SCORE [(-10) to + 10]	Reviewer Initials	Date
Excellent	E	10	Architectural												
Very Good	VG	5	Interiors												
Acceptable	A	0	Structural												
Marginal	M	–5	HVAC												
Poor	P	–10	Plumbing												
			Civil												
			Fire Protection												
			Electrical												
			Landscape Arch.												
			Estimating												
			OVERALL												

Remarks:

1421a-12/94

Figure 4.4   Example A/E performance form

Negotiation with the Top-Ranked Firm

A concise description of the Negotiation Process has been presented by American Council of Engineering Companies (ACEC) and appears as follows:

The Final Selection

When the final selection is made by the agency head, the contracting officer is authorized to begin negotiations with the top-ranked firm. The negotiations are conducted pursuant to the procedures set forth in the Federal Acquisition Regulation (FAR). Usually, the firm is requested to submit a fee proposal listing direct and indirect costs as the basis for contract negotiations. Contract negotiations are conducted following an evaluation of the fee proposal and an audit when the proposed design fee is more than $100,000.

If a fee is not agreed upon within a reasonable time, the contracting officer will conclude negotiations with the top-ranked firm and initiate negotiations with the second-ranked firm. If a satisfactory contract is not worked out with this firm, then this procedure will be continued until a mutually satisfactory contract is negotiated. If negotiations fail with all selected firms, the contracting officer may return to the original contractors' submittal or re-advertise the project. The negotiation process will then continue until an agreement is reached and a contract awarded. As a practical note, it is rare that a contract is not successfully negotiated with the top-ranked firm. For more information, the reader should also review Chapter 12, Managing the Civil Engineering Enterprise.

—American Council of Engineering Companies (ACEC), www.acec.org

FEE-BASED SELECTION

An alternate method of selection of an engineer of professional services is referred to as "fee-based selection." This selection method is based upon the client (alone) or the client and an engineer creating a unilateral scope of work and negotiating a fee for these services. This fee may be a flat rate or a percentage of the overall cost of the project.

This method, which can be used by public agencies and private clients to select an A/E firm, requires that the client prepare either an RFQ or a request for proposal (RFP). While an RFQ describes the project in general terms, the RFP includes a detailed SOW. These RFQs or RFPs are then either advertised or sent to potentially qualified A/E firms and selection is made from responding firms.

THE 6 PERCENT FEE LIMITATION ON FEDERAL DESIGN CONTRACTS—EXCERPTS FROM ACEC

A detailed description of the 6 Percent Fee Limitation has been presented by American Council of Engineering Companies (ACEC) and appears as follows:

Fee Limitations on Design Contracts

Since 1939, federal construction agencies have been required by law to limit the fee payable to an architect or engineer to 6 percent of the estimated construction cost. Presently, there are at least four statutes that prescribe limitations on architect-engineer fees and apply to all civilian and military construction agencies with the exception of the U.S. Department of State.

Federal agencies have interpreted the statutory fee limitations as applying only to the part of the fee that covers the production and delivery of "designs, plans, drawings, and specifications." The agencies, therefore, consider that the 6 percent fee limitation does not apply to the cost of field investigations, surveys, topographical work, soil borings, inspection of construction, master planning, and similar services not involving the production and delivery of designs, plans, drawings, and specifications. Most direct federal awarding agencies have, as a part of their supplement to the FAR, a list of those items exempt from the 6 percent fee limitation.

—American Council of Engineering Companies (ACEC), www.acec.org

WRITING ENGINEERING PROPOSALS

A critical component of the civil engineer's tool box is problem solving. Clients sometimes request engineering and technical support without a clear understanding of their needs. Conversely, some clients have a very clear understanding of their needs but are unsure how to address them. Regardless of the specific client situation, it is imperative that the engineer have a clear understanding and demonstrated skill to communicate problem solving. These key components to problem solving include:

- Identifying the client's particular problem/s
- Possessing background knowledge, the ability to work as a team member, and preparing a clear and comprehensive SOW
- Understanding the client's requirements and constraints
- Having the ability to communicate clearly
- Formulating technical alternatives
- Providing the client with alternative evaluation and/or selection
- Performing engineering design including engineering plans, specifications, and cost estimates
- Offering construction assistance, construction monitoring, or construction management,
- Providing start-up assistance and/or operations and maintenance assistance
- Creating a realistic project schedule

More details on these components appear below.

Problem Identification

An engineering project is typically born as a problem or challenge to the client or operating group. Often, the problem is not clearly understood and rarely is the need articulated well in the form of clearly explained subtasks, tasks elements, schedule, or budget. For the purposes of these discussions we will use the terms "client" and "operating group" synonymously, since the operating group within an industry or agency will actually be the client anyway.

The engineer generally has some idea of the problem from initial communication with the client, or possibly a RFP, and then gains more understanding after a site tour. A client site tour is often referred to as a "site walk." If the client intends to solicit outside commercial engineering support, a site walk may be accomplished with representatives from competing engineering firms, potentially in groups. The client can then address the questions from all the competitors so the competition is fair for all parties. If, by chance, the site walk is performed with the representative(s) from only one firm this could be good or bad news. A long-term client may perform singular-firm site walks with an engineering firm that is trusted. Otherwise, singular site walks can be an indication that the client is simply seeking another proposal or an alternate idea from a competing engineering firm. Experience shows that the success rate, sometimes referred to as the "hit rate" for this scenario is lower than the "hit rate" for a site walk performed with groups.

Once the engineer has a basic understanding of the problem they can begin to articulate the "problem identification." The problem identification statement should generally appear early in the introductory section of the engineering proposal. This statement should be constructed in the engineer's own language and could be enhanced with relevant practical knowledge from other similar projects in the engineer's portfolio. The problem statement will form the foundation of the proposal and should demonstrate to the client that the engineer has a clear understanding of the client's situation and impact on the business activity and operations.

Our experience has shown that a problem identification statement can be greatly enhanced by including a short description of the client's primary objective and any secondary/tertiary objectives if applicable. This "objective" statement should be clear and concise, approximating a paragraph or less. It is also desirable to include an "approach" statement immediately following the objective. This statement will very briefly announce the engineer's overall approach to solving the problem. Again, it is recommended that the statement be clear and concise, approximating a paragraph or less. These objective and approach statements serve as a brief announcement of the overall direction of the proposal.

So far, the engineer has launched the proposal effort and described their understanding of the problem, the objective, and approach. Now is an excellent opportunity to integrate background knowledge into the direction of the scope of work task elements. It may also be an opportune time to suggest optional tasks, beyond the SOW, if they would be applicable and valuable to the client's overall objectives. Background knowledge can consist of other similar projects and/or the application of engineering principles the client is unaware of or did not consider. This is where the engineer demonstrates that they have truly comprehended the client's problem/objective and has analyzed the condition and synthesized potential solutions for the client.

Once the problem analyses and synthesis are complete the engineer should incorporate the client's requirements and other related constraints. The constraints may be as simple as a client choice of color or as complex as the presence of a radiological species on the site. However, these requirements and constraints should be regarded as hard boundaries for the client's project.

A complete well-written proposal should include proposal assumptions. Proposal assumptions are a very important tool for the engineer. In the course of calculating the potential solutions and alternatives to the client's problem the engineer makes numerous assumptions. These assumptions are likely related to the site conditions, complexity of the client's original problem, weather, client contract review periods, client report review periods, resource availability and costs, site accessibility, meeting times, permitting requirements/costs/conditions, and a myriad of other project-related conditions. It is critical to capture these assumptions in the event it becomes necessary to show how task/subtask costs were developed if the project SOW changes or the project schedule is revised.

Background Knowledge, Teamwork, and Scope of Work

The application of the engineer's background knowledge can be weaved through at least three sections of a proposal for work which include the scope of work, the "qualifications" section, and the "project team and personnel resumes" section.

The scope of work is a logical set of tasks and subtasks that will accomplish the client's objective and solve the problem described in the problem identification statement. The work breakdown structure (WBS) is a fundamental tool used in project management and systems engineering. It is likened to a structure resembling a tree that describes the summation of subordinate costs for tasks, materials, and so forth, into their successively higher-level "parent" tasks, materials, and so on. Each element of the WBS includes a description of the parent task to be performed. This technique is used to describe, define, and organize the total scope of work for a project (Norman, et al., 2008). More details on task and subtask identification and the work breakdown structure appear in Chapter 7, Executing a Professional Commission.

A typical SOW often includes a task for synthesizing technical alternatives for the problem. If it's not a task that the client specifically included then the engineer may have an opportunity to add a task that considers alternate, cost-effective solutions.

If technical alternatives are included in the SOW, another typical task element is the "alternative evaluation" of these alternatives. These analyses will most likely include a detailed evaluation of the advantages/disadvantages of each alternative with a corresponding summary on the capital outlay, operation/maintenance costs, permitting requirements, sustainability evaluation, implementability, constructability, and any other criteria important to the client or stakeholders.

Other follow-on tasks can include construction monitoring (or construction management) of the actual construction and implementation of the project. The objective of this task is to have the original designer (the A/E firm) review and verify that the construction conforms to the intended design concepts. Another benefit of this service is to review the contractor's work and be present to comment on any potential change orders the contractor may request. In addition, the A/E firm can also be available to comment on the resource requirements and commitments the contractor provides to the job. The benefit is related to the overall impact to the project schedule and the project delivery date. The engineer generally has a great deal of design and construction experience. If the engineer is on-site for this task, it would be possible to provide a credible opinion on the resource commitments to the project before incurring a potential delay in the final delivery of the project.

Another critical element construction monitoring accomplishes is verification of material specifications. The contractor/builder may use the specifications of the construction materials to their advantage for price reduction to maximize their profits. Substitution of lesser quality materials is usually a disadvantage for the owner with regard to life cycle and/or performance. In addition, if a contractor uses materials of lesser quality this practice may cause a potential liability for the engineer in the overall performance (or failure) of the finished product. Construction monitoring is usually a win/win scenario for the owner and engineer. There will be more on this subject later in this chapter.

Client Requirements and Constraints

A clear understanding of the client's requirements and constraints is an essential building block in the client relationship. The requirements will have a major impact on the scope of services. Comprehending the constraints and explaining how the engineer's proposal will address these constraints will demonstrate critical thinking to the client.

Clear Communication

Clear communications provide the conduit for transmitting the engineer's knowledge and experience of practical application to the client's project. Clear communications are a key component in the client relationship and include verbal, nonverbal, and written skills. More information on this subject may be found in Chapter 13, Communicating as a Professional Engineer.

Technical Alternatives

Clients employ engineers to apply technical knowledge and critical thinking to complex problems and projects. Most projects have many different alternative solutions and these solutions come with a myriad of advantages and disadvantages from project initiation through construction. Civil engineering firms with technical expertise matching the client's needs are best positioned to develop viable technical alternatives to address their client's requirements.

Alternative Evaluation

Many technical alternatives have different cost elements in a specific practical application. For example, one alternative may be significantly more expensive, more reliable, and require much less maintenance versus a simpler alternative that is less expensive. An engineer can present these alternatives in a concise fashion to help the client choose between the alternative that best fits their needs. This is where an understanding of the client's requirements and comprehension of their constraints will enable the engineer to provide superior client service.

Design, Plans, Specifications, and Cost Estimates

Engineering design usually culminates with a set of engineering plans, specifications, and cost estimates sometimes referred to as P, S & E. The level of detail in the engineering plans should be described in detail in the scope of services, assumptions, limitations, and corresponding contract. Depending upon the engineering plans, the set may include civil drawings accompanied with structural, mechanical, electrical, process, and architectural drawings, among others. The specifications may provide details on the materials of construction, preferred vendors or suppliers, interfaces, quality, quantity, type, and compliance with other specifications and specific codes for the general contractor (GC) to procure and install in accordance with the intent of the design. The engineering plans and specifications often include an engineer's estimate for the entire project. The cost estimate may include a large variable: labor costs. In fact, the project should clearly state whether the project requires union labor. The engineer usually has a good idea about material, labor, and installation costs. However, a GC's labor rate can vary according to the local demand for labor or specific classifications of labor. Therefore, the engineer's estimate can only provide general guidance on the total cost of a project. After the project is bid, the construction bids show the real costs.

Construction Assistance, Monitoring, and Management

The engineer is often invited to provide construction assistance or construction monitoring during the GC's installation. The objective of this task is to provide the owner with the materials of construction installed in accordance with the intent of the design.

One critical note for the engineer performing construction monitoring is that the GC (usually) works for the owner. If the engineer observes an inconsistency in materials compared to the specification or a potential installation flaw, the engineer should promptly notify the owner. Frankly, GCs often don't like to have engineers monitor their construction. Extreme caution should be employed so the engineer does not give instructions to the GC. There are many instances where an engineer innocently provided guidance or instruction to the GC in a true effort to provide client service. The GC may have intended (and bid) the project in a different way or provided alternative materials and, therefore, feels a construction claim or change order is warranted since the engineer directed the GC to perform a task differently. These construction claims will come as a surprise to the owner. Remember, owners don't like surprises and claims associated with the engineer's presence on the job can tarnish the engineer's reputation with the client.

Start-Up and/or Operations and Maintenance Assistance

Once the construction of a facility is completed, an owner may have a need for start-up assistance or operations and maintenance assistance. These services may be optional services in the original contract or may be recognized later by the client as necessary. The engineer as the designer is recognized and accepted as the expert and may have an opportunity to provide these services to the client. This is another example where clear communications and client service will "pay off" for the engineer as a reward in the form of additional work.

Scheduling

Work Breakdown Structure—The WBS is a comprehensive classification of the project scope of work that concentrates on the planned project outcomes in a hierarchical fashion. In summary, the WBS is a results-oriented family tree that captures all the work of a project into smaller increments in an organized way. Preparing the project WBS is an important step toward managing the project's inherent complexity and should be developed before the project schedule is prepared (Norman, et al., 2008).

Internal Reviews, Client Reviews—It is important to include internal reviews and client review periods in the overall project schedule. These review cycles are an important component of the project quality system and generally improve the integrity and accuracy of the delivered project. Review periods vary with the complexity of the engineer's product and vary from several days to weeks depending upon the project elements and complexity. In addition, incorporating key client reviews of the deliverable product provides the engineer with an opportunity to achieve a high degree of client satisfaction and allows the owner to have "ownership" in the process.

Response to Comment (RTC) Tables and Resolutions—An RTC table is a tool commonly used by the engineer to communicate a complete understanding of the client's review comments and review comments from others in a comprehensive and organized fashion. The RTC table summarizes all these comments and states "how" they were handled in a column labeled "accepted," "rejected," or "revised," and usually include the final disposition of the comment. This table is an important tool because it provides a summary of every reviewer's comments, some of which may contradict one another. It also simplifies the final checking process since the reviewers do not have to go into the document and track the final disposition of their comments. Additional details on the RTC table and a sample format appear in Chapter 13, Communicating as a Professional Engineer.

Final Production Time—The engineer should consider the final project production time and include this time period in the project schedule. The final product may be a report or data but it does take time to assemble and it should be checked for accuracy and completeness before delivery.

Delivery Time—Upon completion of the final product the engineer can arrange for delivery. Depending upon the ultimate location of the client(s), at least a one- to two-day period should be reserved for final delivery. Electronic delivery of engineering reports or data can be accomplished much more quickly but time should still be reserved for this activity. It is also recommended to follow-through that delivery was achieved and the client can open or read the final product (in the event of electronic delivery).

Reserve Time—An experienced engineer will usually place a one- to two-day reserve period in the project schedule. Experience shows that despite superior planning, events beyond our control can disrupt the project schedule and impact the delivery of the finished product.

Software Programs—Software products should be verified for accuracy before final delivery to the client. It is also recommended to have a brief introductory session where the engineer can present the product and perform an initial demonstration of its utility and capabilities.

THE CONTRACT

The contract is the glue that binds the client and the engineer to the SOW. It is a formal, legal document that usually cites the tasks from the original RFP, the engineer's tasks, schedule, and budget. It includes numerous clauses and terms that usually trump any discussions and verbal agreements made during the entire process. Sometimes clients expect the Engineer to begin the project and tasks for the project before the contract is executed. If a client insists that work begins immediately, it is highly recommended that the Engineer be certain that the contract is in full force and effect. Otherwise, the Engineer may not be compensated for any work performed without a contract. Most contracts include a clause that states "the terms of the contract supersede any other agreements including oral agreements."

Contract review and approval is not simply a formality. Typical contracts are very detailed and complicated legal instruments. For example, if disputes were to occur between the client and the engineer, the case would be tried in the city where the last party signed the contract. Clients usually ask the engineer to review and approve a contract before sending it back to the home office or facility location. Therefore, the last party to sign and date the contract did so at their home office where they likely have legal representation and recognition in their own city. If the engineer were to be challenged and had to appear in court, the travel, meals, lodging, any attorney's fees and preparation would be borne by the engineer. This fact alone might enter into the final decision whether to attempt to challenge a dispute because of the great expense incurred.

Key elements to the contract are the clauses that typically appear in small print on the back side of the contract. These clauses describe the details and conditions for the contract. A standard contract generally will begin with a statement and reason for the contract and list the names of the parties involved in the agreement. There will likely be some additional information such as:

1. The original "request for services" or definition of the scope of services from the originator (client)
2. The original proposal from the consulting engineer and any addenda, meeting minutes, or telephone records relevant to the proposal
3. Any other relevant agreements such as a "master services agreement" (referred to as an MSA)

The contract terms are generally complex, detailed, and are prepared to serve as a legal document governing the arrangement for the parties involved in the contract. The engineer may be tempted to skip over these terms because they are anxious to perform the services, but this could be a costly mistake. These terms are generally listed in numerical order and may range from a few clauses to 50 or more. They outline the business elements of the legal arrangement ranging from the terms for payment to insurance requirements required to receive payment for the professional services rendered. Each client and respective contract is unique so the contract terms should be read carefully by the engineer. The engineer should be confident that they will be able to comply with the contract terms while performing the services, after completion of the work, and when receiving payment for the services.

Some of the typical terms and their meanings are listed for reference as follows.

1. Scope of Services: This clause is one of the most critical ones in the contract. It defines the actual tasks to be performed, the deliverable product(s) resulting from the effort, the budget, the schedule, the project location, the fees, and any other relevant statements describing the services the consulting engineer will provide. The engineer should take great care describing the scope of work and reaching an agreement with the client on an SOW that meets their needs.
2. Standard of Care: The standard of care refers to "how" the consulting engineer will perform their services. This clause describes that standard and generally limits this standard to the level of care and skill normally exercised by similar engineering professionals in the same locale, under similar conditions (such as time requirements, budgets, or task elements in the SOW), at the date the services were performed. This clause is important because it's directly related to the overall quality of the deliverable product that can be produced by the engineering professionals in this area at this time for a given price.
3. Engineer's Responsibility: This clause describes "how and what" the engineer is responsible for and includes a statement that the engineer is an independent contractor which is directly related to providing an independent

opinion and professional ethics. The clause may also state that the consulting engineer will provide qualified staff to perform the work; will employ safe work practices for these staff; will not be responsible for the safety or liability of other staff outside the organization that may be working on the same project; will work cooperatively with the client's employees, consultants, subcontractors, or other staff; and will retain a record copy of the project file and deliverable product for a given period of time.

4. **Client's Responsibilities:** This clause generally states the client will provide all the material, information, reports, and data pertaining to these services without limitation. It may also state that the client will disclose information regarding potentially hazardous situations, chemical compounds, and underground conditions including utility information; past or present information on the general and environmental compliance with local/state and federal regulations; any potential or pending court actions; or any other relevant information on the project or site location. Another client responsibility includes informing the engineer whether any other regulations or requirements should be included in the engineer's scope of services and whether other labor-related conditions may be applicable such as trade union representation or prevailing wage regulations. The clause may also state that the client's employees, consultants, subcontractors, or other staff will cooperate with the engineer.

5. **Insurance:** There are different types and amounts of insurance including automobile, general liability, and more. The insurance clause is important because these requirements may not be available or may impose greater cost that impacts the engineer's fee and profit.

6. **Revisions or Contract Changes:** This clause states that contract changes or revisions may be recommended or required by either party after the project begins by altering, deleting, or adding tasks to the scope of services. If the scope of services is altered, both parties are responsible to renegotiate in good faith to assess an equitable adjustment in the project budget, schedule, and deliverable product and to prepare and approve a project change order or work order reflecting this new adjustment. The clause will also state that if both parties cannot agree and approve a change order that the work will be suspended without penalties to either party and that the engineer is entitled for just compensation for the services performed to date on the project.

7. **Contract Term and Termination:** This clause generally states that the contract will begin at the approval (signature) of the agreement and that the agreement will be in force for a period of time or until the project is complete. It will also likely state that either party may cancel the contract at any given time without cause by providing an advance written notice with a time period varying from two to ten days. The clause also states that the client will compensate the engineer for reasonable expenses and labor charges up to the cancellation of the

contract including any relevant demobilization fees and that the engineer will provide any related file information and partial deliverable products to the client.

8. Force Majeure: If, in the course of performing services, the engineer encounters conditions or causes beyond their control then "force majeure" is declared. Force majeure includes acts of God; acts of a legislative or judicial office; acts by the client's subcontractors; labor disturbance or strikes; floods; hurricanes; fires; war; or severe weather.

9. Site Access: This clause states that the client will provide unimpeded site access to the engineer and subcontractors; space for equipment, materials, or vehicles; utility access including utility services and relevant utility clearances; and any relevant permits.

10. Warranty and Ownership of Waste Products: This clause generally has two subject components. Warranty refers to an overall warranty for the deliverable product produced by the engineer. The engineer is cautioned to limit the deliverable product(s) and the subject clause to the accuracy and timeliness of the verbal and written information provided by the client (or client's contractors, agents) and to also limit the warranty to the general limitations clause of the contract.

The other major subject component regarding this clause is related to any waste products and/or testing materials provided by the client to the engineer. The engineer should be certain to limit the risk of loss of sample materials and the responsibility for disposal of residual sample materials to the owner. The owner is considered the original generator of said materials and is ultimately responsible for proper manifesting and disposal in accordance with federal, state and local regulations. The engineer should be clear that final return of waste materials and ultimate disposal is not included in the original scope of services unless, of course, this task is included in the scope of work.

11. Subcontracting: The subcontracting clause notifies the client that subcontractors may be employed to perform specific tasks for the project and that these staff may require site access or information access as part of the project team. The clause may also mention a mark-up for these services to the client if the engineer contracts with this subcontracting firm. Finally, the clause may also state that the contract agreement cannot be assigned to a third party without the written consent of both originating parties. The part of the clause protects both parties from having to execute a contract with parties other than the original ones in the agreement.

12. Dispute Resolution: This clause outlines the procedure for resolving disputes that may arise out of the interpretation, enforcement, implementation, or performance of the services in the scope of work and contract. Generally, both parties agree that they will attempt to resolve the dispute in a meeting within the existing upper management structure of both parties within a relatively short period of time. If this is unsuccessful,

the next level of resolution is generally by third party or mediation. One mediation procedure often cited is the architect's Construction Industry Mediation Rules or an alternate association. Finally, if this level of resolution is unsuccessful the clause may mention legal claims by filing suit in court in the jurisdiction of one of the corporate headquarters. The caution here is that if a lawsuit is filed both parties may want the suit to be tried in their home town because the courts may favor the local entity. The losing party in the law suit may have to pay travel expenses for the plaintiff and/or attorneys.

The clause on dispute resolution is helpful to provide a roadmap for resolving disputes. This is particularly important because a dispute can often stop or slow the progress on the project which can lead to an array of other problems. It is also helpful to resolve project-related problems at the lowest level of authority possible to keep the client-CE relationships amicable and positive.

13. Governing Laws: This clause generally states that the contract is subject to the laws of the United States and specifically to the laws where either the engineer's or client's home office is located.

14. Severability: This clause usually states that each contract term is separate from one another. For example, it's acceptable for both parties to choose to ignore a specific clause. However, the other clauses of the contract are still in effect.

15. Entire Agreement: This clause generally states that the agreement applies to the immediate task order and any future task orders. It also usually states that any amendments to the agreement shall be in writing. Verbal agreements outside of the written agreement will not be recognized. This clause is important to the consulting engineer because it clearly states that the written agreement is the legal controlling document and that any verbal agreements between the engineer and project-related clients are not recognized. If a client insists on a verbal agreement the engineer is cautioned to prepare a revision clause to the written agreement and have it approved by the client before proceeding with the verbal direction.

16. Risk Allocation: This clause attempts to discuss risks and rewards and tries to balance claims against the engineer to a reasonable proportion for responsibility for the actual cause of any losses. The clause often has multiple parts including:

    a. Limitation of Liability: This subpart clause generally states that the engineer will only be responsible for potential damage or loss payments up to a preset dollar limit (in fees ranging from $10,000 to $100,000) or the limit of the actual contract. The limit usually applies to any potential losses or claims in connection to the contract and it usually has some time period specified when it's possible to seek these claims generally varying from three months to a few years. The clause may also state that claims may only be filed against the organization and not filed against any specific individuals providing some security to the engineering employees for the company. In addition, more recent clauses sometimes limit the percentage of fees that may be based on an estimated percentage of the engineer's contribution toward the resulting problem or loss. This means that if an engineer had a very small role on a project and a corresponding small estimated contribution toward a failure of a large construction project, then the engineer would be responsible for a similar small percentage of the fees to correct the problem or loss. In the recent past, some consulting engineers have performed very limited technical services (with correspondingly small fees, such as a few thousand dollars) and have been sued for large sums, thinking the engineering firm had large assets or large insurance coverage. This clause attempts to level the playing field and potentially expose the engineer to a fair and reasonable amount of exposure when working on large, intricate, or expensive projects for small portions of the actual project. The final contract may have a similar clause protecting the client in the event the engineer suffers losses from a client- or subcontractor-related cause.

A Note on Liability: The most important concepts of this contract clause are for the engineering firm to limit the monetary extent of liability in the project and to attempt to limit the liability to potential causes of damage to the scope of services performed by the engineering firm or some fee like \$10,000, if possible. Typically contracts present reasonable liability coverage to the monetary limits of the engineering services or \$50,000, whichever is less. Limiting liability exposure to an originating cause within the firm allows the firm to be protected by their own insurance.

However, some clients will only accept contracting terms using their own preapproved contracts. Under this condition it is especially important for the engineer to carefully review the liability clauses. We have seen liability clauses by large public and private clients where this clause asks the engineering firm to accept liability for their own subcontractors, and other contractors and subcontractors working on the site. These clauses are particularly onerous because the engineer has no oversight for these other firms and may risk the viability of the entire engineering firm for mistakes made by others.

b. Indemnification for engineer clause: This subpart clause generally states that the client indemnifies the engineer and their employees, officers, shareholders, or agents from any lawsuits, damages claims including attorney's fees potentially caused by the client's negligent performance of services under the contract.

c. Indemnification for client clause: This subpart clause generally states that the consulting engineer indemnifies the client, and their employees, officers, shareholders, or agents from any lawsuits, damages claims including attorney's fees potentially caused by the consulting engineer's negligent performance of services under the contract.

17. Ownership of Instruments of Service: This clause generally states that the client owns the documents but the consulting engineer retains the right to retain a copy for their files. This clause is sometimes challenged when there may be an occasion when the client may not pay for the engineering consulting services in a timely manner and the engineer feels that it is necessary to hold back the finished product until payment is received. In addition, it is advised that both the A/E firm and the client work closely on this clause to avoid misunderstandings in data packages, usability, readability, availability, and presentation.

18. Late Payment/Assessments: A late payment and assessment clause in the contract is important because of the time value of money and the potential capitalization of (potentially) many simultaneous projects within the engineering firm at any given time. The typical engineering firm prepares and sends invoices monthly. This process captures all the project-related costs for the previous month including labor, project expenses, and overhead. Once these costs are computed a draft invoice is typically prepared for the engineering project manager's review and approval which in total may take an additional two to three weeks. Often the total time elapsed by the time the client receives the invoice from the previous month is six to eight weeks. Now the invoice has to proceed through the client's review and approval process which in turn can take an additional three to six weeks.

Often the total time elapsing from a professional at the engineering firm working on the project on Day 1 until the firm receives payment from the client is 8 to 12 weeks under optimum circumstances. Under these conditions, the typical engineering firm is paying at least 8 to 12 weeks' interest for the firm's entire portfolio of projects which can lead to increased overhead expenses thereby increasing the firm's cost and making the firm less competitive. A late payment clause can be a good tool to remind the client about the importance of paying promptly and hopefully the assessment component of this clause will not need to be accessed.

BUDGETING

The project budget is like the fuel for the jet aircraft with the client and the engineer onboard working together. Proper budgeting and budget tracking is critical for both parties—the project and the careers depend upon it. Some key points to remember:

- Budget by tasks, subtasks, and staff codes
- Include subcontractors, oversight and management costs, and any mark-ups
- Include project-related expenses like mileage, per diem, or other related items
- Review assumptions for hidden costs like permitting fees that might be missed
- Include other related items like delivery expenses or report binders, color copies, and the like

ENHANCING THE ENGINEERING FIRM'S PROBABILITY FOR A SUCCESSFUL PROFESSIONAL ENGAGEMENT

"ASCE's Body of Knowledge Technical Outcome 8—Problem Recognition and Solving" is probably one of the most appropriate examples of engineering and business applications for the engineering firm. As part of the client proposal process the engineer and the firm clearly need to apply the six levels of cognitive achievement shown in the following list (American Society of Civil Engineers, www.ASCE.org).

ASCE's Six Levels of Cognitive Achievement Presented in the *Civil Engineering Body of Knowledge for the 21st Century:*

- Knowledge
- Comprehension
- Application
- Analysis
- Synthesis
- Evaluation

These are the key variables in crafting an effective proposal for the client except for two additional variables that the civil engineer may find him- or herself working on for his or her entire career:

- The client relationship and
- Effective project budgeting

This concept can be further explained by the graphic depicted in Figure 4.5.

The civil engineering firm has an opportunity to enhance their chance for winning the client's project and securing a professional engagement by completing the proposal in an effort that matches ASCE's six levels of cognitive achievement discussed. Reaching this high level of cognitive achievement, building a positive client relationship and including an effective project budgeting will usually "seal the deal" for the engineer and the client.

Working Examples of RFPs

The authors have the delightful experience of teaching at a California State University to senior civil engineering students. Specifically, the Senior Project course (CE-190) is a requirement for the CE curriculum. It is a "hands-on" course about real problems in the Northern California area. The projects are typically suggested by public agencies like Rapid Transit (RT), Redevelopment Agencies, Parks and Recreation or State/County Engineering Departments. We would like to share an applicable experience with our readers for one of the semester projects. A local county prepared a

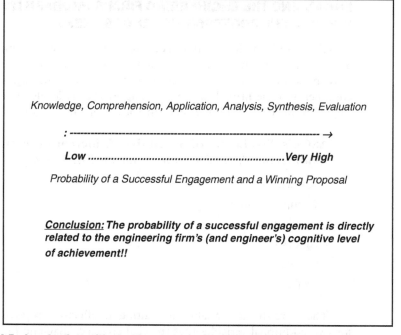

Knowledge, Comprehension, Application, Analysis, Synthesis, Evaluation

: --- →

Low ....***Very High***

Probability of a Successful Engagement and a Winning Proposal

Conclusion: *The probability of a successful engagement is directly related to the engineering firm's (and engineer's) cognitive level of achievement!!*

Figure 4.5   Levels of cognitive achievement for the CE professional and CE firm

Draft RFP for a real problem with a local Wastewater Treatment (WWT) Authority for the decommissioning of an inadequate WWT plant and the design of a new conveyance line to an existing downstream regional facility. The students tackle the problem in groups of four to five students from start to finish, culminating in a 90 percent complete report, then Final Report, and a public presentation as follows:

- Review the Draft RFP and prepare a student team as a fictitious engineering firm
- Conduct a site visit
- Meet with the client
- Prepare a proposal including task descriptions, schedules, level of effort (LOE) budgets, assumptions, and qualifications
- Respond to client comments on the proposal (the clients are volunteer mentors and practicing registered CEs from local consulting engineering firms under the guidance of the University Professor and a lab class instructor)
- Conduct a project kick-off meeting
- Prepare a report outline including objective and approach statements consistent with the proposal
- Prepare a 90 percent report

- Meet with the client on the 50 percent report and respond to client comments on the report
- Prepare a 100 percent report
- Meet with the client on the 100 percent report and respond to client comments on the report
- Prepare a 30-minute presentation on the final report results and give the presentation in a University auditorium open to the public.

An example PowerPoint presentation may be found at www.wiley.com/go/ cehandbook.

Typical Civil Engineering Example RFP

An example RFP for a typical CE problem appears in Appendix A. The problem was an actual one experienced by a small county Wastewater Collection and Treatment Authority. A response in the form of a "Senior CE Student Proposal" from a California State University is presented immediately after the RFP (Appendix B) by one of the CSUS senior CE students groups that called themselves Global Hydraulic Engineers, Inc. The CE students taking the CSUS CE course were under the guidance of a University Professor, an adjunct laboratory instructor, and practicing registered CEs acting as the client/owner for the Authority. The proposal responses therefore were subjected to critical review and revision and contain the essential elements for a professional reply to an actual simulated RFP. In addition, a sample Final Feasibility Report consistent with the RFP is also provided for review (Appendix C) by a different CSUS senior CE student group that called themselves CVision Engineering.

SUMMARY

This chapter describes the request for proposal process, followed by contracting details for performing the project, and enhancing the engineering firm's probability for a successful professional engagement. In addition, as part of a Senior Civil Engineering student project course, a sample RFP is included in Appendix A for reference. This RFP is then followed by a sample proposal for the work in Appendix B, and finally a sample Feasibility Study Report in Appendix C, all prepared by senior CE students as part of their CSUS Senior Project.

Overall, this chapter emphasizes that the engineer can increase the probability of submitting winning proposals and winning more jobs by thoroughly applying the six levels of cognitive achievement: knowledge, comprehension, application, analysis, synthesis, and evaluation to the client's problem. The consulting engineer may also find that they will continually improve two other variables for their entire career: the client relationship and effective project budgeting.

REFERENCES

American Council of Engineering Companies, www.acec.org.

Norman, Eric S., et al. (2008). *Work Breakdown Structures: The Foundation for Project Management Excellence*, John Wiley and Sons, Hoboken, NJ. ISBN: 978-0-470-17712-9.

American Society of Civil Engineers, www.asce.org.

Introduction
Ethics
1 2 3
Professional Engagement
4
History
5
What Engineers Deliver
6
Engineer's Role in Project
Development
7
Permitting
Leadership Managing
Emerging Technology
A B C D E F
Having
a Life
17
8 9 10 11 12 13 14 15 16
Executing a Professional
Commission
Sustainability
Client Relationship
Globalization
Legal Aspects
Communicating

Chapter **5**

The Engineer's Role in Project Development

Big Idea

Civil engineers play many different roles in the project development process.

> Engineering problems are under-defined, there are many solutions, good, bad and indifferent. The art is to arrive at a good solution. This is a creative activity, involving imagination, intuition and deliberate choice.

—Ove Arup

Key Topics Covered

- Background
- Participants in the Process—The Players
- The Flow of Work
- Predesign
- Design
- Design During Bid and Construction
- Postconstruction Activity
- Summary

Related Chapters in This Book

- Chapter 2: Background and History of the Profession
- Chapter 3: Ethics
- Chapter 4: Professional Engagement
- Chapter 6: What Engineers Deliver
- Chapter 7: Executing a Professional Commission
- Chapter 8: Permitting
- Chapter 11: Legal Aspects of Professional Practice
- Chapter 15: Globalization
- Chapter 16: Sustainability
- Chapter 17: Emerging Technologies

(*Continued*)

Related to *ASCE Body of Knowledge 2* Outcomes

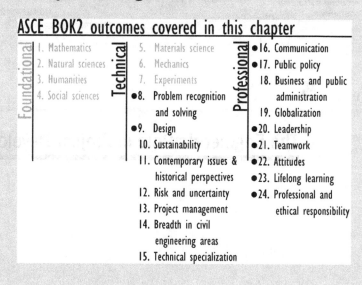

BACKGROUND

The architectural, engineering, and construction (AEC) industry always has operated on the "virtual" organization principle and is infamous for its fragmentation. Constructed products involve a staggering number of players. These include private owners, developers, government agencies, engineers and architects, other designers, builders, product and material suppliers, real estate agents, lending institutions, and inspectors among others. In the United States, where the AEC industry historically has made up 9 to 10 percent of the gross national product (GNP), the majority of firms employ fewer than 20 people.

The industry is design-intensive because most projects are one-of-a-kind. Often with limited local knowledge, AEC professionals must produce unique products with stringent cost, schedule, and quality standards. Much of the design for any particular constructed product is performed by separate individuals and firms. Furthermore, project team members may not be focused on a shared goal. Clients and owners come in a variety of types and sizes, some with considerable design and construction experience and, more frequently, others with none. The owner typically thinks in terms of quality, as well as short- and long-term costs. Architects and engineers have a different perspective; often they are motivated by the desire to avoid mishaps and to minimize their costs relative to billable hours.

Civil engineers must work with a staggering number of determinant and nondeterminant processes, a vast array of participants, and the need to evaluate outcomes and manage risk in order to develop and evaluate prospective designs. This chapter examines the civil engineer's role in project development. The chapter discusses the people involved in moving projects from ideas and needs to completion. It explores the many roles civil engineers play in the design and the project delivery process, as well as the deliverables connected to that process.

PARTICIPANTS IN THE PROCESS—THE PLAYERS

As discussed in Chapter 2, Background and History of the Profession, the first civil engineers had to develop knowledge and skill in a wide range of fields: irrigation, palace and tomb building, weapons, and fortification. Today's civil engineer also must develop considerable expertise, but this knowledge tends to be more technically specialized. What has remained constant for millennia is a need for civil engineers to be problem solvers, innovators, analysts, critical thinkers, and communicators.

Another uninterrupted theme throughout the history of civil engineering is the involvement of three key players in the project delivery process:

- Owner or client
- Designer (engineer or architect)
- Builder or contractor

For a description of the roles that these participants play, see the accompanying textbox—*Participants in the Design Process.*

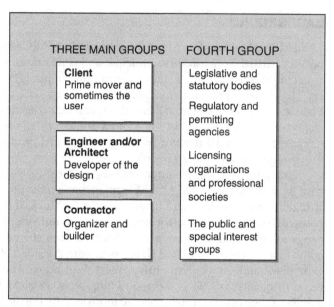

Figure 5.1 Players involved in project delivery

These three main groups—owner or client, designer (engineer or architect), and builder or contractor—coexist and interact with a fourth group composed of legislative and statutory bodies, regulatory and permitting agencies, licensing organizations and professional societies, and the public and special interest groups. (See Figure 5.1.) Chapter 8, Permitting, describes the important role played by these participants in the design process.

Participants in the Design Process: Owners, Design Professionals, and Contractors

Designing and constructing projects frequently involves hundreds of participants; however, there are three principal players in every project: the owner, the design professional, and the contactor.

Owners

Although the list of potential construction project owners is nearly infinite, the short list includes governments (federal, state, and local), districts (school, irrigation, and reclamation), for-profit and nonprofit corporations, partnerships, and individuals. Construction projects occur when a representative of one of these groups seeks to mitigate a need or realize an idea. The owner's role generally is to provide the site; finance both the design and construction of the improvements; give timely, accurate feedback to the design professional; and operate the facility. Owners commonly engage the services of a design professional to conceive the design and produce the construction documents. Owners place contractors under contract to execute the work described in the construction documents.

The Design Professional and Design Consultants

Design professionals—primarily architects and engineers—offer a wide variety of services to project owners. Traditionally, they have created projects and produced construction documents and contract administration on behalf of owners, under service agreements called design contracts. The tumult in the design profession in the last couple of decades has prompted architects and engineers to diversify the services they offer for a fee. Many now include facility lifecycle analysis, recycling and management, [sustainable design and energy efficiency], as well as practice management in the range of services they provide.

The number of different design professionals involved in producing construction drawings varies according to the type and complexity of the project. Individuals or very small organizations generate most drawings for homes. In some states, laypersons may design homes and duplexes without a design professional's license. Developing the design for a hospital, performance center [high rise building], or manufacturing facility, in contrast, may require many highly specialized design professionals who, after rigorous examination, have been licensed by the states in which they do business.

The core participants commonly responsible for the design of building construction projects include architects and landscape architects, and geotechnical, civil, structural, mechanical, and electrical engineers.

Architects, whose authority to design projects derives from state licensing boards, have the daunting task of identifying their clients' problems during the predesign phase and describing their solutions to them, using pictures and words, during the design phase. Architects conceive the physical attributes of a project and incorporate local land-use ordinances and applicable building code requirements into their designs. Their interests and professional responsibilities are focused primarily on how a project looks (aesthetics) and how it works as a product (that is, will it protect its users from the elements and from injury during catastrophic events such as earthquake and fire? Does it fit effectively into its environment? Does it fulfill the owner's needs?).

The number of specialists and the variety of services that design consultants offer is substantial; however, architects commonly hire structural, mechanical,

(Continued)

and electrical engineers for significant portions of building work—areas of specialty for which they frequently do not have the training, license, or personnel. Large design firms, however, frequently have in-house engineering capability, which gives them more market share, greater efficiency, and more control over the design process. Such organizations are commonly referred to as architect/engineer (AE) firms. Regardless of the size and organizational structure of the office, the overall responsibility and liability for the design of a project reside with the architect, who becomes known as the prime design professional (the "prime" designer or contractor is the term given to the entity that signs a contract with the project owner).

Although many civil engineers are qualified to prescribe the treatment required to prepare soil for a project, geotechnical engineers are registered professional engineers who are required to devote several more years to practice and/or additional education after becoming licensed civil engineers before they can legally call themselves geotechnical engineers. They are hired by owners to investigate a project site and produce a comprehensive evaluation of its soil conditions, which are recorded in a geotechnical report. Geotechnical engineers [and in some cases environmental engineers] commonly investigate the past uses of the site and its hydrology, identify its soil types, determine whether and to what extent a site is contaminated, and delineate any procedures that the contractor must follow to prepare the soil for its intended role. For example, soils must be made stable and competent to bear the weight of structures and vehicular traffic for years, and soils may be used to encapsulate solid waste and to line excavations and earthen structures that will contain water.

A host of participants in the design and building process use the geotechnical report. The structural engineer uses the report to design the foundation of a structure; the landscape architect uses the report to develop the specifications for the planting and irrigation of landscaped areas; and the contractor and subcontractors use the report to determine the costs of earthwork (such as excavation, soil preparation, pile-driving, and foundation work) and evaluate the risk associated with it.

The principal concern of geotechnical engineers is how the soil will perform over time with the planned activities imposed on it. Their contracts with the owner normally require them to prepare the geotechnical report, and monitor, inspect, and approve earthwork while it is being performed. Additionally, the geotechnical engineer resolves issues that arise in the course of construction, such as the mitigation of contaminated soil that might not have been apparent during the site investigation. Beyond these functions, they do not typically get involved in design.

Civil engineers typically produce most of the construction documents related to engineering construction (streets and highways, sewer and water treatment plants, harbors, dams, bridges, and utilities). They must be licensed by the state in which the work is performed. On commercial building projects, the civil engineer plays a relatively limited design role, normally taking responsibility for on-site grading, drainage, and paving plans and specifications; for off-site improvements (driveways, gutters, curbs, and sidewalks along a public thoroughfare); and for the design of certain on-site underground utilities (sewer lines, fire system supply, storm drainage, domestic water supply). Civil engineers often cite the standard specifications of the city, county, or state in which the project is located, particularly in the design of off-site improvements. These specifications are frequently tried-and-true specifications that are developed by the state departments of transportation (which invest considerable funding in research) and are often wholly adopted by public works departments at the local level.

Structural engineers specialize in the design of foundations (piles, caissons), substructures (habitable portions of a structure that are below ground, such as basements), and superstructures (the portion of the project above grade, or above the water in the cases of bridges built across bays, lakes, and rivers). Like civil engineers, structural engineers are licensed by the state in which they do business, but they are frequently required to have specialized education and training beyond that of a civil engineer. Structural engineers—frequently hired by architectural firms for their expertise—are focused on the performance of the structural system under various loading conditions that fall into two classifications: static and dynamic loading. Static loading comprises dead loads (gravitationally imposed loads resulting from the weight of the structure and its permanent equipment) and live loads (mobile loads that are not necessarily present at all times). Furniture, snow, hydrostatic pressure (the pressure at any point exerted on a surface by a liquid at rest), and a building's occupants are examples of live loads. Dynamic loads, such as seismic activity and wind can occur suddenly, and vary in intensity, duration, and location.

Structural engineers are responsible for protecting the lives and property of project users in a cost-effective way. Although their focus is on the performance of a structure under the loading conditions just mentioned, they should also be aware of the aesthetics of the project.

Mechanical engineers involved in building project design are responsible for plumbing, sewage and piping systems, and for heating, ventilating, and air conditioning systems (HVAC). Mechanical engineers commonly form consultant agreements with the A/E to develop and describe the plumbing and HVAC systems in buildings, which are designed to ensure the comfort and health of building occupants. Plumbing and sewage systems provide an adequate source of water for human consumption and sanitation, and effectively dispose of wastes generated in the building. The heating, ventilating, and air conditioning equipment is used to control environmental comfort factors such as the temperature of the ambient air in a building, the mean radiant temperature of the surrounding surfaces, the relative humidity of the air, the pureness of the air, and air motion. HVAC and

(Continued)

plumbing systems in building projects present a significant design challenge, particularly in the distribution of conditioned air and piping through the structure. The involvement of mechanical engineers in the design process increases dramatically when they are involved in industrial construction projects, such as refineries, manufacturing facilities, chemical plants, and waste and water treatment plants. Indeed, they may hold the prime design professional role on these projects. Mechanical engineers concentrate on the performance of the systems they design.

Electrical engineers are involved in the design of a variety of construction projects, including massive power generation and distribution systems for state and federal governments, cogeneration power plants, and building construction projects, to name a few. As with the other engineers, electrical engineers must be licensed by the state in which they conduct business. In building construction projects, these engineers design the electrical service and communications systems on the site, as well as the site lighting, usually at the request of the A/E. They also design the service and distribution systems inside the structures. In addition, electrical engineers must design and clearly spell out the type and location of the electrical equipment and cabinetry and the means of distribution and controlling the power. Those engineers who work for the local utility company frequently control the design of the off-site system (the portion found in public utility easements). Electrical engineers focus on the proper sizing of the system, the location of the equipment, the distribution of the power, and the safety of the end user.

Landscape architects, also licensed by the state, specialize in developing ornamental landscaping plans, which includes selecting trees, shrubs, ground cover, and grasses, and designing the irrigation system required to support them. The landscape architect's work may also include some site improvements (such as walkways, garden structures, screens, fencing, and water features, all of which are referred to generally as *hardscape.*) Landscaping plays an important role—not only for the visual beauty it brings to a project, but for the beneficial effects that well-conceived and -executed design can have on the energy consumption of a building, as well as on air and water pollution.

Contractors

Although the term "contractor" is loosely applied to anyone who earns income from constructing things, sole proprietorships, partnerships, corporations, and joint ventures are the common legal entities that assume responsibility and liability for constructing projects under contract with the owner. Many states regulate contractors through licensing boards, which assure the health, welfare, and safety of the public through education, testing, and, where applicable, the enforcement of state license laws.

There are distinct categories of contractor:

- *Engineering contractors* construct engineering projects such as highways, bridges, and industrial construction projects.
- *General building contractors* produce residences, multiple-family projects, commercial and civic buildings, and/or retail spaces.
- *Specialty contractors* focus on one portion of a project, such as plumbing, sheet metal and air conditioning, roofing, insulation, tile, floor coverings, and elevators.

The contractor who signs a construction contract with an owner is called the *prime contractor*. The prime contractor, for a variety of reasons, frequently hires specialty contactors for portions of the work, who become subcontractors under the construction contract. Plumbing, mechanical, and electrical specialty contractors are commonly hired in this fashion.

—Keith Bisharat. (2008). *Construction Graphics*, John Wiley & Sons, pp. 5–7.

THE FLOW OF WORK

The typical project moves through several phases: predesign, design, bid, construction, occupancy, and eventually adaptive-reuse, and decommissioning and/or demolition. Civil engineers can be involved in any of these phases. (See Figure 5.2.)

PREDESIGN

In predesign, or planning, clients enter a "discovery" phase where needs and wishes are explored. If a client is large, client staff may be responsible for preparing a general plan of action and outlining requirements. Often large clients have constructed previous projects and have a wealth of experience on which to draw. A client without in-house capacity may hire a consultant, usually an architect or civil engineer, to help evaluate the need to build. These consultants may hire additional consultants to support their efforts. Table 5.1 gives examples of professional services available in predesign.

At this initial stage, the entity responsible for the evaluation forms a working organization and identifies the information needed. Data should include:

- History of the events leading up to the decision to build
- Purpose and function of the project
- Policy decisions
- Timescale for the project
- Cost limit, or budget, of the project

- Details of the site and services
- Basic details of building requirements
- Comparable best practice

The result of this effort is a *Statement of Need,* which identifies the need for a new or remodeled facility based on business objectives or public policy and enables the client to gain internal approval for the project. In private corporations, upper management grants this approval. In public organizations, governing bodies, such as boards or councils, give the go-ahead.

The Statement of Need concisely states the problem, not the solution. Even after organizational approval has been gained, the client continues to assess needs. Either through the efforts of in-house personnel, with the assistance of consultants, or by a combination of the two, the client evaluates the needs and resources of the organization and generates options to meet those needs. All options should be considered, including the "do nothing" option. Options that do not involve new construction, such as leasing additional capacity or remodeling existing facilities, also need to be contemplated. Relative benefits, drawbacks, and risks need to be analyzed. Feasibility studies may be conducted on particular aspects of the proposed options. Alternatives may be tested to determine their financial, economic, technical, or other advisability.

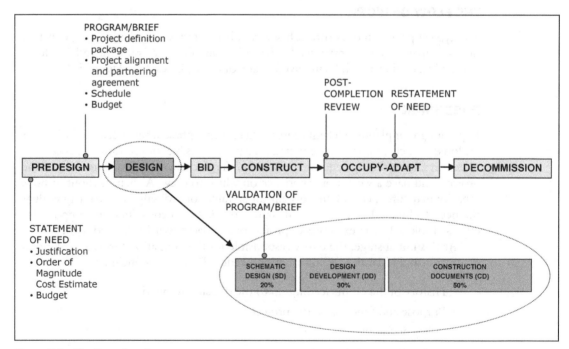

Figure 5.2 The flow of work
(Adapted from Blyth and Worthington, 2001)

Table 5.1 Professional Services Available in Predesign

Profession	Special Skills
General Civil Engineer	• Conducts site assessment • Highlights opportunities, potential problems, and building permit requirements • Coordinates other professionals and prepares design criteria, program, or brief
Environmental Engineer	• Assesses need for environmental impact reports and other permitting requirements • Researches project site's history in relationship to previous uses and hazardous materials (Phase I Environmental Assessment)
Geotechnical Engineer	• Investigates and reports on soils conditions • Proposes initial approach for design of foundations
Structural Engineer	• Advises other professionals regarding load-carrying capacity of existing structures • Evaluates feasibility of conceptual designs
Transportation Engineer	• Transportation planning • Rights of way, current/proposed/use • Existing traffic conditions • Traffic impact analysis, travel demand forecasts
Water Resources Engineer	• Flood Plain Analysis, channel design, stormwater conveyance and treatment, water supply
Land Surveyor Architect	• Provides legal description and topography of site • Conducts site and/or building assessment • Highlights opportunities and potential problems • Coordinates other professionals and prepares design criteria, program, or brief • Assesses Americans with Disabilities Act (ADA) requirements of existing building and/or new improvements
Landscape Architect	• Surveys condition of existing trees and planting • Performs Americans with Disabilities Act (ADA) assessment of existing site improvements
Urban Planner	• Interprets planning regulations • Advises on probable outcome of development proposal, changes of use, or new development
Contractor	• Reviews conceptual design for constructability (buildability) • Evaluates availability of materials and labor • Establishes initial construction schedule
Cost Estimator	• Forecasts project costs including design, construction, and other fees and expenses

Workshops with users may be conducted to establish whether the project represents value for money and meets both organizational and functional needs. Key activities include:

- Confirming that options have been identified
- Agreeing upon which option(s) to pursue
- Identifying potential problems with items in the budget and agreeing on a total budget
- Establishing a timeline
- Carrying out a risk assessment
- Defining clear objectives
- Preparing a *Program* or *Brief*

The *Program, Statement of Need, Basis of Design (BOD)* or *Brief* (as it is called in the United Kingdom) captures the essence of the project. It converts organizational and business language into building terms and fixes functional relationships and major elements of the design. During the proposal phase discussed in Chapter 4, the program or brief forms a key component of the Request for Proposal (RFP). It sets out project parameters used by the client to instruct and select the design or design-build team. It also forms the basis on which to judge the relative merits of proposals submitted by these teams. The program or brief frequently is written with the help of a designer. Table 5.2 depicts a process for developing this document.

Each requirement included in the completed document should be described as being:

- Complete: fully describes the functionality desired
- Correct: is compatible with larger project (system) objectives and accurately reflects users' needs

Table 5.2   Steps Used in Creating a Program or Brief

Elicit	Analyze	Document	Verify
• Write vision and scope	• Consider need to build	• Write purpose and functions of project	• Inspect requirements document
• Identify project objectives	• Evaluate physical context of project	• Record business case	• Cross-check functional performance requirements
• Define procedure for developing requirements	• Evaluate nonphysical context of project	• Establish cost and schedule limits	• Define user acceptance criteria
• Identify key users	• Prioritize requirements	• Adopt Statement of Need template	
• Select champions	• Model requirements in terms of cost and schedule	• Identify sources of requirements	
• Establish focus groups	• Apply Quality Function Deployment	• Create requirements traceability matrix	
• Set up support organization	• Establish roles for project team members	• Prepare job duty statements	

(Adapted from Blythe and Worthington)

- Feasible: possible to implement given the overall project and its environment
- Necessary: is really needed for conformance to overall projects requirements or a standard
- Prioritized: assigned a priority that indicates its relative importance
- Unambiguous: is written in simple, straightforward language so that all readers arrive at a single interpretation
- Verifiable: can be shown to be accomplished

Getting requirements right early results in significant payoffs: improved product quality, savings of time and budget, and better client relations. The concept depicted in Figure 5.3 is known to most firms involved in delivering complex products and is sadly familiar to clients who have experienced cost overruns and schedule delays. Project team members can exercise maximum control over the project's final outcome during the earliest phases. With the passage of time, the ability to exert a positive influence over the end product diminishes. On the other hand, mistakes and omissions in design briefs can lead to higher costs, increased litigation, schedule delays, and lower quality of the final constructed product.

The importance of early project definition has been recognized for many years. As early as the 1970s, Preiser and Peña in the United States made important contributions to the field of *facility programming*. Even earlier in the United Kingdom, a 1960s investigation at the Tavistock Institute addressed briefing from the perspective of communication within the construction industry (Higgin and Jessop, 1963). Others in well-established professional institutions, academia, and client organizations have continued to investigate the subject:

- The Royal Institute of British Architects (RIBA) divides the briefing process into four phases: Stage A, Inception; Stage B, Feasibility; Stage C, Outline Proposals; and Stage D, Scheme Design.
- A monograph produced by the Building Research Establishment (BRE), *Better Briefing Means Better Buildings,* presents an outline that serves as a checklist of important considerations.
- John Worthington and others at the Institute of Advanced Architectural Studies (IoAAS) at the University of York have conducted professional level courses on design briefing and have gathered a large amount of empirical data drawn from its close association with DEGW, an international firm that consults on the planning, design, and management of workspace. See Table 5.3.
- An investigation conducted by Professor James Murray at the University of Reading viewed the brief as a communication tool. Case studies indicated that clients shared four project-related concerns and additional areas crucial to the success of the briefing process. See Table 5.3.
- The University of Salford's Construct I.T. Centre of Excellence have benchmarked the effectiveness of information technology (IT) and identified best practice in briefing and design. This study found that most briefing activities use terminology very specific to the AEC industry, thereby making

communication with clients difficult. Additionally, the study noted that a significant gap exists between the leading-edge IT and best practice.

- Dr. Iain MacLeod of the Department of Civil Engineering at the University of Strathclyde has indicated that establishing a *Requirements File* assists in managing the briefing process. At the outset of the design work a requirements file is opened which contains: (1) a statement of what the client requires the design team/consultant to achieve; (2) a statement by the client delineating performance criteria and constraints; (3) a *design requirements statement,* a comprehensive list of the project requirements drawn up by the design team; and (4) a requirements checklist, based on the design requirements statement, which then is used in the design review process.

- Edward T. White, while on sabbatical from Florida A&M University, conducted a rich study called *Design Briefing in England.* His showed that a good brief does not ensure good design, but a good design is very difficult to produce with a poor brief.

Clearly, the subject of design definition continues to gain the attention of academics and practitioners alike. The emergence of international standards such as ISO 9000 and ISO 14000 and increasingly complex design requirements assure sustained interest. As depicted previously in Figure 5.2, the final program or brief should be validated by the design team selected in the request for proposals/qualifications process (RFP/RFQ) to assure that client requirements have been understood. This graphic illustration shows the criticality of establishing the "basis of design" in the early phases of the project when it is cost-effective to make revisions conceptually, on paper. Conversely, as the project timeline progresses toward completion it becomes significantly more expensive to integrate design revisions once construction is well underway. For more information, see the textbox *Design Programming Primer.*

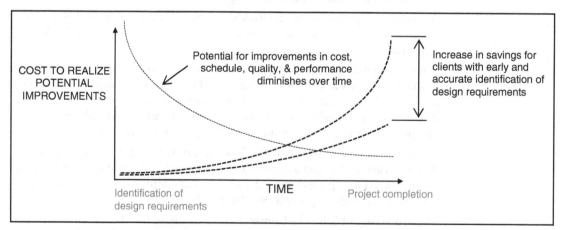

Figure 5.3 Importance of getting requirements right
(Based on the work of Boyd C. Paulson, Stanford University)

Table 5.3 Key Programming (Briefing) Issues

Common Client Concerns	Conditions Necessary for Success	Attributes of an Effective Brief	
• Early indication of cost • Early indication of schedule • Functional constraints—spaces and services • Environmental issues—planning and site	• Client and design team clear points of contact • Decision-making authority of client rep. • Effective client/design team organization • Communication media clearly understood by all	• Limited number of key objectives expressed • Layered and iterative, reflecting design stages • Fluid until the last possible moment • Fixed and frozen once complete	• Requirements backed up with hard data • Performance measured • Innovation balanced with established use • Users and building demands balanced • Related to chosen procurement process • Concise and clear

(Compiled from Murray et al., and Blythe and Worthington)

Design Programming Primer

At the beginning of any project, design requirements are identified. The program or basis of design (in the United States) or design brief (in the United Kingdom) is a statement of requirements that ideally should contain everything a designer needs to know about a client's proposed project. It anticipates functionality, aesthetics, project costs, schedule, quality, safety, and so forth. It also sets the tone for communication among project participants.

One of the early practitioner/authors to address programming in the United States was William Peña of Caudill, Rowlett and Scott (CRS) Architects. In 1969, he directed the first publication of his *Problem Seeking* to clients and planning officials within institutions, corporations, and various public bodies. Soon practicing architects and architectural students discovered the booklet, and in the late 1970s the second edition of *Problem Seeking* joined a multitude of other new publications on programming methods. In 1994, Hellmuth, Obata + Kassabaum, Inc. (HOK) acquired CRS (then CRSS) and eventually undertook publication of the fourth edition of Peña's work. The fourth edition, published in 2001, represents a range of principles developed by Peña as well as other practitioners at CRS and HOK.

Peña purports that "Programming IS analysis. Design IS synthesis." The program, or problem statement, is the last step in problem seeking (programming) and the first step in design. The problem seeking method clearly was a breakthrough more than 30 years ago, and Peña's book continues to

(Continued)

be read by most practitioners. However, as discussed in a later section, the possibilities introduced by digital design and a faster-paced world may necessitate the expansion of these original concepts.

Another well-respected design methodology innovator is Wolfgang Preiser, whose initial work, *Facility Programming*, appeared in 1978. Preiser contributed to the development of the emerging field of facility programming by introducing a generic programming process. This facility programming process included the major actors in the process, the principal beneficiaries of the process, and a discussion of who would pay for the process. The book's last chapter anticipates the potential for a computer-driven database that provides decision support and links facility programming to design and postoccupancy evaluation. Through the use of this database, this 1978 edition also identifies the possibility for designers to apply lessons learned from successes and failures in building performance to future buildings.

Preiser's next book *Programming the Built Environment* appeared in 1985. A third book, *Professional Practice in Facility Programming*, published in 1993, expands on the possibility of using databases as part of knowledge-based, expert systems shells interfacing with computer-aided design (CAD). Preiser speculates that facilities management may dominate the creation of appropriate software systems and modules and that these systems will add modules for facility programming and postoccupancy evaluation. Though more than 20 years old, this vision has yet to be realized.

There are many recent U.S. books addressing briefing (Cherry, 1998; Duerk, 1997; Hershberger, 1999; and Kumlin, 1995). These books have been written by experienced practitioners for practitioners. Both Cherry and Hershberger present a general approach to programming that adopts the best of current methods and also offers a text to be used in an educational context. The books by Duerk and Kumlin similarly focus on methods for eliciting information from clients and creating programs that lead to satisfactory design solutions.

Numerous research projects sponsored by the Construction Industry Institute (CII, 2000a) at various universities across the United States also have demonstrated early and accurate project definition to be crucial for successful project outcome. The findings from these research efforts can be found in a variety of publications (CII, 2000b) that address topics such as:

- Scope Definition and Control (Pub. #RS6-2)

- Project Objective Setting (Pub. #RS12-1)

- Input Variables Impacting Design Effectiveness (Pub. #SD-26)

- Work Packaging for Project Control (Pub. #SD-28)

- Adaptation of Quality Function Deployment to Engineering and Construction Project Development (Pub. #SD-97)

- Pre-Project Planning (Pub. #SP39-2)
- Project Definition Rating Index (PDRI) for Industrial and Building Construction (Pub. #IR155-2 and IR113-2)
- Alignment During Pre-Project Planning—Key to Project Success (Pub. #IR113-3)
- Framework and Practices for Cost-Effective Engineering in Capital Projects in the A/E/C Industry (Pub. #IR113-3)

An analysis of these CII publications, and of other sources found in the literature, reveals the importance of project definition.

The programming methodology that Peña and Preiser pioneered has now become "mature," but a strong need for innovation in the identification and management of design requirements remains. Clients exert an ever-increasing influence over the way design professionals perform their tasks (Hansen and Tatum, 1996). A new importance is being placed on collaboration with the client. The American Institute of Architects' (AIA's) document, *The Client Experience,* urges architects to reach beyond traditional roles both as a profession and for their client base. The document identifies the genesis or predesign phase of a new built environment as one of the areas of greatest growth potential for designers. The AIA's current *Architect's Handbook of Professional Practice* devotes the more than 40 pages to project definition.

DESIGN

After the client has developed a program or brief and has selected a designer, the client and designer enter into a contract for professional services. (See Chapter 4, Professional Engagement and Chapter 11, Legal Aspects of Professional Practice.) In addition to being a legal document, the contract is a communication tool. It spells out the:

- Design tasks to be performed
- Parties' (client's and designer's) specific responsibilities during design
- Client approvals required
- Schedule, including start date, end dates, and major milestones
- Budget, including any contingencies

In some public projects, such as water and sewage treatment plants, private civil engineering firms may work collaboratively with public utility departments to produce a design. In other public infrastructure projects, such as highways and bridges, departments of transportation may develop all of the plans and specifications internally. Many contracts divide the design effort into several discrete phases: schematic design, design development, construction documents, bidding, and construction. Frequently, the client's payment of the designer's invoices is linked to successful completion of these design phases. The percentages of the design effort that these various stages (schematic design, design development, construction documents) represent can vary slightly. Depending on the type of project, the three phases may be referred to as:

1. 20% + 30% + 50%,
2. 30% − 60% − 90%, or
3. 35% − 65% − 95%.

Schematic design involves establishing the general project scope, relationships among project components, basic geometry, and client understanding and acceptance. As part of schematic design, the designer also validates the program or brief that the client has provided.

During *design development,* the design concept is elaborated. In other words, major systems are defined, important decisions are documented, and a clear, coordinated description of the project is developed. Gaining the client's understanding and acceptance is extremely important so that preparation of construction documents can proceed smoothly.

Construction documents provide the contractor with sufficient information to build the project and delineate the responsibilities of the two parties who sign the construction contract—the client (owner) and the contractor. They also provide information about the role of the designer, who is not a party to this contract but who has responsibilities during the bidding and construction phases. The construction documents are comprised of drawings and a project manual, made up of bidding requirements and technical specifications.

See Table 5.4 for a summary of the purpose, activities, and deliverables associated with the various project phases.

DESIGN PROCESS

Though the three phases of design often are depicted in an organized, linear manner for the sake of clarity, in reality the actual design process is far more iterative. There are many books written on the creative design process. Sometimes describing the actual act of design seems like trying to capture lightening in a bottle. Christopher Alexander, a British architect who taught for years at the University of California, Berkeley, wrote two classics that address the involvement of users (end users and clients/owners) in the design process: *A Pattern Language* (Oxford University Press, 1977) and *The Timeless Way of Building* (Oxford University Press, 1979).

Involving users can reap significant rewards. Most civil engineering projects involve considerable stakeholder participation, especially during the early phases of design. Frequently owners hire facilitators to help manage large public meetings where

Table 5.4 Design in Project

	PREDESIGN	DESIGN			BID	CONSTRUCTION	POSTCONSTRUCTION
		Schematic Design (20%)	Design Development (30%)	Construction Documents (50%)			
Purpose	Establish project scope, budget, and schedule Obtain necessary permits Select prime designer	Establish basic geometry and relationships among project components	Develop clear, coordinated description of project, including major systems	Provide contractor with sufficient information to build the project	Assist client in selecting contractor	Assist client in assuring that contractor is building per contract documents	Work to improve design/construct process and position firm to acquire new work
Activities	Assist client in: analyzing project requirements, gaining funding approval, performing environmental impact report(s) obtaining geotechnical reports and preparing request for proposal (RFP)	Verifying information contained in RFP Organizing info and synthesizing possibilities Involving other designers and subconsultants Securing client understanding and acceptance	Defining major systems Conducting workshops: *Value engineering* *Lifecycle analysis* *Sustainability review* Documenting important decisions Freezing client's design changes	Delineating responsibilities of the client (owner) and the contractor Supplying an appropriate level of detail in plans and specifications *Conducting Constructability review*	Conducting prebid conference Answering potential bidders' questions Assisting client in evaluating bids	Making field observations that fulfill requirements for level of attention and testing cited in contract and specifications Resolving problems and discrepancies as they arise	Conducting Postoccupancy review
Deliverables	Design criteria, program, or brief Order of magnitude cost estimate Schedule RFP and scope of work (SOW)	Schematic plans Outline specifications Schematic cost estimate	Design development plans Design development specifications Design development cost estimate	Complete plans Complete Project Manual including Procurement Requirements, Conditions of the Contract, and Technical Specifications Opinion of probable construction cost	Addendum/a	Responses to Requests for Information (RFIs) Reviews of submittals and shop drawings	Postoccupancy Review Report

stakeholders express their concerns. Involving stakeholders early is advantageous because their apprehensions are known and mitigations can be developed in a timely manner. Discovering late in the design process that stakeholders may mount a campaign against a project can become extremely expensive, both to the owner and their engineers. (See textbox, *Collaborative Design*.)

Collaborative Design

In order to become more responsive to clients, design professionals have recognized the need for better communication among the disciplines. In the late 20th century, "constructability" or "buildability" became a way of bringing useful insights from builders into the design process. As ideas about collaboration evolve, the locus and extent of collaboration is expanding to include greater participation not only of professionals but also of stakeholders.

Arriving at an approved design that actively involves stakeholders requires several key components, including:

- A high-level person in the client organization who champions "buy-in"
- Clear guidelines regarding scope, budget, and schedule
- A project organization with explicit roles and responsibilities
- Stakeholder-focused building site committee(s) committed to bringing closure to sometimes difficult issues
- Management and project teams willing to embrace the collaborative process
- Recognition up front by all parties of the impact in time and cost for rework and changes

As a prelude to beginning actual design, a design workshop should be held to introduce the building site committee(s) to the design team. A diagram that explains the process that will be used should be presented. The diagram should convey the idea that the building site committee(s) will be making most of the big decisions during the schematic design and design development phases and that committee involvement will taper off during the construction documents phase. Additionally, an easily read Gantt (bar) chart should be presented that depicts critical review points and presentations to be made to the client and regulatory organizations.

Finally, open communication is key to collaborative design. "I never saw that before!" is a stakeholder comment that strikes fear in the hearts of design project managers and signals potentially serious, negative impacts to the project schedule.

"While the word *design* is used to signify the individual act of conceptualization that puts an idea on the back of an envelope, the same word is used to signify the often long and unusually collaborative process of carrying out the detailed calculations that flesh out the first sketch and thus make it possible to put specific dimensions and manufacturing instructions on formal drawings."

—Henry Petroski, (1997). *Remaking the World: Adventures in Engineering*, Vintage Books

The design process works with information as well as "flashes of insight" on many levels. In pursuit of appropriate and acceptable solutions, designers must process:

- Client requirements
- Technical variables
- Physical, budgetary, and schedule constraints
- Permitting and code issues
- Political realities

Design Analysis

- Throughout an iterative process of examination and criticism, the design emerges. An effort to understand thoroughly the problems to be solved speeds this process. A careful analysis should include:

- Program analysis: convert the information the client has provided into understandable and usable information

- Site analysis: visit the site and organize information into a common scale and format

- Zoning and code analysis: concurrently with site analysis, translate zoning and code issues into building form

- Documentation of existing conditions: establish clear and accurate documentation of existing conditions

- Scheduling: examine the need for project phasing, fast-track sequencing, and time required for permits

- Cost: analyze the project budget and allocate funds to aspects of the project necessary for overall success

- Construction industry practice: evaluate local building practices to know availability of materials and labor and to understand standard processes

(*Continued*)

> • Design precedents: assess previous projects for relevant precedents and similar program, site, context, size, cost, and other design issues
>
> —AIA. (2007). *Architect's Handbook of Professional Practice*, John Wiley & Sons, pp. 522–523.

If design begins with analysis, it proceeds with synthesis. Whatever the method, designers must move past the base data they have amassed. Through a combination of sketching, talking, calculating, and thinking, designers must reach sufficient understanding to form a concept. Tim Brown, CEO of IDEO, the firm known for such innovative designs as the iMac, offers a methodology for making this transition that he calls "design thinking." See Figure 5.4, which outlines the three phases of design

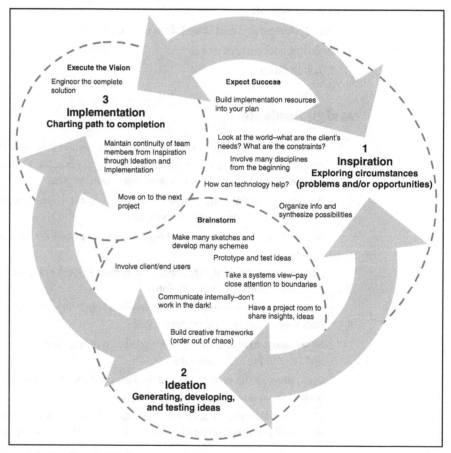

Figure 5.4   Design thinking
(Adapted from Tim Brown, "Design Thinking," *Harvard Business Review*, June 2008.)

thinking. The three phases of design thinking are *inspiration*, *ideation*, and *implementation*. Designers are encouraged to:

- Explore the circumstances
- Generate, develop, and test ideas
- Chart the path to completion

Throughout the design process, and particularly in the construction documents phase where most work is in "production" mode, a *quality control plan* should be implemented. This could include internal quality control review, independent technical review, and in important cases, peer review. For more information about managing work quality, see Chapter 7, Executing a Professional Commission and the textbox, *Importance and Value of a Comprehensive Quality Control Plan*.

Importance and Value of a Comprehensive Quality Control Plan

Quality Control Plans (QCP) are an important aspect of any successful project. The first step in creating a comprehensive QCP is reviewing previous project experience and coordinating with the project manager, principal in charge, and other appropriate staff members. Important elements of a QCP include: a knowledgeable project manager adept at implementing a QCP, an experienced QC Team capable of reviewing contract documents, and an extensive and successful QCP outline.

Once the QCP is drafted, the next step is implementation. Implementation of a comprehensive QCP includes the following: a clear and concise organizational chart outlining roles and responsibilities of the QC Team, proper scheduling of review, including ample time for each review, and good project management practices to ensure the reviews are completed.

A comprehensive QCP can reduce the risks associated with incomplete or poorly completed work products. It can increase the overall quality of the work performed by providing an additional step between the completion of the work and the delivery to the client, which enables a final review for any missing items or incorrect standards. Both of these attributes will lead to a better end product, which will increase client satisfaction and maximize the client's desire to solicit the designer for future work. Repeat business is paramount for sustaining a successful and profitable organization, and any steps that can be taken to ensure this should be implemented.

—Tony Quintrall, HDR

DESIGN DURING BID AND CONSTRUCTION

The work of civil engineers typically does not end with the completion of the construction documents. Most clients rely on their prime designers to help them through the bid phase. As part of the bid process, civil engineers may be responsible for including Division 00—Procurement and Contracting Requirements in the project manual. (See Chapter 6, What Engineers Deliver.) Division 00 includes:

- Advertisement for bids
- Invitation to bid
- Instructions to bidders (contractors)
- Prebid meetings
- Land survey information
- Geotechnical information
- Bid forms
- Owner-contractor agreement forms
- Bond forms
- Certificate of substantial completion form
- Certificate of completion form
- Conditions of the contract
- Procedure for answering bidders questions

Whether acting in the capacity of prime designer or subconsultant, civil engineers usually attend prebid meetings to acquaint prospective bidders with the project. They also answer bidders' questions during the time allotted for the bid and group those responses and clarifications in an *addendum* or *addenda*, if more than one installment is required.

Following contract award (the owner and contractor enter into a contract), civil engineers may be responsible for:

- Attending a preconstruction conference
- Responding to field questions, called requests for information (RFIs)
- Making field observations
- Reviewing submittals, including shop drawings

Making field observations fulfills the requirements for level of attention and testing cited in contract and specifications and encourages quality. The civil engineer's jobsite presence also can head-off problems, because contract documents are never perfect. Additionally, owners may request civil engineers to validate contractors' payment requests (invoices) for accuracy by comparing work or materials in place with percent complete invoiced.

Submittal and shop drawing review is the final element of design review. Reviewing submittals, such as concrete mix design, and shop drawings, such as steel fabrication drawings, assures that the design detailing by the contractor conforms with the intent of the design. However, due to time and budgetary constraints, time to review is limited. The process used by civil engineers for reviewing submittals and shop drawings should be referenced in the general conditions of the construction contract and discussed at the prebid and preconstruction conferences.

Shop Drawing Review Process

1. A/E identifies shop drawings required and establishes schedule for review and resubmission

2. Contractor reviews each shop drawing and establishes acceptability as to:
   - Means
   - Methods
   - Techniques
   - Operations and sequences of construction
   - Safety precautions

3. A/E reviews each shop drawing and determines conformity to:
   - Design intent
   - Compliance with contract documents (plans and specifications); may ask contractor to "revise and resubmit"

4. Contractor appraises A/E of any changes from what was specified in contract documents—A/E may or may not accept

5. Contractor should pay A/E if shop drawings vary considerably from original design (difficult to accomplish because contractor does not have contract with A/E)

6. A/E returns any shop drawings that they have not required

—John Bachner. (1991). *Practice Management for Design Professionals: A Practical Guide to Avoiding Liability and Enhancing Profitability*

In order to reduce liability, architects and engineers use special language when reviewing submittals and shop drawings. Typical language might include:

Review is limited solely to the purpose of checking for conformance with Civil Engineer's design intent and conformance with information contained in the Contact

Documents. Review is not conducted to determine the accuracy and completeness of other information such as dimensions, completeness, installation instructions, or performance of equipment or systems supplied by the contractor, all of which remain the contractor's responsibility. Review neither extends nor alters any contractual obligations and shall not relieve the contractor of responsibility for deviation from the requirements of the Contract Documents.

Additionally, when returning the submittals and shop drawings to the contractor, civil engineers can select among several options, which conclude:

- Accepted as Noted
- Revise and Resubmit
- Rejected
- Not Reviewed, Submittal not Required by Contract Documents
- Reviewed for Project Closeout Requirements Only

POSTCONSTRUCTION ACTIVITY

Some sophisticated client organizations conduct their own design reviews throughout the design process. They also document how closely architectural and engineering firms design to budget by comparing the opinion of probable construction cost provided at the end of design with the bid submitted by the successful contractor. They also track the number of RFIs issued by the contractor as an indication of the quality and completeness of the plans and specifications. As Figures 4.2, 4.3, and 4.4 in the previous chapter indicate, the Veterans Administration (VA) scores the prime designer and subconsultants at the end of schematic design, design development, and construction document design phases; the VA then uses this information when selecting architects and engineers to perform new work.

Though not always done, most design organizations could benefit from a Post-Occupancy Review with the client and end users.

Finally, though difficult to achieve, many organizations—both academic and commercial—are exploring the use of *design performance measures* (DPMs). The belief is that measurement of design performance will lead to improved designs. See the textbox, *Design Performance Measures* for additional information.

Design Performance Measures

Numerous research projects at various U.S. universities over the last two decades have demonstrated that early and accurate project definition is crucial for successful project outcome. The findings from these research efforts can be found in a variety of Construction Industry Institute publications

(CII, 2000a) (CII, 2000b) that address topics such as: Scope Definition and Control; Project Objective Setting; Input Variables Impacting Design Effectiveness; Work Packaging for Project Control; Adaptation of Quality Function Deployment to Engineering and Construction Project Development; Pre-Project Planning; Project Definition Rating Index (PDRI) for Industrial and Building Construction; Alignment During Pre-Project Planning—Key to Project Success; and Framework and Practices for Cost-effective Engineering in Capital Projects in the AEC Industry.

An analysis of these CII publications and of other sources found in the literature reveals the importance of project definition. However, most publications stop short of actual linkage of design quality to measurement of performance. When addressed, design performance measures (DPMs) are usually cost-based. Such measures can be product-oriented, such as lifecycle cost analysis, or process-oriented, such as construction change orders as a percentage of total project cost. Few DPMs address qualitative aspects of design, such as client and user satisfaction, innovation, or aesthetic appeal. This may be the case because practitioners do not perceive the need, or they find the problem too difficult to solve, or they do not wish to enable discussions regarding subjective design factors with clients. Additionally, there is limited funding available for academics to pursue this line of investigation.

Within the U.S. architectural, engineering, and construction industry, development of DPMs is nascent. However, researchers have identified performance measures that explicitly represent project objectives, such as those that might appear in a brief. A guiding principle in defining DPMs is the identification of a critical variable that measures, reflects, or significantly influences a particular performance objective. In most instances, a high-level performance objective will need to be delineated by multiple metrics that influence its overall satisfaction. The following discussion of current research on developing DPMs is divided into two sections: cost-based methods and other approaches.

Cost-Based DPMs

Life-cycle cost is a relatively straightforward performance objective to delineate. However, others, such as energy efficiency, may be more difficult. A group at the Lawrence Berkeley National Laboratory recognizes that metrics cannot stand on their own as they are linked to design assumptions and/or operating conditions. In order to evaluate performance against a benchmark, a Building Life-cycle Information System (BLISS) has been developed. During design, data from this model can be used to simulate performance. Later, simulated performance can be compared to actual building performance. An aggregate Life-Cycle Cost performance objective (Hitchcock et al., 1998) is shown below.

(Continued)

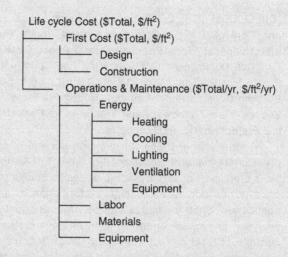

At least two organizations at the National Institute of Standards and Technology (NIST) concern themselves with DPMs. The Building and Fire Research Laboratory (BFRL) focuses on developing measurement methods, fundamental data, simulation models, and life-cycle environmental and economic analysis tools to support sustainability in design. The BFRL has created software called BEES (Building for Environmental and Economic Stability) for designers to select among alternative building materials and products based on environmental and economic performance (NIST, 2002). Following the attacks of September 11, 2001, NIST's Office of Applied Economics has developed financial models to be applied in design to optimize investments in "protective," i.e., anti-terrorist, strategies (Marshall, 2002).

At Stanford University's Center for Integrated Facility Engineering (CIFE), investigators acknowledge that the AEC industry is experiencing a profound change that brings with it the need for productivity improvements, leaner organizations, and more consistent and rigorous performance metrics. Several near-term metrics have been proposed, some cost-based and others having different quantitative measures. Schwegler et al. (2001) describe these as: quality of design documentation (ratio of drawings or 3D objects to dollar value) of work; individual team task performance (transactional data provided by project extranets); assembly complexity (number of simultaneous activities occurring during construction); design iterations (experience in manufacturing has shown the benefits of increasing design iterations but AEC industry practitioners frequently "freeze" design at a low number of design iterations in order to control soft costs); and response time for requests for information (RFIs) and shop drawing review (extremely long response times frequently have been shown to result in major unplanned changes).

Other Approaches for Developing DPMs

In 1999 the High Performance Structures (1999) group at the Massachusetts Institute of Technology was commissioned to design a new building to house Civil and Environmental Engineering (CEE). Both students and faculty rated the design solutions on a set of predetermined design measures that included: (1) flexibility (upgradeability, adaptability, expandability); (2) aesthetics (character, human comfort/security, proportion and scale, material palette, lighting, landscaping); (3) high level of engineering performance (showcase, HPS components); (4) environmentally friendly (impact to environment, energy efficient, material flow, choice of materials); (5) accessibility (ease of entry, multiple circulation paths, controlled access, location of space, communication); (6) constructability (construction efficiency, low impact); and (7) maintainability (cleanability, repairability, and site maintenance). These elements are weighted and then graded on a scale of 1, 2, or 3. Scores on individual elements were aggregated to create an overall score for the design.

Within the Construction Engineering and Management Group in the College of Civil and Environmental Engineering at the Georgia Institute of Technology, a Project Definition Matrix has been developed over time and has been used successfully in workshop settings with industrial clients. This three-dimensional model is based on: six layers of stakeholder perspectives (owner, vendor/suppliers, construction, design, user/operator, external parties); twelve performance parameters (contextual compatibility and response, functional performance, physical performance, cost, time, quality/reliability, safety/security, risk, constructability, maintainability, health, sustainability); and six types of internal and external influences (project characteristics, project objectives, project scope, physical context of the project, non-physical context of the project, project risks). These largely qualitative considerations assist in aligning client and designer expectations (Vanegas, 2001).

A U.S. Defense Department goal is for all military construction to use principles of sustainability, addressing issues such as siting, water and energy efficiency, minimization of pollution, indoor environmental quality, etc. Consequently, the U.S. Army Civil Engineering Research Laboratory (CERL) has developed the Sustainable Project Rating Tool (SPRT). SPRT will be used as a standard measure to rate the sustainability of Army building and infrastructure designs (Flanders et al., 2000).

Researchers at Stanford's Civil and Environmental Engineering Project Based Learning Lab (PBL2) have used metrics to measure cross-disciplinary learning in distributed AEC teams that can relate to improved design quality. The methodology involves a four-tiered classification based on cognitive and situative learning theories. The four tiers are: island of knowledge, awareness, appreciation, and understanding. The approach used has

(Continued)

relevance for planning deployment of automated briefing systems (Fruchter and Emery, 2000).

Given the fragmented nature of the U.S. construction industry, the lack of DPMs is not surprising. In a single project, design usually is performed by a panoply of consulting firms. However, the subject of design definition and quality will continue to gain the attention. The emergence of international standards such as ISO 9000 and ISO 14000 and increasingly complex design requirements assure sustained interest of academics and practitioners alike in improving the briefing process. At the 2001 CIB Congress in New Zealand, 33 papers focused on the "demand side," or stakeholders' perspective, of the construction delivery process—the highest number ever.

Need for a Different Approach

The publications and studies emphasize how crucial good design definition is and clearly point to the necessity for better tools and strategies for handling design requirements. Classifications and checklists are helpful in conceptualizing what a brief is and what it should include. However, by definition, they are generic and cannot make allowances for each project's unique nature. These approaches make a very iterative process look predictable and linear.

SUMMARY

This chapter details the many roles civil engineers play in pre-design, design, and construction. Civil engineers can be involved from the very initial stages of the project inception through project closeout, owner occupancy, and later adaptive reuse or decommissioning. Coplayers in the project development process—clients, civil engineers and other design professionals, contractors, and regulatory agencies—possess diverse perspectives and agendas. Much of design, especially the early phases, requires civil engineers to take abstract ideas and convert them into tangible deliverables. Because of the complexity of today's projects, encouraging involvement of professionals and stakeholders through a collaborative design process can reap significant rewards. Developing and implementing a quality control plan also yields tangible benefits.

The next chapter—Chapter 6, What Engineers Deliver—discusses the deliverables connected to the project delivery process.

REFERENCES

Alexander, Christopher, et al. (1977). *A Pattern Language*. New York: Oxford University Press.

_____. (1979). *The Timeless Way of Building*. New York: Oxford University Press.

American Institute of Architects. (2007). *Architect's Handbook of Professional Practice*. Joseph A. Demkin, ed., John Wiley & Sons, New York.

Bachner, John. (1991). *Practice Management for Design Professionals: A Practical Guide to Avoiding Liability and Enhancing Profitability*. John Wiley & Sons, New York.

Blyth, Alastair, and John Worthington. (2001) *Managing the Brief for Better Design*. SPON Press, London.

Building and Fire Research Laboratory. (2002) www.bfrl.nist.gov/goals_programs/EBP_goal.htm.

Brown, Tim. (2008). "Design Thinking," *Harvard Business Review*, June 1, 2008.

Cherry, Edith. (1998). *Programming for Design: From Theory to Practice*. John Wiley & Sons, New York.

CIB World Building Congress 2001. (2001). *Construction Industry Board, Conference Proceedings*, 2nd to 6th April, Wellington, New Zealand.

Construction Industry Institute (CII) 2000a. http://construction-institute.org. The University of Texas at Austin, Austin Texas.

Construction Industry Institute (CII) 2000b. http://construction-institute.org/services/catalogue/catframe.htm. The University of Texas at Austin, Austin Texas.

Duerk, Donna P. (1997). *Architectural Programming: Information Management for Design*. John Wiley & Sons, New York.

Flanders, Stephen N., Richard L. Schneider, Donald Fournier, and Annette Stumpf. (2000). www.cecer.army.mil/earupdate/nlfiles/2000/sustainable2.cfm

Fruchter, Renate, and Katherine Emery. (2000). "CDL: Cross-Disciplinary Learning Metrics and Assessment Method."*Proceedings of ASCE -ICCCBE-VIII Conference*, Stanford University, August 2000.

Hansen, Karen Lee, I.A. MacLeod, I.M. Tulloch, and D. R. McGregor. (1996). "*BriefMaker*: A Design Briefing Tool Developed on the Internet," International Conference on Trends in Civil and Structural Engineering Design at Strathclyde University, Glasgow. Civil Comp Press, Edinburgh, August 1996.

Hansen, Karen Lee, and C.B. Tatum. (1996). "How Strategies Happen: A Decision Making Framework." *ASCE Journal of Management in Construction*, 12:1, January/February 1996, pp 40–48.

Hershberger, Robert G. (1999). *Architectural Programming and Predesign Manager*. McGraw-Hill, New York.

High Performance Structures Group. (1999). www.moment.mit.edu/Hps/98-99/designfiles/documents.

Hitchcock, Robert J., Mary Ann Piette, and Stephen E. Selkowitz. (1998). "Documenting Performance Metrics in a Building Life-cycle Information System."*Proceedings of the ACEEE '98 Summer Study on Energy Efficiency in Buildings*, Lawrence Berkeley National Laboratory (LBNL-41940), June 1998.

Kumlin, Robert R. (1995). *Architectural Programming: Creative Techniques for Design Professionals.* McGraw-Hill, New York.

Marshall, Harold E. (2002). "Economic Approaches to Homeland Security for Constructed Facilities." Keynote address, Tenth Joint W055-W065 International Symposium on Construction Innovation and Global Competitiveness, University of Cincinnati, September 2002.

Murray, James P., R.M. Gameson, and J. Hudson. (1993). *Creating Decision-Support Systems, Professional Practice in Facility Programming*, ed. Wolfgang FE Preiser. Van Nostrand Reinhold, New York, NY, pp 427–452.

O'Reilly, J.J.N. (1987). *Better briefing means better buildings.* Garston, UK: Building Research Establishment (BRE).

Peña, William M., and Steven A. Parshall. (2001). *Problem Seeking: An Architectural Programming Primer*, John Wiley & Sons, New York.

Preiser, Wolfgang F.E. (1993). *Professional Practice in Facility Programming.* Van Nostrand Reinhold, New York.

Rutherford, James H., and Thomas W. Maver. (1994). *Knowledge-Based Design Support. Knowledge-Based Computer-Aided Design*, eds. G. Carrara and YE Kalay. Elsevier, New York.

Salisbury, Frank. (1990). *Architect's Handbook for Client Briefing.* Butterworth Architecture, London.

Schwegler, Benedict R., Martin Fischer, et al. (2001). *Near-, Medium-, and Long-Term Benefits of Information Technology in Construction.* Center for Integrated Facilities Engineering (CIFE) Working Paper #65, July 2001.

Vanegas, J. (2001). "The Project Definition Package: A Cornerstone for Enhanced Capital Project Performance." *Proceedings of the 2001 World Congress of the International Council for Research and Innovation in Building and Construction (CIB)*, Wellington, New Zealand (paper in conference CD ROM)

White, Edward T. (1991). *Design Briefing in England.* Tuscon: Architectural Media Ltd.

Introduction
Ethics
① ② ③ Professional Engagement
④ ⑤ What Engineers Deliver
History
Engineer's Role in Project
Development ⑥ ⑦ Permitting Leadership Managing Having a Life Emerking Technology Ⓐ Ⓑ Ⓒ Ⓓ Ⓔ Ⓕ
⑧ ⑨ ⑩ ⑪ ⑫ ⑬ ⑭ ⑮ ⑯ ⑰ Sustainability
Executing a Professional
Commission Client Relationship Legal Aspects Communicating Globalization

C h a p t e r **6**

What Engineers Deliver

Big Idea

Civil engineers convert abstract ideas into physical realities through their efforts. The output of these engineering services includes engineering reports, feasibility studies, plans and specifications and construction administration.

> We know where most of the creativity, the innovation, the stuff that drives productivity lies—in the minds of those closest to the work.

> —Jack Welch

Key Topics Covered

- Background
- Contract Documents
- Drawings
- Specifications
- Drawings and Specifications—Final Thoughts
- Technical Reports
- Calculations
- Other Deliverables
- Summary

Related Chapters in This Book

- Chapter 2: Background and History of the Profession
- Chapter 3: Ethics
- Chapter 4: Professional Engagement
- Chapter 5: The Engineer's Role in Project Development
- Chapter 7: Executing a Professional Commission—Project Management
- Chapter 8: Permitting
- Chapter 11: Legal Aspects of Professional Practice
- Chapter 15: Globalization
- Chapter 16: Sustainability
- Chapter 17: Emerging Technologies

(Continued)

Related to *ASCE Body of Knowledge 2* Outcomes

ASCE BOK2 outcomes covered in this chapter

Foundational
1. Mathematics
2. Natural sciences
3. Humanities
4. Social sciences

Technical
5. Materials science
6. Mechanics
7. Experiments
●8. Problem recognition and solving
●9. Design
10. Sustainability
11. Contemporary issues & historical perspectives
12. Risk and uncertainty
13. Project management
14. Breadth in civil engineering areas
15. Technical specialization

Professional
●16. Communication
●17. Public policy
18. Business and public administration
19. Globalization
●20. Leadership
●21. Teamwork
●22. Attitudes
●23. Lifelong learning
●24. Professional and ethical responsibility

BACKGROUND

As the design solution evolves through pre-design and the schematic design, design development, and construction documents phases, the final form of the contract documents begins to take shape as discussed in Chapter 4, Professional Engagement; Chapter 5, The Engineer's Role in Project Development; and Chapter 12, Managing the Civil Engineering Enterprise. In the traditional design-bid-build process represented by Figure 5.2, The Flow of Work, the prime designer is responsible for developing the majority of the documents that form the basis of the construction contract between the owner and the contractor. In building projects, the architect is usually the prime designer, and in large civil projects, the civil engineer is usually the prime designer. These prime designers typically hire numerous subconsultants to assist with the design. Contractors also hire many subcontractors and material suppliers in order to perform the construction. Figure 6.1 depicts the typical contractual arrangements in design-bid-build project delivery. Chapter 11, Legal Aspects of Professional Practice, discusses other forms of project delivery.

Aside from the sheer number of contracts required, the most notable aspect of the contract relationships depicted in Figure 6.1 is that there is no contract between

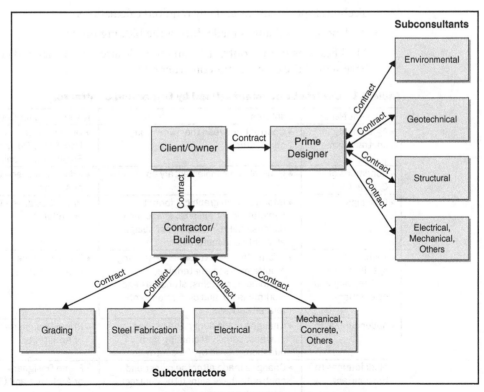

Figure 6.1   Contractual relationships in design-bid-build project delivery

the prime designer and the contractor. Thus, the prime designer and subconsultants must prepare documents, paid for by the client, that the client and stakeholders can understand and that the contractor can use to build the project. This may seem more than a little challenging, and it is!

This chapter examines the kinds of deliverables for which civil engineers are responsible. These include drawings or plans, specifications, and technical reports, as well as other documents such as calculations, meeting minutes, construction reports, reviews of various submittals, and schedules.

CONTRACT DOCUMENTS

The term "contract documents" is used widely in the Architecture, Engineering and Construction (AEC) industry. These are the documents on which the contract for construction is based. Contract documents include more than the drawings and technical specifications. They consist of:

- Agreement/Contract Forms
- Conditions of Contract (General and Supplementary Conditions)
- Drawings
- Technical specifications and any required calculations
- Addendum/a (changes made during the bidding process)
- Modifications to the contract (changes made after the owner and contractor have signed the contract for construction)

Table 6.1   Contract Documents (Used by Owner and Contractor)

Document Name	Contents	Responsible for Development
• Agreement/ contract forms	• Contract between the owner and contractor	• Owner's Attorney with input from Prime Designer—Civil Engineer or Architect
• Conditions of Contract	• General and supplementary conditions	• Prime Designer—Civil Engineer or Architect
• Drawings	• Information in graphical format depicting location, size, shape, and dimensional relationships of design elements and materials	• Prime Designer—Civil Engineer or Architect
• Technical specifications and any required calculations	• Information in text format describing requirements for materials, equipment, systems, standards and workmanship, and performance of related services	• Prime Designer—Civil Engineer or Architect
• Addendum/a	• Changes made during the bidding process, usually stemming from questions raised by the contractor	• Prime Designer—Civil Engineer or Architect
• Modifications to the contract	• Changes made after the owner and contractor have signed the contract for construction	• Prime Designer—Civil Engineer or Architect and Owner

Table 6.1 depicts the kinds of information contained in these documents and the parties responsible for their development.

Though technically not considered part of the contract documents, another set of documents, referred to collectively as *procurement and contracting requirements,* are necessary for selecting a contractor using competitive bid methods. The procurement and contracting requirements, general and supplementary conditions, and technical specifications are collected in a single document assembled by the prime designer. This document commonly is referred to as the *Project Manual.* The agreement/contract forms, project manual, drawings, and addenda comprise the *Bidding Documents,* as shown in Figure 6.2. More information regarding the agreement/contract forms and the different approach used in design-build project delivery is included in Chapter 11, Legal Aspects of Professional Engagement.

There is an important difference in the information that is included in drawings (or plans) and technical specifications—see Table 6.2 for a comparison. The most obvious difference is that drawings largely contain information in graphical/geometric format and specifications contain information in text format. The level of detail

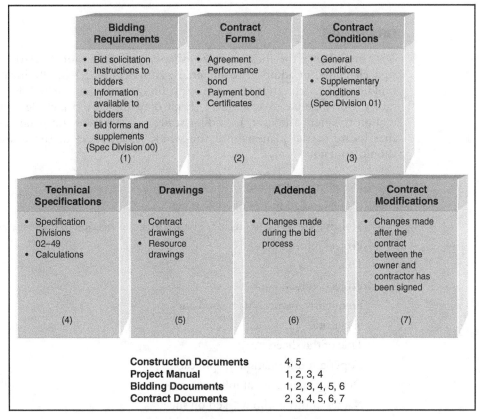

Figure 6.2 Documents are the formal building blocks of project delivery

Table 6.2 Information Contained in Construction Drawings versus Technical Specifications

Construction Drawings	Technical Specifications
• Design requirements represented graphically	• Design requirements represented verbally
• Size, shape, and relationship of elements provided	• Properties and characteristics of elements provided
• Location of elements depicted	• Installation requirements for elements established
• Products or materials shown wherever located	• Products or materials described once
• Products or materials shown generically	• Products or materials identified specifically
• Quantity indicated	• Quality indicated
• Few requirements for testing noted	• Requirements for testing clearly spelled out

included in the drawings and specifications corresponds to the needs of the client, of permitting and regulatory agencies, and of the contractor. The following sections discuss further the content and organization of drawings and specifications.

DRAWINGS

Drawings depict the location, size, shape, and dimensional relationships of design elements, in addition to materials. A drawing set typically includes site and building plans, elevations, sections/profiles, details, and schedules (matrices or tables, not time-based schedules). Each drawing should include sufficient information to orient the user, including scale, a north arrow on plans, and key plans that locate partial plans within the whole. Each sheet in the drawing set also should contain:

- Designer information
  Names and addresses of consultants
  Seal and signature of engineer or architect, as needed in most states
- Project information
  Title
  Project address
  Frequently, owner's name and address
- Sheet title
  Title of the sheet
  Copyright information, if applicable
- Drawing management information
  Names of those who worked on sheet
  Names of those who checked sheet

Table 6.3 Standardized Drawing (Sheet) Sizes

Designation	Architectural Drawing Sizes		Manufacturing Drawing Sizes*	
	(mm)	(inches)	(mm)	(inches)
A		9 × 12		8.5 × 11
B		12 × 18		11 × 17
C		18 × 24		17 × 22
D		24 × 36		22 × 34
E		36 × 48		34 × 44

*American National Standards Institute (ANSI)

- Issue information

  Date(s) of issue, including revisions

  Purpose of issue—bid, permit, construction

- Drawing identification

  Sheet letter

  Sheet number

Drawings can be produced in a variety of sizes. Standardized sheet sizes are shown in Table 6.3.

A uniform method for formatting sheets has been adopted by various organizations, such as the American Institute of Architects (AIA). For example, each sheet is divided into three modules: (1) sheet title block, the information listed above, usually on the right side of the sheet; (2) graphic area, based on a hidden modular grid; and (3) perimeter, or border, with alpha numeric grid coordinates.

There also are standards for organizing the types of information contained on individual sheets. The U.S. National Computer Aided Design (CAD) Standard (NCS) recognizes ten sheet types:

0. *General*—notes, symbols, legends
1. *Plans*—site, building, generally horizontal sections
2. *Elevations*—vertical views of surfaces
3. *Sections*—vertical cuts across plans
4. *Large-scale views*—plans, elevations, sections of larger-scale components
5. *Details*—plans, elevations, sections of smaller-scale components
6. *Schedules and diagrams*—tables, matrices
7. *User-defined*—miscellaneous
8. *User-defined*—miscellaneous
9. *Pictorial representations*—isometrics and perspectives

Additionally, drawing sets are organized by disciplines, each with a letter prefix, such as "C" for civil engineering, "S" for structural engineering, "A" for architecture, and so forth. Information in each discipline is ordered by grouping like information into subdivisions, such as the ten listed above. Thus, drawing organization within each discipline progresses from the general to the specific. The figure *Typical Drawing Numbering System* and the textbox *Content in Drawing Sets* contain more information about the organization and the content of drawing sets.

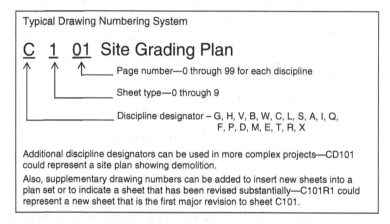

Typical Drawing Numbering System

C 1 01 Site Grading Plan

Page number—0 through 99 for each discipline

Sheet type—0 through 9

Discipline designator – G, H, V, B, W, C, L, S, A, I, Q, F, P, D, M, E, T, R, X

Additional discipline designators can be used in more complex projects—CD101 could represent a site plan showing demolition.

Also, supplementary drawing numbers can be added to insert new sheets into a plan set or to indicate a sheet that has been revised substantially—C101R1 could represent a new sheet that is the first major revision to sheet C101.

Content in Drawing Sets

While the following order reflects the National CAD Standard, much of what is described pertains to drawings that follow other organizations.

Cover Sheets

Cover sheets are, simply, title pages, with the project title, the owner's name, the names of the design professionals involved, a pictorial of the project (frequently a perspective drawing), and similar content.

G: General Information

Site data, location map, energy compliance calculations, building code summary, project square-foot calculations, key plans, general notes, abbreviations, and an index of sheets all belong in the general information category.

H: Hazardous Materials

Hazardous materials occur in a variety of goods and materials, primarily in older structures. Asbestos, for example, occurs in a variety of materials (predating the 1970s), including floor coverings; plaster; piping insulation; ceiling tiles; plaster, floor, wall, and ceiling insulation; and myriad other materials, thus making sheet organization somewhat problematic in this discipline. The way the NCS is currently established, demolition of an HVAC or piping system in an older building would be described in mechanical drawings using the Level 2 discipline designator (see sheet identification paragraphs upcoming). The likelihood that asbestos would be encountered in the duct or piping insulation and sealed joints in older buildings is high. Sitework involving hazardous materials—hydrocarbons, for example—are likely to be described in the civil drawings. Just how this discipline will develop remains to be seen; however, the idea is to identify the location of hazardous materials so that their potential for harm is mitigated.

V: Survey/Mapping

These pages contain relevant survey and map information. It is normally the owner's responsibility to provide the design professional with accurate information related to the real estate being developed. Vicinity maps, common on drawings sets, and general layout information are recorded in this sheet set.

G: Geotechnical Information

Providing information in a drawing on site soil conditions makes sense from several viewpoints. A number of people make use of the geotechnical report in their analysis of work requirements, including the owner, the prime contractor, and subcontractors. Though including this information in a drawing set is cumbersome, it is perhaps justified by the ready availability of the information.

W: Civil Works

The civil works category is, for all practical purposes, the same as civil drawings (the next one); however, it was added at the behest of government agencies that are responsible for civil work that encroaches upon the property of multiple landowners. A municipal pipeline project that impinges upon numerous landowners, for example, would be the appropriate project type to describe in this category.

C: Civil Drawings

In a drawing set that describes civil construction projects, such as highway and street improvement projects, it is not necessary to distinguish the discipline from others—the entire project is "civil drawings." When the discipline is distinguished, however, it is generally when civil drawings form a part of a commercial building set. Civil engineers design and describe the off-site improvements (curbs, gutters, and sidewalks along public thoroughfares), on-site grading and paving requirements, and underground utilities for building construction projects. Their work is recorded in the "C" sheets—civil sheets.

L: Landscape Drawings

Landscape drawings generally include planting plans and schedules (lists of plants, shrubs, and grasses), the irrigation system required to support the

(Continued)

plants, and hardscape (trellises, fences, site benches, patios, walkways, etc.) described in plan, elevation, section views, and details.

S: Structural Drawings

These drawings describe the elements, components, and assemblies of structural systems, and the manner in which they are connected, for a variety of projects. They are the construction equivalent of the skeleton, tendons, and muscle matter of the human body; in fact, "tendons" is a term used to describe the steel cables that are inserted in components, such as precast, prestressed concrete piles and girders, and in cast-in-place post-tensioned concrete beams, bridge decks, and floor slabs. Architects frequently determine the basic structural system, since the functional arrangement of a project and the required aesthetic treatment may dictate column spacing and therefore spans; however, the calculations and detailed structural design parameters are the bailiwick of the structural engineer.

A: Architectural Drawings

Architectural drawings are the heart and soul of a building project—virtually all other disciplines act in support of the architect's design, which is described in these drawings. To some extent this is due to the architect having overall control as the prime design professional and the uniqueness of building projects; however, in highway projects, for example, where design standards for construction are common and projects are co–developed and managed by district and central DOT offices, the various engineers involved act more as equals. There are as many approaches to design as there are architects—some conceive projects from the exterior and fit the functions within a shell; others determine the appropriate functional relationships of a building and develop the shell from them. No matter the origin of the design, the other disciplines take their cues from the architect's drawings, which are the most wide-ranging drawing set. Depending on the charges to the architect, the drawings can include master project planning and building design (frequently for multiple buildings) from basic systems or shells to complete buildings with interior details, furniture, and even fabric design.

I: Interiors

As just noted, architects might be given the responsibility to design a complete building or building shell for an owner. When the latter occurs, it is often because a developer or owner has anticipated that there will exist a demand for space within the building when it is complete and has undertaken to construct it on a speculative basis. Commercial real estate brokers monitor the construction and lease activity of buildings, and earn fees for facilitating lease agreements between building owners and tenants. Under certain lease agreements, tenants have the responsibility to design and construct their office space, within parameters established by the building owner, and will commission interior or building architects to produce the necessary design documents. It is these drawings, as well as drawings used in subsequent lease activity in the project, that are inserted in a drawing set under the interiors category.

Q: Equipment

Equipment drawings run the gamut of equipment that might be used in a project, from bank vaults, teller equipment, and ATMs to library, theater, videoconferencing, and commercial cooking, bakery, and laundry equipment. Some of the drawings required in this division can be complex, as, for example, in the case of bank vaults, which are subject to compliance with federal legislation governing their construction.

F: Fire Protection

To aid fire departments in their fire-fighting efforts, fire codes require comprehensive fire protection plans from project owners. Access and egress to the site; on-site street widths and radii; the location of hydrants, water mains, trees, overhead power lines, utility service disconnects; and anything else that could affect the success of fire-fighting efforts are subject to review by fire districts. Fire suppression systems, which are systems that actively fight fires (automatic sprinkler systems of a variety of types), as opposed to systems that simply detect or prevent fires, belong in this division as well. In addition to having some of the problems associated with other mechanical systems, fire suppression systems are complicated hydraulic systems whose performance is sensitive to minor changes in design. They are carefully reviewed in the design phase and are actively monitored during and after construction by the fire districts having jurisdiction in the community.

P: Plumbing

Plumbing systems are designed and described by mechanical engineers and recorded in plans, elevations, and sections, as well as in isometric schematics and fixture schedules. The basic parts of the system include drain waste and vent piping, hot and cold water supply, and fixtures.

D: Process

Process refers to systems that support the conversion of raw materials into a commercial product. Complicated forests of piping, controls, and storage facilities, process facilities are worthy of a distinct division in drawing sets. Refineries, canneries, and wineries are examples of process facilities—the

(Continued)

latter being an example of projects for which building construction drawings and process facilities might be combined.

M: Mechanical

Mechanical drawings describe the location, size, and type of equipment for distributing, filtering, humidifying/dehumidifying, cooling and heating air, as well as the distribution and control systems required in a project.

E: Electrical

Electrical drawings describe the electrical service (utility-provided wiring, metering, main switches, and grounding), distribution (panelboards, switchgear, and wiring emanating from the boards), branchwork (circuitry), and devices used in a project. As with other drawings, the electrical engineer uses plans, sections, details, and schedules to describe the project.

T: Telecommunications Drawings

Changes resulting primarily from widespread computer use, as well as developments in telecommunications technology, have resulted in a dramatic increase in the attention given to telecommunications systems. The NCS has provided room for additional developments by creating a separate division for these systems.

R: Resource

Resource drawings consist of any drawings that are created prior to and sometimes during construction, as well as "measured" drawings—drawings that describe existing conditions that are used in the development of remodeling plans, among other types. As to subject matter, these drawings contain whatever information might be required for a remodeling or refurbishing project—structural, mechanical, and other plans are among the possibilities.

X: Other Disciplines

This category is a miscellaneous division. Any participant—an acoustical consultant, for example—could produce the necessary drawings for atypical kinds of work.

Z: Contractor Drawings

Shop or fabrication drawings are among the types of drawings that are the responsibility of the contractor, hence, the division "contractor drawings." Subcontractors or manufacturers use shop drawings to demonstrate to their shop personnel, the contractor, and to the design professional how an assembly or component—described in general terms by the architect or engineer—will be produced. Structural steel, trusses, fire suppression systems, and vertical transportation are examples of the kinds of work that are detailed in shop draw-ings. Shop drawings are the responsibility of the contractor; however, it is gener-ally the architect who provides the list of work items that require shop drawings as a part of the submittal process. The drawings are reviewed initially by the contrac-tor, and then are sent to the design professional, who reviews them for compliance with design intent. Although some controversy has arisen as to the timing (prior to permit approval) of certain submittals, as well as the liability associated with shop drawing approval, the design professional is interested in understanding generally how the contractor plans to execute portions of the work.

O: Operations

This category exists for the benefit of facilities management personnel, who have the responsibility for maintaining the facilities of a company or institution as well as for modifying facilities to suit changing needs. Drawings generated by facili-ties management employees or by design firms that describe proposed changes to the facility find a home in this category.

—Keith Bisharat. (2008). *Construction Graphics*, John Wiley & Sons, pp. 26–28.

SPECIFICATIONS

The AIA states that specifications are written requirements for materials, equipment, systems, standards and workmanship for the work, and performance of related ser-vices. In their book *Construction Specification Writing: Principles and Procedures*, Rosen and Regener list the following topics best covered in specifications:

- Type and quality of every product, from simple material to system
- Quality of workmanship during manufacturing, fabrication, application, installation, and finishing
- Requirements for fabrication, erection, application, installation, and finishing
- Regulatory requirements, including applicable codes and standards
- Overall and component dimensional requirements for specified materials, manufactured products, and equipment
- Specific descriptions and procedures for product alternates and options
- Specific requirements for administration of the contract for construction

Those who use drawings and specifications—owners/clients, plan checkers, engi-neer/architect field representatives, estimators, contractors, subcontractors, material suppliers, and inspectors—benefit from standardized formats. Standardization

Table 6.4   U.S. and Canadian Drawing and Specification Standards

Drawings	Specifications
• U.S. National Computer Aided Design (CAD) Standard (NCS)	• Construction Specifications Institute (CSI) MasterFormat™
• Construction Specifications Institute (CSI) Uniform Drawing System (UDS)	• Construction Specifications Institute (CSI) SectionFormat™
• American Institute of Architects (AIA) CAD Layer Guidelines	• Construction Specifications Institute (CSI) PageFormat™
• National BIM Standard Project Committee— National Institute of Building Sciences (NIBS)	• American Institute of Architects (AIA) MASTERSPEC®

enables AEC professionals to communicate more easily with one another, minimizes confusion, and saves time. Some large public owners, such as the U.S. Army Corps of Engineers, and private companies, such as The Boeing Company, develop and adopt their own standards. In such cases, designers and contractors must familiarize themselves with owner-defined standards.

Table 6.4 lists the organizations largely responsible for creating the standardized systems of drawings and specifications used throughout the United States and Canada.

SPECIFICATION FORMAT

The Construction Specifications Institute (CSI) has developed a specification numbering and formatting system, called MasterFormat™, which is used widely. Prior to 2004, CSI-standardized specifications had 16 divisions, 01 through 16. Because of increased project complexity and the desire to address the needs of the market, CSI's 2004 MasterFormat™ was increased to 50 divisions, 00 through 49. It also changed the basic specification numbering system from five to six digits. See the following figures *Typical Specification Numbering System* and *MasterFormat™ Division Numbers and Titles*:

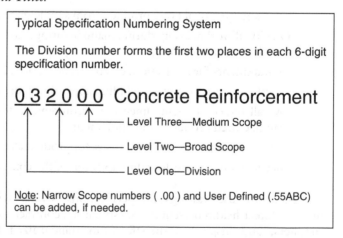

MasterFormat™ Division Numbers and Titles

PROCUREMENT AND CONTRACTING REQUIREMENTS GROUP
Division 00   Procurement and Contracting Requirements

SPECIFICATIONS GROUP
GENERAL REQUIREMENTS SUBGROUP
Division 01   General Requirements

FACILITY CONSTRUCTION SUBGROUP
Division 02 Existing Conditions
Division 03 Concrete
Division 04 Masonry
Division 05 Metals
Division 06 Wood, Plastics, and Composites
Division 07 Thermal and Moisture Protection
Division 08 Openings
Division 09 Finishes
Division 10 Specialties
Division 11 Equipment
Division 12 Furnishings
Division 13 Special Construction
Division 14 Conveying Equipment
Division 15 Reserved
Division 16 Reserved
Division 17 Reserved
Division 18 Reserved
Division 19 Reserved

FACILITY SERVICES SUBGROUP
Division 20 Reserved
Division 21 Fire Suppression
Division 22 Plumbing
Division 23 Heating, Ventilating, and
                     Air Conditioning
Division 24 Reserved
Division 25 Integrated Automation
Division 26 Electrical
Division 27 Communications
Division 28 Electronic Safety and Security
Division 29 Reserved

SITE AND INFRASTRUCTURE SUBGROUP
Division 30 Reserved
Division 31 Earthwork
Division 32 Exterior Improvements
Division 33 Utilities
Division 34 Transportation
Division 35 Waterway and Marine Construction
Division 36 Reserved
Division 37 Reserved
Division 38 Reserved
Division 39 Reserved

PROCESS EQUIPMENT SUBGROUP
Division 40 Process Integration
Division 41 Material Processing and Handling
                     Equipment
Division 42 Process Heating, Cooling, and Drying
                     Equipment
Division 43 Process Gas and Liquid Handling,
                     Purification, and Storage Equipment
Division 44 Pollution Control Equipment
Division 45 Industry-Specific Manufacturing
                     Equipment
Division 46 Reserved
Division 47 Reserved
Division 48 Electrical Power Generation
Division 49 Reserved

Source: Construction Specifications Institute (csinet.org)

The specifications numbering systems developed by the Construction Specification Institute (CSI) and the American Institute of Architects (AIA) provide for inclusion of *Procurement and Contracting Requirements* and *General Requirements (General Conditions of the Contract)*. Table 6.5 lists typical information included in Divisions 00 and 01.

Table 6.5   Information Contained in Divisions 00 and 01

Division 00 Procurement and Contracting Requirements	Division 01 General Requirements
• Advertisement for bids	• Summary of work
• Invitation to bid	• Price and payment procedures
• Instructions to bidders	• Product substitution procedures
• Prebid meetings	• Contract modification procedures
• Land survey information	• Project management and coordination
• Geotechnical information	• Construction schedule and documentation
• Bid forms	• Contractor's responsibility
• Owner-contractor agreement forms	• Regulatory requirements (codes, laws, permits, etc.)
• Bond forms	• Temporary facilities
• Certificate of substantial completion form	• Product storage and handling
• Certificate of completion form	• Owner-supplied products
• Conditions of the contract	• Execution and closeout requirements

CSI also has created SectionFormat^{TM}, which presents a unified way of depicting information contained in each specification section. Each specification section is divided into three parts:

- Part 1—General

  An extension of Division 01—General Requirements unique to this specification. Describes work covered by the specification, as well as administrative and procedural requirements such as submittals and quality assurance.

- Part 2—Products

  Details regarding materials, products, equipment, systems, and quality control.

  Describes products to be incorporated into project, such as mix design and off-site fabrication.

- Part 3—Execution

  Preparatory and on-site actions to be taken.

  Describes erection/application/installation, as well as field quality control and manufacturer's field services.

Typical information included under these headings is included in Figure 6.3. A complete package of typical construction documents including the project manual, drawings, and specifications is depicted in Figure 6.4.

SECTION XXXXXX

SECTION TITLE

PART 1—GENERAL

1.1   SECTION INCLUDES
       A.   Element of Work ⎤
                           ⎬ Next level of detail
       B.   Element of Work ⎦
1.2   RELATED SECTIONS
1.3   ALLOWANCES
1.4   UNIT PRICES
1.5   ALTERNATES
1.6   REFERENCES
1.7   DEFINITIONS
1.8   PERFORMANCE REQUIREMENTS
1.9   SUBMITTALS
1.10   QUALITY ASSURANCE
1.11   DELIVERY, STORAGE, AND HANDLING
1.12   PROJECT CONDITIONS
1.13   SEQUENCING AND SCHEDULING
1.14   WARRANTY
1.15   MAINTENANCE

PART 2—PRODUCTS

2.1   MANUFACTURERS
2.2   MATERIALS
2.3   [MANUFACTURED UNITS][EQUIPMENT][COMPONENTS][ELEMENT OF WORK]
2.4   ACCESSORIES
2.5   MIXES
2.6   FABRICATION
2.7   SOURCE QUALITY CONTROL

PART 3—EXECUTION

3.1   EXAMINATION
3.2   PREPARATION
3.3   [ERECTION][APPLICATION]INSTALLATION]
3.4   FIELD QUALITY
3.5   MANUFACTURER'S FIELD SERVICES
3.6   ADJUSTMENT AND CLEANING
3.7   DEMONSTRATION
3.8   PROTECTION
3.9   SCHEDULE

END OF SECTION

XXXXXX - #

Figure 6.3   Abridged version of CSI's SectionFormat™

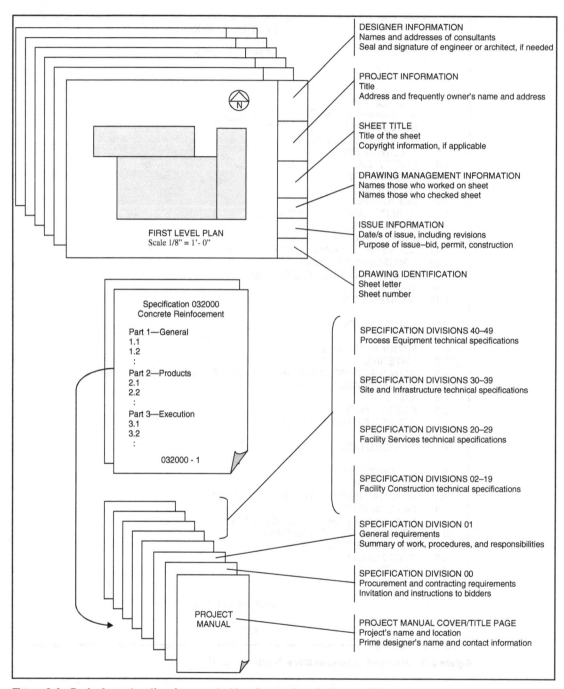

Figure 6.4 Typical construction documents (drawings and project manual) format

METHODS OF SPECIFYING

There are four widely recognized methods of specifying. These are:

- *Descriptive Specifying*: Exact properties of materials and methods are described in detail, without referring to specific manufacturers or suppliers.
- *Reference Standard Specifying*: Standards developed by trade organizations, institutions, and government organizations are cited within the three other specification methods.
- *Proprietary Specifying*: Actual brand names, model numbers, and other unambiguous information define the materials, methods, or systems; may be made less restrictive by naming two or three manufacturers or by providing "or equal" terms.
- *Performance Specifying*: The end result desired is described and the contractor, manufacturer, and/or fabricator supplies the solution that meets the defined criteria; designer must include a provision for appropriate tests to measure performance

Each type of specifying has advantages and disadvantages. These are summarized in Table 6.6.

Some Authors and Publishers of Reference Standard Specifications

AISI (American Iron and Steel Institute)

ANSI (American National Standards Institute)

ARMA (Asphalt Roofing Manufacturers Institute)

ASHRAE (American Society of Heating, Refrigerating, and Air-Conditioning Engineers)

AASHTO (American Association of State Highway and Transportation Officials)

ASTM International (formerly American Society for Testing and Materials)

AWI (Architectural Woodwork Institute)

AWS (American Welding Society)

BHMA (Builders Hardware Manufacturers Association)

ACI (American Concrete Institute)

ICC (International Code Council)

NECA (National Electrical Contractors Association)

(Continued)

NEMA (National Electrical Manufacturers Association)

NFPA International (formerly National Fire Protection Agency)

NIST (National Institute of Standards and Technology)

NRCA (National Roofing Contractors Association)

OSHA (Occupational Health and Safety Administration)

PHCC (Plumbing, Heating, Cooling Contractors Association)

SAE (Society of Automotive Engineers)

SDI (Steel Door Institute)

SMACNA (Sheet Metal and Air Conditioning Contractors' National Association)

SSPC (Systems and Specifications for Painting and Coatings)

Table 6.6 Advantages and Disadvantages of Specifying Methods
(Adapted from Harold J. Rosen and John R. Regener. (2005) *Construction Specifications Writing: Principles and Procedures*, 5th edition.)

DESCRIPTIVE SPECIFYING		REFERENCE STANDARD SPECIFYING	
Advantages	Disadvantages	Advantages	Disadvantages
Describes exactly what the designer intends	Requires designer to describe design intent carefully; "wordsmithing" necessary	Clearly states which standards of production, workmanship, and quality apply	Can be cited incorrectly, causing confusion at best or errors at worst
Is applicable to all conditions and circumstances	Results in long specification documents	Is based on well-tested and accepted work developed by experts	Can be used inappropriately in place of descriptive specifications
Permits free competition—omits brand names	Is time consuming to produce and may require more time in evaluating submittals	Saves designer time by not having to "reinvent the wheel"	May be difficult to enforce; contractor must interpret the standard
Provides good basis for bidding; desired work results clear	May be too elaborate for small projects	Can be used on most projects, except very small projects	May be obsolete or based on too low a standard
PROPRIETARY SPECIFYING		PERFORMANCE SPECIFYING	
Advantages	Disadvantages	Advantages	Disadvantages
Controls product selection strictly; contractors know exactly what is expected	Prefers some manufacturers and suppliers over others	Spells out design intent only; requires contractor to deliver systems that work	Requires definition of all attributes, requirements, criteria, and testing
Bases details closely on data supplied by manufacturers or suppliers	Reduces or eliminates competition	Delegates technical responsibilities to the contractor	Delegates technical responsibilities to the contractor

PROPRIETARY SPECIFYING		PERFORMANCE SPECIFYING	
Advantages	Disadvantages	Advantages	Disadvantages
Reduces specification production time	May use products or materials with which the contractor has no or bad experience	Can result in shorter documents	Can be time consuming to produce
Simplifies bidding by narrowing the competition	May create the possibility of increased cost due to lack of competition	Encourages the development of new technologies and permits free competition	May be too elaborate for small projects

DRAWINGS AND SPECIFICATIONS—FINAL THOUGHTS

Division 01—General Requirements typically spells out whether the drawings or specifications take precedence if the information in one conflicts with that of the other. In a dispute, however, specifications often are given greater significance than drawings for several reasons. First, specifications may show design intent more clearly than drawings. And second, those involved in the legal system—attorneys, judges, and juries—as well as construction managers are more familiar with interpreting text-based rather than graphics-based documents.

The best way to avoid conflicts between drawings and specifications is to limit duplication of information. Drawings and specifications work together to tell the whole story, but their purposes are different. Making global changes to drawings and specifications is difficult because they are created using different software; there is no "search and replace" command to replicate changes through both sets of documents. Many designers request the reader/bidder to ask questions if there's an apparent discrepancy between the drawings and the specifications.

Finally, the information presented in both types of documents should be clear and concise. Many different people use the drawings and specifications. The language selected by civil engineers and other consultants should be comprehensible to owners, technical specialists, construction field personnel, and government agencies alike. Like any professional deliverables, drawings and specifications also should be correct and complete.

TECHNICAL MEMOS AND REPORTS

Engineers are often given assignments in problem solving that may be part of an overall larger project. However, some problem solving may be on a large scale and may be associated with an operating component of a company or an organization. The size of the problem and its potential impact usually dictate whether the assignment will produce a technical memo or a technical report. A technical memo is usually less formal than a report but still has much of the same content. A technical report will generally include several additional sections beyond a technical memo including an executive summary, references, and/or appendices. In addition a technical report typically has greater detail and depth of the information than a technical memo.

Probably the most important difference between a technical memo and a technical report is the intended audience. A technical memo can be shorter and include less detail because the audience may be colleagues or peers within the organization, senior management or other related departments within the organization. In such cases, the audience is usually familiar with the problem, technical approaches, and analytical details.

An example of a technical report format is included in Chapter 13, Communicating as a Professional Engineer. In addition, a sample short technical report titled "The Benefits of Green Roofs" may also be found in Appendix E. Additional sample technical reports may be found in the Appendices. For example, Appendix A includes a sample Request for Proposal for a Pipeline Routing Study, Appendix B includes an example engineering proposal to accomplish this requested scope of services in Appendix A, and Appendix C includes a sample feasibility report to address the RFP found in Appendix A.

The United States Environmental Protection Agency (U.S. EPA) has several excellent reference documents for engineers including a guidance document for conducting feasibility studies (FSs). This guidance document is titled, "Guidance for Conducting Remedial Investigations and Feasibility Studies under CERCLA, Interim Final" (October 1988) EPA 540/G/89/004, OSWER 9355.3-01. This is an excellent reference for conducting FSs and includes good graphics and tables as examples.

Another useful tool for engineers is a cost estimating guidance document. This U.S. EPA reference can provide useful information but the engineer will need to update specific details relevant to the location, and contemporary material and labor rates. The reference document is titled:

"A Guide to Developing and Documenting Cost Estimates During the Feasibility Study," U.S. Army Corps of Engineers and U.S. Environmental Protection Agency EPA 540-R-00-002 OSWER 9355.0-75 www.epa.gov/superfund July 2000

CALCULATIONS

Engineers are also given assignments in problem solving which may require engineering calculations. The calculations are usually associated with an operating component of a company or an organization or could be part of a repair or replacement for a necessary component. Engineering calculations are often requested by a professional colleague from a technical branch like "plant operations" of a company or agency. The requestor generally has a great deal of knowledge on the overall performance and utility of the system and simply needs some technical assistance from an engineer.

Calculations should be set up much like engineers are taught in their degree programs. The problem statement should be clearly defined and a deliverable product should be agreed upon with the requestor. The original assignment may be given to the engineer in a hallway conversation or in a more formal kick-off meeting. If the engineer is unclear about the assignment or problem, some additional communications will be necessary to ferret out the specific information needed to proceed. A typical format for "engineering calculations" is outlined below:

- Problem Statement
- Objective and Approach
  - The objective statement should be carefully worded to reflect the originator's request in the engineer's own words. The objective statement is usually a sentence, or maybe several, and can include a primary objective, secondary objective, or others as necessary.
  - The approach statement should reflect "how" the engineer plans to approach the problem. It can include details on required tools, any required testing or analyses, required resources, materials, restrictions on ongoing operations, required health and safety details (the H&S plan will likely be included later but mentioned here), and any other pertinent details related to the approach. State whether the operations (if this is part of an operating system) will need to be shut down or restricted in any way during the assessment, maintenance, and repair period.
- Known Information
  - A description and relationship of the specific component within the system should be included, as well as photos or sketches with dimensions and impacts if possible.
  - It is suggested that the engineer include a sketch or diagram if it will help crystallize the problem and related details.
  - For clarity, the "problem statement" should be connected with the objective/approach and known information.
- Technical References
  - Technical references on materials of construction, reference documents, existing engineering drawings, manufacturer's catalogue information, performance objectives, specifications, operating requirements, and other related information also should be included. This information should be made as detailed as necessary to meet the objectives, health and safety requirements for personnel, and operations requirements.
- General Scope of Work Section
  - This section should show the general tasks that are required to accomplish the task.
  - There may be an assessment task to gather information related to the component failure, replacement, or upgrade.
  - If there is an assessment task, an evaluation task, design task, or calculation task will likely follow.

- Depending upon the complexity of this problem, completing this series of tasks with product evaluation, specification details, ordering, pricing, or custom development and manufacture may be necessary.
- A clear statement should be made regarding the requestor's desired product, how it should be transmitted, and how it should be delivered.
- The task effort should include some time for development and planning of a schedule and budget (or time estimate). The engineer is highly cautioned to include assessment time, reference time, and QC time in the overall estimated time to complete the effort. Often the requestors do not understand the interrelationships of components within an overall system and can possibly "oversimplify" the problem statement in an effort to speed up the delivery of the required product. Conversely, the engineer is highly cautioned not to "overcomplicate" the problem when communicating with the requestor. An experienced engineer will find that there is a delicate balance here and may request some mentoring or coaching from more experienced colleagues when addressing this issue.

- Calculations
  - The calculations should be clear and concise showing the reason for the calculation related to the problem statement, objective/approach, known information including drawings or photos, and general scope of work.
  - The engineer should write clearly, include references, show detail, include assumptions, verify conditions, initial (or sign) and date each sheet. The engineer should have these calculations peer-reviewed and all calculations should be quality checked (QC checked) by a competent professional in the field of endeavor.
  - If the document will be released for public files, the specific state where the work is accomplished may require the engineer to stamp the calculations.
- Transmitting the Final Deliverable
  - The engineer should prepare a letter of transmittal to the requestor that includes the relevant information agreed upon in the kick-off meeting. Sample letters and transmittals are included for reference in Chapter 13, Communicating as a Professional Engineer.
  - The engineer should keep file copies of the information for future reference especially if the laws governing professional engineers in the state where the work is accomplished require the engineer to do so.

OTHER DELIVERABLES

The civil engineer often has numerous responsibilities during the construction phase of a project and frequently is responsible for: (1) facilitating the permit process; (2) responding to requests for information (RFIs) from the contractor; (3) reviewing

Table 6.7 Useful Construction Phase Terms

Term	Meaning
Submittal	Shop drawings, material data, and sample required primarily for the engineer and architect to verify that the contractor has purchased the products required in the plans and specifications. Concrete mix design calculations are an example.
Shop drawing	Drawing or set of drawings submitted to the prime designer for review by the general contractor. Produced by the contractor, supplier, manufacturer, subcontractor, or fabricator. Typically required for prefabricated components, such as structural steel, trusses, precast, elevators, etc.
Steel fabrication drawings	Shop drawings showing the fabrication of structural steel components. Depict steel joint connections and detailed dimensional information for all parts and assemblies, including tolerances. Calculations generally included.
Request for information (RFI)	Questions directed to the prime designer by the general contractor to gain clarification or to confirm the interpretation of a detail, specification, or note on the construction drawings. Used to secure a documented directive and often results in a change to the scope of requiring changes in project budget and/or schedule.

fabrication (shop) drawings and other submittals; (4) attending meetings and writing meeting minutes; (5) making recommendations to the owner regarding construction progress payments; and (6) documenting installation instructions. Table 6.7 contains further information related to construction terms.

SUMMARY

The civil engineer and other designers are responsible for a dizzying array of deliverables. The challenging and rewarding aspect of the project delivery process is that these designers ultimately create something from nothing. As later chapters show, client relations, project management, and communication are key to this process. Ultimately, the delivery of the civil engineer's stock in trade—studies, reports, calculations, plans, specifications—enables a sort of alchemy to take place. These project deliverables result in build/no-build decisions and have been used to construct almost all the built environment that surrounds us. The need for their accuracy and completeness and clear communication cannot be overstated.

REFERENCES

American Institute of Architects. (2008). *The Architect's Handbook of Professional Practice*, 14th edition. Joseph A. Demkin, AIA, executive editor. John Wiley & Sons, Inc. Hoboken, New Jersey. ISBN 978-0-470-00957-4.

Bisharat, Keith A. (2008). *Construction Graphics: A Practical Guide to Interpreting Working Drawings*, 2d edition. John Wiley & Sons, Inc. Hoboken, New Jersey. ISBN 978-0-470-13750-5.

Construction Specifications Institute. (2005). *The Project Resource Manual—CSI Manual of Practice*, 5th edition. McGraw-Hill, New York. ISBN 0-071-37004-8.

Madan, Mehta, Walter Scarborough, and Diane Armpriest. (2010). *Building Construction: Principles, Materials, and Systems—2009 Update*. Pearson Prentice Hall, Upper Saddle River, New Jersey. ISBN-13: 978-0-135-06476-4.

Rosen, Harold J., and John R. Regener. (2005). *Construction Specifications Writing: Principles and Procedures*, 5th edition. John Wiley & Sons, Inc., Hoboken, New Jersey. ISBN 0-471-43204-0.

Introduction

Ethics

Professional Engagement

1 2 3 4 5 6 7 8 9 10 11 12 13 14 15 16 17 A B C D E F

History

What Engineers Deliver

Emerging Technology

Engineer's Role in Project
Development

Permitting

Leadership   Managing

Having
a Life

Executing a Professional
Commission

Client Relationship

Sustainability

Globalization

Legal Aspects   Communicating

Chapter **7**

Executing a Professional Commission—Project Management

Big Idea

Effective project management knowledge and techniques are essential for conducting project operations. The project manager is at the heart of the project and must be aware of all activities related to the initiation, planning, execution, monitoring and control, and closure of the project.

Of all the things I've done, the most vital is coordinating the talents of those who work for us and pointing them toward a certain goal.

—Walt Disney

Key Topics Covered

- Introduction
- The Basics of Project Management
- The Major Parties on a Project
- Project Sectors
- Project Teams
- Project Initiation
- Project Estimates
- Project Management Plan Components
- Staff Selection Guidelines for the PM
- The Project Manager's Responsibilities
- Project Risk Management
- Design Coordination
- Summary

Related Chapters in This Book

- Chapter 3: Ethics
- Chapter 5: The Engineer's Role in Project Development
- Chapter 6: What Engineers Deliver
- Chapter 9: The Client Relationship and Business Development
- Chapter 12: Managing the Civil Engineering Enterprise
- Chapter 13: Communicating as a Professional
- Chapter 14: Having a Life
- Chapter 15: Globalization
- Chapter 16: Sustainability
- Chapter 17: Emerging Technologies

(Continued)

Related to *ASCE Body of Knowledge 2* Outcomes

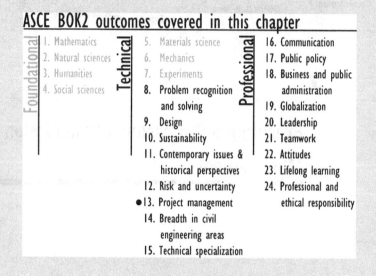

ASCE BOK2 outcomes covered in this chapter

Foundational
1. Mathematics
2. Natural sciences
3. Humanities
4. Social sciences

Technical
5. Materials science
6. Mechanics
7. Experiments
8. Problem recognition and solving
9. Design
10. Sustainability
11. Contemporary issues & historical perspectives
12. Risk and uncertainty
● 13. Project management
14. Breadth in civil engineering areas
15. Technical specialization

Professional
16. Communication
17. Public policy
18. Business and public administration
19. Globalization
20. Leadership
21. Teamwork
22. Attitudes
23. Lifelong learning
24. Professional and ethical responsibility

INTRODUCTION

Project Management Background

It's appropriate to begin a discussion on project management with a brief description of the historical development of the management of projects. Peter Morris has provided a detailed account of the development of project management as a distinct discipline (Morris 1994). Other authors (Kerzner 1998) and Morris postulate that project management as a discipline was spawned in the mid-20th century.

However, mega-projects date to a distant time; the Roman Coliseum was built by four different contractors (Morris 1994). Early undertakings by the Celts, Egyptians, Greeks, Romans, and Chinese involved the entire community and served to cement secular and religious authority. During the great Gothic period, expression of the devotion to God was more important than timely project delivery but, as economies and technologies became more sophisticated, so did the use of contracts to realize projects. By the 18th century, those who designed projects were separate both contractually and organizationally from those who built them.

The economic development of the Victorian era led to huge infrastructure projects and industrialization. Authors whose early theories on scientific management emerged at the beginning of the 20th century are still quoted. These include: Taylor and Gilbreth (time and motion studies) and Gantt (production scheduling). Weber also established his theories on bureaucracy during this period. But it was not until the 1930s that an academic writer proposed the use of a coordinator who might be used to administer a task involving several functional areas (Morris 1994). Morris views this addition of a separate mechanism to integrate the various entities making up a project as *the inception of modern project management*. He further proposes that the rise of modern project management between the 1930s and 1950s is related to:

- Development of systems engineering in the U.S. defense and aerospace industry
- Engineering management practices in process engineering
- Developments in management theory, particularly in organization design
- Evolution of the computer, enabling many project management tools

Kerzner, on the other hand, holds the more typical view that project management began in the 1950s and 1960s (Kerzner 1998). A major advancement came in 1958 during the development of the POLARIS missile program by the U.S. Navy, helped by the Lockheed Missile Systems division and the consultant firm of Booz-Allen & Hamilton when PERT was developed.

During that period the literature abounds with journal articles that proposed various models of organizational redesign which would lead to better control over resources, and thus, better project control. The result was an organizational structure with multiple layers of management. The 1970s saw a focus on organizational behavior. Firms and researchers pondered how to get desired productivity from these

**Table 7.1 Evolution of Design toward Collaboration and Integration
(Adapted from Hughes, 1998)**

Modern	Postmodern
Taylorism	Systems Engineering
hierarchical/vertical	flat/layered/horizontal
specialization	interdisciplinary
rational order	messy complexity
centralized control	distributed control
experts	meritocracy
tightly coupled system	networked system
micro-management	blackboxing
hierarchical decision making	consensus-reaching
bureaucratic structure	collegial community
incremental	discontinuous
closed	open

elaborate command and control structures. With advances in information technology, by the 1980s the emphasis shifted to project management quantitative tools, such as "performance evaluation and review technique," or PERT, and "critical path method," or CPM (Render and Stair 1982, Wiest and Levy 1974). As a result of business process re-engineering (BPR) in the 1990s and the changing nature of technology, many firms have fewer layers of management and relatively flat organizational structures. The presence of multidirectional, cooperative work flow necessitates better communications. Additionally, the advent of the "virtual" organization, for instance, a multifirm organization (an example would be a prime contractor and key subcontractors) formed around a project that will be disbanded upon project completion, strongly underscores the importance of integration in project management.

Literature on project management reflects the need for flexibility and the changing context in which projects exist. The phrase "management by projects" may best capture the situation today. Thomas Hughes, in *Rescuing Prometheus: The Story of Mammoth Projects* (1998), has captured the essence of these changes by comparing the modern and postmodern approaches to design (Table 7.1).

A Discipline, But Not a Theory

Although books on project management abound, their primary focus is prescriptive. That is, authors outline what to do to optimize cost, schedule, quality, profitability, and so forth. Works on project management *theory* are much less abundant. In fact, project management appears to be a blend of ideas drawn from other disciplines. The Project Management Institute's guide to accepted knowledge and practices of the profession includes nine project management "knowledge areas," each of which has a robust theoretical underpinning (PMI Standards Committee 1996). These areas are:

- Integration management
- Scope management

- Time management
- Cost management
- Quality management
- Human resources management
- Communications management
- Risk management
- Procurement management

To this list can be added:

- Organization structure
- Organizational behavior

Shehnar and Dvir approach project management theory from a different perspective, that of the innovation literature (Shehnar and Dvir 1996). These authors observe that one of the basic deficiencies of project management theory may be that little distinction is made between project type and project managerial problems. They propose a two-dimensional typology of projects that maps system (project) scope against technological uncertainty as a way of distinguishing project types as illustrated in Figure 7.1.

System scope refers to the notion that there are different hierarchies inside a system or product with different levels of activities. Shehnar and Dvir define these categories of project scope as: *assembly* (components and/or modules combined into a single unit); *system* (collection of interactive elements, performing independent functions and fulfilling a specific need); and *array* (widely dispersed collection of different systems functioning together to achieve a common purpose). The technology axis ranges from *low-tech* to *super high-tech*. At one end of the spectrum is technology that exists and is readily available and acceptable (e.g., that used to build roads); at the other end is technology that does not exist at project initiation (e.g., the Apollo moon landing).

Shehnar and Dvir's typology appears to offer a useful way to move toward an integrated theory of project management. The authors suggest other typologies that might be developed. Possible dimensions include: the fit between project type, project management style, and project effectiveness; project environment, such as economic, political, social, geographic, and cultural; and the degree of difficulty in articulating user/customer requirements and the point in the project lifecycle at which these requirements are identified.

Given the factors mentioned above, most projects vary significantly along several critical dimensions rendering a universal typology meaningless. From a broader perspective, acknowledging that all projects have typology with varying phases of complexity can be quite useful. When the Project Manager recognizes this fact, he or she can gain insight into timing and location of potential *hotspots,* such as interface between players, integration, communication, and so forth that may manifest itself later in the project.

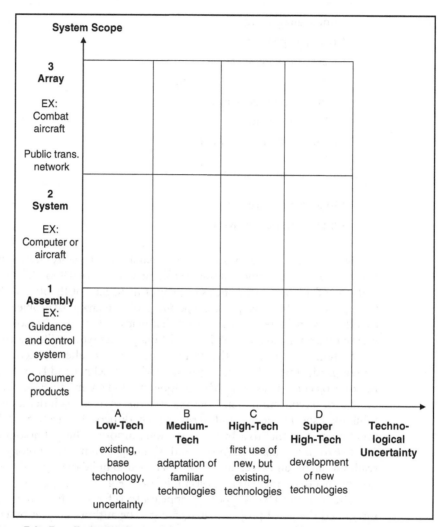

Figure 7.1 Two-dimensional typology

A Glimpse into the Life of a Project Manager

Charlene just had a full morning of budget and projection meetings for her business unit and now she was exhausted as it approached 12:30 pm. She actually got up almost an hour earlier than usual to get bagels and snacks for the management team since it was her turn. Her Engineering Manager, Steve, needed a complete rundown of the previous quarter's invoicing for her four clients (and all the other client invoices in the unit) plus the anticipated staff level of effort (LOE) resource projections and subcontractor needs for the next quarter. She stayed up late the night before to get the figures into the new spreadsheet requested by her boss through the Denver corporate office. Charlene left the meeting with seven priority tasks on her "to do" list that needed attention by the end of the day. On the way to her office she ran into Frank who reminded her about the overdue deadline for her comments and recommendations for two new engineer candidate employees she recently interviewed. The interviewees were anxiously awaiting a response from the firm since they both now had competitive offers from other companies and a public agency.

As Charlene returned to her office she rested her eyes on the pleasant photo on her wall calendar but she suddenly remembered it's month end and she also needed to enter her time into the time sheet database or her paycheck would be delayed. She hadn't had a free moment all week to enter her time and she was now seriously out of compliance with the policy. She knew she would receive the computer automated reminders and a special "spanking" phone call from the Engineering Manager on this one, since Federal regulations require daily time entries. Her main concern now was the three urgent e-mail alerts on her screen and the blinking light on her voicemail suggesting a missed call from one of her clients. She normally would have seen the e-mails come in on her PDA but Steve did not permit PDAs in the quarterly meetings because of the disruptions. Charlene quickly scrolled to the urgent e-mails as she bit her lip unintentionally.

Apparently Jim, the field crew lead technician, had a minor health and safety incident in the field. Jim hit an obstacle on the construction site with the new truck and had a flat tire and bent wheel rim. In the process of changing the tire the new staff engineer nicked his thumb on the bent rim and required five stitches. Unfortunately, Charlene, as the PM for this project, had to fill in the Accident Form 3592 by the end of the shift and report it to the Engineering Manager and the Director of Corporate Health and Safety. She began to feel her neck ache indicating the first sign of one of her dreaded headaches.

Charlene pulled up Form 3592 on her computer as she called Jim on his cell phone. After four rings and no answer, she left a message on Jim's voicemail to check in and see if he needed any assistance. She then called Sara, his colleague and trainee. Sara answered after one ring anxiously waiting Charlene's call. Sara stated they were driven to the emergency room by the client since their new truck was now disabled. The client's PM, Wendy, was really nice and understanding over the past 18 months on this project—they even exchanged birthday cards. However, Sara said there was one minor problem as she mentioned

(Continued)

the Jim's blood stains on the client's car seat. Charlene sounded surprised and slightly irritated about this mishap but quickly recalled they did not have a chance to equip the new truck with a standard company first aid kit. The Engineering Manager would not be happy about this situation and would likely be a lot more irritated than the client with the stained car seat. Charlene hung up the phone and went to find Steve to inform him of the disabled vehicle and stranded technicians at the hospital.

As Charlene rounded the corner she noticed Steve's lights were out signaling his likely lunch appointment. She phoned him immediately to tell him the news. The Engineering Manager picked up after the second ring and was very quiet on the other end as she recanted the events thus far. Steve apologized to his clients at the lunch table as he called for the bill. He understood that he was likely going to the project site or the hospital to pick up the field crew, Jim and Sara, as he quickly headed back to the office. Jim chose to go to the project site to arrange for a tow truck and thank the client, Wendy, for taking Jim and Sara to the emergency room. He was thinking about what solvent might take Jim's blood stains out of the seat fabric in Wendy's personal car.

Charlene took her personal car to the emergency room to pick up Sara and Jim. They were waiting by the exit door as she pulled up and gave a hug to Jim showing her delight that he was okay after the stitches and unanticipated tetanus shot. Jim apologized for the mistake and took the blame for not having the first aid kit in the new truck. It really wasn't Jim's fault, or Charlene's fault either, since the kit was on backorder and expected in next week. Charlene actually passed within a few blocks of her house as she longed for a long hot bath, a snack, and a glass of wine. When they arrived back at the office, Charlene told them both to go home for the day. In her mind, she thought the minor accident would likely be decided as "avoidable" by the Corporate Health and Safety Manager and that Jim would have to retake the defensive driving course as a reminder and sort of a punishment. She also recalled a recent presentation she heard about "how we pay" for mistakes in this morning's long meeting.

Charlene then returned to her office to tackle the unopened e-mails and voicemails as it approached 4 pm. Her headache was in full force accompanied by a stomach pain now, long after lunch as she searched for the Tylenol and an emergency snack bar in her bottom desk drawer. She scanned here-mail origination time, titles, original sender, and distribution and chose the second unopened e-mail to open immediately as she dialed her voicemail box for that news. The deep voice on the voicemail message confirmed that the message was from her largest telecommunications client Civil Construction Vice President, Jonathan. He said he was surprised that there was a new proposed ASTM testing technique required on the construction materials for all new facilities under construction in the eastern region. He was concerned that it could potentially cost over $25,000 in unanticipated expenses to his firm. He ended the concise message with an irritated tone and requested a callback soon.

Charlene quickly composed a brief e-mail response (knowing Jonathan's personal habit of checking fresh e-mails over unopened ones) and then picked up the phone for Jonathan's cell. They shared the same cell provider, Jonathan's company. Charlene wondered whether this could have had any impact on her firm winning the multi-million dollar design and construction monitoring contract. Jonathan picked up the call on the first ring knowing it was Charlene while he simultaneously received confirmation of the pending e-mail reply from her. Charlene opened with a warm greeting and then apologized for missing Jonathan's earlier call and previous e-mail. Jonathan returned the greeting warmly and stated he only had a few minutes since he was at the airport waiting for his flight home. Charlene didn't bother providing Jonathan excuses on "why" she could not respond earlier and simply got right to the point and asked whether he had a chance to read yesterday's e-mail notification and pick up her previous phone message. Jonathan stated he had been in strategic planning meetings for the last two days regarding funding for his company's corporate challenge and the current flurry of construction. Charlene mentioned that she had just learned about these new testing requirements and she had already planned to equip her field staff with the new testing equipment and that, overall, this test will replace a similar but slower technique thereby saving the client money and that construction may be able to proceed a bit faster. Charlene thought to herself how she really appreciated her CE QA team for this information. Jonathan laughed and thanked Charlene for being prepared and promised more good news when he had more time at their regular bi-monthly meeting next week. As Jonathan finished his final words Charlene heard a flight announcement in the background. Taking the nonverbal cue Charlene said she looked forward to the upcoming meeting. She quickly said "happy anniversary" and closed with "stay well," a little saying they said to each other often. Jonathan said thanks again and closed with "you stay well, too." Jonathan wondered how he found such an intelligent, responsive PM and one who could keep track of his anniversary date almost better than he could. Charlene thought her new "client key information" database was working well, knowing that two weeks ago Jonathan let it slip out that his ten-year wedding anniversary was at the end of the month.

Charlene took a deep breath, opened the second e-mail as she prepared to call Steve to check on the progress of the disabled new truck. She saw the second e-mail was related to a civil design question for her third largest client. She composed a brief reply on her understanding of the design aspect, included her lead designer on the reply message, and suggested that more recent information

(*Continued*)

could be provided by the lead designer. She knew this client well and suspected her information was sufficient and was going into the client's month-end report. However, she thought it was better to provide more information than less and marked this client as a follow-up phone call in two days. Meanwhile, Steve picked up the call on the second ring, said the new truck went to the dealer's shop and he was on his way home. She communicated her brief accomplishments and that she would be at work a bit longer finishing up. She saved Jonathan's promise of a "surprise" for her next face-to-face meeting with Steve when she needed to explain "how" the truck incident happened.

Charlene addressed the three urgent e-mail alerts and the urgent VMX and now turned to her time sheet, the required Health and Safety Form 3592, Frank's second request for her recommendation on the two interviewees, and the seven things on her "to do" list. The time sheet was the easy one and only required ten minutes and then she pulled up Form 3592 on her screen. She filled out the urgent part knowing that some information would not be available until the injured party, Jim, returned to work. The accident was fully covered and Jim had not had an accident in his last five years with the business unit so she wasn't too concerned. Completing these activities lifted a weight off her mind as she thought the Tylenol was kicking in as her headache receded. She remembered the blood stain in her client's (Wendy's) car as she researched a solvent for blood and a car detail company that could accomplish the task. She placed a quick call to Wendy's cell to ask her if she would like a rental car for a day. Wendy turned down the nice request but Charlene did get a commitment for a date and time to get Wendy's car in for a detail cleaning and a special wash and wax treatment. Wendy was so nice. Charlene thought this fix would require some flowers, too. Charlene thought of the "ways a PM pays when a team member makes a mistake on a project."

Charlene finished a brief message to four of the seven items on her "to do" list as she forwarded them off to her technical team for finishing up the information. She addressed the other three items quickly, and sent off the e-mail and attachments to Steve, as was requested of her. Charlene thought to herself that she actually was quite efficient when nobody was around, the phone didn't ring, and e-mails didn't come in. The quietness of the office was not surprising as it approached 6:30 pm. She realized she had to leave for her dinner appointment with her best friend. She put on some soft rock music on the quick trip home before changing and heading out for the evening.

Charlene thought to herself that her original plan of becoming invaluable to her employer and clients was a good one. She recognized that nobody could foresee all the potential project impacts and she was proud of her accomplishments for the day. On the way to dinner, she placed one more call to Jim to check on his hand injury and empathize with his pain.

There Are Three Ways a Project Manager Pays When Making a Mistake!

An engineering project manager experiences similar ways to pay for a mistake even if the mistake is not the fault of the PM or the firm:

1. Money: When a mistake or a problem comes between a client and an engineer, the most obvious form of payment is money. The payment can take the form of additional services, materials, or supplies paid by the eor shared with the client.

2. Time: The time element is usually experienced with a slip in the schedule or a delay in delivery of the final product or services. The schedule slip will likely be accompanied with additional costs borne by the engineer and/or client

3. Pain: In a professional work atmosphere the third element that may accompany a mistake, depending upon the seriousness and circumstances, is emotional pain. In this instance, the emotional pain may be embarrassment or a sting to one's character or reputation. If the mistake is in the form of a health and safety incident, it may also be accompanied with actual physical pain as well.

THE BASICS OF PROJECT MANAGEMENT

Definition of a Project

A project is an endeavor that is undertaken to produce results expected from the requesting party. To be a bit more specific, a project is an endeavor that will:

- Accomplish a specific client need or goal
- Include related activities guided by a leader or project manager
- Be composed of cross-departmental personnel or unique experts capable of providing unique services to the project
- Be performed for a fixed duration of time

Generally, the requestor or owner does not have the expertise or the time (or both) to take on the endeavor on their own. The project will likely have a fixed or estimated life of its own which can vary from days to years. For example, a project to repair a software program may take days and a project to build some of the world-renowned civil engineering projects depicted in Chapter 2, may take decades. Once the project is complete, the operating component and maintenance will likely revert to the owner's operations and maintenance personnel as the project team is disbanded. However, some contemporary engineering project owners offer the designers and/or builders a chance to provide services for short-term start-up operations and occasionally long-term facility operations.

Scope/Schedule/Budget Triangular Relationship

A project's scope of work, often referred to as the "scope," is a definition of the tasks required to complete the project to meet the client's needs. The complete scope of work for a project requires a schedule and budget. If the owner does not supply a schedule as part of the scope, an experienced PM will include a schedule for a project and an experienced negotiator will know that the schedule is part of the negotiations. For example, if an owner wants a 40-story hotel and convention center designed and built in 48 weeks, this effort will include a significantly more concentrated effort for managing the design team, materials procurement/delivery, and the construction team than an identical project built over a two-year period.

A project like this was built in Las Vegas -on a fast-track basis with the construction crews framing and placing concrete for each floor over a one-week period and then moving to the next floor. The casino owner had evaluated the increased construction costs for the accelerated construction compared to the standard construction cost. The owner then considered the casino revenue lost on the longer 100-week standard construction period and the costs associated with the shorter 48-week construction period. The casino owner realized that the additional hotel and casino revenue significantly exceeded the costs for the accelerated construction. The owner determined they would attempt building one floor a week for the high-rise hotel and convention center. This example illustrates how the budget is uniquely tied to the scope of work and schedule for each project, as illustrated in Figure 7.2.

This relationship of scope, schedule, and budget can be thought of as a connected triangle where each side represents an essential component of the "project" managed by the project manager (PM).Remembering the connected sides of this triangle

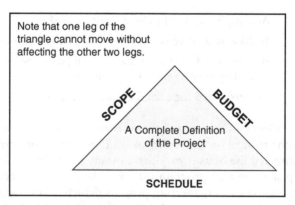

Figure 7.2   Scope, schedule, and budget relationship

enables the PM to see that these three components are interrelated and that one side cannot move or be adjusted without affecting either of the other two components.

Quality during this stage of project development is another important issue that needs to be addressed. The client will likely have an idea about the quality of the project and will include their thoughts on quality in the scope details. The important point is that quality should connect the scope, schedule, and budget. If the final project does not meet or exceed the owner's expectation for quality, the project likely will be rejected by the client. There will be more discussions on quality later in this chapter.

THE MAJOR PARTIES ON A PROJECT

The most important parties in a project are the owner, designer, and builder/contractor. These parties have the most invested in the project and stand to gain or lose the most if the project is performed correctly on time and within budget. A cooperative team approach is best when performing projects although the builder/contractor is often brought into the project at a later stage. It is recommended to have several builder/contractors comment on the early design phases of the project to get valuable feedback on the constructability of the proposed design.

A cooperative team approach with the owner, designer, and builder/contractor will often avert adversarial relationship problems later in the project.

The Owner's Role

The owner's role in the project is critical to setting operational criteria for the details on the scope of work. The owner should also identify:

- Their preferred level of involvement
- Their proposed review process
- Levels of approval based on dollar amounts, scheduled completion, or budget completion
- Any special equipment
- Any important company policies regarding safety, security, labor requirements, and so forth

In addition, the owner should also set parameters on total cost, payment of costs, major milestones, and the required project completion date.

The Designer's Role

The engineering designer should specify the number, type, and details of the:

- Proposed design alternatives
- Number, type, and amount of computations

- Number, type, and amount of drawings
- Level of detail for the specifications

The engineering designer can reduce the level of ambiguity and increase the level of understanding with the owner and builder/contractor by providing a clear understanding in written form with exhibits, tables, and figures on the design detail.

The Contractor's Role

The contractor should clearly state (or be asked to state) that they will:

- Perform all the work in accordance with contract documents prepared by the engineering designer
- Furnish all labor, equipment, material, and know-how necessary to build the project as designed
- Prepare regular budget and status reports for the owner
- Develop project quality reports for the engineer and owner
- Develops accurate budget and schedule controls

In summary, the contractor should act as if they are an owner and report information clearly and effectively to keep the project on schedule and within budget.

A Brief Summary of the Basics

Project management is defined as the art and science of coordinating people, equipment, materials, money, and schedules to complete a specified project on time and within approved cost, and includes the functions of planning, organizing, staffing, directing, and controlling.

PROJECT SECTORS

Project sectors relate to the market sector the project will service. The building sector generally includes the residential, commercial, or industrial market depending upon the client. The infrastructure sector includes private or, more likely, public owners and involves transportation, water, and other systems needed for a functioning society. The process sector includes the industrial and/or commercial markets. For example,

Building Market

Projects will likely be composed of:

- Commercial, educational, office, hospital, residential buildings
- Prime designer: *Architect*

Infrastructure Market

Projects will likely be composed of:

- Transportation (roads, bridges, airports, waterways, and water treatment facilities)
- Prime designer: *Engineer*

Process Facilities Market

Projects will likely be composed of:

- Chemical plants, oil refining, pharmaceutical, pulp and paper, electrical generating
- Prime designer: *Engineer*

There is a large array of approaches for delivering projects to the owner in these diverse sectors. Project delivery methods are influenced by whether a project is competitively bid or negotiated. Many engineering firms will not even submit proposals for competitively bid projects, where the focus is mainly on lowest price rather than the engineering firm's qualifications.

The amount of risk a client is willing to accept also greatly influences the choice of a project delivery method. There are risks associated with rapid construction (fast-track). If the client desires an accelerated service, with design solutions in the field, the client must be willing to accept a fairly high level of risk. In fast-track delivery methods, a detailed schedule showing critical dates for design packages and procurement is essential.

Project size is another import factor in selecting the appropriate project delivery method. For example, more elaborate methods, such as integrated project delivery outlined below, may not be appropriate for small projects; but larger, more complex projects may benefit greatly from such approaches.

Chapter 11 discusses project delivery methods in detail, however, a brief list of available methods is included here for reference:

- Design-bid-build (DBB) requires two separate contracts: owner-prime designer and owner-contractor. If the prime designer and contractor enjoy a solid working relationship, the owner's risk exposure is moderate; but if multiple conflicts arise between the prime designer and contractor (who do not have a contract with each other), the owner could be exposed to a high level of risk, including cost over-runs and schedule delays. This situation inevitably leads to a high level of risk for the prime designer also.
- Design-build (DB) requires only one contract from the owner's perspective: owner-design build entity. The design build entity can be a partnership or joint venture between the prime designer and builder. Alternatively, either the designer or builder can take the lead, putting the other in a subordinate position. The DB project delivery approach is meant to reduce the owner's risk and to shift that risk to the design build entity.

- Integrated project delivery (IPD) requires one contract: a multiparty agreement. In IPD, the owner, prime designer, builder, and possibly other significant project participants (a key subcontractor, for example) sign a single contract. In IPD risk is meant to be shared equally by all signatories to the contract.

- Construction management (CM) at risk requires two contracts: owner-prime designer and owner-CM. In this delivery method the construction manager sign contracts on the owner's behalf with the general contractor and key subcontractors. In CM at risk the CM is assuming some the owner's risk.

- Construction management (CM) as agent typically requires three contracts: owner-CM, owner-prime designer, and owner-contractor. In this project delivery method, the CM acts as the owner's agent and may manage both design and construction; however, the owner assumes the primary risk for design and construction.

PROJECT TEAMS

Project teams are composed of members vital to the success of the project and must include a leader to guide overall efforts. The team leader is the PM and the PM must rely on the team for technical expertise. The PM also acts as a coach, answers questions, clears the way for progress, and makes sure desired outcomes are understood by all team members.

The PM influences a diverse set of individuals with sometimes competing goals, needs, and perspectives. The PM needs to motivate these individuals since often the members may be assigned to the project from other departments within the firm. The PM often deals with multiple teams within the organization including administrative, budget office, health and safety, and various engineering disciplines. The design team organizations may include architects, subconsultant engineers, and CAD specialists. Construction teams have a different culture and may have a shorter-term perspective on the project.

The PM must possess team management skills, since these various subteams need to be an integral part of the project organization, referred to as the "project team." The project team must have a well-defined mission with goals and trust instilled by the PM. As part of the team management skill, the PM will likely conduct team building exercises, such as the project kick-off meeting, where they will stress that:

- All participants need to use effective communications and should confirm that information given and received is clearly understood

- All participants have a common customer

- Team success will allow continuity of project team

- Remembering key words "responsibility," "honesty," "kindness," "respect," and "communications" is critical

- The positive aspects of the project lead to success and perhaps future work for the same client

With regard to the project team, the PM likely will request or pick members in accordance with the client's requirements for the scope, budget, and schedule; acquire resources for the project; develop processes for decision-making; and develop a leadership style that is respected and accepted by the whole project team.

PROJECT INITIATION

Once an engineering business unit actually has won a project after completing the initial proposal effort, interviews, and contracting, the actual phases of work for CE projects begin, as depicted on Figure 7.3.

The phases of delivering the project include:

- Project definition, referred to as "predesign," which includes an approximate 10 percent design that shows the layout and footprint and possibly an elevation

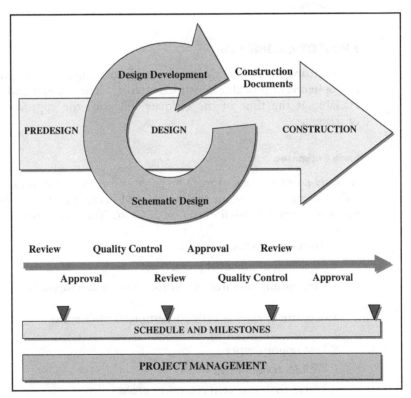

Figure 7.3 Delivering the project

- Design phase, which is composed of three phases referred to as schematic design, design development, and construction documents. The schematic design will be about 10 to 20 percent and include input from the owner, the PM, and the design engineer. The design further develops to 60, then 90 percent completion, sometimes referred to as a "check print," for the client's review and comments and possibly the builder/contractor for a constructability review
- Once the 100 percent design, which becomes part of the contract documents. The Construction Documents typically include the plans, owner's statement, a complete set of reviewed and stamped engineering plans, and a set of specifications including general specs, technical specs, contract requirements, a bid sheet, and a proposed construction schedule. With the exception of the plans, this information frequently is bound into a 'book' called the *Project Manual*.
- Occupancy, which refers to occupancy of the project site by the owner.
- Adaptive re-use and decommissioning, which are activities that can take place years after project completion, but which may be addressed in the current design.

PROJECT ESTIMATES

Project managers frequently are involved in creating a series of project cost estimates, or "opinions of probable construction cost." The amount and type of information available at the time of the estimate influence the approaches taken and levels of accuracy.

Early Estimates

An early estimate is an estimate prepared before completion of detailed design Early estimates are an important tool for the PM because these very preliminary cost estimates are often the basis for business decisions. These early estimates are also used for:

- Asset development strategies
- Screening of potential projects
- Committing resources for further project development

Inaccurate or careless early cost estimates can lead to:

- Lost opportunities
- Wasted development effort
- Lower-than-expected returns or profit

There are also classifications of estimates/re-estimates including:

- Sponsor's (client's) study (conceptual design)
- Preliminary engineering (preliminary design)
- Schematic engineering (Schematic Design)
- Detailed engineering (design development)
- Final design (construction documents)

There are several primary factors that should be considered when preparing and reviewing estimates, or comparing them to one another, including:

- Standardization of the process to allow for comparison of items
- Alignment of objectives between client and designer so there's an early and common understanding on the cost of a project
- Selection of appropriate methodology for building size, location, material selection, or other methodologies
- Collection of project data and historical costs
- Organization of estimate into desired formats because some funds will likely come from different sources or even organizations
- Documentation of basis costs and accuracy of the estimate are important qualifiers
- Review and checking to verify content and accuracy
- Feedback from project implementation for future efforts to perform continuous improvement.

There are benefits of alignment with cost estimates throughout the life of a project including:

- Establishing an understanding of the product or service received for the cost paid
- Determining the level of effort associated with the project and can establish a budgetary target for various components of the project
- Establishing work processes and a staffing plan
- Highlighting critical issues early in the project so alternatives can be arranged or high-cost elements can be assessed more closely
- Improving and documenting scope definition for the record as the project progresses
- Assisting the client's understanding of what is included in the estimate and what is not included

- Establishing responsibilities of parties involved and respective budget allocations and responsibilities
- Creating cohesiveness between project team and client in a way that builds trust and commitment

There are some critical questions that should be considered when preparing early estimates, such as:

1. What is the current level of definition for the scope of work? For example, is it enough to know there will be a 10,000-square-foot building or would an estimate be more accurate if the definition included more information such as two floors, one standard elevator and a freight elevator, two pairs of bathrooms, steel roof, masonry construction, two 400-square-foot kitchen units, and two 300-square-foot conference rooms?

2. What level of detail is the customer expecting, a five-line early estimate or a five-page estimate?

3. What resources are required for this deliverable, when is it due, and what is the appropriate format?

4. How will this early estimate be used and what decisions will be made from it?

There's one additional tool that can be very helpful for the engineer regarding the preparation of cost estimates. Cost estimating will likely involve several specialists on the project team and there will likely be a cost estimate kick-off meeting. Here's a checklist of issues for the estimate kick-off meeting:

1. What are the client's main issues, cost, quality, time, or other concerns?

2. What level of accuracy and what format is the client expecting?

3. When is the estimate due, when will the materials be purchased, and when will the work be performed?

4. What is the budget for the preparation of this estimate and how much effort is required?

5. Are there any customer-furnished items such as land, grading, security, utilities, materials, fencing, or other items?

6. Are there any special tax issues or credits, or funding requirements that could have an impact on the project cost?

7. Are there similar projects in the nearby vicinity that this project may be compared to?

8. What is the availability for labor and any specialty trades required for this project?

9. Are there any special permitting issues for this project that could affect the cost or schedule?

10. Does the client have any specific information sources the project team should contact or avoid?

This checklist is project specific, client and time dependent, and includes only the basic checklist items. However, the checklist will provide the engineer with a starting point for cost estimation discussions with the client.

Project Budget Estimates

Project budget estimates levels of accuracy significantly vary with the level of completion of design. As depicted in Figure 7.3, the project moves through the predesign phase, through schematic and design development, on to the construction documents, and to construction. The project becomes increasingly defined and refined. Each level of design generally has a client review component where there is an opportunity to conduct detailed discussions on the project thereby crystallizing the client's image and expectation in the engineers' and project team's minds. Therefore, it follows that if there's little or no design the "level of accuracy" for the budget estimate will be much higher than with more refined engineering design and project detail. The general level of accuracy for project budget estimates appears below:

- No design work: Accuracy ±50 percent
- Preliminary design: Accuracy ±30 percent
- Design development: Accuracy ±15–20 percent
- Construction documents: Accuracy ±5–10 percent

Of course an owner may have other techniques for estimating their project costs, especially if they perform those projects on a regular basis, such as with small restaurants, commercial buildings, or civil structures. These owner techniques are often evaluated by the following:

- Parametric techniques such as square footage, cubic footage, linear footage, or some other common comparative measure, or
- Historical costs based on past projects, especially if they are in the general vicinity and time frame.

From an engineer's point of view design budgets and the business unit's compensation for these projects can be estimated by several means including:

- Lump sum based upon similar projects or experience
- Salary cost multiplied by a fixed multiplier
- Cost plus a fixed fee (payment) for the service
- Cost plus a variable fee (incentive-based payment) for the service
- A fixed percentage of construction

From a contactor's point of view construction budgets and the contractor's compensation for these projects can be estimated by several means including:

- Fixed price if the contractor has a solid database to work from and a high level of confidence
- Cost reimbursable plus a fixed fee

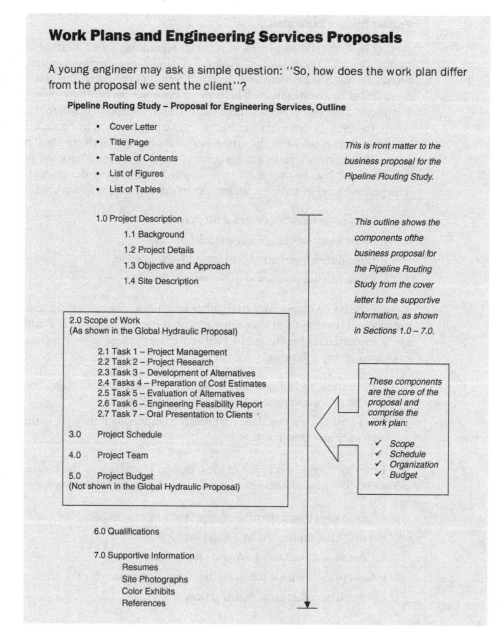

Work Plans and Engineering Services Proposals

A young engineer may ask a simple question: "So, how does the work plan differ from the proposal we sent the client"?

Pipeline Routing Study – Proposal for Engineering Services, Outline

- Cover Letter
- Title Page
- Table of Contents
- List of Figures
- List of Tables

This is front matter to the business proposal for the Pipeline Routing Study.

1.0 Project Description
    1.1 Background
    1.2 Project Details
    1.3 Objective and Approach
    1.4 Site Description

This outline shows the components of the business proposal for the Pipeline Routing Study from the cover letter to the supportive information, as shown in Sections 1.0 – 7.0.

2.0 Scope of Work
(As shown in the Global Hydraulic Proposal)

    2.1 Task 1 – Project Management
    2.2 Task 2 – Project Research
    2.3 Task 3 – Development of Alternatives
    2.4 Tasks 4 – Preparation of Cost Estimates
    2.5 Task 5 – Evaluation of Alternatives
    2.6 Task 6 – Engineering Feasibility Report
    2.7 Task 7 – Oral Presentation to Clients

3.0   Project Schedule

4.0   Project Team

5.0   Project Budget
(Not shown in the Global Hydraulic Proposal)

These components are the core of the proposal and comprise the work plan:

    ✓ *Scope*
    ✓ *Schedule*
    ✓ *Organization*
    ✓ *Budget*

6.0 Qualifications

7.0 Supportive Information
    Resumes
    Site Photographs
    Color Exhibits
    References

The core components of a proposal that actually discuss the work to be done, the organization and team to do it, the schedule of activities, and the budget to accomplish the effort become part of the work plan. The work plan is contained within the engineering firm's proposal to do the work. In essence, the initial work plan is the proposal minus the introduction and the qualifications and reference materials (marketing materials) to sell the work to the client. The marketing materials included in the proposal usually demonstrate that the proposer has completed similar projects successfully, has the qualified staff to accomplish the work—on time and within budget. So, the marketing material in the proposal has a very important role in helping the client(s) visualize a successfully completed project with a capable team to perform the work, while the proposal's schedule of activities, the team and team organization, schedule, and budget are the starting points for developing the work plan.

The PM begins preparing the work plan first as a "scope of work" within the proposal by analyzing previously gathered material the firm has been collecting since its business development personnel began following the proposed project. This preliminary work plan should reflect the scope of work outlined in the client's proposal. The PM will combine this information with all the background material prepared by the client. The PM will then:

- Become familiar with the owner's objectives and overall project needs
- Formulate an objective and approach
- Identify additional information required
- Organize the review process into three categories:
  - Scope
  - Budget
  - Schedule

The work plan can be strengthened by adding the owner's perspective and specifically adding the authorized representatives as an integral part of the project team. The owner's representatives serve two purposes:

1. To provide information and clarity to the project requirements. Optionally these representatives may actually review and approve all team decisions, and

2. To define quality in addition to scope, cost, and schedule. Optionally, if the owner does not have the expertise in-house to perform this function, they may hire an independent third party to act on their behalf. A construction manager can also accomplish this task for the owner.

If representatives are a part of the project team their detailed role and level of involvement needs to be articulated and well understood to avoid problems later.

PROJECT MANAGEMENT PLAN COMPONENTS

Plan Purpose

Once the project scope of work, schedule, and budget are defined, it's time to get to work. A very useful document to guide the efforts and provide necessary procedures for the project team is the Project Management Plan (PMP). The PMP is like a road-map that provides important and general information and project procedures to the project team and stakeholders. The PMP is generally distributed to all team members at the project kick-off meeting, discussed in detail and updated as necessary as the project progresses. The PMP should be a live document, updated on a regular basis and controlled by a revision number and date. PMP revisions should be an agenda item discussed at each project meeting.

PMP Components

- General project information:
  - Purpose
  - Objective
  - Approach
- Major client, stakeholder, and general information
  - General contract information
  - Contract type (prime contractor, subcontractor, design-build, or multiple prime)
  - General contact information including e-mail and phone
  - Preferred method of contact
- Organizational structure
  - Organization chart
  - Project manager or project engineer
  - Key Disciplines (technical disciplines such as civil engineering, planning, environmental sciences, civil design, and administrative services such as document production, contracting, accounts receivable, and so forth)
  - Key team members and leaders in each discipline area, as well as their roles within the project organization, especially as it relates to client and vendor or subcontractor interaction
- Scope of work summary, or work plan
  - Summarize key tasks and subtasks
  - Show relationships of tasks to key disciplines
  - Show subcontractors, other services, materials, and other resource requirements

- Project schedule
  - Present project schedule
  - Present flow chart of activities that relate disciplines to tasks and schedule, generally shown as an accompanying CPM or PERT chart
- Project budget
  - Present a summary spreadsheet displaying tasks, project team member names, summary of labor hours, subcontractors, materials, or other resource needs. Relate this information to the overall project budget.
- Communication plan
  - Establish project communication protocols for client, project team, stakeholder, and partner contact. Protocols should describe the preferred method of contact and frequency for the client, stakeholders, and key partners (written, teleconference, e-mail contacts, and frequency such as daily, weekly, or monthly).
  - Set up regular schedule and reserve dates and time slots for periodic work sessions and project review meetings. The importance of project progress meetings cannot be overstated. These meetings are vital to ensure that a cross-feed of project information is occurring so that problems can be identified and addressed, or even anticipated and eliminated before they become impediments to the performance of work related to the project.
  - Describe the project filing system, including hard copy and electronic file formats.
- Project health and safety plan
  - Minimum plan content is described below. Plan should be upgraded as required by complexity of the field tasks, capability of the staff, and overall risk assessment.
  - Plan should include information such as:
    - Task activities including soil borings with a drill rig, site recon, enclosed spaces, chemical exposure, and so forth.
    - Potential hazards such as heat stress, poisonous plants, slip, trip and falls, ladder use, and the like.
    - Routes of exposure such as inhalation, skin exposure, puncture wounds, and so on.
    - Suggested safety equipment such as safety shoes, hard hats, safety glasses, sunscreen, or appropriate clothing.
  - Site map and routes to emergency care
    - Name and phone number for police and emergency care
    - Map to first aid facility

- Suggested field staff partnering system (minimum of two staff members working together as partners if possible). For single staff activities it is recommended to have a phone call in procedure such as one phone call in the morning and one in the afternoon to report progress and safety status.
- Miscellaneous: First aid kit in vehicles, cell phone or radio availability, flashlights, flares, and so forth
- Health and safety plan to be reviewed and approved by senior engineering staff, certified safety professional (CSP), or certified industrial hygienist (CIH) professionals
- Project quality plan
  - Engineering tasks should be checked and verified by other qualified professionals in the discipline area.
  - Quality assurance (QA) personnel should be listed for each major discipline in the organizational chart mentioned above.
  - Quality assurance personnel should be involved from the initial proposal or scoping activities and throughout the project. QA personnel should not be expected to review and verify project information at the conclusion of the project.
  - Engineering documents that include conclusions and recommendations are generally required to be conducted by a registered engineer, stamped and dated by the registered professional. The recommended view should reflect requirements by the Board of Professional Engineers in the state in which the work is contracted.

In summary, the PMP should be a useful document often consulted by the PM, project team members, client, and other project stakeholders for project guidelines and procedures. This document will be a live document that should be discussed in the periodic project meeting and updated as necessary, tracked with a revision number and date so members can keep abreast of developments.

STAFF SELECTION GUIDELINES FOR THE PM

At some point in the project lifecycle the PM will likely be engaged in the process of staff selection for the project. Staff selection will usually involve numerous individuals, all with a vested interest, including the PM, the individual staff members, each discipline manager, the client or client's representative(s) among others. The experienced PM will often begin soliciting interest for staff before the project contract is signed or the kick-off meeting is conducted. Generally, the goal for the PM will be to select staff for the project based on criteria as illustrated in Figure 7.4. This criteria includes:

- Project and client needs
- Staff availability

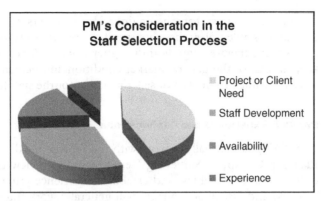

Figure 7.4   Staff selection process

- Previous experience and qualifications
- Staff development
- Project budgets and/or staff rates

Once the PM has completed the project management plan, the PM can begin the negotiation process for specific staff individuals.

Project and Client Needs

Of course one of the key elements in this decision will be the client's needs and the best choice for the project staff within each discipline that the project requires. For example, if the project has complex civil design features it would be a good choice to select very experienced civil designers rather than more junior staff. Conversely, if there were common drafting or AutoCAD requirements, then the PM might consider a staff engineer who was eager to learn new AutoCAD skills under the tutelage of another experienced engineer willing to mentor the junior engineer.

Staff Availability

The PM often encounters one important but basic question when negotiating for staff resources on a project: Which staff is "available" to work on a particular project? In the private sector, many companies often desire to operate with the minimum staff for the general market conditions to get the work done. In other words, the private sector generally operates a little light on staff with a higher work load per individual to maintain billability and cost competitiveness. Private sector managers usually like this scenario because they prefer to periodically place overtime requirements on their staff than the opposite condition where they might have to reduce work hours or reduce the number of staff on payroll. This general market condition can fluctuate over time and if long work hours accompany a high work load with corresponding higher profits, then these managers will consider hiring additional people.

In the public sector, the engineering managers generally find the workloads a little more stable unless there are specific conditions like new regulations or new programs that generate a significantly higher work load. So, in summary, the PM should keep abreast of the general market condition and demand for experienced staff and remain flexible in a buyer's or seller's market for the staff to conduct the project.

Previous Experience and Qualifications

The PM's goal is usually to work with the appropriate discipline manager (such as the Planning Group, or Surveying Department) to review the scope of work, task requirements/deliverables, and establish experience criteria for staffing the project. The PM and discipline manager will generally work together to prequalify the staff that have previous experience while also considering individual staff schedules or other commitments, staff career development goals, and/or cross-training between other clients or similar projects. For example, it might make sense to train a junior engineer on a particular project if they were particularly interested in working for a specific client.

Staff Development

Most public agencies and private engineering consulting firms perform annual appraisal and development evaluations on their staff. The first part of these evaluations is generally the appraisal portion that communicates the staff person's performance to the individual. The second portion of these evaluations generally includes the staff's development goals, indicating where the individual would like to direct their career. During these evaluation discussions, the staff individual generally expresses the type of work or experience they would like to obtain over the next year or two to develop their career. This process is a good one because it allows the discipline managers to learn the individual staff's desires while the PM learns where some staff might want specific experience.

Project Budgets and/or Staff Rates

A new PM quickly learns how to accomplish tasks and prepare deliverables for the client or finds alternate work. As part of this education, the PM gets exposed to different ways to accomplish the same goal. For example, the PM may have an option to place a very experienced engineer on a specific task and compare this level of effort and cost versus using a lesser experienced engineer with some senior mentoring and comparing this level of effort and overall cost to the project. There are some more specific examples of this situation under "Tracking Work/Work Breakdown Structure" in this chapter. The point here is that the PM can create alternate scenarios using a mix of staff to accomplish the same goals but come up with different cost impacts to the project. This particular skill can come in very handy for the experienced PM where this applied knowledge may allow the PM to cut costs, create development or coaching scenarios for eager staff, or increase profits for the firm or agency as needed.

In summary, the PM and discipline managers work together to discuss project assignments and corresponding deliverables, confirm staff interest and availability, discuss staff development goals and client requirements, and obtain firm commitments. If the project is delayed for some reason or the staff member is not available, then the PM and discipline managers generally will work together to find alternative solutions. Regular communication between the PMs, discipline managers, department managers, and staff members are good to keep one another updated on developments that may affect project commitments or staff availability.

THE PROJECT MANAGER'S RESPONSIBILITIES

The most important task of the PM is to maintain the Scope/Schedule/Budget triangular relationship depicted in Figure 7.2. The PM must follow the contract terms with the client (PMI 1996). Of course, the responsibilities do not stop there when there are so many interrelated tasks to manage to accomplish this main goal. These other tasks include:

- Maintaining ethical conduct, as discussed in Chapter 3
- Conducting problem recognition and solving, as discussed in Chapter 4
- Maintaining a high-quality product, as discussed in Chapter 4
- Maintaining the engineer's role in project development, as discussed in Chapter 5
- Producing engineering deliverables, as discussed in Chapter 6
- Managing the permitting requirements, as discussed in Chapter 8
- Maintaining the client relationship as discussed in Chapter 9
- Conducting oneself as a leader, as discussed in Chapter 10
- Conducting the project in accordance with the contractual and legal aspects, as discussed in Chapter 11
- Preparing and reviewing the invoices, as described in Chapter 12
- Managing the civil engineering enterprise, as discussed in Chapter 12
- Communicating as a professional, as discussed in Chapter 13
- Having a life, as discussed in Chapter 14

The PM's Time Commitment

The PM function can require an hour or two per week or can actually be managed by a project management office with numerous staff tracking the key components, depending upon the size and complexity of the project. If the project is relatively simple or has a small budget, the PM may actually perform the PM functions as well as all other the task functions. These simple projects may be something as routine as a Preliminary Site Investigative Report or a Spill Prevention, Containment and Contingency (SPCC) Plan, where the engineering budget hours may only require 10 to 100 engineering/technical support staff hours. These relatively small project budgets don't really allow adequate time or budgets for a distinct PM task.

For medium to large projects, the PM is likely just to provide PM functions, depending upon the client and the business unit that employs the engineer. PM services and corresponding budgets typically require from about 2 to 10 percent of the overall project budget, depending upon the client and the business unit. As the projects get more complex and the project budgets increase, the PM budgets will likely become a smaller proportion of the overall project budget.

Complex or large projects typically providing technical services greater than 2,000 engineering/technical support staff hours may be managed by an entire PM office. This office includes staff services such as project management and client services, accounts receivable/payable, quality control, legal support, health and safety, technical writing, project tracking and/or other specific project staff as necessary. Medium-sized engineering projects ranging from 100 to 2,000 engineering/technical support staff hours may require the same services; but these services may be provided by shared matrix-managed staff or staff capable of performing more than one function, such as accounts receivable/payable, technical writing, and project tracking.

The *project definition* is then completed by assessing the detailed technical services required to perform the project. For example, a water resources management project may require services like civil engineering, hydraulic modeling, geotechnical engineering, environmental permitting by environmental support staff, site restoration and biological services, landscape design services, technical writing, editing, mapping, geographic information system services, project tracking, accounts payable, contracting and legal services. These technical services comprise the core disciplines that the PM will integrate into the staff matrix.

The PM, usually in combination with the business unit's management, will then assess how the project will be conducted, the staff services that the business unit can provide, and any outside resources that may be required. The key point is that the PM, in conjunction with the business unit management staff, should evaluate the appropriate technical and client services for the project and then assess how these services can be best provided to fit the client's needs and budget. These aims sometimes are addressed in the organization breakdown structure (OBS).

This discussion leads us to the work breakdown structure and the business unit/organizational support.

Work Breakdown Structure

The *work breakdown structure* (WBS) defines the work to be accomplished and divides it into identifiable components that can be managed. The WBS does not yet define responsible parties (RPs) accountable for performing the work. But people are a necessary and important component of this process. After the WBS has been prepared, the next step is to identify the RPs, sometimes referred to as "resources" in the business unit/organization to perform the work. This component of the project is a critical one because the PM and/or the business unit's management team usually assess the overall schedule and availability of the staff and their capabilities to accomplish the tasks identified. For example, an engineering task might be accomplished by 100 staff support hours with senior engineering oversight of 10 to 20 support hours. Alternatively, this same task might be accomplished by about 70 senior engineering staff hours with 8 hours of senior engineering oversight. This evaluation becomes one that considers budget efficiency, staff training, coaching/mentoring, or even staff availability (more on this later).

One of the other input decisions for selecting the business unit's staff is to identify technical disciplines and individual staff capabilities responsible for the WBS functions. Selecting the individual staff members links the WBS to the organization and completes the *project framework*. Costs then can be linked to individual functions within the WBS through a cost breakdown structure (CBS).

Once the project framework is complete, the project schedule can be prepared. A tool useful to the PM who needs to develop a project schedule is the *critical path method* (CPM) (PMI, 1996). CPM was originally developed by the DuPont Company in combination with Remington Rand as consultants in the mid-1950s. A similar method, referred to as the *performance evaluation review technique* (PERT), was developed by the U.S. Navy in 1957. Both scheduling techniques are referred to as a network analysis that defines interrelationships of activities with corresponding scheduling of cost elements and resource availabilities. The CPM and PERT techniques require a detailed understanding of the identified tasks and their interrelationships to one another in a logical sequence (Anderson 2009).

After the PM generates the WBS, this network diagram can be linked to a schedule using either the CPM or PERT technique. This handbook will not go into the details of network diagramming since there are many references available on this subject already. However, the PM should understand the basics of these techniques. Their primary purpose is to develop a baseline for the project and measure project progress. Project scheduling techniques can be as simple as bar charts, Gantt charts, Microsoft project scheduling software, or much more involved software programs. In engineering consulting it is important to pick a scheduling software program that best fits the client's needs, the capability of the PM and the project team, the complexity of the project, and the project budget (Kerzner 2010).

In summary, the phases for the development of project plan are:

- Project Definition
- Project Framework, Organization Breakdown Structure
- Work Breakdown Structure
- Cost Breakdown Structure
- Project Scheduling (CPM and PERT)
- Project Tracking, Evaluation, and Control

Some Definitions

- OBS—Organization(al) breakdown structure: identifies organizational, rather than task-based, relationships
- WBS—Work breakdown structure: defines and groups a project's discrete work elements
- CBS—Cost breakdown structure: links to WBS and classifies costs into cost units, elements, and types
- CPM/PERT—Critical path method/performance evaluation review technique: links WBS, and possibly the CBS, to a schedule

Concepts Used in Project Progress Tracking

Project progress can be tracked using various indices. These tracking methods vary but basically compare the work completed in relation to the original budget and original schedule. Some definitions follow:

- Earned work-hours = (budgeted work-hours) × (percent complete)
- Percent complete = $\dfrac{\text{Sum, earned work-hours of tasks included}}{\text{Sum of budgeted work-hours of tasks included}}$
- Cost Performance Index (CPI)
  - CPI = $\dfrac{\text{Sum of earned work-hours of tasks included}}{\text{Sum of actual work-hours of tasks included}}$
- Only tasks for which budgets have been established are included
- Change order would need to be prepared for new work
- Schedule Performance Index (SPI)
  - SPI = $\dfrac{\text{Sum of earned work-hours to date}}{\text{Sum of scheduled work-hours to date}}$
  - Compares amount of work performed to the amount scheduled to a point in time
- If CPI or SPI is greater than 1.0, then:
  - Trend is favorable and can be "cumulative" or for a "defined period"
  - Trend can be plotted on a graph
  - Trend can be applied to design and construction
- If CPI or SPI is less than 1.0, then:
  - Trend is unfavorable, i.e., poor project performance

- Trend can be plotted on a graph
- Trend can be applied to design and construction
- Earned Value System

  BCWS = budgeted cost of work scheduled (Original Plan)

  ACWS = actual cost of work scheduled (Actual)

  BCWP = budgeted cost of work performed (Earned)

  ACWP = actual cost of work performed (Actual)

  CV = cost variance

  CV = Earned – Actual

  SV = schedule variance

  SV = Earned – Planned

  BAC = Budget at completion (Total cost for the project)

  EAC = Estimate at completion

  EAC = ACWP + estimated effort still needed to complete the work

  VAC = Variance at completion

  VAC = BAC – EAC

  - If a positive number, indicates an increased potential for profit or accelerated schedule; negative indicates potential cost overrun or schedule slippage.

Tracking Methods

The *percent complete* method is probably the most common method for typical civil engineering project tracking. This method simply compares the percentage of work completed and plots this figure on a graph and compares it to the percentage of time for the task to be completed (Wiest 1974). The *y*-axis portrays the percentage of work completed and the *x*-axis portrays the percentage of time for the task to be completed. Both axes should be scaled the same so that the point representing 25 percent complete versus 25 percent of time completed forms a 45 degree line bisecting the 90 degree angle of the *x*- and the *y*-axis. A typical graph representing project percentage completion is illustrated in Figure 7.5. This particular example is very typical and illustrates how project teams often start out rather slowly where the percentage of work completed lags behind the percentage of time on the schedule. (PMI, 1996) At some point the project team either gets a time extension or significantly ramps up construction to catch up to the percentage of time completion on the *x*-axis. These graphs can be created for each task or collectively by calculating the total project completion percentage data. One caution about this method of project tracking is a tendency for the

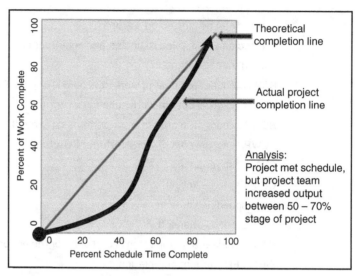

Figure 7.5 **Typical percent complete graph**

project team members to overestimate their percentage of work completed. A good cross-check on this estimate is to follow-up this question with:

"What is your estimate to complete the task?"

The answer to this query should provide a better indication of the remaining work to be completed for each task.

An alternate project tracking method is called the *CPI*, which calculates the sum of earned work-hours of tasks included divided by the sum of actual work-hours of tasks included. This index simply shows that if a project team member completes the assignment for less than, or equal to, the number of hours assigned the ratio will be just greater than or equal to 1.0. A ratio of 1.0 or greater indicates a positive performance and completion within the assigned task budget. If the ratio is less than 1.0, this indicates a negative performance and completion over the assigned task budget with a likely negative impact on the budget.

The *SPI* is a similar tracking method which calculates the sum of earned work-hours to date divided by the sum of scheduled work-hours to date. This index shows the amount of work-hours performed to date compared to the amount of work-hours scheduled to be completed to date. If a project team member completes more hours on the project assignment than the amount of scheduled work-hours to date, the ratio will be greater than or equal to 1.0. A ratio of 1.0 or greater indicates a positive performance and completion of the amount of hours scheduled or greater than the amount of work-hours scheduled. If the ratio is less than 1.0, this could indicate a negative performance and likely incomplete assignment. This answer will require a more definitive answer to how much work had actually been completed. It is possible that efficient team members could have completed the assignment in less than the amount of scheduled hours thereby surprising the PM with a positive budget impact.

Using the earned value system, a PM can calculate a cost variance or a schedule variance for the project. This method compares the budgeted cost of work scheduled (in the original plan) compared to the actual cost of work scheduled (actual cost). A cost variance is calculated by simply subtracting the actual cost from the earned cost. Of course the PM will have to be sure the task assignment is on schedule and compares these costs for the same stage of task completion for this calculation to be accurate. A schedule variance is calculated in much the same way, by subtracting the actual cost of work scheduled from the budgeted cost of work scheduled.

PROJECT RISK MANAGEMENT

Another way of looking at the project tracking methods discussed above is to consider them as tools used to analyze, anticipate, and address the risks inherent in any project undertaking. Project Risk Management, however, is a separate process unto itself, and is responsible for responding to both negative (adverse) and positive events in the lifecycle of a project. When defining risk from a health and safety perspective, it could be defined in terms of the possibility of suffering some type of harm or loss. Project risk, however, involves not just negative outcomes, but potentially positive outcomes. Opportunities may arise from how a project is conducted. (Winch 1997.)

The first step in managing risk is identifying it, usually by performing a cause and effect analysis of the phases of the project to evaluate the "what if" scenarios—both worst and best case—if project tasks are not performed according to plan. The way to obtain this information is often very prosaic—interview or ask the people performing the work what most worries them about what could go wrong within a given task or portion of the project.

The next step involves quantifying the risk, by using statistical tools or techniques to identify what risks have a fairly high likelihood or occurring, and which ones are not likely to ever happen at all. Once this list has been developed, additional quantification of the potential costs to the project, both to schedule and budget, must also be performed. The outcome from this quantification exercise should result in a short list of manageable opportunities (positive risk) and negative threats.

The approach to responding to these risks is straightforward—either avoid the risk, mitigate for it, or make the management decision to accept the consequences. How to respond to them is a bit more complex. In order to achieve the desired outcome for a given project risk, the project manager needs to work with his project team to develop contingency plans and alternative strategies or "workarounds" that directly address the risk. Upper management may become involved in the form of deciding to purchase insurance or bonds, if the risk of the project rises to such a level that it warrants it. Ideally, especially for large, complex projects that may have multiple and diverse risks entailed in the performance of the work, a risk management plan should be prepared that identifies, at least conceptually, the corrective actions that will be performed in the project to respond to a risk as it occurs. This may seem a bit akin to "crystal ball gazing," but thinking about the risks a given project may face prior to the actual performance of the work can go a long way toward heading off these threats before they ever can affect the project in the first place—it's referred to as "mitigation of risk."

DESIGN COORDINATION

For purposes of this discussion, the PM and the design team leader (DL) are two different functions and individuals. The DL will be a responsible member of the project team working for the PM. After the DL develops a work plan that includes the scope, schedule, budget, deliverables (reports, drawings, specs, and so forth) and a work breakdown structure, there are numerous other management functions that should occur including:

- Managing scope growth (creep): This is commonly referred to as "scope creep," which includes incremental changes that alter the original scope. The DL should have a process for controlling these generally small incremental changes. The DL has the control to record these changes and decide when these revisions actually alter the scope with a schedule impact or cost impact. This will be a delicate balance because the PM and the DL are trying to develop a relationship with the client while trying to hold his company's cost to the original effort estimated. The problem is that the project won't function effectively without controls for scope creep. In addition, the contract for the work will have legal implications if the schedule and scope of work aren't controlled.
- Project team meetings: Team meetings should occur at regular intervals throughout the project—a typical frequency is weekly, throughout the project.
- The DL usually sends out "Meeting Agendas" before the meeting and defines the interface of the various disciplines on the project. The DL or an alternate party should record meeting notes. (Sample forms for these meeting notes are included in Chapter 13, Figure 13.3 for reference. It is the DL's responsibility to assure that meetings stay on track, record key decisions, record action items, record responsible individuals to perform those actions with deadlines and finally stay productive.
- Weekly/monthly reports: Regular weekly or monthly reports may actually be part of the contract documents. These reports can act as a record of change as the design progresses through the project lifecycle.
- Distribution of documents: Document distribution must be timely and conducted with a sense of urgency. If document distribution is not conducted effectively, it will increase work load and bog down the project.
- Authority/responsibility checklist: Tracking action items and the responsible parties for accomplishing these actions is also the responsibility of the DL. It is also wise to track the tasks and the responsible individual identified for accomplishing these tasks. The DL may also use a checklist of duties for design.

Team Management

The DL is responsible for team management in three distinct areas:

- Within project team
- Between team and client
- Between team and engineering organization's management

Communication between and among the design team, other project team members, and client is critical to achieve a clear understanding of goals and objectives. These communication problems seem to become amplified as clients emphasize the need to stay on schedule. We understand "why" some clients want to keep on track. It's because of the rate of expenditure. Some clients believe in the "open window" concept, which is: The longer the window is open the more dollars fly out. Often there's no real deadline except for the client's desire to cut back on the rate of spending and move closer to construction and overall completion. But, even with effective communication, problems can arise. Some typical problems include:

1. Differing outlooks, perspectives, priorities
2. Fear of completing one job because of the unknown afterward
3. Scope creep by the owner
4. Lack of understanding or coordination among team members
5. Lack of confidence in the PM or DL
6. Unsure of path forward for some team members

To reduce these potential problems the DL or PM should continually emphasize commitment, clarity, and unity. This is where the leader's perception and people skills become very important.

Evaluation of Design Effectiveness

Evaluating the effectiveness of an overall design is difficult to quantify. The opinions of the effectiveness would likely be judgmental and subjective depending on whether the evaluator was an owner, user, regulator, or engineer. One way to quantify design effectiveness is through the number of design revisions or adherence to the original schedule or design budget. There are subcriteria that may also be evaluated such as:

- Constructability: The ease with which the contractor can build the project reflects, at least partially, the quality of the final construction documents. Constructability can be evaluated from the perspective of construction managers (CM) specialists, or contractors. Often, constructability is quantified based on the number of requests for information (RFIs) coming from the field and on the resultant change orders in a given project.

- Flexibility: This refers to the capability of using alternate materials that generally meet the same specification and criteria but possibly offer a more attractive price or availability.

- Labor skill/availability: The largest component of most construction jobs is labor. Elaborate or intricate design concepts will likely require a higher-skilled labor force transcending into a much higher cost to the owner. The designers should consider this component in the overall design process to keep the overall construction price within expected ranges.

- Substitution: Experienced designers likely have experience with substituting materials due to temporary shortages or price spikes. Importing steel or decorative tile from a foreign country may have unexpected impacts on the schedule or project budget. Performing a design that can easily handle substitutions can provide the owner with potential negotiation power when the time comes to place the order for these special materials.

In summary, evaluating the effectiveness of a design is judgmental. But creating a design that includes criteria such as constructability, flexibility, skill, and compensation rates for labor and substitution will provide the owner with a degree of comfort like a hidden insurance policy.

From Design Engineer to Project Manager

Bright and motivated engineers ultimately are faced with a dilemma: continue on their path of growing technical expertise, or jump to the professional fast track of business development and project management. For many young engineers, the concept of a project manager is quite nebulous. Business and management classes are not typically part of the engineering curriculum (which is unfortunate), and project management involves skills that are not common to the engineering practitioner. The transition from design engineer to project manager can be challenging.

So what exactly does an engineering project manager do? Simply put, the project manager is responsible for everything. That by no means implies that the managers do everything, or even that they are an expert at everything. It does mean that when anything goes wrong, the responsibility lies with the manager. Responsibility for "everything" includes: business development (one has to get the project in the first place), project delivery (on time, within budget, and with the satisfaction of all project goals), internal staff performance, consultant and subconsultant performance, human resources (staffing, teamwork, dispute resolutions, hiring, and firing), client success (the ultimate measure of project delivery), and time management. Because of the breadth of these responsibilities, project managers are given a tremendous amount of "rope" within most firms. The trick, of course, is to avoid hanging one's self.

Project Management Skills

A manager must have a team, and the project team is often quite diverse, even for relatively small projects. An engineering project will often require specialists in multiple branches of engineering as well as planners and architects. Beyond the immediate engineering challenges that must be resolved, a project manager must deal with business issues, political issues, interpersonal issues, and public relations—all requiring skills that engineers are less than famous for.

Formal engineering training includes a study of mathematics, the physical sciences, computer programming, specific engineering fields, and general education. Often, the only opportunity for early formal training in the soft skills associated with project management comes with "general education." Candidly, engineers tend to avoid coursework that involves the intensive practice of written and oral communication, business, and accounting. Postgraduate training of engineers in these fields is typically not encouraged by employers or embraced by employees. Engineering project management, therefore, requires that an individual be enthusiastic about deriving personal and professional growth from less formal types of education.

The mentality of an engineer, although useful for the design of a bridge or a machine, is not immediately conducive to project management. Engineers think in absolutes. A system either meets the design criteria or it does not. They take pride in their work and are meticulous about it. Engineers are paid to provide the best possible solution, but often interpret this as a charge to provide the one correct solution. They are not paid to be flexible (allowing a building to collapse "only a little bit" is not an option). Politics and bureaucracy are anathema to an efficient engineering project. Unfortunately, all of these tendencies are counterproductive to effective project management. Flexibility is, of course, critical. There are always better ways of doing things, and the need for a timely project delivery often outstrips the need for a perfect project. Moreover, the engineer is not the only important project stakeholder. Politics and bureaucracy are unavoidable because engineering projects often involve large amounts of public money. Effective communication with nonengineers becomes critical.

Project management requires skills in business development, leadership, and persuasion, among others. Business development means marketing. Successful business development comes from knowledgeable project managers that are familiar with their clients' needs, wants, and constraints. Marketing is actually quite straightforward if one's past project performance is excellent. A successful manager needs to inspire confidence and project competence. All commitments should be met without excuses, and solutions should be presented to a client, never problems. If one can honestly stand behind a track record that demonstrates these qualities, then repeat business will always follow.

(Continued)

The leadership needed to be a project manager does not come from a promotion and it is not granted from above. It comes from leading by example, not by pushing from behind. Leadership is manifested in project management through interactions with the project team, flexibility, discipline, and advocacy for the team. A leader must be both flexible, for nothing ever goes exactly as planned, and, at the same time, disciplined. Certain standards, such as the project schedule, must be set and maintained, or a project will not succeed. Finally, leadership means advocating for the needs of the project team. The client is not always right. Ignoring that fact can mean alienating the people who have the ability to keep the client well satisfied in the long term.

More than anything, project management is about persuasion. A mentor frequently reminds me that his job is to get people to do what they really did not want to do. The need to be persuasive is far reaching: We must persuade our project team to get the job done correctly and on time; we must persuade company principals to provide needed resources; we must persuade our clients that our solutions are effective; and we must persuade agencies and companies to approve our clients' projects. This requires excellent communication skills as well as leadership and mutual trust.

Training

Formal training for engineering project management can be hard to come by. Large firms and agencies sometimes have established internal programs, but they often amount to a stack of manuals and a few seminars. Most successful project managers learn through the school of hard knocks. A few schedules and budgets are blown, lucrative contracts are lost, and fellow employees are alienated. If an individual can avoid making major mistakes more than once, they succeed. If not, they go back to designing, but often at a new firm.

From Design Engineer to Project Manager—Continued

Assuming that one did not have the foresight to take extra business and communication courses in college (and most of us did not), other training resources must be utilized. First, one should obtain the appropriate tools. Project management software ranges from terrific to terrible. A project manager needs tools that provide all the relevant project and staffing data, without overwhelming the user with irrelevant information. Proper support from accounting personnel is important. Second, there is no replacement for voracious reading and self-study. Everyone has an opinion about how to manage, and many have written books on the topic (including, or course, the author of this text). At best, one can glean some useful ideas from each of these opinions. Project management style needs to vary with the individual project manager, the particular engineering field, and the particular firm involved, and therefore must be carefully crafted by each individual. There are no design manuals for how to manufacture a project manager. Finally, a project manager must learn to stop talking and listen. Watch carefully at your next big project meeting—the most impressive person in the room will likely sit quietly through most of the meeting. Take note when they do talk, as it will probably be in regard to something of substance.

Is It Worth It?

Project management can be rewarding, but the profession offers its own set of challenges. For example, from a client's perspective, projects are either great or terrible, and they are not always great. The responsibility for a failed project lies entirely with the project manager. Economies fluctuate, and that means that a manager must occasionally fire and layoff coworkers and sometimes friends. Because of these responsibilities, a good project manager is respected, but a certain emotional distance is necessarily maintained from most coworkers. Project managers (and companies, for that matter) that forget that they are working in a business cease to exist.

For some ambitious engineers, this discussion is discouraging, for project management involves a lot of skills that require additional development and training and are outside the typical engineer's comfort zone. Despite the challenges, a career in management is worth considering because it means career and personal advancement. One becomes responsible for business development, and truly has a major stake in the project ownership. Project managers are typically involved with a project from its inception through its completed construction. This long-term personal stake in a job generates tremendous professional satisfaction. Some engineers even enjoy practicing the interpersonal skills that are required. Finally, great satisfaction is derived from doing a job that most of one's peers cannot do.

What Now?

For those passionate about leading their profession, engineering project management offers a unique opportunity. Ready to make the leap? The first step in preparing yourself is to broaden your skills. In particular:

- Practice writing and public speaking
- Take a business class and read everything you can find
- Take on a leadership role somewhere (it doesn't have to involve engineering)
- Broaden your engineering resume

> • Most importantly, spend some time working in construction or manufacturing. Learn to build what you are designing.
>
> —Matt Salverson, Ph.D., PE, Dokken Engineering and California State University, Sacramento

SUMMARY

Effective project management skills are essential for conducting efficient project operations. The project manager is at the heart of the project much like a captain of a ship is often on the bridge. The PM must maintain close communication with the Client ands also be aware of all activities related to the initiation, planning, execution, monitoring and control, and closure of the project. Infusing the project team and task efforts with the Client's vision of the final deliverable product requires skill and experience but will help assure a final quality product delivered on time and within budget. A bonus award for the PM, the project team, and the organization would be the Client's next job, a letter of appreciation, and/or a good recommendation based on a project well done.

REFERENCES

Anderson, David R. et. al (2009) *Quantitative Methods for Business.* South-Western Cengage Learning. Mason, OH. ISBN: 13:978-0-324-65181-2

Hansen, Karen Lee. (1998). "Integration of Methods for the Effective Administration of Processes in the Construction Industry." Presented at Expo Construccion '98. February 1998, Mexico City. Vanir Construction Management, San Jose, California.

Hobday, Michael. (1998). *Product Complexity, Innovation and Industrial Organisation: CoPS Working Paper.* June 1998. CoPS Publication No 52.

Kerzner, Harold. (1997). *Project Management: A Systems Approach to Planning, Scheduling, and Controlling.* John Wiley & Sons, New York.

Kerzner, Harold. (1998). *Project Management: A Systems Approach to Planning, Scheduling, and Controlling*, p. 73.

Kerzner, Harold. (2010). *Project Management Best Practices: Achieving Global Excellence.* (The IIL/Wiley Series in Project Management). John Wiley and Sons, Inc. Hoboken, NJ. ISBN: 978-0-470-52829-7

Morris, Peter W. G. (1994). *The Management of Projects.* Thomas Telford, London.

Render, Barry and Stair, Jr., RalphM. (1982). *Quantitative Analysis for Management.* Allyn & Bacon Inc., Massachusetts.

Shenhar, Aaron J., and Dov Dvir. (1996). "Toward a typological theory of project management," *Research Policy*, Vol. 25, Issue 4.

The PMI Standards Committee. (1996). *A Guide to the Project Management Body of Knowledge (PMBOK)*. Project Management Institute.

The PMI Standards Committee. (1996). *The Global Status of the Project Management Profession*. Project Management Institute.

Wiest, Jerome D., and Ferdinand K. Levy. (1974). *A Management Guide to PERT/ CPM*, Prentice-Hall of India Private Limited, New Delhi.

Winch, Graham. (1997). "Thirty Years of Project Management: What Have We Learned?" Presented at British Academy of Management, Aston, 1996 (revised March 1997). Bartlett School of Graduate Studies, University College London.

Introduction
Ethics
Professional Engagement
1 2 3 4 5
History
What Engineers Deliver
Emerging Technology
6 7 8 9 10 11 12 13 14 15 16 17 A B C D E F
Engineer's Role in Project
Development
Permitting
Leadership Managing
Having
a Life
Executing a Professional
Commission
Client Relationship
Legal Aspects
Communicating
Globalization
Sustainability

Chapter **8**

Permitting

Big Idea

Permitting is a "key element" in the design process and is part of the "basis of design" for engineering projects. A civil or environmental engineering construction project could have potentially tens or hundreds of applicable federal, state, and local regulations or Presidential Executive Orders. Civil engineers should recognize and understand these regulations and incorporate the essential elements of compliance into the project statement and scope of work.

What good is a house, if you haven't got a decent planet to put it on?

—Henry David Thoreau

Key Topics Covered

- Introduction
- Accept Requirements for Permits
- Respect the Staff Implementing the Permits
- Initiate the Permitting Process Early
- Managing Permits
- Streamlining Permits
- Sample Permit Table
- Summary

Related Chapters in This Book

- Chapter 3: Ethics
- Chapter 5: The Engineer's Role in Project Development
- Chapter 11: Legal Aspects of Professional Practice
- Chapter 13: Communicating as a Professional

(*Continued*)

Related to *ASCE Body of Knowledge 2* Outcomes

ASCE BOK2 outcomes covered in this chapter

Foundational	Technical	Professional
1. Mathematics	5. Materials science	16. Communication
2. Natural sciences	6. Mechanics	● 17. Public policy
3. Humanities	7. Experiments	18. Business and public administration
4. Social sciences	8. Problem recognition and solving	19. Globalization
	9. Design	20. Leadership
	10. Sustainability	21. Teamwork
	11. Contemporary issues & historical perspectives	22. Attitudes
	12. Risk and uncertainty	23. Lifelong learning
	13. Project management	24. Professional and ethical responsibility
	14. Breadth in civil engineering areas	
	15. Technical specialization	

INTRODUCTION

Project permitting hardly was addressed in previous engineering handbooks, much less made a chapter title. But ask any engineer today what their major headaches are and environmental permitting will be in the top five. There is no more frustrating problem, and one for which current handbooks offer so little practical advice. Avoiding frustration and delay associated with project permitting requires three important tools:

1. An attitude that accepts the requirement for permits and the expertise and opinions of the agencies and staff that are trained to implement them.
2. An approach that includes project permitting as an important early step in the project schedule.
3. A plan to streamline permits by preparing them in a comprehensive and consolidated manner.

Engineers traditionally were trained to believe that they held all responsibility and authority for a project's implementation and success. In practice, engineers experience confusion, conflict, and frustration when confronted with the need to obtain environmental permits. The conflict arises because: first; a regulatory agency (not the engineer) has authority over the project approval and implementation; second, the regulatory agency (not the engineer) can control (and substantially delay) the project schedule; third, agency staff are often from a land use, cultural resources or biological background and may seem to have personality and communication methods that contrast with those of engineers. The engineer may perceive a regulator to be less technically competent.

The common reaction is for the engineers to try to deny the permitting authority and proceed forcefully. Some will blame the agency for delaying the schedule and endangering the project. Some will rail against the subjective nature of the permit requirements and complain how difficult the agency staff can be. Some will spend time attempting to rationalize the directions of the regulatory agency. None of these responses will lead to a successful project, a successful relationship, or reduce the stress and anxiety of an engineering career.

A more successful approach to permitting includes the following steps:

- Accept the requirement for permits and actively seek out what may be required
- Respect the agency staff implementing the permits
- Identify the permitting requirements and initiate the permitting processing early in the project—at least 12–18 months in advance of implementation for a simple project.

ACCEPT THE REQUIREMENTS FOR PERMITS

Environmental permitting prior to 1965 was practically nonexistent. In the latter 1960s the public became increasingly concerned with environmental pollution and enacted a series of laws to reduce further degradation and restore previous conditions. Most of the current environmental permitting requirements stem from the following foundation laws:

- The National Historic Preservation Act 1966
- The National Environmental Policy Act 1969
- The California Environmental Quality Act 1970
- The Clean Air Act 1970
- The Clean Water Act 1972
- Federal Endangered Species Act of 1973

From these foundations sprung rules and regulations designed to codify and detail necessary compliance at both the state and federal level. Different levels of government may have different permitting requirements, some of which may be similar, but many of which will be stricter than the underlying federal law. The laws describe the public health and trust; the permit process is the mechanism used to implement these definitions of public welfare and environmental protection.

It is important to acknowledge that these laws were created by a process that included input by the people of the United States to protect the public trust. If the project engineer believes the permit requirements are unnecessary, or ignorant, then they should work to revise or change the law. Engineers, clients, and the public should not forget that the laws were enacted to represent the will of the majority in an effort to preserve our heritage, history, environmental quality, endangered species, and quality of life. These concepts are similar to the engineer's ethics as presented in Chapter 3, Ethics and Chapter 11, Legal Aspects of Professional Practice.

In reality, permitting requirements are a significant project input for the design process. The permitting conditions and requirements are as important as other design requirements because the project may not be built until these requirements are fulfilled. The challenge for the engineer is to identify the permit requirements and schedules, and incorporate this information into the data collection, requirements, and design basis for the project to keep it on schedule and within budget. This information is also presented in more detail in Chapter 5, The Engineer's Role in Project Development.

RESPECT THE STAFF IMPLEMENTING THE PERMITS

Engineers require years of college education, an advanced degree, and continuing education to gain a professional license. The engineering curriculum includes advanced classes in calculus, physics, statics, chemistry, and more. Engineers deal in facts. Engineers are paid more than planners, cultural resources specialists, and biologists. Therefore, engineers are smarter, right? This attitude becomes a dangerous stumbling block for many projects.

Generally, the permitting staff has as many years of advanced education, continuing education, and professional licenses as engineers. In many cases they were required to take the same advanced classes in calculus, physics, and chemistry as engineers, in addition to training in planning principles, application of the law, economics, and principles of theoretical ecology. In other words, agency staffs have the same level of education, training, and experience in their profession as an engineer does in theirs. Therefore, engineers should show the same respect to nonengineering professionals as they would an engineering colleague.

One area to show constructive respect is to listen and read carefully the agency requirements. Understanding the underlying purpose of the permit or agency can be a positive way to show the regulator that you have "done your homework." Be cautious in interpreting or reinterpreting the requirements. Permit requirements may seem subjective, rather than factual, which puts the engineering mindset at an immediate disadvantage. Agency staffs generally have the authority and responsibility to interpret the rules, and it is wise to listen to them carefully.

Interacting with the agency staff is a key area to employ collaboration techniques to establish a win-win scenario. Distrust between project proponents and engineers and the regulatory staff is common for many of the reasons described above. The tendency is for engineers to minimize their interaction with the regulatory staff. On the contrary, using every opportunity to increase time with the agency staff will generally increase trust and improve the permitting relationship. This is one area where social engineering is as important as technical skill in reducing potential schedule and cost impacts. More information on this topic is presented in Chapter 13, Communicating as a Professional Engineer.

INITIATE THE PERMITTING PROCESSING EARLY

Building permits are generally a "back end" process. Once the project design and schedule are complete, applications are filed for building authorization in a process that extends less than 30 days. Environmental permits, by contrast, may require 12 to 18 months to apply for and obtain approval, *and likely will require changes in the project design or schedule*. Permits may require field investigations or surveys over multiple years before permit applications can be prepared. The project engineer should anticipate that a complex civil or environmental engineering construction project may need to comply with tens or hundreds of applicable federal, state, and local regulations or Presidential Executive Orders. The engineer needs to recognize and understand these regulations, incorporate the essential elements of compliance into the project statement, scope, and schedule of work. Some examples of permits that require extended preapplication times follow:

- *The Endangered Species Act* is intended to prevent the extinction of species by the actions of the federal government. The Act prohibits the "take" of endangered species. "Take" is defined very broadly to include both direct mortality and modification of the habitat supporting the species. All species indentified and listed are covered. These can include widely known species like eagles and wolves, and also not so well-known species such as the elderberry beetle, the San Francisco garter snake and the desert tortoise. The habitat features that support endangered species can be a patch of sand, a rainwater pond that fills once in ten years, or some plant community like coastal scrub. In order to determine that a project will not endanger a species, the applicant needs to determine if the species is present, if the habitat characteristics that support the species are present, and in some cases survey the locations and health of populations that are outside the project area. For species that are only present during a short period of the year or perhaps only every other year, surveys nearly always will require at least two years of field work. An example of this is vernal pool fairy shrimp that occurs in temporary rainwater pools every two to three years. Other fairy shrimp species occur across the United States and more are likely to be listed as endangered. For projects that occur where fairy shrimp may be present, one to two years of survey work and application preparation, followed by at least 135 days of consultation by agency will be necessary.

- *The Historic Resources Preservation Act* requires that projects not damage or disturb historic resources. Historic resources can include bone, stone, burial grounds, artifacts, or other evidence of previous human inhabitants. Historic resources can now also include historic structures as little as 50 years old. Another more recent extension of the law includes prehistoric resources, such as fossils and fossil evidence. Cultural and prehistoric resources evaluations follow a sequential path of reviewing existing records and literature, performing reconnaissance-level surveys, geomagnetic surveys or potholing, and at the most extreme, excavation, recovery and curation of resources prior to developing the project. A literature search and reconnaissance survey can take as little as four weeks, but excavations, recovery, and curation can require years to complete.

- *The Clean Water Act* requires that projects not degrade surface waters. Under Section 404 of the CWA, regions identified as "waters" and "wetlands" are also protected. A formal determination of the location and extent of wetlands is required before allowing project impacts. Some wetlands determinations are complex and may require multiple wet or flowering seasons to complete. Section 401 prohibits discharges or changes to water courses that can degrade water.

- *The National Environmental Policy Act (NEPA)* requires that federal agencies evaluate the potential environmental impacts of a project before implementing, approving, or funding them. Any project that uses, for example, **Federal Highway Administration (FHWA)** or **U.S. Army Corps of Engineers** funding will require an Exemption, an Environmental Assessment, or Environmental Impact Statement (EIS) to evaluate the potential impacts of the project and propose methods to avoid, minimize, or compensate for them. In some cases, impacts may be unavoidable, but necessary to achieve a more important objective. NEPA documents generally evaluate 13 different potential impact areas including air quality, cultural resources, biological impacts, water quality, land use, noise, and socioeconomic impacts. An EIS can require one to two years to prepare and publish. An EIS is made available for public comment and if impacts are identified that were not evaluated in the EIR, the process may need to be repeated one or more times. The final decision based on the NEPA document can be legally challenged on the basis of incomplete information, unsupported conclusions, or a high degree of uncertainty on the analysis, extending the time needed for approval. For further information it is suggested that the reader read *A Citizen's Guide to the NEPA, Having your Voice Heard*. It provides an excellent summary and may be found by researching "NEPA" and the Council of Environmental Quality or at the following web address: http://ceq.hss.doe.gov/nepa/Citizens_Guide_Dec07.pdf

- *California Environmental Quality Act (CEQA)* is California's equivalent of NEPA and requires evaluation for projects authorized, permitted, or funded by state agencies be analyzed for environmental impacts. The CEQA document is an Environmental Impact Report (EIR). In some cases where both federal and state agencies are expected to authorize the project, a combined EIR/EIS may be prepared. Some agencies such as the California Energy Commission may prepare a "CEQA-equivalent" document instead of a CEQA document. The necessary time and cost of a CEQA document are similar to the NEPA document and can be likewise challenged by the public.

- Many federal or state agency permits may specify that a CEQA or NEPA document be completed *before* permit authorization is effective. For example, a NEPA document is required before the United States Army Corps of Engineers can consult with the United States Fish and Wildlife Service to extend Section 7 authorization for a state or privately led project. The California Department of Fish and Game requires that a CEQA document be completed before authorizing work under a Streambed Alteration Agreement.

Identifying Permits with the U.S. Environmental Protection Agency

The EPA can be a valuable resource for the engineer to identify environmental permits required for a proposed engineering project. The EPA's foundation is protecting public health and the environment. Most federal laws proposed and accepted by Congress and signed into law by the President do not have sufficient

(*Continued*)

detail for immediate implementation. Congress authorizes the EPA to write the regulations that explain the details necessary to implement the environmental laws. Also, some laws come into effect as Presidential Executive Orders (EO). These EOs can establish additional conditions or performance standards, or influence the regulatory schedule.

For example, EO 12898 (signed by President William J. Clinton in 1994) is titled "Federal Actions to Address Environmental Justice in Minority Populations and Low-Income Populations." Its goal is to focus federal attention on the environmental and human health effects of federal actions on minority and low-income populations with the goal of achieving environmental protection for all communities. The EO directs each agency to develop a strategy for implementing environmental justice to avoid disproportionately high and adverse human health and/or environmental effects on minority and low-income populations. The Order is also intended to promote nondiscrimination in federal programs that affect human health and the environment, as well as provide access to public information and public participation. The Major Laws and Executive Orders promulgated by the U.S. EPA are presented below:

U.S. EPA website, 2009 www.epa.gov/lawsregs/laws/index.html

Some Major Laws and EOs that Should Be Considered

- Atomic Energy Act (AEA)
- Clean Air Act (CAA)
- Clean Water Act (CWA)—(Originally The Federal Water Pollution Control Amendments of 1972)
- Comprehensive Environmental Response, Compensation and Liability Act (CERCLA, or Superfund)
- Emergency Planning and Community Right-to-Know Act (EPCRA)
- Endangered Species Act (ESA)
- Energy Policy Act
- EO 12898: Federal Actions to Address Environmental Justice in Minority Populations and Low-Income Populations
- EO 13045: Protection of Children from Environmental Health Risks and Safety Risks

- EO 13211: Actions Concerning Regulations That Significantly Affect Energy Supply, Distribution, or Use
- Federal Food, Drug, and Cosmetic Act (FFDCA)
- Federal Insecticide, Fungicide, and Rodenticide Act (FIFRA)
- Federal Water Pollution Control Amendments—see Clean Water Act
- Marine Protection, Research, and Sanctuaries Act (MPRSA, also known as the Ocean Dumping Act)
- National Environmental Policy Act (NEPA)
- National Technology Transfer and Advancement Act (NTTAA)
- Nuclear Waste Policy Act
- Occupational Safety and Health (OSHA)
- Ocean Dumping Act—see Marine Protection, Research, and Sanctuaries Act
- Oil Pollution Act (OPA)
- Pollution Prevention Act (PPA)
- Resource Conservation and Recovery Act (RCRA)
- Safe Drinking Water Act (SDWA)
- Superfund—see Comprehensive Environmental Response, Compensation and Liability Act
- Superfund Amendments and Reauthorization Act (SARA)—see Comprehensive Environmental Response, Compensation and Liability Act
- Toxic Substances Control Act (TSCA)

Laws and EOs that Influence Environmental Protection

Another way for the engineer to cross-check the environmental laws and EOs that might apply to a specific engineering project are to check the EPA list by business sector. In this case the engineer would simply look up the specific industry that is considering engineering services.

Most business sectors are affected by a number of major environmental statutes and regulations. EPA's website offers a regulation's text, history, statutory

(Continued)

authority, supporting analyses, compliance information, or related guidance for numerous industry sectors. The nature and scope of activities vary across facilities in any single sector. The information provided on the EPA website will not necessarily apply to all facilities within that specific sector. The key industry sectors are:

REGULATORY INFORMATION BY BUSINESS SECTOR

- Aerospace
- Agriculture
- Automotive
- Chemicals
- Computers/Electronics
- Construction
- Dry Cleaning
- Electronics/Computers
- Energy
- Federal Facilities
- Fishing
- Food Processing
- Forest Industry
- Furniture
- Garment/Textiles
- Healthcare
- Leather Tanning and Finishing
- Local Governments/Municipalities

- Lumber/Pulp/Paper
- Metals
- Mineral/Mining/Processing
- Paper/Pulp/Lumber
- Pesticides
- Petroleum
- Pharmaceuticals
- Plastics/Rubber
- Power Generation
- Printing
- Pulp/Paper/Lumber
- Retail
- Rubber/Plastics
- Shipping, Shipbuilding, and Repair
- Solid Waste
- Textiles/Garment
- Transportation
- Tribes

Identification of Applicable Permits

The successful engineer will assign experienced project staff or permitting experts when identifying applicable permits for a specific task effort or project. This permitting function has become a specialist's job and the engineer should not be tempted to simply consult the Internet for simple answers where the employer's or client's best interest, legal liability, and funds are at stake. For example, would it be responsible for an inexperienced, energetic individual to design a bridge by using the Internet?

MANAGING PERMITS

Recognizing that environmental permits are not a "back-end" process, a successful engineer will list project permitting as a first order task, beginning with proposal or contract review. Following are some suggestions for specific considerations during the proposal, contract review, construction, and postconstruction stages of the project.

- During Proposal or Contract Review—review carefully:
  - Any reference to client or project-required environmental policies or regulations
  - Any requirement to appoint an individual with specific responsibilities or qualifications (Environmental Manager) who will be responsible to implement and monitor environmental compliance
  - Any limitations or approvals needed for specific materials. For example, it is increasingly common in California for agencies to specify that only ultra-low sulfur fuel vehicles be used and that machine idling times be limited. Agencies increasingly specify that no pesticides or herbicides be used without prior agency approval. The potential consequences of these limitations on cost or schedule need to be identified in the proposal and cost estimate.
- Upon Project Implementation:
  - Consider appointing a dedicated Environmental Permitting Manager from the kick-off of the project.
  - Make Environmental Permit Review and Compliance a standing part of project meetings.
  - Incorporate your corporate environmental policy into the requirements of the project.
  - Monitor and record implementation of environmental policies and procedures that show compliance with permits. For example, important data to show permit compliance include separating waste streams and implementing recycling programs, minimizing waste streams through recycling, documenting regular inspections, worker awareness trainings, ride-sharing programs, and similar initiatives.
  - Remain aware of materials, services, and subcontractors that provide environmental benefits for consideration in permits. While not explicitly required in many contracts, such measures are part of many permits, and can be a crucial discriminator or "value added" feature of an engineering or construction agency.
  - Remain aware of and train workers so that project participants know they are representatives of the client and the project and that behavior and environmental actions both on and off the job site are important to project success.

- Invite the agency staff to inspect the site or review project procedures. In fact, this can be the engineer's first step in a collaborative process. It may also seem like inviting enforcement, but in the event of an enforcement action, the judge will generally look with favor on a project proponent that shows a willingness and desire to comply. Any fines or enforcement actions are likely to be waived or reduced as a result. As a practical matter, agency staff rarely has the time to perform additional reviews or site inspections. This is another opportunity to "reach out" to your agency counterpart to share the information on the current condition of the project. These little actions will build rapport and the reputation for the engineer as a team player in this complicated process.

The Collaborative Process

Inviting the agency staff to inspect the site or review the draft work plan or project procedures can be the engineer's first step toward working with the agency in a collaborative process. Engineers should consider every interaction with agency staff an opportunity to establish a closer working relationship with the agency. Building collaborative relationships with regulatory agencies will benefit the client and the engineer!

Communicating with Permitting Agencies

Nearly all permits require a final report, and some require periodic monitoring and reporting for one, five, or ten years. Communicate with the regulating agencies on the progress of these requirements and submit these reports in a timely manner. If there is an established format, the engineer should consider asking the agency staff person if they would like to view a working draft. If there is no established outline or format, then the engineer should consider submitting a draft outline for agency review and comment.

- Nearly all permits require a letter of contract completion or closure to be filed.
- The project engineer should seek a letter of concurrence or other evidence that all project permit requirements have been completed.

Table 8.1 titled, "How Permitting Is Integrated with Project Phases," describes permitting activities for large civil engineering projects. This summary applies for all phases of the project from initiation and proposal efforts through project implementation, construction, completion, and closeout. The general activities are described for each major phase of the project accompanied by the general timeframe and deliverable product from each phase of activity. It's noteworthy that environmental permitting can typically take 12–18 months but may require from one to five years or more, design from one to three years, and construction can take months to many years to reach completion and closeout. The next section describes how to streamline permits for effective project management.

STREAMLINING PERMITS

There are ways to streamline permits. As noted above, most complex civil and environmental engineering projects will require multiple permits from multiple agencies. Preparing the permit materials together, with close collaboration between design engineers and the permitting staff can make the process more efficient.

To streamline project environmental permitting:

1. Prepare one combined project description listing the facts and figures of the project. This should include the purpose and objectives of the project, the proposed project footprint (including electrical transmission, gas, water supply, sewer, or other utilities) with clearly established project boundaries, city and county boundaries, and other key landmarks. An access plan will need to be created that shows transportation routes, proposed ingress/egress, and parking locations. The project description should include a proposed schedule, estimates of construction staffing and effort, as well as estimated water use, waste generation, and plans for recycling, disposal, and waste management. The project owner, operator, and construction contractor should be identified. Generally, the construction contractor will not be known and can be entered in permit applications as "TBD" (to be determined). However, the need to identify these parties for the permits forces a timely determination of who will be responsible, and in whose name permits should reside. For example, it may be desirable to have construction waste permits and construction stormwater permits in the name of the contractor *if the regulating agency allows it*.

2. Do not attempt to skip Step 1. Attempting to initiate project permitting without a *written* project description is likely to result in frustration and substantial rework and delay. This is not to ignore that projects will change during design. However, it is generally more efficient to amend or revise the project description subsequently, then to proceed with permit applications that lack a comprehensive project description.

Table 8.1 How Permitting Is Integrated with Project Phases (Medium to Large Civil Engineering Projects) (Adapted from Conrad Bridges, Vice President HDR [retired])

1	2	3	4	5
3–6 Months	6–12 Months	3 to 60 Months	1 to 3 Years	1 to ?? Years
Initiation	Funding	Environmental Review	Design	Construction
• Project Need Identification • Project Scoping Study • Feasibility Study • Preliminary Design • Multiple Alternates • Identify Required Permits, Agencies, Time Frames • Cost Estimate • Schedule	• Confirm Permit Requirements • Cost (Total) Engineering; Construction; Real Estate; Right of Way(s); Utilities; Mitigation(s)	• NEPA • CEQA (State EIR) • Refined Preliminary Design • Environmental, Technical, State Reviews Biological; Air; Noise • Public Meetings • Agency Review(s) and Approval/s • Reconfirm Permit Requirements	• Final Technology Decision, e.g., Selection of Bridge Type • Cost Estimate(s) • Design Submittals for Owner's Approval 30% 60% 90% 100%	• Advertisement • Bid • Award • Construction • Fulfillment of Final Permit Requirements • Closeout
Approval to Proceed	*Funding Authorization*	*Permits*	*Contract Documents*	*Completed Project*

3. Attach the comprehensive project description to each permit application, in lieu of describing details in an application form. In this manner the same information is submitted to all the regulating agencies and can be tracked and amended efficiently. One useful technique is to amass hard facts and figures (area of disturbance, expected megawatts of generation, estimates of waste generation). Using a separate table for factual information allows quick revisions without the need to re-do word processing and document formatting.

SAMPLE PERMIT TABLE

Table 8.2 lists common regulations and permits required for a hypothetical gas-fired energy project in California as an example of the environmental permits required for a typical project. Natural gas-fired energy projects involve the engineering design, permitting, and installation on industrial or private properties. The projects include construction of access roads to the facilities, engineered foundations for the facilities, installation of the structures as well as transmission lines to and from the facilities. There are numerous federal, state, and local permits required through all phases of project initiation through construction and operation (as shown in Table 8.1).

Projects involving roads, buildings, bridges, outfalls, or levee improvement will require similar or analogous permits, for which this table may be a useful beginning. Projects involving hazardous materials or remediation of toxic waste sites will have a similar list, predicated on the Comprehensive Environmental Response, Compensation and Liability Act (Superfund) or Resource Conservation and Recovery Act (RCRA). The engineer should prepare an analogous table for each project early in project scoping.

SUMMARY

Environmental permitting for civil and environmental engineering projects has become a much larger part of project management than was traditionally the case. For this reason, the experience of senior engineers may not be as helpful as an open collaborative attitude to comply with requirements that were nearly absent historically. The primary sources of frustration and project delay associated with environmental permitting are a lack of understanding of the need for permits, a lack of having a project management plan to address them, and not implementing methods to acquire permits in a streamlined and efficient manner. Permitting should be regarded as a "key element" in the design process and be part of the "basis of design" for engineering projects. The permitting task should be delegated to qualified staff and addressed at an early phase of the project to allow time to complete preapplication investigations and modifications to the design, if required.

Table 8.2 Natural Gas-Fired Energy Project—State Permit Requirements (California)

Permit	Agency	Time Frame	Cost	Requirement
PRECONSTRUCTION				
Application for certification pursuant to Warren Alquist Act	California Energy Commission (CEC)	18–30 months	Cost reimbursement, generally $1–$3M	Required to allow construction of energy generation facilities in California. CEC takes "one stop" authority before all other agencies.
Determination of compliance and authority to construct	Regional Air Pollution Control District	Just after AFC filing	Initial fee + application fee based upon heat input (~$160,000)	Required to allow discharge of air pollutants (oxides of nitrogen, sulfur, carbon particulate, etc.)
401 Water Quality Certification	Regional Water Quality Control Board	4–6 weeks	$500 filing fee. Based on capital cost of improvements	Required to demonstrate no adverse effect to water quality and beneficial uses.
Section 10 Consultation and Habitat Conservation Plan	U.S. Fish and Wildlife Service	At least 120 days	No filing fee	Required to demonstrate that endangered species will not be adversely affected, and prepare compensation plan for any impacts. Potential cost can be 2% of project cost.
Section 2081 Consultation with Memorandum of Understanding	California Department of Fish and Game	At least 120 days	No filing fee	Required to demonstrate no impacts to *species declared endangered by the state. Potential mitigation costs of 1% of project cost.*
Section 7 Consultation with Biological Opinion	National Marine Fisheries Services	At least 120 days	No filing fee	Required to demonstrate no adverse impacts to anadromous threatened and endangered species, where a federal agency has permitting authority. Section 7 is required for the USACE to authorize fill of wetlands projected by federal law.
Clean Water Act Section 404 Permit	U.S. Army Corps of Engineers	45–120 days	No filing fee	Required for activities that would fill waters or wetlands, including vernal pools, creeks, drainages or ditches.

Permit	Agency	Time Frame	Cost	Requirement
PRECONSTRUCTION				
Historic Preservation Act, Section 106 Permit	State Historic Preservation Office	30–120 days	No filing fee	Because issuing a Section 404 authorization is a *federal* action, the USACE will require compliance with Section 106. The Historic Preservation Act now generally includes both prehistoric (fossils), cultural (native American artifacts and burials) and historic (buildings or structures more than 50 years old) evaluations.
Streambed Alteration Agreement (California Fish and Game Code Section 1600)	California Department of Fish and Game	45–120 days	Generally $500 filing fee	Required for construction or operation activities that would affect the area within the "bed and banks" of any "stream" as defined by NSMA
General Plan Amendment	Local county	45–120 days	$500 filing fee, based on capital cost of project	Required if the proposed project changes the designated use of the property (e.g. from agricultural to industrial or residential)
Conditional Use Permit	Local county	45–120 days	$500 filing fee, based on capital cost of project	In lieu of a General Plan Amendment; required if the proposed project does not conform to the allowed uses of the property.
Encroachment Permit for pipelines	Caltrans	Within 60 Days	$70 per hour for review and inspection	Required for water, gas, or electrical lines to cross or parallel roadways.
Encroachment Permit for pipelines	Local county	2 weeks (work with project engineer)	Varies depending on type of roadway and work	Required for water, gas, or electrical lines to cross or parallel roadways.
Encroachment Permit for pipelines	Railroad	To Be Determined (TBD)	To Be Determined (TBD)	Required for water, gas, or electrical lines to cross or parallel railroads.

(Continued)

243

Table 8.2 (Continued)

Permit	Agency	Time Frame	Cost	Requirement
PRECONSTRUCTION				
CONSTRUCTION PERMITS				
Trenching and Excavation Permit	Cal/OSHA	24 hours	TBD	Information needed to demonstrate compliance with environmental and safety requirements.
Permit for the erection of a fixed tower crane	Cal/OSHA	24 hours	TBD	Information needed to demonstrate compliance with environmental and safety requirements.
Single Trip Transportation Permit	Caltrans	2 hours	$16	Information needed to demonstrate compliance with environmental and safety requirements.
Single Trip Transportation Permit	Local county	Same day	$16	Information needed to demonstrate compliance with environmental and safety requirements.
Annual Trip Transportation Permit	Caltrans	2 weeks	$90	Information needed to demonstrate compliance with environmental and safety requirements.
Transportation Permit	Local county	Same day	TBD	Information needed to demonstrate compliance with environmental and safety requirements.
Construction Stormwater Permit	State Water Resources Control Board	1 week	Fee Likely Required– TBD	Required to demonstrate that measures are implemented to prevent stormwater contaminating surface or groundwater
Construction Permit	City or county	4–6 weeks	Review and inspection costs	Information needed to demonstrate compliance with environmental and safety requirements.

Permit	Agency	Time Frame	Cost	Requirement
OPERATIONS PERMITS				
Risk Management Plan (RMP)	Local County Emergency Management	Contact local county for accurate time estimate required.	Fee Likely Required–TBD	Needed to demonstrate that project materials (compressed gases, lubricants, fuels, explosives) will not create an unacceptable hazard to the local public
Hazardous Material Permit	California Highway Patrol		$100 initial fee/$75 annually	
Hazardous Materials Business Plan (HMBP)	Local County Emergency Management	2–4 months	$10,000–$20,000 to prepare the HMBP	Obtained prior to commencement of operations
Hazardous Materials Storage Permit (Must be submitted with HMBP)	Local County Emergency Management	2–4 months	~$1,000–$2,000 plus nominal filing fee	Obtained prior to storage of HM
Risk Management Plan (RMP)	CUPA (Local County)	4–6 months	$30,000–$35,000 to prepare document	EPA requires an RMP be prepared prior to project operation.
Industrial Wastewater Permit	Local county	TBD	TBD	Required by local agency.
Solid Waste Permit	Local county	TBD	TBD	Required by local agency.
Spill Prevention Control and Countermeasures Plan (SPCC)	CUPA (Local county)	2–4 months	$10,000–$20,000 to prepare the plan	Needed prior to storage of oil at the site
Hazardous Waste Generator Permit	CUPA (Local county)	1 month	Nominal filing fee	Obtained prior to construction
Industrial Stormwater Permit		TBD	TBD	Required by local agency.
User Agreement	Local city	TBD	TBD	Required by local agency.
Sewer Discharge Permit	To local wastewater jurisdiction	TBD	TBD	Required by local agency.
Title 22 Engineering Report	State Department of Health Services	TBD	TBD	Required by local agency.

By using the tools described in this chapter, the successful engineer can distinguish himself/herself by maintaining constructive working relationships with the regulatory agencies, avoiding and minimizing environmental permitting delays and maintaining cost control on projects.

REFERENCES

Construction Industry Research and Information Association. 1994. *Environmental Handbook for Building and Civil Engineering Projects.*

Liu, D. H.F., (Ed). (1974). *Environmental Engineer's Handbook.* http://ceq.hss. doe.gov/nepa/Citizens_Guide_Dec07.pdf, 55-page PDF document.

U.S. Environmental Protection Agency, Laws and Regulations, 2009 www.epa.gov/ lawsregs/laws/index.html#env

Introduction
Ethics
Professional Engagement
1 2 3 4 5 6 7 8 9 10 11 12 13 14 15 16 17 A B C D E F
History
What Engineers Deliver
Emerging Technology
Engineer's Role in Project
Development
Permitting
Leadership Managing
Having
a Life
Executing a Professional
Commission
Client Relationship
Legal Aspects
Communicating
Globalization
Sustainability

Chapter **9**

The Client Relationship and Business Development

Big Idea

Almost all engineering projects begin with a client who has a need to be filled with a budget and schedule in mind. Building the relationship with the client is the foundation of business development.

> A customer is the most important visitor on our premises, he is not dependent on us. We are dependent on him. He is not an interruption in our work. He is the purpose of it. He is not an outsider in our business. He is part of it. We are not doing him a favor by serving him. He is doing us a favor by giving us an opportunity to do so.
>
> —Mahatma Gandhi

Key Topics Covered	Related Chapters in This Book
• Introduction	• Chapter 3: Ethics
• The Foundation of a Lasting Relationship	• Chapter 4: Professional Engagement
• Building upon the Relationship—The Superstructure	• Chapter 5: The Engineer's Role in Project Development
• Maintaining the Relationship	• Chapter 7: Executing a Professional Commission—Project Management
• Cultivating Business Opportunities	• Chapter 11: Legal Aspects of Professional Practice
• Business Development	• Chapter 12: Managing the Civil Engineering Enterprise
• Conflict Management	• Chapter 13: Communicating as a Professional Engineer
• Summary	• Chapter 14: Having a Life
	(Continued)

Related to *ASCE Body of Knowledge 2* Outcomes

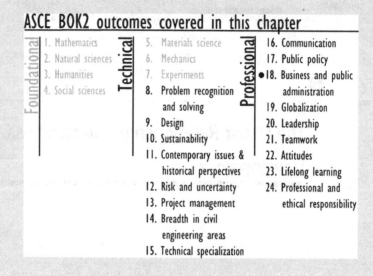

ASCE BOK2 outcomes covered in this chapter

Foundational
1. Mathematics
2. Natural sciences
3. Humanities
4. Social sciences

Technical
5. Materials science
6. Mechanics
7. Experiments
8. Problem recognition and solving
9. Design
10. Sustainability
11. Contemporary issues & historical perspectives
12. Risk and uncertainty
13. Project management
14. Breadth in civil engineering areas
15. Technical specialization

Professional
16. Communication
17. Public policy
● 18. Business and public administration
19. Globalization
20. Leadership
21. Teamwork
22. Attitudes
23. Lifelong learning
24. Professional and ethical responsibility

INTRODUCTION

In Chapter 1 the fact that ABET adopted Engineering Criteria 2000 (EC2000) and took a completely new approach to engineering education was discussed. The new focus was to identify *outcomes* of engineering education focusing on what is learned rather than what is taught. The 11 outcomes of civil engineering education identified are:

1. *Mathematics, science, and engineering*—an ability to apply knowledge of mathematics, science, and engineering
2. *Experiments*—an ability to design and conduct experiments, as well as analyze and interpret data
3. *Design*—an ability to design a system, component, or process to meet desired needs
4. *Multidisciplinary teams*—an ability to function on multidisciplinary teams
5. *Engineering problems*—an ability to identify, formulate, and solve engineering problems
6. *Professional and ethical responsibility*—an understanding of professional and ethical responsibility
7. *Communication*—an ability to communicate effectively
8. *Impact of engineering*—the broad education necessary to understand the impact of engineering solutions in a global and societal context
9. *Lifelong learning*—a recognition of the need for, and an ability to engage in, life-long learning
10. *Contemporary issues*—a knowledge of contemporary issues
11. *Engineering tools*—an ability to understand techniques, skills, and modern engineering tools necessary for engineering practice

In reviewing these criteria from the professional practice perspective it quickly becomes apparent there may be a key missing element. This element is "the client." The client may be a paying customer if the engineer is in private practice, a member of the public if you're in public service, or a commander if you're in the military. Regardless of the engineer's business relationship with their respective "clients" there are key components to successful relationships which can enhance the professional's career, improve the engineer's enjoyment of the practice, and help gain more respect for the profession.

The key components to a successful client relationship are:

- Trust
- Respect
- Managing Expectations—It's all about relationships

- Follow-through
- Effective Communication
- Scope/Schedule/Budget—Maintaining this triangular relationship by applying effective project management skills
- Understanding the client's business model and/or their perspective and his/her stakeholders
- Anticipating the client's needs before their request—Help crystallize the client's initial thought process in an effort to comprehend their needs
- Quality
- Going above and beyond the competitors
- Personalize your delivery

Engineers should recognize that clients create the need for engineering services regardless of whether you are in private engineering consulting, in public service, or in the military. "Your" client may be an industry, a paying private customer, or an army General but, it will be someone, numerous people, a community, a state, or the Nation that will depend upon the needed engineering service. *Client service will distinguish a good engineer from an excellent engineer. And, the client relationship is at the heart of client service.*

THE FOUNDATION OF A LASTING RELATIONSHIP

The client-engineer relationship is "at will" because both parties may leave the relationship at any time provided there was no express contract for a definite term governing the relationship. In the professional world there is a standard of conduct where patience, courtesy, and professionalism are always expected. An illustration of the relationship appears in Figure 9.1.

Trust, respect, commitment, follow-through, and communication are the foundational elements of the client relationship:

- *Trust:* In the course of the relationship the engineer will learn a lot about the client's business. This information about the client's business model may include:
  1. Key partners or subcontractors
  2. Raw materials used in their process
  3. The client's clients and/or competitors
  4. Market segment pricing and/or profit margins
  5. Regulations governing their business activities including commerce and operations
  6. Other business elements for a successful profitable business

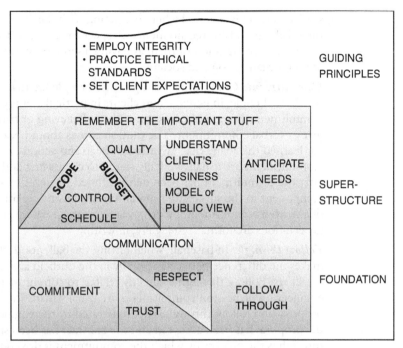

Figure 9.1 Building relationships

One can imagine that any client would be cautious and guarded sharing this valuable information. Often the contractual agreement will have clauses restricting the engineer to confidentiality and limiting the sharing of this information. Therefore, trust is a very important element in the foundation of the client-engineer relationship.

For example, in environmental engineering and particularly the hazardous waste management practice, the engineer learns about the waste produced from an industry's manufacturing operations. Working backward and applying reverse engineering, the engineer could learn about the raw materials used in a process, material suppliers, the actual process operations, labor requirements, and potentially marketing information. This information could all be provided to a competitor or potentially used in some way against the client if trust were absent in the relationship. And, in fact, some industrial and chemical manufacturing clients do require confidentiality agreements before engineering tasks proceed. Of course, engineers are also bound by professional ethics regarding sharing private information regarding the Client's business.

- **Respect:** Respect for the individual is a critical component to successful interactions and building relationships. It is the foundation of strong relationships. Respect is demonstrated by recognizing a person's abilities or qualities especially as a component of an organization or company. If an engineer is in public service, respect is essential when interacting with the public and acknowledging their feelings and/or requirements. Respect toward another individual demonstrates your recognition of them as having a sense of worth and a valuable place in their organization and society.

- *Commitment:* Commitment is a promise or a pledge that you provide to someone. People, in general, and clients in particular enjoy hearing commitments but are often cautious about believing or trusting someone who gives a verbal commitment. The caution comes about because almost everyone has learned the adage that "words are cheap and actions are valuable." Commitments are usually verbally stated to clients first and are truly valued after the commitment has been demonstrated. Commitment is an essential component to the successful foundation of a positive relationship. An example of a written commitment is a proposal or contract where promises and expectations are formalized legally, in writing.

- *Follow-through:* In baseball, when hitting the ball, good "follow-through" makes the difference between a pop fly in the outfield and a home run into the bleachers. Follow-through is directly related to commitment because in the engineering profession commitments are usually made in a meeting or a proposal and follow through is demonstrated by the engineer in the field after the proposal has been accepted and the contract has been signed. Follow-through is the process by which the commitment is driven home to the client (or to the public) and the initial promise is demonstrated in real-time delivery. Follow-through, like commitment, is essential to the successful foundation of a positive relationship.

- *Communication:* There is an entire chapter devoted to communication in this book but, let's just say that to achieve effective communication it is essential to send and receive "the complete message." We believe the strength of the relationship between the "engineer and the client" is directly proportional to the efficiency of their communication.

The Fundamental Rule of Client Relationships: NO SURPRISES !!!

BUILDING UPON THE RELATIONSHIP—THE SUPERSTRUCTURE

Once a solid relationship is present, the engineer can look forward to building upon this foundation to create a lasting relationship. This process works well for client relationships as well as with personal friends and colleagues. The additional elements that follow can help create a mature, lasting client relationship. In engineering terms, this phase of relationship building is like constructing a superstructure upon a firm foundation.

These elements will come into play in a relationship some time after the foundation of the relationship has developed, and they include maintaining the scope/schedule/budget relationship, understanding the client's business model and stakeholders, anticipating the client's needs, delivering quality, and remembering the important stuff.

- *Maintaining the scope/schedule/budget relationship:* This relationship of scope, schedule, and budget can be thought of as a connected triangle where each side represents an essential component of the "project" managed by the project manager (PM). The project will be described in a picture and/or text representing the scope of work. The client will have a schedule and budget in mind for ultimate completion and delivery of the project. Many times the client will need assistance defining these terms and the engineer may help (and negotiate) this definition by providing assistance to the client. Once defined, however, the engineer will maintain a positive client relationship by adhering to scope, schedule, and budget constraints.

- *Understanding the client's business model and stakeholders:* Understanding the client's business model will provide a distinct advantage to the engineer when building a strong lasting relationship with the client. This understanding goes beyond the simple grasp of knowledge the client completes his/her task and receives compensation or recognition for it. This understanding includes knowing the competitors, key business terms and concepts, the client's management structure, products, issues, and other related components of the client's business.

- *Anticipating the client's needs:* Anticipating the client's needs before they make a request can help crystallize their thought processes. If the timing is right, this effort can help shape the client's project while the flexibility to define the scope of work still exists. In the public sector, anticipating the public's needs can also help shape the project or allow the engineer to develop a response to any public concerns.

- *Delivering quality:* The quality of the deliverable product will generally be defined in the scope of work. If there is any question about the overall quality of the deliverable, the engineer should clarify the quality issues when discussing the scope of work and schedule. The quality of the deliverable will be related closely to the schedule, the level of effort, and corresponding fee to produce the deliverable. Greater detail about quality is discussed in Chapter 7, Executing a Professional Commission—Project Management.

- *Remembering the important stuff:* The "important stuff" could be anything that is important to the client. If the engineer is a public servant, there may be issues that are politically sensitive. Important things might include quality, schedule, the client's birthday or favorite holiday; the form or package of the deliverable; big issues like security, or little ones like whether the client likes their middle initial on the transmittal letter. In order to enhance the client relationship, the engineer should observe, listen, remember, and employ the important stuff in the relationship. Remembering the important stuff and incorporating this into the deliverables will set the smart engineer apart from others.

MAINTAINING THE RELATIONSHIP

Maintaining a superior client relationship means going above and beyond the competitors. Clients can acquire engineering services from many sources. High scores in "client service" have become of paramount importance for obtaining and maintaining the "outstanding" rating category. Going above and beyond the competitors most likely will involve personalized service like making deliveries by hand, exhibiting flexibility when scheduling meetings to the client's preferred date or location, providing extra copies, or sending technical articles related to the client's business. Whatever the service is, it should be performed with thought and the intention of benefitting the client.

The Value of Keeping Current Clients Happy

"When I was first stock piling knowledge, I took on Philip Kotler's Kotler on Marketing, a seminal college textbook. Easily the hardest book I've ever slogged through, it practically bored me to tears. I would have vastly preferred falling asleep on the airplane, or going to bed earlier, or just watching the tube.

But I knew the loving thing was to keep trying, to keep reading, to keep working, and that it would pay off. Or, at least, I hoped so.

A few weeks later I finished the book on a plane, with no idea how I would ever use its knowledge. Forty-eight hours later, I was quoting an important section to a thousand people at a marketing conference, telling them that the cost of attracting a new customer is five times the cost of keeping a current customer happy, and that customer profitability tends to grow with the length of customer tenure (long-term customers tend to buy more, they recommend the company more, and they cost the company less). And, the cost to maintain them is lower because they don't complain as much as unhappy customers.

These insights gave me fresh perspective on what marketing is all about: retaining your customers. Until that time I had always talked about accumulating new customers through marketing."

—Tim Sanders, *Love Is the Killer App*, pp. 198–199

Personalize Your Delivery

Personalized delivery for client deliverables can be an actual delivery made in person by the engineer or submitting the deliverable before the due date. This personal component can set the engineer apart from other competitors.

In addition to exceeding the competitors' performance, several principles contribute to the successful maintenance of the client relationship. These are: choosing clients carefully, setting client expectations, maintaining ethical and moral standards, and earning a profit.

- *Choosing clients carefully:* Clients select engineers, but engineers choose to be selected. The chances of maintaining a successful relationship with a client are enhanced if client and engineer are well matched.

  Clients are responsible for administering the process designed to select an engineer. Ideally, the client's selection process starts with the client's clear understanding of the engineer's strengths as a specialist and commitment to a particular market sector. But client selection can begin before the engineer ever meets with the prospective client as a "client." It could start with an informal meeting at a conference or at a convention or by introduction at a local community meeting. The initial meeting may be at a luncheon seminar related to a general subject overview or at a presentation at the client's office.

  Obtaining some basic information or doing some research on a prospective client will enhance the engineer's odds of being selected later. Some questions the engineer might ask are:

1. What are the current issues related to this market sector and to this client in particular?
2. Are these issues clearly understood by the engineer and are they in his/her practice area?
3. Has the engineer worked with more than one other client in this market sector on a similar matter?
4. Does the engineer have relevant experience in this area?

   Additionally, the engineer should assure that anyone referred to his or her office is a priority because this individual is reaching out and is seeking a valued relationship.

- *Setting client expectations:* Engineers often dive into projects as soon as the client has agreed to work with them. It's common that the engineer gets immersed in a project developing engineering solutions that he/she may tend to lose their clients in the details. Probably the most crucial client expectation to manage is communication. At the outset, client and engineer should establish the frequency, type, and standards of communication desired. Without this understanding, clients may form their own expectations about how the relationship will work. This should be the first expectation to be clarified.

Other details on communications may include:

1. Good times to contact the client

2. When the engineer is available

3. The preferred method of contact, e.g, phone, text messages, or e-mail

4. How to handle urgent messages

- *Maintaining ethical and moral standards:* The engineering profession is one of the most highly regarded professions due to the ethical and moral standards held by engineers. Maintaining these standards is the engineer's duty under the business and professional codes, the canons of ASCE, ACEC, NSPE (and other highly respected engineering organizations), and each state's standards for professional engineers. Ethical treatment of clients is essential for maintaining a positive client relationship.

- *Earning a profit:* In private enterprise maintaining a profit isn't just a principle, it is a requirement. However, there may be short periods of time when a private firm "invests" in a proposal effort for a major project or a key client and actually may not maintain a profit for a month or two or three. But in general, a profit is required for businesses to stay in business. Thus, both clients and engineers need to understand that the engineer's ability to make a profit on a project is a key to maintaining a strong client relationship.

CULTIVATING BUSINESS OPPORTUNITIES

Business can come from many sources. Among these are: participation in professional associations; company press releases; journal and magazine articles; printed brochures; websites; newsletters; awards and competitions; and advertising. These approaches vary in cost and effectiveness. Advertising often is cited as being one of the most expensive and least effective ways of cultivating new business for engineering firms.

Several less expensive approaches are available involving efforts by company employees. These include developing common ground with the client; networking; volunteering; speaking engagements; and asking for referrals:

- *Developing common ground with the client:* Learning if there is some common ground that can be built upon with clients is important. This common ground may be enjoying the same sport, enjoying the same food at a luncheon engagement, hobbies, family, or other areas of common interest. All of us often display pictures of our loved ones or favorite vacation spots in our work spaces. A simple observation can lead to an interesting conversation, if time permits and it seems appropriate.

- *Networking:* Networking is an integral component of business development. Networking can take place through professional organizations such as the

ASCE, ACEC, NSPE, at business conferences and seminars, through alumni associations and other social clubs, basically anywhere.

- *Volunteering:* Volunteering for public service projects offers an opportunity to share common ground with the public and potential customers. There's a tremendous need for volunteer services at the local shelter for the homeless, creek or beach cleanup activities, and for those willing to participate in running or walk-a-thons to raise money for good causes. A great deal of personal as well as professional satisfaction can flow from volunteer activities.

- *Speaking engagements:* Another potential area for engineers to gain exposure is by offering to speak on a contemporary issue at a civic organization, school, university, or event. Public speaking offers an opportunity for the engineer to expose local problems that require public attention or to highlight engineering services that are potentially available to solve problems. A public speaking engagement that is done well and timely may create valuable opportunities. Many private firms as well as public agencies will entertain short presentations over lunch time for their staff eager to learn about engineering solutions to contemporary problems.

- *Asking for referrals:* Finally, after completing a successful job or project for a client, the engineer should consider asking for a reference. This request can first be in the form of a feedback questionnaire from upper management in the same company (or agency). Engineers are interested in improvement on their services and feedback from the questionnaire can provide valuable information on how to perform better the next time. Assuming the questionnaire is positive it may allow an opportunity to ask the client if they would mind providing a reference for other potential clients in the future.

Networking Primer

After initially meeting people it's common to inquire about common areas of interest. Establishing these common areas may start with initial questions involving subjects like:

- Where you're originally from
- Whether you have a family or where your children go to school
- Your daily commute route
- Your interests outside of work
- Your latest vacation or weekend outing
- Or other subject areas
- Note: Use caution and common sense when asking personal questions!

(Continued)

> Once these areas have been discussed it's also common to learn how alike we all are and that we may know some of the same people. This initial introductory dialogue can become the foundation for the next potential encounter with this individual or a mutual acquaintance. In these initial encounters you have an opportunity to introduce yourself and your profession and potentially to learn about possible engineering opportunities.

BUSINESS DEVELOPMENT

Although all employees can be involved in furthering the interests of their companies, most firms formally vest responsibility for business development with certain employees. In smaller firms the principal(s) will be involved in cultivating new business. In larger firms, there may be a Director of Business Development with accompanying staff. Often project managers and section managers are expected bring work into their firms.

An example of a model for business development for private engineering consultants is presented in Figure 4.1 This business development (BD) process begins with identifying an initial problem or a project lead. If the firms deems the opportunity worthwhile, the potential project enters the tracking phase, usually managed by a senior engineer in the firm who monitors the pulse of this specific project. This process can take days or more likely months, providing an opportunity for the firm to consider initial strategies when entering the positioning phase. It's in this phase where the firm will identify the outstanding accomplishments, tools, or personnel that set this firm apart from competitors.

A "Request for Proposal" (RFP) is an invitation for firms to evaluate the client's problem and to assess whether they can provide a viable, cost-effective solution. The engineering firm must decide whether it should "invest" valuable time and effort into the proposal efforts. There are books written on creating winning proposals; and needless to say, creating a thorough proposal takes a great deal of technical, administrative, and support effort.

Proposals in the defense industry for sophisticated fighter jet aircraft or rockets can be volumes 5 to 10 feet thick or more to provide a complete answer to the government's request. Typical engineering proposals may be 5 to 10 pages or 1 to 5 volumes, depending on the complexity and project budget. The proposal phase may be a few days or a few months, depending upon the client. However, most engineers feel they never had enough time before the deadline, which accentuates the importance of the tracking and positioning phases preceding the RFP.

These expenses are not related to a specific client or project so they must be charged to the firm's overhead expenses. Firms try to limit their overhead expenses because high overhead drives their services rates up for all customers; and high rates can make the firms uncompetitive. Most firms try to limit these proposal overhead expenses to 1 to 4 percent of the total project fees they anticipate receiving if they win the project. Therefore, firms will often evaluate the client's engineering cost estimate and compare it to their own cost estimate before making a go/no-go decision on the proposal. However, there may be times when a firm goes "all out" for a proposal beyond their own set guidelines to restrict other competitors from establishing a foothold with a business sector or a specific client or to establish a foothold in a new market.

The next phase, client/agency review, may take a few days or a few months again, depending upon the scope of the project and urgency of the need. After the review is complete the interview and selection phases follow. The interview phase is typically referred to as the "shortlist interview," which requires considerable organization and preparation for the private consulting firm. The interview may last from 20 minutes to hours in the client's office. Typically, the client will "drill" the engineer with questions related to their approach, project team or staff, schedule, or budget. These interviews may take anywhere from 2 to 50 hours or more for each of the project participants in the interview, which in turn drives up the engineer's overhead cost.

After the interview, the client/agency will pick the winning team and make announcements to all the participants. One of the most common questions entertained by the client/agency from the losing engineering firms is: "So, how did our firm rank in the process?" Experience has taught the seasoned engineer that the client/agency response is often: "Your firm came in second, but we wish you lots of luck on the next one." Often, in reality the winning firm is number 1 and "all other firms" are ranked as number 2. This secret and odd response to the ranking question often occurs because these competitors don't usually ask one another how they ranked, and the client/agency doesn't want to discourage or restrict future responses from the firms. However, there are clients/agencies that share their ranking process with the competitors but typically not in writing. The engineering firm can gain a great deal of insight into improving their proposal process if they request feedback from the client interview panel.

The final phase before the long-awaited project kick-off meeting is the contract negotiation process. Experienced firms often request a sample contract in the very beginning of the BD process because the terms of the contract may prevent them from proposing in the first place. Some clients/agencies may want to assign "all risk" to the engineering firm for the entire client's staff and other contractors on the project site, or they may request some sort of guarantee for the project before it's even started. There could be other contract terms requiring some sort of insurance protection or payment structure that the engineering firm cannot withstand. This negotiation phase may be an arduous process and may result with the "winning" firm ultimately rejecting the work. This forces the client back to their interview list. Alternatively, the process may be smooth, leading swiftly into the project. Therefore, the negotiation phase is one that requires special attention to detail and experience on project and contract management.

Work Smarter . . . Not Harder!

CONFLICT MANAGEMENT

In the professional world, providing thorough and direct communication is an effort to avoid conflict. However, reality sometimes sets in with difficult issues like tight schedules, budget constraints, contract issues, competition for resources, subcontracting issues, and more. Even the most experienced project manager and experienced team can encounter unexpected issues that result in a conflict between the engineer and client or owner. Some basic conflict management skills can help manage these conflicts, repair the damage, and hopefully salvage the relationship.

Some typical, real-life experiences affecting the engineer and client are listed below. Whether the client or the engineer could have anticipated these (typical but imaginary) events before signing their agreement is doubtful:

- The client PM left the client's firm and the engineer had to temporarily freeze the project until the client could re-assign a new PM.

  The Result: The project was on hold for two months and the engineering PM had a struggle to re-assemble the team and incur re-start costs that could not be billed to the project. Therefore, meeting the client's needs caused the engineering firm to lose revenue and profit on this project.

- The federal government exercised the "stop work clause" in the contract because of priority funding for an international conflict.

  The Result: The engineering firm disbanded the project team, incurred major overhead costs (and lost profits) to keep the employees on the payroll while they were being re-assigned to other projects.

- A subcontractor had a safety incident and one of their employees had a minor laceration that required first aid. The client recognized that this was the second event in six months and temporarily stopped all field work on the West Coast affecting over 50 projects in the entire program to re-examine the Health and Safety Plans for the engineer and subcontractors. The engineer's PM had to redirect the project staff to keep them billable during the three-month evaluation period.

  The Result: The engineering firm lost revenue and the client lost confidence in the engineering firm's field safety protocols, even though the firm was not directly responsible for the incident. Finally, the engineering firm did not win the follow-on multi-million dollar contract one year later.

- The engineering firm's PM left the firm and the firm had two weeks to re-assign another less-experienced PM to the project.

  The Result: The client canceled the contract and the original PM re-contracted the client with the new engineering firm.

- On another similar project with a different engineering firm, the firm's PM left the company. The firm hired another more-experienced PM to retain the client and the project.

The Result: The firm incurred higher overhead costs because the new PM was compensated at a higher rate and was hired through an agency with expensive placement fees. The engineering firm was able to keep the client. However, the firm lost revenue and profit because the project staff (hours and effort) was temporarily reduced by the client until a new engineering PM was hired.

- The engineering firm entered into a contract with a chemical analytical laboratory to perform analytical testing for potentially toxic and hazardous compounds. A large elaborate and expensive data collection field effort was planned and conducted. During a very busy summer period, the analytical laboratory performed the analyses but later realized that the samples were analyzed past their prescribed holding times. The data would not be accepted by the regulating agency and was considered invalid.

The Result: The analytical laboratory sincerely apologized for the error and offered to re-analyze new samples for a significantly reduced fee. As a result the client missed the report deadline and was angry with the engineering firm and the laboratory. The engineering PM was able to negotiate with the regulatory agencies for additional time so the client would not be fined. However, the engineering firm suffered a significant cost impact due to the need to re-collect field samples and then have them re-analyzed at a different laboratory for a higher fee. The engineering firm decided to seek restitution from the analytical laboratory.

The point of all these typical, imaginary real-life experiences described above is that despite all reasonable care taken by the engineering PM and the client PM, these situations were not anticipated. The engineering PM and the client PM may have common goals on the project, but each PM has a different employer with different stakeholders that can place them in direct conflict with one another.

The engineer should attempt to resolve these conflicts as soon as possible and at the lowest management level possible. Frequent and direct communication on these issues is highly recommended, as described in Chapter 13. The contract or agreement most likely includes language and direction on pursuing solutions to these problems, but often at a greater expense to one party—frequently the engineer. Specific contract issues and terms are discussed in Chapters 4 and 11 for reference.

In addition, applying the 4 Cs of conflict resolution may be beneficial, if the atmosphere for resolution presents itself and can be fostered. If either party attempts to integrate their legal representatives into the resolution, the PMs usually lose control of possible alternatives. Often negotiations get very difficult and can get more aggressive at this point, and salvaging the relationship becomes more problematic.

The 4 Cs of Conflict Management

Experienced PMs often implement the 4 Cs of conflict resolution as part of the effort to reach an equitable agreement. Preselecting the targeted strategy of the resolution is important because only two of these strategies offer an equitable solution. A more detailed discussion on this management technique appears in Chapter 13, Communicating as a Professional Engineer.

Collaboration—A collaborative strategy in conflict management is when two parties come up with a completely new idea that pleases them both. Win/Win

Compromise—A compromise in conflict management is when two or more sides agree to accept less than they originally wanted. Draw/Draw

Co-existence—Both parties agree to disagree. Lose/Lose

Capitulation—One party completely gives up resisting the other. Lose/Win (Maybe)

Some Tips about Building Relationships

- Keep in touch with your clients! In sales jargon, we actually refer to this as "touching," which means staying in contact by phone, e-mail, or meetings on a regular basis. This can be as small as asking about their family, remembering or referring to something they are interested in or their hobbies or a business problem they previously shared with you. It never hurts to take notes. Think about sending them articles that may be of interest that you have come across in your reading. It is also advisable to call sometimes just to "check in" without an agenda.

- Be curious about your clients. People generally like to talk about themselves. A connection opportunity may be the pictures or mementos on their desk. They are there for a reason. You never know what will help you make that "connection" that can lead to further business or referrals.

—Janet Riser, V.P. and Executive Manager at Janney, Montgomery and Scott LLC

SUMMARY

The fundamental foundation of client relationships is trust, respect, commitment, follow-through, and communication. Building upon the foundation of this relationship includes anticipating the client's needs, understanding the client's business model or public's views, excellent scope, schedule, budget control in contractual deliverables and projects, an appreciation for the client's expectations on quality, and remembering the client's important stuff. Guiding principles for maintaining relationships include integrity, ethical standards, and learning to set expectations. Client service will distinguish a good engineer from an excellent engineer. And, the client relationship is at the heart of client service. The fundamental rule of client relationships is: NO SURPRISES (except good ones)!

REFERENCES

Bachner, John. (1991). *Practice Management for Design Professionals: A Practical Guide to Avoiding Liability and Enhancing Profitabilty.* John Wiley & Sons, Inc., New Jersey.

Jones, Tricia S. and Brinkert, Ross. (2008). *Conflict Coaching: Conflict Management Strategies and Skills for the Individual.* Sage Publications, Inc, ISBN-10: 1-412-95083-X.

Kidde, Barbara and Fetterberg, Fred. (1992). "Succeeding with Consultants: Self-Assessment for the Changing Nonprofit." The Foundation Center. (http://foundationcenter.org/) ISBN-10: 0-879-54450-3.

Sanders, Tim. (2009). *Love Is the Killer App.* Three Rivers Press, New York.

Introduction
Ethics
1 2 3 Professional Engagement
History 4 5 What Engineers Deliver
Engineer's Role in Project 6 7 Permitting Leadership Managing Having Emerging Technology A B C D E F
Development 8 9 10 11 12 13 a Life 14 15 16 17 Sustainability
Executing a Professional Client Relationship Legal Aspects Communicating Globalization
Commission

Chapter **10**

Leadership

Big Idea

There are four component parts to effective leadership that include business strategy and economics, technical skill, and public affairs/marketing. An effective leader possesses skills and acumen in all four areas. Leadership style should be adjusted to fit the situation.

> Genius is one percent inspiration and ninety-nine percent perspiration.

> —Thomas Edison

Key Topics Covered	Related Chapters in This Book
• Introduction	• Chapter 3: Ethics
• Leadership Styles	• Chapter 5: The Engineer's Role in Project Development
• Tools for Leadership and Management	
• Four Quadrants of Effective Leadership	• Chapter 11: Legal Aspects of Professional Practice
• Public Service Leadership (for Government Employees) or Marketing Leadership (for Consulting Engineers)	• Chapter 13: Communicating as a Professional Engineer
• Secret Recipe for an Effective Leader	• Chapter 14: Having a Life
• Summary	

(Continued)

Related to *ASCE Body of Knowledge 2* Outcomes

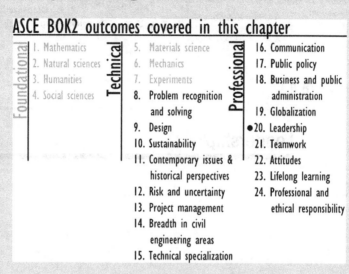

ASCE BOK2 outcomes covered in this chapter

Foundational
1. Mathematics
2. Natural sciences
3. Humanities
4. Social sciences

Technical
5. Materials science
6. Mechanics
7. Experiments
8. Problem recognition and solving
9. Design
10. Sustainability
11. Contemporary issues & historical perspectives
12. Risk and uncertainty
13. Project management
14. Breadth in civil engineering areas
15. Technical specialization

Professional
16. Communication
17. Public policy
18. Business and public administration
19. Globalization
● 20. Leadership
21. Teamwork
22. Attitudes
23. Lifelong learning
24. Professional and ethical responsibility

INTRODUCTION

Leadership is both the act of guiding and the process of showing the way to proceed toward a common goal (Northouse 2010). A leader takes risk, exhibits initiative, inspires and motivates others to follow and perform an activity. In a social order, such as a community or an employer's office, a leader must have the self-confidence to lead others, intelligence and the forethought to plan, and the respect and trust of followers. Engineers can make good leaders because they are generally intelligent and have the ability to plan. If the engineer possesses the additional ingredients of a dynamic personality and strong character, they can gain the respect and trust of their followers.

A Good Leader:	"Characteristics" of a Good Leader:
• Inspires • Motivates/Recognizes • Sets a Good Example	• Trustworthy • Intelligent • Courageous

LEADERSHIP STYLES

Leadership style is the manner and approach of providing direction, implementing plans, and motivating people. Kurt Lewin (1939) led a group of researchers to identify different styles of leadership. This early study has been very influential and established three major leadership styles. The three major styles of leadership identified by Lewin are (U.S. Army Handbook 1973):

- Autocratic
- Democratic
- Delegative or Free Reign

Effective leaders are aware of all three types of leadership styles and adapt these styles to the individual and the situation. It's possible to use all three styles on the same individual in different situations, and it's possible to use all three styles on different individuals for the same situation. Note that the leader is still responsible for the decisions and performance of the team members regardless of the style used. Let's look at the definitions of these leadership styles.

Autocratic Leadership

An *autocratic leadership style* is one where the leader tells the team members or followers what to do, how to do it, and when to do it without any input from their followers. Typical instances for using an autocratic style might be when there is an emergency situation that requires immediate action like, "Call for help!" or when the leader has all the information to solve a particular problem and requests a specific tool or action. Generally the leader desires (and requires) little input from the team members (Army Handbook 1973). New or unskilled team members may be accustomed to this leadership style but experienced and skilled team members will not appreciate being dictated to and told what to do and how to do it.

Democratic Leadership

A *democratic leadership style* occurs when the leader invites the team members or followers to provide input into the decision-making process respecting them and validating their input into the overall process (Army Handbook 1973). This style might be used when the team members have valuable information regarding the process and the leader has other knowledge or information regarding the process (Northouse 2010). The leader cannot be expected to know everything and relies on knowledgeable employees. Generally, the leader maintains the final authority but considers the team members' input. This style displays the leadership strength of the leader and usually generates respect from the team members. There are time requirements for including the team members' input, so this style may not be employable during urgent situations or where the team members have little working knowledge to share on the process.

Delegative Leadership

A *delegative leadership style* is one where the leader provides broad guidance to the team members and allows the team members to decide what to do, how and when to do it, and where to do it (Army Handbook 1973). This style is often employed with very senior and skilled members when the team members can correctly analyze the situation, determine what needs to be done, and how to do it (Northouse 2010). The leader understands that they cannot do everything so they set priorities and delegate certain tasks. However, one of the pitfalls, especially for new leaders, is to simply assume that all the team members are skilled and fully capable. In such situations, the leader needs to "check in" with the team members to verify their level of self-confidence and capability. There's more information on this communication process in Chapter 13, Communicating as a Professional Engineer. This style may be employed when the leader enjoys full trust and confidence in the team members.

> There is no one leadership style that can be employed for all team members under all circumstances.

> Conversely, it is unlikely that one leadership style will be appropriate for each individual employee all the time.

There are numerous factors that leaders should consider when choosing a leadership style for a team member and a specific situation. The key is for the leader to be aware of, and consciously choose, the most appropriate style—including consideration of their individual "natural" style. Some factors are listed below:

- Urgency of the task
- Competence level and skill level of the team members
- The level of respect and trust between the leader and team member
- The leader's knowledge level and team members' knowledge level of the task
- Task type—meaning the level of structure (existing procedures) associated with the task, complexity of the task, number of components, number of steps, extent of knowledge required, and technical and cognitive abilities needed to accomplish the task
- Relationship with and knowledge of team members
- Other critical factors

Additional information may be found on the following website: http://donaldclarkplanb.blogspot.com/2006/06/drucker-leadership-training-is.html

So, the message is that leaders should adjust their leadership style to fit the circumstance and the employee.

> A leader's natural style may have components of all three leadership styles described above.

Quotes from Famous Leaders

- "If your actions inspire others to dream more, learn more, do more and become more, you are a leader." John Quincy Adams
- "Setting an example is not the main means of influencing another, it is the only means." Albert Einstein
- "A leader is one who knows the way, goes the way, and shows the way." John C. Maxwell
- "Innovation distinguishes between a leader and a follower." Steven Jobs
- "To be a leader, you have to make people want to follow you, and nobody wants to follow someone who doesn't know where he is going." Joe Namath

—www.thinkexist.com

Additional Thoughts on Leadership

Leadership is a relationship between the leader and the team member(s). A leader can't lead without followers. The leader has to demonstrate continually respect, trust, and courage to maintain leadership and inspire and motivate the followers. In general, leadership is also *nonhierarchical*. Just because the supervisor has a title, that doesn't make him or her a leader, but it does make them the boss. "Supervision without leadership" presents challenges for those at organizational levels both below and above such individuals. Leaders exist at all levels within an organization (Goffee and Jones 2009).

Each of us holds the power to inspire confidence in others!

In a single work session, two dedicated team members can accomplish much more than twice the work as each working individually!

TOOLS FOR LEADERSHIP AND MANAGEMENT

Most engineering managers are leaders and they generally have common tools in their tool boxes independent of whether they are involved in engineering consulting or public service. We'll focus on the responsibilities of a "manager" and later discuss how this relates to being a "leader" and how an engineer can be both a manager and a leader. A manager assumes responsibility for a business unit or an element of a government service like an engineering department or an organization. For purposes of this discussion, let's agree to use the term "business unit." The manager's overall responsibility usually includes:

- Planning functions of the business unit for short- and long-term operations
- Monitoring the daily operations of the business unit, the organization of the team often including the business unit's marketing and sales functions (if it's an engineering consulting organization), or the public service/public affairs function (if it's a government business unit)
- Leading the strategic planning functions of the company
- Controlling the income revenue and cost elements of the business unit, which is referred to as the Profit & Loss (P&L) responsibility (if it's engineering consulting) or the budget responsibility (if it's government)

So, in essence, managers *plan, organize, lead,* and *control* (POLC) to accomplish the mission of the business unit (DuBrin 2000).

Planning

The planning function includes an analysis of where the business unit is going in the short- and long-term periods. The plan may be referred to as a "business plan," a "marketing plan," an "operations plan," a "strategic plan," or "some other plan." But, the manager's plan will include a strategy, much like a chess player, that describes where the business unit should be going to respond to the market or to the public's needs. The manager will most likely have a "vision" that will guide and shape the plan.

Organizing

Organizing activities are related to assessing the needs of the business unit with regard to personnel, equipment, services, and resources to operate the unit to accomplish the mission. Organizing also includes the responsibility to accommodate the staff with a safe, comfortable work environment, the required tools and equipment to accomplish staff job responsibilities, training to enhance and increase staff development, output, and efficiency (DuBrin 2000). Managing the organization can be as simple as ordering new computers or software or as complicated as finding a new cost-effective space to house the projected growth (or decline) of the unit.

Leading

The effectiveness of the leadership function separates "managers" from "leaders." A leader will lead the strategic planning function of the business unit in a way that matches the vision of the executive management team (DuBrin 2000). Hopefully, the vision communicated in the strategic plan matches the perceived direction of the market for the consulting engineer leader or the direction of the public's views for the government engineer leader. This is where the leadership ability distinguishes the manager from the leader.

Controlling

Controlling an engineering business unit is probably a lot like flying a light aircraft from a quiet valley, over the mountains and through a storm. The first thing a new pilot would likely do is:

- Inspect the aircraft (on the ground)
- Study the operator's manual
- Learn about the gauges and navigation system

- Run through some mock scenarios (on the ground)
- Take lessons from a qualified, experienced pilot
- Take more lessons
- Run a solo flight
- Then be ever cautious

Similarly, a new manager coming into an existing business unit would likely:

- Learn about the business unit, the staff, the goals, clients, and deliverables
- Study the existing operating procedures, guidelines, and requirements
- Learn about the revenue and costs for the unit and compare this to the output
- Run some mock scenarios aligned with the business plan or strategic plan
- Counsel a qualified, trusted engineering manager familiar with the unit
- Manage the unit solo while carefully watching the gauges under the guidance of the qualified, trusted engineering manager
- Then be ever cautious

Whether you're a new pilot or new engineering manager, managing a business unit will likely involve some smooth operations, some rocky terrain, and some strong storms. It's important to understand the business unit's plan, the organizational staff, resources, partners and output deliverables, and the appropriate leadership technique for the staff and conditions of the business climate. "Controlling" and managing the unit will likely stretch the engineering manager's skills until he/she gains the appropriate leadership skills

FOUR QUADRANTS OF EFFECTIVE LEADERSHIP

Remember, the leadership styles, as described earlier in this chapter, are *autocratic, democratic,* or *delegative.* The Manager generally focuses on the direction, manipulation, and deployment of the organization staff and resources. The manager often uses feedback from the past while trying to forge ahead. It's similar to driving a car while focusing on the rear view mirror but occasionally glancing ahead. The manager will guide the business unit from last month's staff reports, revenue and cost reports, while trying to stay on track with the business plan.

Leaders often display strong character, strength, fortitude, and courage to lead the way toward the vision in the strategic plan. The leader paves the way described in the strategic plan before the followers can see the vision the leader embraces. The leader is focusing on the path ahead while glancing backward to see the previous destination and gauge progress. Effective leaders do not forget that they need to control the velocity and direction of the business unit.

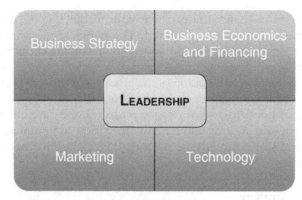

Figure 10.1   Four quadrants of effective leadership

Management is needed to run the operation and stay on track with the realities of cost, revenue, and resource usage. Leaders are needed to prevent stagnation, to encourage continuous improvement and innovation to compete in the market. Leaders also create new ways to serve clients and/or the public more effectively.

So, what's the right mix of management and leadership? Are there specific areas in which a leader can excel? Let's look at typical areas of engineering management where engineers excel at leadership. These leadership specialty areas are depicted in Figure 10.1, and are referred to as the four quadrants of leadership:

- Strategic Leadership
- Financial Leadership
- Technical Leadership
- Marketing Leadership (for consulting engineers) and Public Service Leadership (for government engineers)

These four quadrants represent essential developmental skill areas in addition to the other tools previously mentioned in the manager's tool box for the leader to practice honing throughout their career (Altier 1999). Effective leadership requires a delicate balance in all four areas whether the engineer is in private practice or public service. Continuous improvement and development of these skill areas make engineering a career and a practice and not just a job.

Strategic Leadership

Strategy involves concentrated thought, planning, direction, team organizational skills, and vision. Strategic leadership requires concentrated thought and keen observation that evolves into a vision and action to realize that vision. This means that the consulting engineer leader will observe the surrounding conditions in a particular market and concentrate on where the market is going in order to create a vision of the products or services required. The public service employee will perform a similar function by observing the surrounding conditions within the boundaries of a state/city or political environment and concentrating on the needs and desires of the public in order to create a vision of the products or services required for that entity.

Strategic leaders understand the balance between managing the business unit based upon past performance measures and carefully investing the unit's time or profit into creating and realizing the vision communicated in the strategic plan. If we could stroll through the strategic leader's mind we would observe the thought patterns on:

Strategy

1. Where do we want to be and where do we need to be?
2. How can I articulate the vision so staff will understand the criticality of realizing that vision?
3. How much time and how many resources can we invest into realizing this vision?
4. Is there any way to accelerate the investment process and just how much can we afford to devote to our future?

Public Service Considerations

1. Do I truly understand the public opinion?
2. What is the view of this business unit to the public?
3. How can I gauge the public opinion and respond to it?
4. Is the current operating procedure meeting or exceeding our mission?

Market Considerations

1. Do I truly understand my client's needs and expectations?
2. Do I understand where my client's market is going and can the business unit respond to that direction?
3. What can the unit do to increase the client base and diversify?
4. Can I cross-sell my services to clients from other departments within the organization?

Management

1. Can I achieve the operating goals and simultaneously invest in the vision?
2. Are there any major pieces of equipment or other resources that are almost completely depreciated that could break down and significantly impact the unit's overhead or expense cost?

3. What are my operating and investment costs for public affairs (community involvement and donations) and marketing and are they in line with the revenue and other competitors?

4. Are we achieving the unit's profit or service goals and the corporate or agency's goals? Can the output be improved?

These questions and corresponding answers can help to position the business unit strategically and can result in successful profitable operations or public service merit awards.

Financial Leadership

Financial leaders know and comprehend the value of the services and products produced by the business unit in relation to materials, overhead costs, direct costs, and expenses. The leader understands how these costs and expenses are related to one another, how to reduce these cost impacts, and how to increase output. The leader also understands that the unit's exemplary service (or profit) provides the fuel for the unit to pay the salaries, provide training, pay the rent, fund special events like parties and picnics, all of which are necessary to maintain the engine to continue accomplishing the work.

Financial leaders understand the balance between the cost of business and revenue required for investing in the vision in the strategic plan. If we could stroll through the financial leader's mind we would observe the thought patterns on:

General Financial Concepts

1. Does staff understand the cost of doing business and how they may help lower overhead expenses?

2. Are our unit's marketing costs and overhead costs in line with the competition and the market?

3. Does the unit have adequate cash flow, reserve capacity, a relationship with a financial institution to secure quick loans at reasonable costs?

4. Are the unit's direct costs in line with the indirect costs?

5. Am I receiving a just compensation for the unit's products and services and are there ways to increase mark-ups for outside vendors or materials?

Risk-Related Financial Concepts

1. Does the unit employ adequate risk and uncertainty analyses to protect the unit's viability?

2. Have the key unit managers received important training on negotiating, interviewing, sensitivity, prevention of sexual harassment, defensive driving, and other mandated training by the personnel department?

3. Does the unit have adequate safety and health practices to protect the employees and is the unit protected/insured from lawsuits on related issues?

Project Management

1. Do the unit's project managers understand the budgets we create have a tremendous impact on the unit's overall performance and viability?

2. Do the PMs understand that if a PM loses $50,000 on a particular project that this represents the total profit on a $1 million project if the profit were set at 5 percent?

3. Are the PMs carefully watching for scope creep and prompt invoicing to maintain the unit's revenue stream?

4. Do the PMs understand the cost of money for delayed invoicing and unbilled labor?

5. Do we have adequate reserve funds for unanticipated impacts?

6. Do the PMs understand when an issue is out of the contract's scope and how to create a change order?

7. Do the PMs understand our competition's strengths and weaknesses compared to our own unit?

Technical Leadership

Technical leaders display technical acumen, excellent proficiency in engineering and ingenious application of solutions to client problems. In addition, the technical leader usually possesses strategic and financial leadership qualities, too. These leaders are often on the cutting edge of new technology and developments and include services like:

- Research and Development
- Process Development
- Technical Services
- Customer Support
- Specific Quality Cells (for example, the Geotechnical Engineering Quality Cell or the Environmental Engineering Quality Cell)
- Product Development, among others

Technical leaders often publish their works in magazines or periodicals to gain professional recognition in their field of expertise. These published leaders can then offer their prospective clients the latest journal article and may gain technical points for receiving future work.

Technical leaders understand the balance between the investment in technical innovation and investing in the vision in the strategic plan. If we could stroll through the technical leader's mind we would observe the thought patterns:

Innovation

1. Are the unit's technical leaders aware of the most recent developments in the areas of expertise locally, nationally, and internationally?
2. Does the organization encourage critical thinking and an atmosphere conducive to process improvement and innovation?
3. Does the organization recognize individuals who have demonstrated innovation or award programs for encouragement?
4. Can the technical leaders identify innovative technologies that fit potential pilot or full-scale applications for projects in the unit's practice area?
5. Are there any grant funds or private funds available for furthering the application of new technologies on the unit's practice area?

Organization

1. Does the organization attract top technical talent and if not, what can be done to improve this condition?
2. Does the organization have a training and/or mentorship program?
3. Does the organization encourage and compensate colleagues for pursuing secondary degrees or other specialized coursework?
4. Does the organization encourage colleagues to communicate potential problems early in the project when the cost to correct a situation is usually lower?

Vision

1. Do the technical leaders display critical thinking and offer innovative perspectives to today's challenges?
2. Do the technical leaders display strategic thinking for application adopting technology to today's challenges?
3. Do the technical leaders comprehend and accept there are realistic limits to the investment of technology for the organization?
4. Are these leaders willing to accept their roles in the daily tasks to run the organization while maintaining their roles as technical leaders?

Marketing Leadership

Marketing Leadership generally applies to consulting engineers as business development activities. In public service, engineers working for the government also display leadership but generally with a different purpose. Both leaders are discussed below.

PUBLIC SERVICE (FOR GOVERNMENT EMPLOYEES) OR MARKETING LEADERSHIP (FOR CONSULTING ENGINEERS)

For the purposes of this discussion, the information presented for consulting engineers is related to "marketing" the firm and its employees and the information presented for government employees is related to public service. The critical elements for each are very similar and relate to the outreach efforts to communicate the service capabilities of these respective organizations. Marketing and public service leaders understand the value of the services and products produced by the business unit and recognize that if their services do not meet the expectations of the users, the existence of the business unit will be jeopardized. Regardless whether it's private industry or public service, these leaders are continually challenged with producing more results in less time with improved service. The leadership in this particular area is very dependent on accurate visual aids, good presentation skills, and clear communications.

Public service and marketing leaders understand the balance between the cost of presentation and outreach and consistency with the vision in the strategic plan. If we could stroll through the public service and marketing leaders' minds we would observe the thought patterns on:

Outreach

1. Do the public service and marketing leaders make themselves available for outreach efforts and do they understand the criticality of this effort?
2. Can the leaders prepare and present effective presentations that communicate the message in a concise and deliberate manner?
3. Do the leaders display respectful and tactful responses to questions and comments from the clients or public?
4. Do the leaders participate in open forums, strategy sessions, and workshops in a manner that exhibits genuine concern for the stakeholders?
5. Do these leaders relate well to people from all levels of society?

Marketing and Presenting the Concept

1. For public service, do the leaders understand the costs associated with democratic participation, public awareness of the business unit, and the subject of the presentation?
2. For consulting engineers, do the leaders understand the costs associated with the costs of marketing, strategic positioning, sales presentations, proposal preparation, short-list interviews, getting the project, and maintaining the client relationship?
3. Can the leaders express the concepts well and empathize with the impacted listeners' concerns?

4. Do the leaders understand how to plan, prepare, and execute the public presentations?

5. Do the leaders understand the value of customer service?

SECRET RECIPE FOR AN EFFECTIVE LEADER

This chapter has provided a lot of discussion on the tools for effective managers, leadership styles, and the foundational strengths of effective leaders. So in an effort to describe a virtual effective leader a "perfect recipe" has been devised. When engineering managers carefully shop for the right individual leader(s) to fit their specific environment, they are more likely to achieve success for their organization.

Ingredients of a Good Leader:

- Start with an energetic individual with a thirst for learning. Now add the following:
  - A liter of strong character
  - A liter of humility
  - Two liters of organization
  - A kilogram of intelligence
  - A pinch of empathy
  - A kilogram of courage
  - A liter of confidence
  - Preheated trust and ethics
  - A liter of reliability mixed with follow-through
  - A genuine smile
- Now, mix the above ingredients together with demonstrated communication skills, add strong listening abilities, and sensitivity training.
- Simmer over even-heated mentorship with years of experience, continued training, advanced technical/business degree, and dedication to common objectives.
- Garnish with politeness, business attire, and a genuine smile (yes, a fresh one).
- Serve warm to the public (or clients) after a brief introduction in a pleasant environment.

SUMMARY

The three major styles of leadership are:

- Autocratic—telling or demanding
- Democratic—interactive and consensus building
- Delegative—or free reign

The management style should be adapted to the specific situation and the individuals being managed. Effective managers employ the tools of planning, organizing, leading, and controlling to accomplish the mission of the organization. *A manager/leader carefully balances the use of feedback from previous quarterly results while strategically leading the business unit to the future.*

Dynamic individuals can become leaders in one or more business component area including strategic leadership, financial leadership, technical leadership, and public service/marketing leadership.

REFERENCES

Altier, William J. (1999). *The Thinking Manager's Toolbox: Effective Processes for Problem Solving and Decision Making.* Oxford University Press. ISBN: 0-195-13196-7.

DuBrin, Andrew J. (2000). *The Active Manager: How to Plan, Organize, Lead and Control Your Way to Success.* January 2000. South-Western College Press, Chula Vista, CA. ISBN-13: 978-0-324-02740-2.

Goffee, Rob and Gareth Jones. (2009). *Clever: Leading Your Smartest, Most Creative People.* Harvard Business Press, Watertown, MA. ISBN 978-1-422-12296-9.

Northouse, P. (2010). *Leadership: Theory and Practice.* Sage Publications, Thousand Oaks, CA. ISBN: 978-1-412-97488-2.

U.S. Army Handbook. (1973). "Military Leadership."

Introduction

Ethics

1 2 3 4 5 6 7 8 9 10 11 12 13 14 15 16 17 A B C D E F

History

Professional Engagement

What Engineers Deliver

Engineer's Role in Project
Development

Permitting

Executing a Professional
Commission

Leadership Managing

Client Relationship

Legal Aspects

Communicating

Having
a Life

Emerging Technology

Sustainability

Globalization

Chapter **11**

Legal Aspects of Professional Practice

Big Idea

All civil engineers, whether employed in public or private practice, need to know the legal consequences of their actions.

> Lawsuit: A machine which you go into as a pig and come out as a sausage.

> —Ambrose Bierce

Key Topics Covered

- Introduction
- U.S. Legal System
- Statutes
- Common Law
- Contract Law
- Contracts in Project Delivery
- Risk Management
- Insurance and Bonds
- Dispute Resolution
- Alternative Dispute Resolution
- Affirmative Action, Equal Opportunity, and Diversity
- Summary

Related Chapters in This Book

- Chapter 3: Ethics
- Chapter 4: Professional Engagement
- Chapter 5: The Civil Engineer's Role in Project Development
- Chapter 6: What Engineers Deliver
- Chapter 13: Communicating as a Professional Engineer

(Continued)

Related to *ASCE Body of Knowledge 2* Outcome

ASCE BOK2 outcomes covered in this chapter

Foundational
1. Mathematics
2. Natural sciences
3. Humanities
4. Social sciences

Technical
5. Materials science
6. Mechanics
7. Experiments
8. Problem recognition and solving
9. Design
10. Sustainability
11. Contemporary issues & historical perspectives
● 12. Risk and uncertainty
13. Project management
14. Breadth in civil engineering areas
15. Technical specialization

Professional
16. Communication
17. Public policy
18. Business and public administration
19. Globalization
20. Leadership
21. Teamwork
22. Attitudes
23. Lifelong learning
24. Professional and ethical responsibility

INTRODUCTION

The purpose of this chapter is to give civil engineers a basic overview of the U.S. legal system and to highlight some potential legal danger zones. The chapter discusses key legal issues that affect design professionals and suggests what can be done to avoid legal pitfalls. Specifically, the chapter addresses: tort law; contract formation and clauses; contract structures used in various project delivery methods; litigation; alternative dispute resolution; the role of insurance and indemnification; and affirmative action, equal opportunity, and diversity.

Regardless of scale, constructed facilities—dams, highways, airports, water treatment plants, office buildings, shopping malls, industrial facilities, houses—use the *project* as an organizing device. In the construction industry, projects tend to be arranged in a way that epitomizes the fragmented form of hierarchical network organization, in which contracts are highly specialized. The project team members are assembled temporarily, for just as long as is necessary, from the array of players outlined in Chapter 5, The Engineer's Role in Project Development. The usual approach relies on price-based contracting methods (market transactions). This often results in adversarial relations between participants, in which information is concealed and information flows are disrupted. In other words, the Architectural, Engineering, and Construction (AEC) industry is rife with miscommunications and misunderstandings. Such an environment creates the need to be aware of the legal consequences of one's actions and knowledgeable enough to know when the advice of an attorney should be sought.

U.S. LEGAL SYSTEM

As depicted in Figure 11.1, the U.S. legal system is divided into two distinct branches: *Statutory Law* and *Common (or Civil) Law*. Statutory law concerns itself with the rules of behavior—statutes—enacted by a legislative body. Common law is divided further into two branches—*Contract Law* and *Tort Law*. Not surprisingly, contract law is based on contracts. Tort law is based on *precedents*, or prior rulings.

Criminal and common law proceedings can arise from the same conduct. One of the more infamous examples involves the famed football player O.J. Simpson, who was found innocent of murdering his wife under an applicable criminal law statute, but was held liable for her death in a civil proceeding.

Generally, American legal principles related to contracts and torts are derived from the English Common Law in what can be termed as the "Anglo-American Common Law" or "Common Law system." In contrast, the "Civil Law system" is typically used to signify a legal system based on the Roman Law system. Civil law–based legal systems are found in Continental Europe and in Louisiana in this country. Creating the potential for confusion is the fact that the American legal system is divided into two basic components, criminal proceedings and civil matters. The latter can include both civil actions based on contract obligations and tort (typically negligence) actions. Actions on warranties have elements of both contracts and torts.

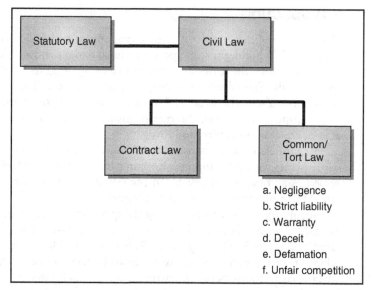

Figure 11.1 U.S. legal system

Statutory law reflects policies established by a legislative body on the national, state, or local level. Legislative bodies enact statutes making certain conduct criminal. However, statutes are routinely enacted that affect contract obligations and potential tort liability. For example, a state statute may make a pay-if-paid clause (e.g., a prime designer must pay a subconsultant within a specified period of time after having received payment from the owner) unenforceable on contracts performed in that state. Whether the action is in contract or in tort, court precedents are relied upon to interpret contract language, for example, to determine whether certain conduct (action or inaction) creates or exposes a party to tort liability for damages in a civil (not criminal) action. Even in the area of criminal law, courts rely upon legal precedent in determining the intent of the language in a particular statute.

Contracts are the primary means to define the parties' respective obligations or duties and rights in a commercial transaction. One basic principle of the Anglo-American Common Law system of contracts is the concept of "freedom of contract." That is, the parties to a contract are basically free to agree to the terms of a contract unless those terms violate a public policy, usually set forth in a statute. For example, in some states the parties are free to agree to establish by contract a specific statute of limitations on actions for beach of that contract, for instance, a one-year period to sue on a payment bond. However, many legislatures will limit the parties' freedom to contract away or limit certain rights.

STATUTES

Legislative bodies such as the U.S. Congress, state legislatures, county boards of supervisors, and city councils enact statutes or ordinances affecting civil or commercial matters as well as criminal conduct. An example of a statute affecting a potential commercial arrangement is a licensing statute controlling the right to practice engineering in a particular state. That type of law may include provisions affecting civil obligations, for example, the validity of a contract to provide engineering services *and* may also include a criminal sanction, for example, a fine, for failing to comply with the licensure requirements.

While statutory law is sometimes viewed as including only criminal law, that is a too narrow definition. Statutes do define criminal conduct as serious crimes, frequently involving violence; these are called *felonies*, commonly punished through incarceration. Less serious crimes are called *misdemeanors*, usually punished by a relatively small fine, such as parking tickets. Alleged infractions of criminal laws or ordinances are prosecuted by *the people*, for instance federal, state, or county attorneys general. Judgment is rendered by a *trier of fact*—a judge and/or jury—who also determines the penalty.

COMMON LAW

Though civil engineers can be prosecuted under a criminal statute for criminal negligence, among other crimes, most civil engineering danger zones exist in contract or tort actions based upon Anglo-American Common Law principles, as modified by particular federal, state, or local statutes/ordinances. Actions in contract or tort are brought by companies or individuals, rather than the state. Generally, these actions are termed civil actions, as distinguished from criminal actions. Even though labeled as civil actions, they are based upon the common law rather than the civil law system found on the European continent and in Louisiana in the United States. A civil action is an adversarial system in which an aggrieved party—the *plaintiff*—takes action against the party alleged to be at fault—the *defendant*. One key feature of a civil action in the American legal system is the concept of *discovery*, which allows each party to claim the right to examine the other the party's evidence and witnesses prior to court proceedings.

As noted above, the common law has two main divisions: *Contract Law* and *Tort Law*. Contract law's main legal concerns involve whether a contract has been broken, or *breached*. The meaning of certain words and phrases establishes the enforceability of various terms and conditions. If contracts could not be enforced, commerce would come to a standstill.

Tort Law concerns itself with *torts*, civil wrongs for which a court of law will grant a remedy. *Civil wrongs* presuppose *norms of behavior* by which society expects people to abide. The *triers of fact* determine:

- What these norms of behavior are
- Whether or not they have been violated
- What the remedy or damages should be

Tort Law

Tort law also presupposes that individuals have a *duty* to uphold these norms of behavior. Therefore, tort also encompasses the idea of *breach of duty*, for which the law will grant a remedy.

In establishing norms of behavior and related issues, courts look at prior rulings, called *precedents*. Some national design professional associations have legal funds used to appeal unfair or unreasonable rulings to prevent them from becoming standards. Left unchallenged, such decisions could become the equivalent of laws, not only in the jurisdiction where the issue is resolved, but also where the precedent is reviewed. New tort law emanating from California can be adopted by civil courts throughout the nation, including federal courts.

This situation has a significant influence on the way engineering and architecture are practiced. For many years engineers and architects (A/Es) were protected by the concept that they owed a duty of care only to their clients; a third party could not claim negligence. This concept has substantially eroded. Tort law holds that design professionals owe a duty of care to anyone who could be damaged physically or monetarily.

According to John Bachner, author of *Practice Management for Design Professionals: A Practical Guide to Avoiding Liability and Enhancing Profitability,* issues most likely to affect civil engineers in tort law include:

- Negligence
- Strict liability
- Warranty
- Deceit
- Defamation
- Unfair competition

Negligence

Importantly, an error or omission is not necessarily negligence. In order for a plaintiff to win a negligence claim, five conditions must be proven as fact:

1. Defendant was required to abide by standard practice.
2. Defendant owed a duty of care to the plaintiff(s).
3. Defendants breached that duty of care.
4. There was a causal connection between the breach and alleged injury.
5. The injury was real.

Condition 1: Standard of Practice

In claims filed against design professionals, plaintiffs do not usually need to establish that a standard of practice applies. Historically, professionals have been bound to abide by standards of practice. What is important is to establish what the standard of practice was at the time it was allegedly breached.

> Standard of practice definition: "The ordinary skill and competence exercised by members of a profession in good standing in the community at the time of the event creating the cause of action."
>
> - Ordinary skill and competence: "ordinary"—professionals are not required to provide the highest level of skill
> - Members of a profession in good standing: peers providing same or similar services
> - In the community: if very specialized, could be national or global; otherwise local, but it must be established
> - At the time of the event creating the cause of action: needs to be evaluated when alleged negligent act was committed

Condition 2: Duty Owed

Absence of a contractual relationship, *privity*, is not a defense. However, the duty must be foreseeable.

Condition 3: Breach of the Standard Practice

Plaintiffs and defendants hire their own expert witnesses, who are supposed to serve the trier of fact (judge and/or jury). Although their role is not one of advocacy, some expert witnesses become "hired guns." Opposing council can defeat or impeach their testimony by pointing out errors, contradictions, weaknesses; however, this can be difficult because most attorneys lack an intimate understanding of civil engineering. Expert witnesses should be required to perform the necessary research, but this does not necessarily happen. The problem can be overcome by working with one's own experts, but there may not be sufficient time. Also, frequently, a jury is more difficult to educate than a judge, so plaintiffs often request jury trials.

Condition 4: Causal Connection

Showing that a civil engineer may have breached the standard of practice is not sufficient to win a suit. The plaintiff must show that the negligent act was the *actual cause* or *proximate cause*. The *actual cause* can be established by the *but for* or *main factor*: but for the defendant's negligence, the damage would not have occurred. The negligent act was the main factor in events resulting in damage. Establishing *proximate cause* requires consideration of three factors: the number of possible causes, the presence of intervening causes, and the foreseeability of the consequences. The plaintiff would need to show that the defendant's actions were closest to the sole cause of the problem.

Condition 5: Real Injury

Finally, plaintiffs must demonstrate the value of the losses in order to establish compensatory damages, in the form of actual losses and/or pain and suffering. If there have been no actual damages, under the principle of *unjust enrichment* no award can be made.

Strict Liability

Strict liability most frequently applies to product manufacturers. Negligence does not need to be proved. The plaintiff simply must show that:

- The product had a defect
- The defect existed when the product was acquired by the purchaser or user
- The defect caused or contributed to an injury
- Under, normal use, the product failed

In some courts, a house can be considered a product; but generally, most constructed facilities are not. A problem can arise when plans, specifications, and reports are viewed as products. Clearly, very few, if any, plans and specifications are flawless. To limit exposure to strict liability laws, in contracts and correspondence between the civil engineer and his or her clients, using the term *instruments of service* is preferable to *plans and specifications*.

Warranty

A warranty promises that things are exactly as they are represented. When civil engineers warrant their work, they are opening the door to the doctrine of strict liability. There are two types of warranties: express and implied. Express warranty exists when language is present such as "I guarantee or warrant that . . . "; when the accuracy of a test is attested to; or when a civil engineer assumes complete responsibility for the accuracy of his or her statements, for instance, makes unequivocal statements. Implied warranty is a problem in jurisdictions where plans, specifications, and reports are treated as products.

Deceit

A typical example of deceit occurs when a salesperson withholds or misrepresents information while a customer is making a purchase decision. The salesperson makes false statements with the intent of altering the customer's position. However, civil engineers may be subjected to charges of deceit when negligence cannot be proved.

Defamation

Civil engineers can be exposed to charges of defamation when they publish written derogatory statements about others that may expose plaintiffs to public hatred and ridicule. Again, the plaintiff must suffer injury and damages. The spoken version of defamation is *slander*. The oral statement must be made before witnesses who understand that the statement is disparaging.

Unfair Competition

Charges of unfair competition arise when a plaintiff believes that a civil engineer's actions have interfered with his or her commercial activities. There are several versions of unfair competition:

- Commercial disparagement—a published statement that is false, injurious, made with malice, interferes with the plaintiff's commercial relationships, and results in special damages (e.g., cost of redesign)
- Interference (malicious) with contractual relations—an intent to interfere and actual interference with contractual relations
- Interference with prospective economic advantage—same as above, except that no contract exists

Unfair competition also may be mentioned in some professional organizations' codes of ethics. (For more information, see Chapter 3, Ethics.)

> Joint and several liability: Defendants can be held liable both collectively and individually for all damages, regardless of their involvement or degree of fault.

Statutes of Limitation and Repose

Several other concepts are very important when considering the law's influence on civil engineering practice. An effective argument civil engineers can use to fend off claims is that the *statute of limitation* or *statute repose* has expired, that is, that the claim is time-barred. Under a statute of limitation, the clock starts running once a defect has been discovered. If the state's statute of limitation is four years, the aggrieved party has four years to file a suit. Frequently, this can be many years after the design has been completed. A statute of repose begins after construction is substantially complete. Thus, if a defect is found five years after substantial completion and the statute of repose is four years, a claim for construction defects would be banned.

Under some circumstances, designers could have lifelong liability. Professional organizations, such as the ASCE and AIA, monitor developments in various courts around the country and generally mount appeals to decisions that might become dangerous precedents. Conversely, many courts wish to protect statutes of repose to ensure protection to consumers of professional engineering and construction services.

> A *statute of limitation* begins once a defect has been discovered.
>   A *statute of repose* begins after construction is substantially complete.

CONTRACT LAW

A contract is a legally binding agreement that sets forth each party's responsibility to each other. No project should be pursued without a written agreement. To do otherwise would leave the assignment of responsibilities to assumption and key provisions to the vagaries of memory. The contract is a key aspect of client-design professional communication. Violations of responsibility and/or obligation can result in legal action.

There are many terms used to categorize contracts:

- *Bilateral/unilateral*—involving two parties/involving one party
- *Enforceable/unenforceable*—containing all necessary elements/not containing all necessary elements (e.g., statute of frauds requiring certain contracts to be in writing to be enforceable or statute of limitations has passed)
- *Void*—missing one or more elements, perhaps due to an oversight
- *Voidable*—giving a party the right to call the contract void
- *Express/implied*—agreeing explicitly/relying on parties' actions toward one another
- *Written/verbal*—text or orally based

For a Contract to Be Binding, Six Elements Must Be Present

1. Agreement—acceptance of an offer
2. Consideration—agreed upon or perceived value
3. Legality—adherence to public policy (enforceability)
4. Authentication—signature or corporate seal (attestation)
5. Capacity—signatories sane and have authority to represent business entity
6. Legal purpose—subject of contract must be legal

Contract Formation

Several issues need to be considered when forming a contract. These include assumption of liability, professional liability insurance, disparate bargaining power, and indemnifications.

Assumption of Liability

Liability in tort law revolves around negligence; liability in contract law centers on whether a contract provision has been breached. If a civil engineer *agrees in a contract to act in a nonnegligent manner,* a negligent act would create both tort and contractual liability. A good approach is to avoid provisions that obligate more assumption of responsibility than common law requires. This would include provisions that require civil engineers to perform at the "highest professional level." The standard of practice requires only "ordinary skill and competence."

Professional Liability Insurance

Assumption of liability has major implications for professional liability insurance coverage. Most policies exclude coverage of liabilities assumed in the contract, a fact that should be discussed with clients. If well-informed by the civil engineer, clients may be persuaded to drop such problematic clauses, pay the added cost for such coverage, or may become interested in including additional services (e.g., field observation).

Disparate Bargaining Power

Disparate bargaining power occurs when one party has an unfair advantage over another during contract negotiation. A *contract adhesion (to be "stuck")* can result when one party to the contract appears to have little power in relationship to the other. This can happen, for example, when a very large client offers the only opportunities for commissions in a small town. A civil engineer, who is forced to accept an onerous contract clause in order to secure a commission, can follow-up with a letter to the client stating that the provision was accepted because the civil engineer needed the work and the principle of disparate bargaining power was at work.

Indemnifications

A good definition of *indemnification* is "to secure against loss or damage." Contract clauses can include indemnifications to protect either party. For example, the client could indemnify the civil engineer for any certifications required. Construction contracts between the contractor and the owner often indemnify owners and design consultants to protect them from being sued by workers and/or visitors injured on the jobsite.

The best measures to take when entering into a contract with a client are to:

- Perform professionally
- Create realistic expectations
- Make clients aware of risks
- Minimize one's own (engineer's) risk
- Obtain liability insurance and be sure of coverage
- Consult a knowledgeable attorney

Contract Wording

Words are never more important than when used in contracts, and common words must be selected and used with care. Words such as "all," "every," "none," "whose," "who," "he/she," "his/her" can be significant. Definitions are needed because interpretations of various words and phrases may differ dramatically. For example, "hazardous materials" may mean different things in federal, state, and/or private work.

Attorneys play an important role in reviewing contracts for appropriate language and conformance with applicable clauses and laws. However, civil engineers should not abdicate their responsibility to attorneys. A contract is a communication tool, and reaching a mutual understanding of responsibilities and restrictions with the client can set the tone for all the work that follows.

Typical Contract Formats

Contracts assume various forms; but at a minimum, regardless of the form, all contracts for civil engineering services should include:

1. Scope of services
2. General conditions
3. Performance schedule
4. Fee proposal

Some of the more commonly used contract formats include: conventional proposals, negotiated terms and conditions, multiple contracts, special contracts for major projects, client-developed contracts, purchase orders, and model contracts.

Conventional Proposals

As discussed in Chapter 4, Professional Engagement, civil engineers often acquire work by preparing a proposal in response to a Request for Proposal (RFP) or Request for Qualifications (RFQ). In qualifications-based selection (QBS), the work scope included in the proposal becomes part of the contract between the client and civil engineer. If the client is using a fee-based selection method, the client's proposal should have (but not always does have) a well-defined statement (scope) of work (SOW). In either case, the SOW is an important component of conventional proposals. However, if the SOW is not well conceived, the civil engineer can be held responsible for an impossible-to-deliver, unilateral workscope. In QBS, the civil engineer's proposal to the client can clarify the SOW; in fee-based proposals, the civil engineer's cover letter can identify areas needing modification.

These conventional proposals also typically include a section under the heading of *General Conditions*. General conditions are nontechnical understandings between the two parties to the contract. They include the business context and mutual responsibilities, such as when payments are due and any limitations of liability. Most design firms have developed standard general conditions.

Whether the selection method being used is two-envelope QBS or fee-based, conventional proposals will include some mention of fee, generally including staff rates/time, computer use, testing, and a schedule of fees and costs for additional services. A schedule for the services provided also will be incorporated into the contract. The schedule should depict milestones, such as completion dates and dates of major deliverables, noted elsewhere in the contract.

Negotiated Terms and Conditions

Negotiation of contract terms and conditions is not strictly related to monetary issues. Often the discussion centers on scope of services needed. Sometimes the client may not agree to everything in the general conditions, such as indemnification provisions. Clients and the civil engineer may want to shift risk; and contemplating the assignment of risk is best done in the earliest stages in order to minimize its effect. Sometimes general conditions are rewritten, or an *addendum* (*Special Conditions*) may be attached to the contract.

Multiple Contracts

Sometimes owners contract separately and simultaneously with the prime (designer or contractor) and the subconsultants or subcontractors. This approach gives the owner more management responsibility and more control. Subconsultants have greater access and may receive more prompt payment. Additionally, the prime designer may reduce "vicarious liability," in other words, exposure to liability stemming from contractual relationships with other design professionals.

Special Contracts for Major Projects

For large projects, standard general conditions seldom suffice. Custom contracts with standard contract clauses frequently are developed, requiring attorney involvement.

Client-Developed Contracts

Large clients can be powerful, and they sometimes attempt to shift risk (liability) to their designers—engineers and architects. Although the attitudes of large clients may be difficult to change initially, disputes may be even more difficult to win later. Some client-developed contracts attempt to shift more liability to the designer than is required by law or custom. Typically, these additional risks are not accepted by insurance companies.

In such cases, the civil engineers may become the client's source of "free" insurance. To offset this increased risk, civil engineers can:

- Read RFPs and/or RFQs carefully—responses can become part of the contract
- Have their attorneys review the client-developed contract very carefully
- Charge a fee premium over the usual amount charged for similar services
- List outstanding issues in a cover letter accompanying the proposal and negotiate later
- Attempt to make the client accept risk

Civil engineers need to proceed cautiously with clients who accept risk too willingly. Clients must be able to honor the changes they have agreed to, that is, they need to have sufficient monetary resources to cover increased risk.

Purchase Orders

Occasionally, clients use purchase orders to hire civil engineers. This typically happens when a public client, who may have a critical need, does not have time to conduct a formal selection process. Public contract codes limit the amount public agencies can commit through the use of purchase orders, so projects using this type of contract tend to be small. Purchase orders are designed primarily to purchase materials and are not really appropriate for professional engineering services. Civil engineers need to exercise good judgment when signing such agreements.

Model (Standard Form) Contracts

Model contracts are developed by professional associations such as:

- American Institute of Architects (AIA)
- ConsensusDOCS LLC (Associated General Contractors (AGC) and 20 other organizations)
- Design Build Institute of America (DBIA)
- Engineers Joint Contract Development Committee (EJCDC)—a consortium of the American Council of Engineering Companies (ACEC), Associated General Contractors (AGC), American Society of Civil Engineers (ASCE), and the National Society of Professional Engineers (NSPE)

The documents produced by these organizations provide a vital function. They offer an economical way for parties to contract with one another without "lawyering up." Each organization has retained attorneys to develop contracts (including general conditions) on behalf of their membership. Most contract clauses have been tested over time, and the documents are revised periodically to reflect changes in law. (See Table 11.1 for a partial list of readily available model contracts and Appendix F for examples.)

Internationally, the International Federation of Consulting Engineers (FIDIC) is a leader in standard form contracts. The FIDIC, headquartered in Lausanne, Switzerland, is a coalition of international, independent consulting engineers. Its forms are widely used in developing countries and are recognized by the World Bank. The Joint Contracts Tribunal (JCT) for the Standard Form of Building Contract publishes documents commonly used in the United Kingdom. The Engineering Advancement Association of Japan (ENAA) publishes contracts, also recognized by the World Bank, used on power plant projects constructed on a design-build basis.

Although these organizations attempt to create contracts that are fair and balanced, it should be noted that there may be some bias in favor of their membership.

Table 11.1 Commonly Used Model Contracts

Origin	Contract Number	Contract Name
American Institute of Architects (AIA)	A191™	Standard Form of Agreements Between Owner and Design/Builder
	A195™	Standard Form of Agreement Between Owner and Contractor for Integrated Project Delivery (2008)
	A201™	General Conditions of the Contract for Construction (2007)
	A295™	General Conditions of the Contract for Integrated Project Delivery (2008)
	A503™	Guide for Supplementary Conditions (2007)
	B101™	Standard Form of Agreement Between Owner and Architect (2007)
	B102™	Standard Form of Agreement Between Owner and Architect without a Predefined Scope of Architect's Services (2007)
	B103™	Standard Form of Agreement Between Owner and Architect for a Large or Complex Project (2007)
	B104™	Standard Form of Agreement Between Owner and Architect for a Project of Limited Scope (2007)
	B108™	Standard Form of Agreement Between Owner and Architect for a Federally Funded or Federally Insured Project (2009)
	B195™	Standard Form of Agreement Between Owner and Architect for Integrated Project Delivery (2008)
	B201™	Standard Form of Architect's Services: Design and Construction Contract Administration (2007)
	B202™	Standard Form of Architect's Services: Programming (2009)
	B203™	Standard Form of Architect's Services: Site Evaluation and Planning (2007)
	B204™	Standard Form of Architect's Services: Value Analysis, for use where the Owner employs a Value Analysis Consultant (2007)
	B205™	Standard Form of Architect's Services: Historic Preservation (2007)
	B206™	Standard Form of Architect's Services: Security Evaluation and Planning (2007)
	B207™	Standard Form of Architect's Services: On-Site Project Representation (2008)
	B209™	Standard Form of Architect's Services: Construction Contract Administration, for use where the Owner has retained another Architect for Design Services (2007)
	B210™	Standard Form of Architect's Services: Facility Support (2007)
	B211™	Standard Form of Architect's Services: Commissioning (2007)
	B214™	Standard Form of Architect's Services: LEED® Certification (2007)
	B352™	Duties, Responsibilities and Limitations of Authority of the Architect's Project Representative, recommended as a reference document when an Architect's Project Representative is employed (2000)
	B503™	Guide for Amendments to AIA Owner-Architect Agreements (2007)
	B727™	Standard Form of Agreement Between Owner and Architect for Special Services (1988)
	B901™	Standard Form of Agreements between Design/Builder and Architect
	C101™	Joint Venture Agreement for Professional Services (2007)
	C401™	Standard Form of Agreement Between Architect and Consultant (2007)

(*Continued*)

Table 11.1   (*Continued*)

Origin	Contract Number	Contract Name
DBIA*	DBIA 501	Design Build Consultant Services Agreement (2010)
	DBIA 520	Owner/Design Builder Preliminary Agreement(2010)
	DBIA 525	Owner/Design Builder Lump Sum Agreement (2010)
	DBIA 530	Owner/Design Builder Cost Plus Fee with Option for GMP Agreement (2010)
	DBIA 535	Owner/Design Builder General Conditions (2010)
	DBIA 540	Design Builder/Design Consultant Agreement (2010)
Engineers Joint Contract Development Committee (EJCDC)	E-500	Standard Form of Agreement Between Owner & Engineer for Professional Services (2008)
	E-505	Standard Form of Agreement Between Owner and Engineer for Professional Services, Task Order Edition (2009)
	E-520	Short Form of Agreement Between Owner & Engineer for Professional Services (2009)
	E-530	Standard Form of Agreement Between Owner & Geotechnical Engineer (2006)
	E-525	Standard Form of Agreement Between Owner and Engineer for Study and Report Phase (2009)
	E-560	Standard Form of Agreement Between Engineer & Land Surveyor for Professional Services (2007)
	E-564	Standard Form of Agreement Between Engineer & Geotechnical Engineer for Professional Services (2006)
	E-568	Standard Form of Agreement Between Engineer & Architect for Professional Services (2006)
	E-570	Standard Form of Agreement Between Engineer & Consultant for Professional Services (2006)
	E-582	Model Form of Agreement Between Owner and Program Manager (2004)
	E-990	Owner Engineer Documents, Full Set
	E-991	Engineer Subconsultant Documents, Full Set
	C 1910-40	Standard General Conditions of the Contract Between Owner and Design-Builder
	C 1910-41	Standard Form of Subagreement Between Design-Builder and Engineer for Design Professional Services
	R-001	Commentary on EJCDC Environmental Remediation Documents (2005)
	R-520	Standard Form of Agreement Between Owner & Environmental Remediator, Stipulated Price (2005)
	R-521	Standard Form of Agreement Between Environmental Remediator & Subcontractor, Stipulated Price (2005)
	R-525	Standard Form of Agreement Between Owner & Environmental Remediator & Subcontractor, Cost-Plus (2005)
	R-526	Standard Form of Construction Subagreement Between Environmental Remediator & Subcontractor, Cost-Plus (2005)
	R-700	Standard General Conditions of the Contract Between Owner & Environmental Remediator (2005)
	R-750	Standard General Conditions of the Subagreement Between Environmental Remediator & Subcontractor (2005)
	R-990	Environmental Remediation Documents, Full Set

Table 11.1 (*Continued*)

Origin	Contract Number	Contract Name
Consensus DOCS**	ConsensusDOCS 300	Tri-Party Collaborative Agreement (Owner, Designer, and Contractor all sign the same agreement —LEAN construction approach, also known as alliancing or relational contracting)
	ConsensusDOCS 301	Building Information Modeling (BIM) Addendum
	ConsensusDOCS 310	Green Building Addendum
	ConsensusDOCS 400	Preliminary Owner/Design-Builder Agreement For use in conjunction with ConsensusDOCS 410 or ConsensusDOCS 415
	ConsensusDOCS 410	Owner/Design-Builder Agreement and General Conditions (Cost Plus w/GMP)
	ConsensusDOCS 415	Owner/Design-Builder Agreement and General Conditions (Lump Sum)
	ConsensusDOCS 420	Design-Builder and Architect/Engineer Agreement
	ConsensusDOCS 421	Design-Builder's Statement of Qualifications for a Specific Project
	ConsensusDOCS 450	Design-Builder/Subcontractor Agreement
	ConsensusDOCS 470	Design-Builder Performance Bond (Surety Liable for Design Costs)
	ConsensusDOCS 471	Design-Builder Performance Bond (Surety Not Liable for Design Costs)
	ConsensusDOCS 472	Design-Builder Payment Bond (Surety Liable for Design Costs)
	ConsensusDOCS 473	Design-Builder Payment Bond (Surety Not Liable for Design Costs)
	ConsensusDOCS 481	Certificate of Substantial Completion for Design-Build Work
	ConsensusDOCS 482	Certificate of Final Completion for Design-Build Work
	ConsensusDOCS 491	Design-Builder's Application for Payment (Cost Plus, w/GMP)
	ConsensusDOCS 492	Design-Builder's Application for Payment (Lump Sum Contract)
	ConsensusDOCS 495	Design-Build Change Order (Cost Plus, w/GMP)
	ConsensusDOCS 496	Design-Build Change Order (Lump Sum)
	AGC 455	Standard Form of Agreement Between Design-Build Contractor and Subcontractor (where the Design-Builder and the Subcontractor share the risk of owner payment)
	AGC 465	Standard Form of Agreement Between Design-Build Contractor and Design-Build Subcontractor (where Subcontractor provides a guaranteed maximum price and where Design-Builder and Subcontractor share risk of owner payment)
	AGC 499	Design-Build Teaming Agreement
	ConsensusDOCS 800	Owner/Program Manager Agreement and General Conditions
	ConsensusDOCS 801	Owner/Construction Manager Agreement
	ConsensusDOCS 802	Owner/Trade Contractor Agreement (Construction Manager is Owner's Agent)
	ConsensusDOCS 803	Owner/Architect-Engineer Agreement (Construction Manager is Owner's Agent)
	ConsensusDOCS 810	Standard Agreement Between Owner and Owner's Representative

*Design Build Institute of America
**Endorsing organizations include: Associated General Contractors (AGC), National Association of State Facilities Administrators (NASFA); The Construction Users Roundtable (CURT); Construction Owners Association of America (COAA); Associated Specialty Contractors, Inc. (ASC); Construction Industry Round Table (CIRT); American Subcontractors Association, Inc. (ASA); Associated Builders and Contractors, Inc. (ABC); Lean Construction Institute (LCI); Finishing Contractors Association (FCA); Mechanical Contractors Association of America (MCAA); National Electrical Contractors Association (NECA); National Insulation Association (NIA); National Roofing Contractors Association (NRCA); Painting and Decorating Contractors of America (PDCA); Plumbing Heating Cooling Contractors Association (PHCC); National Subcontractors Alliance (NSA); Sheet Metal and Air Conditioning Contractors' National Association (SMACNA); Association of the Wall and Ceiling Industry (AWCI); National Association of Electrical Distributors (NAED); National Association of Surety Bond Producers (NASBP); The Surety & Fidelity Association of America (SFAA)

Contract Interpretation

Courts interpret contracts as a whole and attempt to give reasonable meaning to all terms. However, specific negotiated provisions are given more weight than general terms. Words and how they are used in a specific industry are very important (*terms of art*). Courts look at the performance of the individuals involved in the contract in question (*course of performance*), as well as how they have performed in previous contracts (*course of conduct*). Unilateral mistakes and/or unexpressed intentions are not considered part of the contract.

When design professionals are asked to utilize BIM (Building Information Modeling), they would be well-advised to seek answers to the following questions:

- Determine if BIM is required or desired by the owner. Is the use of BIM an evaluation factor in the award of the contract?
- How does the owner expect the contractor to use BIM on the specific project? Public agencies and private owners may have very different goals and requirements.
- What software is required for the project? Will the current software be fully interoperable with other BIM software used by other contributors on the project?
- Does the owner require the BIM documents to be produced in a certain format? Will the software be fully compatible?
- Are resources (equipment and personnel) needed to implement BIM for a particular project or owner?
- What are the legal responsibilities for a party's contribution to the model as well as a party's access to the model?
- What is the standard of care for each party's contribution to or use of the model?
- What are the consequences (cost, time, and responsibility) if the BIM process flags a design conflict?
- What are the procedures and protocols for designating projections derived from a BIM model?
- What legal representations are made as to the dimensional accuracy of the models?
- Who is responsible for any cost, time, and liability related to any design revisions made during a collaborative BIM design process?
- What are the storage and retrieval requirements for electronic files and data?

- What data security issues and various access levels to the BIM models need to be considered?
- Who will own the design rights (intellectual property rights) for certain data generated during the BIM design process?
- What insurance and bonding risks can arise as BIM is used to facilitate providing preconstruction services?
- Will BIM be used for the RFI process?
- What will be the subcontract/purchase order terms and conditions?

Adapted from Federal *Government Construction Contracts,* Second Edition. (2010). Kelleher, Abernathy, Bell, and Reed editors.

"Think Twice" Contract Clauses

1. Certification
    - Problem—contract may require CIVIL ENGINEER to certify that certain conditions exist before, during, or after construction; but certify can be interpreted as guarantee or warrant
    - Solution—if clause cannot be eliminated from contract, include definition of "certify" acknowledging that CIVIL ENGINEER cannot certify conditions whose existence cannot be known with certainty

2. Consequential Damages
    - Problem—CIVIL ENGINEER may be held responsible for damages as a consequence of an event over which the CIVIL ENGINEER had no control; damages could be completely out of proportion with the CIVIL ENGINEER'S fee
    - Solution—establish a general limit of liability in the contract

3. Construction Cost Estimates
    - Problem—the client may view a construction cost estimate provided by the CIVIL ENGINEER as a "guaranteed maximum"; an inaccurate estimate could trigger claims
    - Solution—hire (or have the client hire) a consultant who specializes in preparing construction cost estimates; in the contract, refer to "opinion of probable construction cost" rather than "cost estimate"

(*Continued*)

4. Construction Monitoring
   - Problem—no set of plans or specifications depicts the project completely; the contractor must complete the design according to "custom," which can be subject to debate in court; regular site visits by prime designers, geotechnical engineers, and structural engineers can be beneficial in recognizing and solving problems in a timely manner
   - Solution—construction monitoring should be an additional service included in the contract and the CIVIL ENGINEER should accept only the responsibility that is spelled out in contract
5. Curing a Breach
   - Problem—difficulties can arise when the method for curing a breach is not addressed in the contract
   - Solution—identify mutual responsibilities and what a breach does, and does not, imply
6. Discovery of Unanticipated Hazardous Materials
   - Problem—injured employee or other party can file claim
   - Solution—when earth work or existing structures are involved, add contract clause acknowledging effects of changed conditions
7. Excluded Services
   - Problem—seeks to eliminate a claim that client was not made aware that certain services were available
   - Solution—identify excluded services
8. Freedom to Report
   - Problem—some contractors file claims against CIVIL ENGINEERS when their reports are critical of the contractor's work
   - Solution—client indemnifies CIVIL ENGINEER or CIVIL ENGINEER reports confidentially to client (client disseminates report rather than CIVIL ENGINEER)
9. Indemnification ("to secure against loss or damage")
   - Problem—some forms of client-proposed clauses are more onerous (distasteful) than others
     - Broad form—CIVIL ENGINEER agrees to hold harmless and indemnify client from any and all liability, including cost of defense, arising out of performance of work; requires CIVIL ENGINEER to cover client's costs, even when problem has been caused solely by the client

- Intermediate form—CIVIL ENGINEER agrees to hold client harmless from and against liability arising out of CIVIL ENGINEER'S negligence, whether it be sole or in concert with others; CIVIL ENGINEER may be required to pay 100 percent of damages though has caused only 1 percent

- Limited form—CIVIL ENGINEER agrees to hold harmless and indemnify client against liability arising out of CIVIL ENGINEER'S negligent performance of work; potentially mixes tort law liability with contract obligations making the CIVIL ENGINEER liable both in tort and contract law; liability insurance may only cover tort liability

- Solution—attempt to eliminate such clauses or add clause regarding disproportional payment for liability; have contract examined by legal council and work with professional liability insurer; have contractor's insurance carrier add owner and owner's agents under contractor's liability insurance

10. Jobsite Safety

   - Problem—claims can arise from clauses making the CIVIL ENGINEER responsible for acceptance of stop-work authority

   - Solution—avoid clauses that go beyond the CIVIL ENGINEER'S responsibilities normally required by law because these clauses could void liability insurance coverage; suggest language to be used in the client's contract with the general contractor stating that the contractor agrees to waive liability claims against the owner and owner's agents for injury or loss; refuse engagement if you believe safety matters will not be managed effectively

11. Limitation of Liability

   - Problem—some clients may not see the value of limiting CIVIL ENGINEER'S liability

   - Solution—discuss the issue in terms of risk management and add a risk allocation contract clause; a CIVIL ENGINEER always has a liability limit, which is the amount of money available to satisfy claims; the fee should reflect risk—some projects are more risk prone; also include the dollar amount for aggregate liability

12. Maintenance of Service

   - Problem—when a CIVIL ENGINEER is a subconsultant, the prime contract may be cancelled; someone else may do construction monitoring, which can create problems in interpreting plans and specifications

(Continued)

- Solution—include a contract provision that enables the CIVIL ENGINEER to carry-on work even if owner-prime designer contract is dissolved

13. Ownership of Instruments of Service

- Problem—client may want to own the plans, specifications, reports, boring logs, field data and notes, laboratory test data, calculations, and estimates, which can result in unauthorized reuse; if the jurisdiction views these as products, any defects (errors and omissions) might be treated as product defects, which could invoke the doctrine of strict liability rather than negligence and could obviate professional liability insurance

- Solution—include a contract provision that indemnifies CIVIL ENGINEER against unauthorized reuse and compensates CIVIL ENGINEER for the cost of any defense

14. Record Documents

- Problem—client may want the CIVIL ENGINEER to provide record documents (as-builts) based on information furnished by others; the accuracy of this information is difficult to verify and the CIVIL ENGINEER may be held liable for losses arising from errors

- Solution—include a contract provision that makes clear the potential for inaccuracies and eliminate terms like ''as-built drawings'' or ''corrected specifications,'' which imply ''without error''; use terms such as ''record specifications'' and ''record drawings'' and add a prominent notice on each page of record plans and specs

15. Right to Reject and/or Stop Work

- Problem—client may want the CIVIL ENGINEER to reject a contractor's work or to stop work if corrections are not made; the CIVIL ENGINEER'S role should be based more on observing and monitoring

- Solution—add a contract provision that clearly states the CIVIL ENGINEER'S responsibility; advise client to reject work that does not conform with CIVIL ENGINEER'S recommendations, specifications, and design; if the client insists on a CIVIL ENGINEER stop work provision, include a contract clause that provides CIVIL ENGINEER'S full waiver from any claim or liability, as well as indemnification

Adapted from John Philip Bachner. (1991). *Practice Management for Design Professionals: A Practical Guide to Avoiding Liability and Enhancing Profitability.*

Table 11.2 Clients' Acquisition Strategy
(Adapted from Design Build Institute of America [DBIA] *Fundamentals of Project Delivery* Course Materials, 2009)

Project Delivery System	Procurement Method	Contract Format
• Design-Bid-Build (DBB) • Multiple Prime • Construction Management at Risk • Agency Construction Management • Design-Build (DB) • Design Assist	• Sole Source • Limited Competition—Negotiation • Qualifications-Based Selection • Best Value Selection • Fee-Based Selection	• Lump Sum (Fixed Price) • Cost Plus a Fixed Fee (Cost Plus) • Guaranteed Maximum Price • Target • Unit Price

CONTRACTS IN PROJECT DELIVERY

Clients are faced with many choices. They must choose the delivery system, procurement method, and contract format most appropriate for each project. (See Table 11.2.) These choices directly affect the civil engineer's role, the type of services provided by the civil engineer, and the civil engineer's means of compensation.

Project Delivery Systems

Clients have several options when selecting a project delivery system, as shown in Figure 11.2. These include: Design-Bid-Build (DBB); Design-Build (DB); Construction Management at Risk; Agency Construction Management; Design-Assist; and Multiple Prime.

Design-Bid-Build

Design-Bid-Build (DBB) is still the most common method used for designing and constructing projects. As depicted in Figure 5.2, DBB is a linear process involving separate stages for design, bid, and construction. In DBB, the prime designer and client enter into a contract for design services. When the design is complete, contractors bid on contract documents prepared by the designers. (See Figure 6.2.) After an appropriate period of time, two weeks for small projects and many more for large and/or complex projects, the client enters into a contract with the wining builder. Both Figure 6.1 and Figure 11.3 show the contractual relations involved in DBB, where the prime designer and contractor share no direct, formal contract.

Owners may like DBB because they are able to work directly with the prime designer and are able to exercise control over the entire project, including construction. They also can benefit from having a number of contractors give them prices for the proposed project and being able to select the lowest bid. Also, DBB is allowed in public procurement. The downside of DBB is that the contractor has no input in design, and the process tends not to be collaborative.

Design-Build

In Design-Build (DB) project delivery, the client contracts with one entity for both design and construction. This entity can take the form of a limited partnership, joint

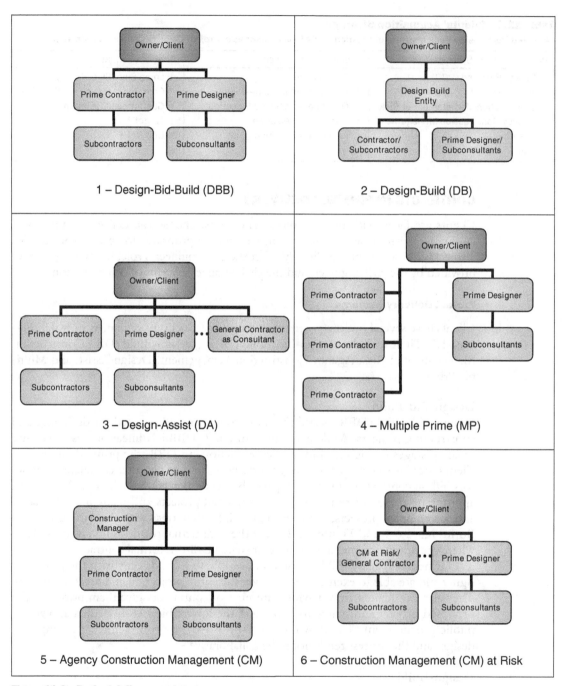

Figure 11.2 Project delivery systems

venture, or fully integrated corporation and can be led by a designer, contractor, or developer. The client is responsible for providing the DB entity with design requirements and performance specifications. Innovation, value engineering, and constructability (the buildability of the design) can be enhanced because these are all in the best interest of the DB entity as well as the client. Schedules can be shortened and costs can be known early.

In DB, the designer and contractor are on the same team and can make unified recommendations to the client. DB takes the owner out of the position between the designer and contractor and consequently results in fewer changes, fewer claims, and less litigation. DB also accommodates complex project phasing and allocates risks to those who can best manage them.

Multiple Prime

Sophisticated clients or clients with in-house construction management capability may use a variation of DBB. In multiple prime project delivery, after the design is complete the client hires multiple contractors instead of one general contractor. Thus, a single project may be divided into various contracts such as site development, steel fabrication and erection, mechanical, electrical, and so forth. Or a client may divide a large project into different zones or phases. The advantage of multiple prime contracting is that it eliminates the general contractor's fees and it is allowed in public procurement. Two primary disadvantages involve the client's duty to coordinate the multiple prime contractors and the potential for liability for delays caused by one of the prime contractors.

Construction Management at Risk

In Construction Management at Risk, the client hires a Prime designer and a construction manager. The construction manager helps to manage the design and then builds the project. The construction manager's involvement enables the client to know construction costs earlier, provides opportunities for value engineering (making changes to maintain or increase value to the client while lowering construction costs), and can reduce schedule time. Also, when budgets are really tight, a guaranteed construction cost can be known earlier.

Agency Construction Management

Agency Construction Management is a variation of Construction Management that is marked by a very different role and level of responsibility for the firm performing as a construction manager. In Agency Construction Management, the construction manager (CM) acts as a consultant to the client and manages the design-bid process, and construction; but the CM does not construct the project. Agency CM adds a layer of management and accompanying fees to the project. However, many clients do not have the specialized knowledge required to oversee design and construction and can benefit greatly in terms of managing scope, schedule, and budget from the involvement of a CM. Agency CM is allowed in public procurement.

Design-Assist

In Design-Assist, a contractor is hired by the client as a consultant during the design phase of DBB. The contractor can assist the designers with constructability reviews and decisions influencing cost and schedule. It can be argued that the amount of money spent by the client will be exceeded by potential savings, but problems frequently arise when the contractor hired to do the consulting is different from the contractor who is chosen to construct the project. They may have valid but different ways of building the project, resulting in changes during construction. Also, in some jurisdictions the contractor consulting prior to bid is not allowed to bid on the project due to conflict of interest laws.

Procurement Method

Private clients have a range of options open to them for selecting their design consultants and contractors. These include: Sole Source; Limited Competition, for instance, negotiated; Qualifications-Based; Best Value; and Fee-Based selection. Due to public contracting laws, public clients are more limited. Ability to use these procurement methods varies from jurisdiction to jurisdiction, for example, federal agency, state, country, municipality, and so forth.

Sole Source

Sole source selection is as it sounds. The client chooses a designer or builder and enters into a contract. This is a common practice in the private sector, especially when only one firm can provide highly specialized services. Generally, sole source selection is not permitted in the public sector, unless issues involving national security or emergencies arise.

Limited Competition—Negotiated

Again, negotiated contracts often tend to be the domain of private, rather than public, clients. In this form of procurement, a list of potential service providers is prepared by the client. This list may be based on firms with whom the client has had previous positive experience. In order to gain an adequate competitive basis, a client selection panel interviews each firm. Based on some established selection criteria, the selection panel enters into negotiations with the top one to three firms and makes its selection based on those negotiations.

Qualifications-Based Selection

In qualifications-based selection (QBS), the selection of firms is based on their qualifications and project approach. As discussed in Chapter 4 - Professional Engagement, QBS frequently is used to select design firms and is open to many (though not all) public clients, as well as private. QBS is a competitive process involving a two-envelope system. The first envelop contains the engineer's response to the RFP, and the second envelope contains the cost for executing the proposed work. Ideally, the client

Table 11.3 Best Value Selection Example (Second Phase)

Firm	Technical Score (60 pts. max)	Price	Price Score (10 pts. max)	Total Score (100 pts. max)
Acme Civil Engineers	49	$1,700,000	35	84
Benefit Civil Engineers	53	$1,650,000	37	90*
Capital Civil Engineers	47	$1,600,000	39	86

*Winning proposal

issues an RFP with a clear definition of project requirements (scope) and well-defined budget and schedule constraints.

Proposing firms respond with the two envelopes and are ranked using the information contained in the first envelop. Those making the short list (typically, from three to ten firms) are interviewed and ranked again. After a firm has been selected based on the merits of their proposal and interview(s), the client opens the second envelop to learn the price. Negotiations with the top-ranked firm ensue. Scope and/or services may be added or deleted until a mutually acceptable price can be reached. If the client and the top-ranked firm cannot reach an agreement, the client may open negotiations with the firm ranked second.

Best Value Selection

Best Value (BV) selection is a two-phase process that combines technical and cost criteria. In BV selection, the client issues a Request for Qualifications (RFQ), and responding firms are ranked based on their proposals. Firms making the short list are then issued an RFP and perhaps given a stipend or honorarium to cover the costs of phase-two proposal preparation. Their second proposals include both their technical approach and cost, which are scored independently. Consequently, a firm with a higher cost could be selected over a firm with a lower cost, if their technical proposal is superior. See Table 11.3.

Fee-Based Selection

Most engineers do like to compete on the basis of cost alone. Fee-based selection, where the contract award criterion is based on price alone, is not really appropriate for the highly judgmental and creative work required of most engineers. The client's review process is fast and uncomplicated. On the other hand, there are few incentives for innovation or exceptional performance on the part of the engineer.

Contract Format

Clients can choose among several different types of contract formats, including Lump Sum (Fixed Price), Cost Reimbursable—Cost Plus a Fixed Fee (Cost Plus), Guaranteed Maximum Price (GMP), Target—and Unit Price.

Lump Sum (Fixed Price)

Lump sum contracts utilize a single price. In order for a civil engineer to sign a lump sum contract, the project scope (SOW) should be very clear. In this type of contract, the risk is borne by the engineer; but the engineer also accrues any savings in time/ fee. Contracts are easy to administer because the need for cost substantiation is limited to determining progress payments. When all parties are performing ethically, price, schedule, and performance are known at project outset.

Cost Reimbursable

Cost reimbursable contracts pass-through the actual cost of the work to the client. These contract types are used typically in negotiated work when the project is technically difficult or unique. Savings are passed on to the client, but price is not necessarily known until the project is complete.

In Cost Plus a Fixed Fee (Cost Plus) contracts, the civil engineer bills the client for the cost of the work performed, plus a fee. The fee is usually a fixed percentage, which is negotiated prior to contract award. Guaranteed Maximum Price (GMP) contracts are a variation Cost Plus; the upper limit of the cost of the project is known initially. As long as project costs remain below the GMP, savings will accrue to the client. There is substantial overhead associated with identifying and reporting costs, and problems can arise regarding wage mark-ups and project staffing levels.

In Target contracts, the client and engineer establish a target price and an incentive fee based on a set of performance criteria at the beginning of the project. The engineer's proposal contains no profit. The budget is reconciled periodically (e.g., quarterly) with the substantiated cost of work. Fee is earned in this period, if the performance and price targets are met. This type of contract format has the potential to lower the client's risk and cost because the engineer has incentives for reducing costs. As in other forms of cost reimbursable contracts, cost reconciliation is a burden.

Unit Price

Most Unit Price contracts have to do with units of materials (such as cubic yards of concrete or aggregate), time (such as hours of design services), or completed elements of construction (such as linear feet of roadway paving). They are intended to price variable quantities. Problems can arise when the units are composite (such as linear feet of pipe—where do the excavation, base-course, hangars, rebar, backfill, and so forth fit in?) and, therefore, are not easily defined. Additional confusion is introduced when units are substituted for payment milestones. Activities such as permit acquisition or procurement of special equipment are not measured in quantities. Also, situations may change. A contract may be priced on labor rates for straight time but may require overtime during the course of the project. Assumptions behind unit prices need to be spelled out carefully to the client in the proposal.

In addition to the understanding and, where appropriate, influencing the client's choice in project delivery system, procurement method, and contract format, civil engineers need to understand the underlying risks associated with all projects they undertake.

Dealing with Contract RISK

1. Execute the contract in a professional manner
   - Be quality oriented
2. Thoroughly educate clients
   - Use general conditions to cover costs
3. Identify problems
   - Impose a fee premium or indemnification
4. Identify unavoidable/unacceptable risks
   - For example, modification of existing structure, hazardous materials
5. Use the contract to close loopholes and avoid traps
   - List services the client has declined
6. Establish extralegal conditions
   - Use dispute resolution (mediation/arbitration), reduce statute of limitations/repose, restrict damages
7. Develop clear contract language
   - Make sure that the contract is easily understood by the client and triers of fact
8. If a client is adamant about not taking prudent measures
   - Walk away

Adapted from John Philip Bachner. (1991). *Practice Management for Design Professionals: A Practical Guide to Avoiding Liability and Enhancing Profitability.*

RISK MANAGEMENT

Constructed projects involve the integration of many technical subsystems and components in a context that often involves complex and sensitive economic, social, political, and environmental decisions. An unbroken flow of information from early debates and strategic project definition through design, engineering, procurement, construction, operation, and decommissioning is practically nonexistent. This flow is difficult to achieve because the nature of decisions and types of decision-makers change considerably over project lifecycles.

Some projects experience relatively stable partnerships between firms. This permits long-term interorganizational networks to be established and integrated information systems to be deployed. More commonly there is greater uncertainty. Risk is managed in hierarchical networks of firms. This form of organization is typified by a management and decision-making environment in which problems are highly interdependent while the people, methods, and organizations involved are extraordinarily independent.

> "Risk" can be defined as danger, chance, and/or exposure to loss or injury.

Dealing with Risk in General

The most effective "preventive" approach to dealing with risk is to perform professionally. This professional approach to practice concerns not just execution of technical efforts but also a focus on eliminating misunderstandings and unrealistic expectations between civil engineers and their clients and among civil engineers and other professionals. Also very important are:

- A good formal contract
- Good project management
- Knowledge of prevalent risks

Civil engineers can attempt to transfer risk, typically via insurance and/or indemnification. Professional liability insurance is the most common mechanism, though many risk exposures are not covered or are covered only in part. Therefore, professional liability insurance affords only partial transfer of risk.

Indemnification ("to secure against loss or damage") of the civil engineer by the client may be appropriate if risk is created by nature of the project and the civil engineer is powerless to control the risk. Indemnification can be full or partial (risks shared). Partial indemnification can be used to cap the civil engineer's liability, but clients also want to minimize risks. Like insurance, indemnifications are imperfect risk transfer mechanisms.

Transferring *all* risks is impossible. Civil engineers can establish a *loss reserves* account by including risk in their fees; but competition in the marketplace can make this difficult. Also, some problems associated with managing risk are not merely financial. Unsuccessfully managing risk can become an emotional drain and can have a negative effect on professional reputation.

Risk retention is a fact of life. Consequently, civil engineers need to understand and abide by the "rules of the road." They should be thoroughly acquainted with their responsibilities and how to execute them. Being careful with selection of clients and projects is extremely important. Well-managed firms have business plans that identify the types of clients and projects to pursue. Some are far more risk prone than others, for example, hazardous materials remediation. Even before proposing on a project, the civil engineering firm should:

- *Do* background research on the client and the *current* project
- Critically *evaluate* your firm's capability and experience with the type of work and project
- *Not rely* on past behavior as an indicator of future
- *Not rely* on assumptions

Also necessary before signing a contract, the civil engineering firm should be sure that the contract includes:

- Appropriate clauses
- Explicit general conditions
- A well-defined workscope
- An accurate schedule
- A clear budget

Other actions to take include insisting on an adequate fee or reducing workscope. However, limiting internal plan review, shop drawing review, or construction monitoring is not advisable. Providing additional quality control and applying realistic work assignment and scheduling procedures also help manage risk—less experienced individuals need oversight. Also, civil engineers should not agree to deadlines that cannot be met and may need to hire additional personnel to meet schedule demands.

Establish a Risk Management Program

At least one person in a civil engineering firm should be assigned the task of risk management. If the firm is small, the *Risk Management Program* could be part of quality control, though it should include more than technical quality control. Establishing a risk management program is an attempt to prevent losses that are not really the fault of the firm. Through a risk management program, all staff can be educated on loss prevention. It is important to balance risk and reward. (See Figure 11.3.)

Documentation also is another key to managing risk. (See Chapter 13, Communicating as a Professional Engineer.) Proper documentation often can result in claims being dropped. Notes, memos, e-mails, and meeting minutes can eliminate some of the worst client statements a civil engineer can hear:

"We never said that . . . "

"You said . . . "

"I've never heard [seen] that before . . . "

Finally, civil engineers should react quickly to the symptoms of problems, maintain open lines of communication, and use emotion intelligence—if someone's body

Figure 11.3 Balancing risk and reward

language or your "gut" is telling you that there is a problem, there probably is. Deal with issues promptly—unlike most wine, disputes seldom improve with age.

Risk Management

Constructing "Reasonably Believable" Edifices: Lessons from Software, Implications for Construction

Rather than speak of "correct" software in the unqualified sense, one should rather speak of reasonably "believable" software. Rather than say software is error free, the most one can usually say is that the conditions under which the next errors will be manifested have not yet arisen. [1]

Introduction

Software development and construction activities are socially distributed across place and time. As with the many project-based businesses, the principal constituent activities involved in software development and construction work are usually performed within a network, populated by different groups (often within distinct organizations) each responsible for various "stages" of the development process. Relationships between these groups are often strained; historically, conflict, confrontation, and adversarialism have characterized the product development process.

Both software development and construction projects often suffer excessive cost and time overruns and produce poor quality end products. There are several aspects to poor quality products in software and construction, these include products that are unfit for their purpose, inappropriate to end users' needs, unreliable (containing a large number of faults), and those that have poor maintenance and upgrade capabilities. Clearly, construction clients and end users of software alike face serious uncertainties and risks that the product they anticipate will not be delivered.

Risk in Project-Based Industries

Uncertainty becomes risk when the perceived significance of the consequences of an uncertain event becomes critical. Therefore, risk can be defined as any uncertain event which has an impact (usually adverse) on the outcome of the project. [2] Some researchers make a slightly different distinction between risk and uncertainty. In their view, risks are those uncertainties which can be identified and quantified.[3] In other words, risk occurs when the outcome of an event or circumstance can be predicted on the basis of statistical probability. For the purposes of this article, risk is discussed using the former, broader definition: Risk involves the likelihood that a system will behave in unpredictable ways, the outcome of which may be undesirable.

Risk is endemic to modern society. It is a key feature of economic growth in a period characterized by rapid technological change, the emergence of a pattern of innovation based on technology fusion [4] and the transition to a knowledge-intensive economy.[3] Risk is a precondition for innovation and is intimately connected to learning and change.[4] During innovation, the generative, uncertain, and complex activities involved in spurring innovative variety are in conflict with the simultaneous requirements for convergence, control, regulation, and standardization.[5] In order to resolve this conflict, a balance needs to be struck between the demands for risk reduction and the need for variation.

Risk is particularly associated with project-based businesses such as software development and construction. In these industries, product development depends on coordinating the activities of a network of socially distributed (both in terms of their disciplinary base as well as their geographical location) actors, each with their own particular strategic aims and objectives. Concerns over the economic, environmental, and technological hazards connected to such businesses have grown alongside scientific techniques of risk assessment, involving the evaluation of ''proneness to failure''[3], and risk management practices (which are often implemented as mere acts of faith).

(Continued)

Networked structures are typical in project-based businesses.[8] In general, networks are prejudiced in favor of innovative activities associated with the generation of a variety of new ideas, as distributed groups work independently (often at different locations) on developing aspects of a technological system. This stimulates uncertainty and generates risk in the product/project environment. These take precedence over the generation of ideas which are linked to standardization and control.[9]

The risk associated with placing the emphasis on generating a variety of ideas in the product/project environment is compounded by the lack of coordination between members of the network. As a result, in the absence of effective risk management practices, networks can often be considered as a source of risk in themselves. One response to this has been to stimulate cultural change in project-based industries in order to introduce synergy among project teams and end the inherent adversarialism that has previously characterized project-based businesses such as software development [10] and construction.[11]

Approaches to Risk in Construction

Within the business of construction, multiple techniques have been adopted to improve risk management in projects. These include PERT (Program Evaluation and Review Technique) used to analyze schedule risk, Monte Carlo simulations, cause and effect diagrams, fault trees, and decision trees, etc. (See Table A below for a list of these techniques.) Project management experts advocate the use of these methods.

Originally risk management was the domain of insurance companies, where risks are bought and sold. Architectural and engineering firms carry errors and omissions insurance policies. Additionally, many owners require surety bonds of their contractors. A surety bond is a financial instrument issued by a third party (surety company) to a first party (owner) which provides a guarantee that a second party (construction firm) will complete a project according to the terms and conditions of a written contract. In the 1980s the surety bonding business experienced several successive years of heavy losses, which has lead to the use of innovative approaches such as neural networks as a way of rating new construction bond applicants.[16]

However, the approach to risk used by construction firms seldom is based on sophisticated quantitative techniques. In an investigation of strategic decision making in large architectural, engineering, and construction firms, Hansen [17] noted that firms tend to ignore remote possibilities, look at a few outcomes rather than all alternatives, and focus on opportunities rather than dangers. Firms developed several ways of managing risk by using incremental approaches, performing cost justification exercises, limiting expenditures to those which could be funded from existing projects rather than overhead, and passing risk to vendors and subcontractors through the use of performance clauses in contracts.

A study carried out in the UK resulted in a similar finding. A survey of expert project management practitioners was undertaken to determine the level of awareness of project risk analysis and management techniques. The sample was drawn from the special interest group concerned with risk of the UK Association of Project Managers (APM). Although the 37 practicing member sample was small, participants were considered to be expert. As indicated by Table A, more traditional techniques such as checklists, Monte Carlo simulation, and PERT were favored.[18]

Regardless of the risk analysis technique employed, the focus on remedial action by optimizing within task performance has tended to lead to sub-optimization of the overall project process. This fragmented approach currently is being challenged by sophisticated owners who want more value from their expenditures on constructed products, by project financing vehicles such as the Private Finance Initiative (PFI), and by European Commission (EC) Directives requiring member countries to implement national legislation which demands greater teamwork in the planning and execution of projects (introduced in the UK by the 1994 Construction (Design and Management), or CDM, Regulations).[19]

Table A   Project Risk Analysis Techniques

Technique	Used currently	Used in past but not now	Aware of but not used	Have not heard of
	%	%	%	%
Catastrophe theory	0	0	56	32
Checklists	76	0	8	4
Controlled Interval & Memory Modelling	8	0	48	32
Decision trees	44	0	48	0
Fuzzy set theory	0	0	64	24
Game theory	8	0	48	24
Influence diagrams	28	0	48	12
Monte Carlo simulation	72	4	16	0
Multiple criteria decision making models	24	0	36	28
PERT	64	4	24	0
Sensitivity analysis	60	4	20	8
Utility theory	4	0	48	36

Note: % indicates percentage of positive response from 37 participants
Adapted from: J.A. Bowers, 1994

(*Continued*)

Therefore, Risk Management, as a distinct activity, is gaining attention in construction. Practitioners and researchers alike advocate a two phased approach using assessment and action.[2][20] The figure below depicts the usual components of the risk management process.

Elements of risk management

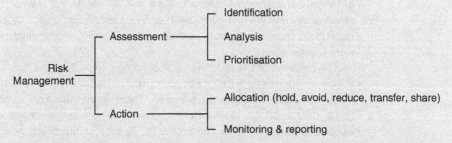

Although techniques for analysis are more robustly developed, the identification of risk is seen as the most critical element of any risk management scheme as risk cannot be managed if it is not identified.

The Changing Nature of Risk Allocation

Recent economic conditions have tended to change the nature of risk allocation in construction. In a recessionary period, the number of business failures generally increases, which leads to the desire to share the risk of financial failure and changing economic conditions.[21] Additionally, due to downsizing, many firms do not have diverse portfolios of projects and operations across which to spread risk. These firms are often interested not in sharing risk but in shifting risk altogether to another party.

A study on perceptions and trends in risk undertaken in the U.S. compares a survey of ENR's (Engineering News and Record) top 100 U.S. contractors with an American Society of Civil Engineers (ASCE) survey.[21] The results indicate that certain risks are associated consistently with either contractor or owner. Risks allocated to contractors in both surveys include those involving: labor, equipment, and material availability; labor and equipment productivity; quality of work; safety; labor disputes; and competence. Risks allocated to owners in both surveys include: permits and ordinances; site access/right of way; defective design (which normally would be transferred to the architect/engineer); changes in work; and changes in government regulations. Apparently, owners and contractors are willing to accept these risks as they consider themselves able to control these risks or take action to curtail or prevent their occurrence.

The trend, however, appears to be toward sharing contractual/legal risks, which in the earlier survey were borne more by owners.[21] Today many large construction firms either retain lawyers or employ them in their home offices.

Contractors may be more comfortable in negotiations and are therefore more willing to share risks associated with change orders, contract delay resolution, and indemnification. Additionally, the use of insurance has increased which provides an addition means of avoiding risk.

Finally, current interest in design build and private finance initiative (PFI) projects also has focused attention on risk management and on achieving objectives in terms of time, cost, quality, and performance as well as future income stream. In these cases the bidding team is under considerable pressure to ensure that risks are allocated correctly.

The management of risk in construction is becoming important as concerns over economic, environmental, and technical hazards have grown alongside scientific techniques of risk analysis. Yet currently available tools for analysis of risk do not seem to have been adopted by the majority of firms. As the next section illustrates, approaches used in software can provide valuable lessons for construction.

Lessons from Software, Implications for Construction

A number of collaborative approaches to software development have capitalized on the awareness that risk can be reduced through effective communication. Among these are the socio-technical systems approach to participative systems design, exemplified by the ETHICS [22] method, and the Scandinavian collective resource approach, see [23]. Common to these many variants is prototyping, whereby continual negotiation among members of the project team iteratively evolves the design, refinement and development of the product under development, for example, Rapid Application Development (RAD) in software.

RAD techniques for software development involve intensive negotiations among heterogeneous and cross-functional teams of specialists. Each specialist in a RAD team is a representative of one of the various functional groups in the product development network. These specialists coalesce on a temporary basis in order to collaboratively and incrementally evolve new software solutions to business problems. Under certain circumstances, this has been shown to lead to software project success through eliminating misunderstandings in communication and enhancing business members' ownership over the software product.[24]

In view of these positive results, a revised risk management scheme for use in construction is proposed. (See the following figure.) In this enhanced model, risk identification is given a more prominent role and techniques drawn from software, such as RAD, are used to encourage communication and co-operation among members of the project network.

(Continued)

Revised risk management scheme

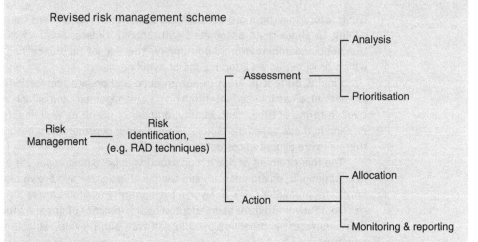

Conclusions

Through enhancing communication between the distributed actors involved over the project lifecycle, it is hoped that networked product development processes will become characterized by openness to new ideas and by access to decision making processes. The by-product of this enhanced communication is the early identification of risk associated with the project process.

References

1. A. Macro, and J. Buxton. (1987) *The Craft of Software Engineering*. New York: Addison-Wesley.

2. Crossland, Rose, Jon H. Sims, and Chris A. McMahon. (1995). "An Object-Oriented Design Model Incorporating Uncertainty for Early Risk Assessment." 1995 ASME Design Engineering Conference, Boston, USA.

3. R. F. Fellows. (1996) "The Management of Risk," The Chartered Institute of Building, Ascot 65, 1996.

4. F. Kodama. (1991) *Emerging Patterns of Innovation*. Boston, Mass.: Harvard Business School Press.

5. OECD, "Employment and Growth in the Knowledge-based Economy,"Paris: OECD, 1996.

6. R. Herbold, "Technologies as Social Experiments. The Construction and Implementation of a High-Tech Waste Disposal Site.," in *Managing Technology in Society*, A. Rip, T. J. Misa, and J. Schot, Eds. London: Pinter, 1995, pp. 185–197.

7. J. Jelsma. (1995). "Learning about Learning in the Development of Bio-technology," in *Managing Technology in Society*, A. Rip, T. J. Misa, and J. Schot, Eds. London: Pinter, pp. 141–165.

8. M. Hobday. (1996) "Product Complexity, Innovation and Industrial Organisation," Research Policy.

9. D. Foray and M. Gibbons. (1996). "Discovery in the Context of Application," *Technological Forecasting and Social Change*, vol. 53, pp. 263–277.

10. S. Easterbrook. (1991). "Negotiation and the Role of the Requirements Specification," The University of Sussex, Brighton, Cognitive Science Research Reports 197, July 1991.

11. M. Latham. (1994). "Constructing the Team," HMSO, London, Final Report of the Government/Industry Review of Procurement and Contractual Arrangements in the UK Construction Industry.

12. G. J. Hodgson. (1995). "Design and build—effects of contractor design on highway schemes," Institution of Civil Engineers and Civil Engineering 108, May 1995.

13. Thurston, D.L. (1991). "Design Evaluation of Multiple Attributes under Uncertainty," *International Journal of Systems Automation: Research and Applications*, vol. 1, pp. 143–159.

14. T. DeMarco. (1982). *Software Systems Development*. New York: Yourdon Press.

15. I. Sommerville. (1992). *Software Engineering*. New York: Addison-Wesley.

16. Kangari, Roozbeh and Moataz T. Bekheet. (1995). "Risk assessment of construction bonds underwriting using neural network technique," *Integrated Risk Assessment: Current Practice and New Directions*, R.E. Melchers and M.G. Stewart, eds. A. A. Balkema, Rotterdam The Netherlands, pp. 139–146.

17. Hansen, Karen Lee. (1993). "How Strategies Happen: An Investigation of the Decision to Upgrade Computer Aided Design (CAD) in Architectural, Engineering, and Construction Firms." Doctoral Dissertation, Department of Civil Engineering, Stanford University, Palo Alto, California, USA.

18. Bowers, John A. (1994). "Data for project risk analyses," *International Journal of Project Management*, vol. 12. no. 1, pp. 9–16.

19. Neale, Brain S. (1994). "Generic health and safety standards—EC Directives and technical policy implications." *ISARC Proceedings* 1994. Elsevier.

20. Boothroyd, Catherine. (1997). "Managing Risk in Construction," Managing Value and Risk for the Client's Benefit, Members Report 96-21-S for the Construction Productivity Network, CIRIA, London.

(Continued)

21. Kangari, Roozbeh. (1995). "Risk Management Perceptions and Trends of U.S. Construction," *Journal of Construction Engineering and Management,* American Society of Civil Engineers, Dec. 1995, vol. 121, no. 4, pp. 422–429.

22. E. Mumford. (1995). *Effective Systems Design and requirements Analysis: The ETHICS Approach.* London: Macmillan.

23. G. Walsham. (1993). *Interpreting Information Systems in Organizations.* New York: John Wiley and Sons.

24. J. E. Millar. (1996). "Interactive Learning in Situated Software Practice; Factors Mediating the New Production of Knowledge During iCASE Technology Interchange," in *SPRU.* Brighton: University of Sussex, pp. 248.

—Jane E. Millar, Ph.D. and Karen Lee Hansen, Ph.D.

INSURANCE AND BONDS

As the frequency of claims increases and balancing revenues and losses becomes more difficult, professional liability insurance has become vital to the practicing civil engineering professional. Professional liability insurance is an essential part of risk management and also may figure in marketing services. Bonds typically are associated with construction; but more design professionals are becoming involved with design-build and integrated project delivery contract structures. Civil engineers should be familiar with the various bonds required by owners.

Liability insurance: A contract under which an insurance company agrees to protect a person or entity against claims arising from real or alleged failure to fulfill an obligation or duty to a third party who is an incidental beneficiary.

Professional liability insurance: Insurance coverage for the insured professional's legal liability for claims arising out of damages sustained by others allegedly as a result of negligent acts, errors, or omissions in the performance of professional services.

Bond: In suretyship, an obligation by which one party (surety or obligator) agrees to guarantee performance by another (principal) of a specified obligation for the benefit of a third person or entity (obligee).

—*The Architect's Handbook of Professional Practice*, p. 985, p. 999, and p. 995.

Professional Liability Insurance Industry

The principal players in the professional liability insurance industry are:

- *Insureds*—those to whom a policy's coverage is extended
- *Insurance agents*—those who specialize in selling various types of insurance and who obtain commissions based on their sales
- *Insurers*—insurance companies that issue policies and establish what policies do and do not cover, limits and deductibles, and the premium that must be paid
- *Actuaries*—compute the odds that a given risk will materialize and the probable cost of that risk
- *Underwriters*—evaluate each applicant to determine the extent to which various risks may occur and what the premium should be
- *Claims Managers*—advise insureds on what course of action to take when a claim arises
- *Reinsurers*—those who insure insurers, providing both back-up (should a large claim be filed) and stability to the industry

Professional liability insurance is outlined in a policy (contract) between the civil engineer (insured) and insurance company. Most policies are purchased from independent insurance agents (brokers). Protection is provided for the professional in case of negligence. Coverage can be augmented with commercial liability coverage that provides more extended overall protection.

The cost of professional liability insurance is influenced by several factors: risk, demand, and level of coverage, among others. Results of the analysis of prevalent risks conducted by insurance company actuaries and underwriters are a prime aspect. In addition, the law of supply and demand is at work. When the supply capacity of insurance is high and demand is low, insurance costs will be relatively low. However, if the demand for insurance increases, insurance costs become more expensive. Level of coverage also affects cost.

Liability Insurance Coverage

Professional liability insurance coverage is designed to create a source of recovery in the event that a civil engineer, or other professional, experiences claims arising out of damages sustained by others allegedly as a result of negligent acts, errors, or omissions in the performance of professional services. Policies do vary in coverage, so the person responsible for purchasing the policy should "read the fine print" and choose an insurance broker with the same care they would exercise in selecting an attorney or accountant. Qualifications, services available, cost, and ability to communicate all matter. Some professional liability insurance companies provide extensive educational and risk management assistance programs; others offer little advice or guidance.

General policy considerations might be:

- What scope of coverage is offered, what endorsements are available to expand the coverage, and what is excluded from the coverage?
- What is the cost of the basic policy and any endorsements?
- Is this coverage part of the firm's overall financial management program?

Specific concerns for professional liability insurance include:

Policy limits : The more protection a policy provides, the more expensive it is. An analysis should be conducted to determine the lowest practical amount of coverage. Another consideration is the amount of coverage required by clients.

Deductible: Usually, raising a policy's limit is relatively less expensive than lowering a deductible. There has to be a balance among the deductible, premium, and level of coverage, considering that a new deductible obligation occurs with each claim.

Cost: The cost of a firm's professional liability insurance is calculated individually by an insurance company's underwriter. The cost of coverage is based on the type of practice (geotechnical, structural, environmental, and so forth), geographical area, project mix, claims history, coverage needs, and resulting risks to the insurer. A firm should ask its insurance agent how its premiums will be calculated—this presents a way to be able to compare policies.

Insurability: Professional liability insurance only covers the professionals for whom it is purchased. Problems arise when owners ask design professionals to indemnify them or require certificates that have the effect of express warrantees or guarantees—none of these is covered by professional liability insurance.

Subconsultants: Prime designers (see Chapter 5, The Engineer's Role in Project Development) routinely retain consultants. Because of this contractual relationship, prime designers are exposed to *vicarious liability* for damages resulting from subconsultants' negligence. Consequently, prime designers need to review the policies of their subconsultants for limits of liability and potential gaps in coverage.

Joint ventures: From a legal point of view, joint ventures are similar to partnerships, even though the joint venture has a more limited purpose. If a professional liability claim is made against a joint venture—or a team or association acting as a joint venture—one or all members can be held liable for any decisions rendered against it. Care needs to be taken regarding the type of professional liability insurance used by the joint venture.

Project professional liability insurance: Project insurance covers the entire project team, even those who practice without insurance. Owners usually pay for project insurance when they desire coverage beyond that normally carried by design firms. When the project scope is greatly increased, this can provide a way to cover consultants who could not otherwise obtain coverage.

Expanded project delivery: Insurance companies have begun to offer coverage for designers involved in design-build or acting as developers. This coverage often is provided through endorsements to existing basic policies. An analysis of this coverage should be conducted to locate and eliminate any potential gaps.

Claims: Claims can be made directly via a demand for money or services based on an allegation of a wrongful act. Claims also can arise from more subjective circumstances, such as the threat of action or a troubling circumstance. Insurance policies should be checked for terms that require timely reporting of claims because not reporting claims on time could jeopardize coverage. A legal term important to know is *barratry,* the fomenting of claims where none exist.

In addition to professional liability insurance, the civil engineering firm may want to carry additional liability insurance, including:

General liability insurance : Civil engineering firms can be exposed to loss from other liability exposures such as slips and falls, libel and slander claims, and property damage to third parties arising from office operations and nonprofessional activities at jobsites. General liability insurance policies usually cover claims involving third-party liability (bodily injury, personal injury, property damage, and so forth). They do not cover professional, automobile, and workers' compensation exposures.

Employment practices liability insurance: In addition to utilizing sound management practices, some firms might choose to purchase employment practices liability insurance to protect against losses arising from employee charges of harassment, discrimination, and wrongful termination.

Insurance may seem deceptively simple; but insurance coverage and cost are influenced by many factors, not the least of which are financial markets. Buying appropriate insurance coverage is a major business decision. For the majority of civil engineering firms, the most devastating professional and business risks stem from litigation based on accusations of negligence in the performance of professional services.

Bonds

Almost all contractors use the services of national surety companies, which provide written bonds guaranteeing the performance of obligations. Like the insurance industry, these companies are subject to public regulation. The most common bonds are bid, payment, and performance bonds, required by the owner. A bid bond guarantees that if a contractor is successful in winning a bid, the contractor will enter into a contract within a specified period of time and furnish any required bonds. A payment bond provides assurance that certain payment obligations associated with a construction project will be satisfied. A performance bond typically is delivered at the time the contract between the owner and contractor is signed. It guarantees that the contractor will perform the work in accordance with the contract documents. The owner usually reserves the right to approve the surety company and the form of the bond. These bond requirements are included in the bidding documents prepared by the prime designer. (See Chapter 6, What Engineers Deliver.)

The class of project (A-1, A, B, and Miscellaneous), the contract amount, and the contract format (lump sum, guaranteed maximum price, unit price, and so forth) combine to determine the cost of bonds. The ability of an entity (contractor, design-build joint venture, partnership, or integrated engineering-procurement-construction firm) to obtain these bonds greatly influences the type and size of projects it can pursue.

Before issuing bonds, surety companies usually undertake thorough investigations. According to Clough, Sears, and Sears, the following are the usual and most important aspects of this type of inquiry:

Essential characteristics of the project under consideration including: size, type, and nature; identity of the owner and the owner's ability to pay; contractor's adequate resources, equipment, expertise, and experience

Total amount of bonded and unbonded uncompleted work in the contractor's current inventory, including work that has yet to be awarded—helps determine if the contractor's working capital, equipment, and organization are becoming overextended

Adequacy of working capital and availability of credit substantiated by financial reports—may prevent the contractor from taking on too much work

Amount of money between the contractor's bid and the next lowest bid, what was "left on the table" —if the spread is more than 5 or 6 percent, there is cause for concern and the surety company may question the soundness of the contractor's bidding practices

Largest contract amount of similar work successfully undertaken and completed to date by the contractor—the surety company is more comfortable if the contractor stays within its realm of expertise; if the contractor wants to enter a new market sector, the surety will advise starting with small projects

Terms of the contract and bonds required, details of how the owner's payments will be made to the contractor, retention (amount of money to be withheld from each of the owner's payments until the project is complete), time allotted for construction, liquidated damages, and required warranties—all affect the contractor's ability to perform the work

Amount of work subcontracted and qualifications of the subcontractors—must possess the necessary organization, financial resources, and experience to carry out the work

After a bonded project is completed, the owner is asked to send a final report to the surety. This report includes a statement regarding the contractor's execution of the contract, changes that were made during the course of the work, and the final total contract amount, which will be used to adjust the amount of the final bond premium.

Professional liability insurance and bonds are both ways of attempting to manage risk. But not even the best risk management plans are guaranteed to yield 100 percent success rates.

DISPUTE RESOLUTION

Clearly, negotiation is the best way to resolve differences. When that does not work, the parties involved have several courses of action open to them. *Litigation,* the filing of a lawsuit, is the most common form of formal dispute resolution. Litigation proceeds in accordance with a complex set of rules. Years may pass between filing a lawsuit, for instance, making a claim, and presenting the case in court. Due to delays and costs, the vast majority of lawsuits are settled before going to court.

Because so much construction litigation involves designers, civil engineers are advised to be familiar with civil litigation procedures and Alternative Dispute Resolution (ADR) practices, such as mediation and arbitration, which are discussed in the following sections.

Civil Litigation

Civil litigation is adversarial and expensive in terms of both time and energy. It can involve considerable emotion because the concepts of battle, winners-losers, and punishment are at play. Civil engineers may find themselves in court for a variety of reasons, such as late or nonpayment by clients or problems arising vicariously from the actions of other design consultants and/or contractors.

There are four stages in civil litigation: Pleadings, Pretrial, Trial, and Post-trial. (See Figure 11.4.)

Pleadings

The purpose of the pleadings phase is to establish the basis of the claim and to determine if the *plaintiff*, the entity filing the lawsuit, has a case. During this first phase of the lawsuit, the *defendant* must be informed that the plaintiff has initiated action. A representative of the plaintiff (a process server, typically sheriff, marshal, constable, or the like) delivers in person to the defendant:

- Summons and complaint, identifying the defendant(s) and the actions being taken against him/her/them
- Name of the court
- Name(s) of the plaintiff(s)
- Name and address of the plaintiff's legal counsel (attorney)

Defendants have a specific number of days to respond, typically three weeks plus or minus several days. If they do not respond, the court enters a *default judgment* against them. In other words, the judge finds in favor of the plaintiff. The defendant's more prudent course of action is to meet with an attorney, who files a formal notice that the summons has been received. The defendant and attorney can then determine

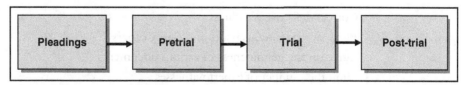

Figure 11.4   Stages in civil litigation

a prudent course of action. Their decision will be based on the legitimacy of the complaint, its potential worth, the plaintiff's ability to sustain an extended legal action financially, the impact of an extended legal action on the defendant and his or her reputation, and the standing of the plaintiff's counsel.

Often both parties attempt to negotiate a settlement and may repeat the process several times if unsuccessful initially. If they are able to reach an agreement, a written document (sometimes called a *stipulation*) is filed with the clerk of the court and the case is recorded as closed. Concurrent with negotiations, the defendant's attorney can file a variety of motions that test the strength of the plaintiff's claims. Among these is the *motion to dismiss*, also known as *demurrer*, which argues that the plaintiff does not have the legal right for a favorable judgment. The defendant's counsel also can file motions arguing that the court does not have jurisdiction over the claim, that the defendant's claim is vague or ambiguous and should be refiled, and/or that portions of the defendant's claim are redundant, immaterial, or scandalous and that they should be removed.

After decisions about the motions have made, the defendant must reveal his or her position with an *answer* to the complaint in the form of a *denial* (*affirmative defense*), a *counterclaim*, or a combination of the two. An affirmative defense alleges the plaintiff's claims essentially are true but that certain explanatory facts have been excluded, such as the plaintiff's contributory negligence or expiration of the statute of limitations. Once filed, a counterclaim becomes a cross-suit to which the plaintiff must answer with a *reply*. The plaintiff must follow the same process that was used initially by the defendant.

Once complete, the complaint, answer, and reply (if the defendant has filed a countersuit) form the pleadings. Based on preestablished procedures, the pleadings can be amended, but eventually the pleadings identify only those issues that can be raised at trial. (See Figure 11.5.)

Pretrial

Following closure of the pleadings phase of litigation, either the plaintiff or the defendant files a *notice of trial* asking the court to put the suit on its calendar. Both the plaintiff and defendant can demand a jury trial. If so requested, the court is obligated to provide a jury. The judge may do so even if the parties prefer not to have a jury trial.

Prior to the trial, the judge calls for a mandatory *pretrial hearing* or *conference*. In an effort to save court time and cost, the plaintiff and defendant's attorneys appear before the judge in his or her chambers and may be ordered to:

- Correct defective pleadings
- Eliminate extraneous issues and clarify others
- Agree on the genuineness of various documents
- Limit the number of expert witnesses
- Identify the scope of *discovery*

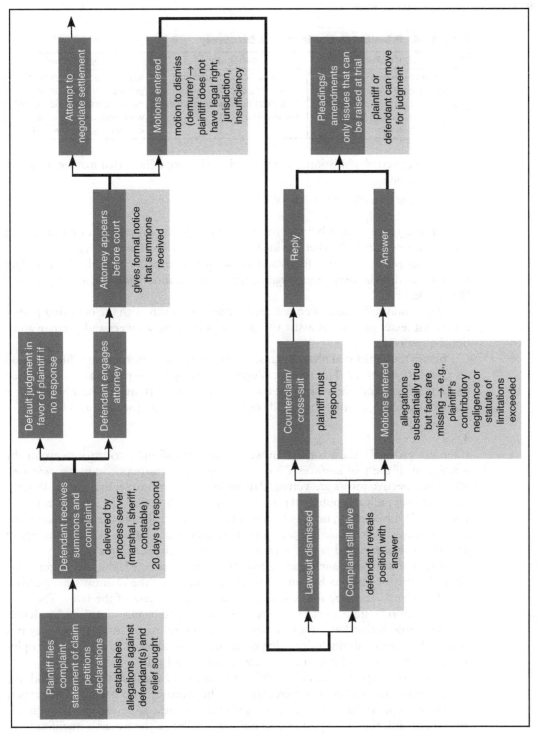

Figure 11.5   Pleadings phase of litigation

Table 11.4 Principal Methods Used in Discovery

Subpoenas Duces Tecum	Interrogatories	Depositions
Issued by the clerk of the court (or the attorney of record) and delivered by a process server to compel the opposing party to provide certain information	List of fact-based questions sent to the opposing party that must be answered under penalty of perjury	Interrogation of opposing party's witnesses under oath—is transcribed word-for-word and can be used as evidence during the trial

- Make pretrial admissions, admitting the existence of facts that may help the other side
- Engage in settlement negotiations

Discovery also occurs before the trial. Through the process of discovery, both sides have access to each other's evidence. Discovery saves court time and expense, as well as narrowing the suit's focus and allowing few surprises at trial. The principal methods used in discovery are: *subpoenas duces tecum, interrogatories,* and *depositions.* (See Table 11.4.)

Depositions often are taken of expert witnesses, usually with the opposing party's expert witnesses present to assist the attorneys evaluate answers and prepare additional questions.

Before the actual trial phase begins, one of the parties' attorneys may file a motion for *summary judgment.* This motion alleges that the opposing party cannot prove that the facts are true and, therefore, that the case has no merit. If the summary judgment is granted, the claim, counterclaim, and/or both are dismissed. (See Figure 11.6.)

Trial

If the suit proceeds, the attorneys answer a *calendar call* and a courtroom eventually is assigned. If the trial involves a jury, the judge and attorneys conduct a *voir dire,* where prospective jurors are vetted. Attorneys question prospective jurors. When a juror's answer indicates potential problems with prejudice, financial interest, relationship to an involved party, or the like, the attorneys can challenge the juror stating the reasons. This is called a *challenge for cause.* Attorneys also have a discreet number of *peremptory challenges* and can have a juror dismissed without citing a reason.

After the jury is impaneled and alternate jurors are selected, the trial begins. The plaintiff's attorney begins by summarizing the facts from the plaintiff's perspective. The defendant's attorney follows with the defendant's view of the facts. Then the plaintiff's attorney starts to call witnesses, who are bound by various rules of evidence. One such rule is that lay witnesses may only testify to matters of fact and may not express their personal conclusions. Expert witnesses are allowed to express their opinions and conclusions. Other rules of evidence are included in Table 11.5.

Like jurors, expert witnesses are subjected to *voir dire.* After taking the stand, the expert witness recites his or her credentials. The attorney who has hired the expert witness attempts to have the court recognize the witness as an expert. Though seldom successful, the other party's counsel may challenge the expert's qualifications.

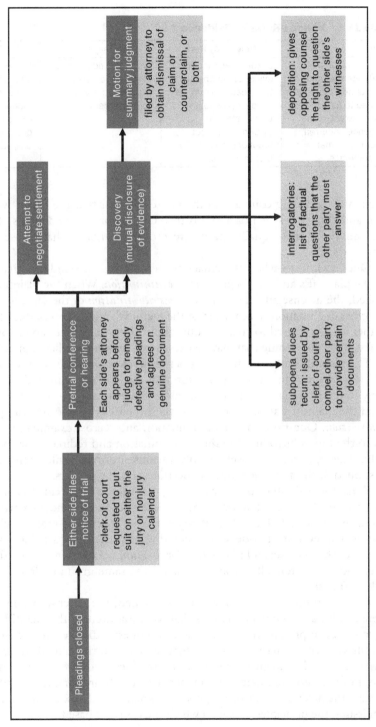

Figure 11.6   Pretrial phase of litigation

Table 11.5 Various Rules of Evidence

Parole Rule	Relevancy Rule	Hearsay Rule	Best Rule
Bars admission of evidence that differs from understandings agreed to formally, as in a contract, if that evidence occurred after the formal agreement(s)	Determines that only evidence outlined in the pleadings may be used; serves to limit *circumstantial evidence*, indirect proof or disproof of a fact in question	Precludes admission of evidence based on what a witness was told by others	Deems that only the best possible form of evidence must be produced at trial, e.g., exhibits, such as documents, should be originals, not copies

Frequently, opposing council and the judge *stipulate* that the expert witness is fit to serve and the witness is not allowed to recite his or her credentials. This can be a disadvantage if these qualifications are more remarkable than those of the other party's expert.

Once a witness—whether factual or expert—takes the stand on behalf of the plaintiff, the plaintiff's attorney begins *direct examination*. When the plaintiff's attorney is finished, the defense attorney initiates *cross examination* of the same witness. On one level cross examination is meant to test the memory and/or knowledge of a witness; on another level, it is used to create doubts about the witness's credibility or to cause the witness to say something that may give the jury a reason to dislike him or her.

The plaintiff's attorney may conduct *redirect examination* of the witness to correct answers given in cross examination that could lead to misunderstandings or to clarify statements that appear to be at odds with evidence developed through discovery. If the plaintiff's attorney redirects, then the defense attorney can pursue *recross examination*. Questions in cross examination and recross examination must be limited to the topics introduced in direct examination and redirect examination, respectively. The opposing counsel can file various motions while witnesses are being questioned, including objections to questions and/or answers.

When the plaintiff rests his or her case, the defense can decide to continue with its case or to settle. If the defense elects to continue, the defendant's attorney can file various motions, including: (1) *motion for a directed verdict*—granted when the judge believes that, even if all evidence presented were true, an impartial jury would decide in favor of the defense; and (2) *motion for a voluntary nonsuit*—allows the plaintiff to begin another, potentially stronger, case on the same grounds after the plaintiff has paid court costs.

If these motions are not made or are denied, the defense begins its case. The defense follows the same procedures that were followed by the plaintiff. After all evidence has been presented, the defense can request a directed verdict. If denied, the plaintiff's attorney can offer evidence that rebuts what the defendant's witnesses have said. Then the defendant's attorney can introduce evidence to contradict what was brought up in rebuttal. Once all evidence has been heard, first the plaintiff's attorney and next the defendant's attorney present *summations*. These summations outline the chief issues, the principles of law on which the cases are based.

After both attorneys have completed their statements, the judge *charges the jury*—reviewing the closing statements, outlining the relevant issues of law, going over witness testimony, and giving advice about how the jury should evaluate what they have heard. Usually the jury is advised that the plaintiff (or defendant if there is a counterclaim) has the burden of proof and that proof is based on the *preponderance of evidence*. Both parties' attorneys can file requests to charge, listing special issues to be considered by the judge. They also can object to the charge given by the judge.

After being charged, the jury enters the deliberations phase of the trial. When a decision is known, the jury delivers its verdict. The losing side can request a *judgment notwithstanding the verdict*, essentially asking the judge to overturn the jury's decision. If the *judgment notwithstanding the verdict* is denied, the losing side may make a *motion for a new trial*. When the jury cannot reach agreement, a *hung jury* results and the case must be retried; but if the verdict stands, the judge directs entry of a *final judgment* in favor of the winning side. (See Figure 11.7.)

Post-Trial

Either party to the case appeal the court's decision for a variety of reasons—errors were made or perhaps the judgment was too high, for example. The appellant (the party appealing the court's decision) must notify the clerk of the court where the trial was conducted and the appellee that an appeal is being mounted. The appellant prepares a *record of appeal*, citing precedents explaining why the appeal should be granted. The appellant must post a bond to cover the cost of any judgment in the event the appellant's assets are lost during an unsuccessful appeal. The appellee also usually files a record of appeal outlining why the original verdict should stand.

Several judges sit on the appeals court. The court's decision is based on the majority opinion. The appeals court may agree with the previous ruling, overturn or modify the judgment, or grant a new trial. If the case presents complex issues of law, the verdict of an appeals court may be appealed to a state supreme or appellate court and, in very rare circumstances, to the U.S. Supreme Court. The aggrieved party may also seek redress in a federal court. (See Figure 11.8.)

ALTERNATIVE DISPUTE RESOLUTION

Alternative dispute resolution (ADR) has been used in the U.S. construction industry since the 1870s. ADR is exactly like it sounds—an alternative to the long and draining process of resolving disputes through litigation. ADR consists of several procedures including Mini-Trials, Dispute Review Boards, but mainly variations of Mediation and Arbitration. All of these approaches have advantages and disadvantages, which are explored below.

Mediation

Mediation is the "newest" form of ADR and often precedes other dispute resolution procedures. Mediation is a conciliatory process that can be voluntary or mandatory.

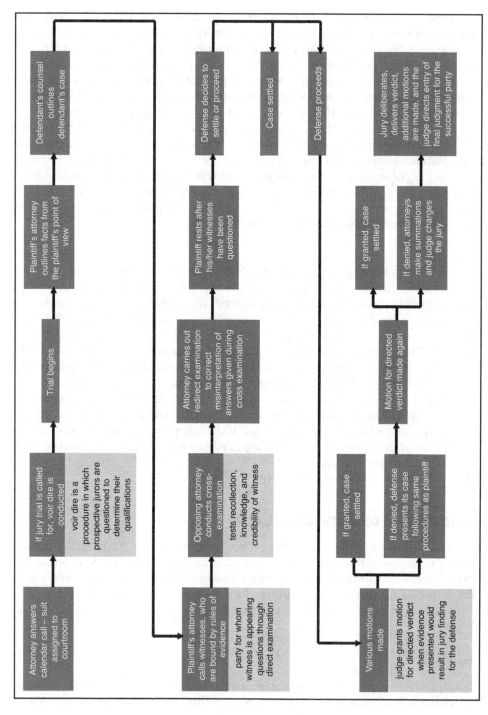

Figure 11.7   Trial phase of litigation

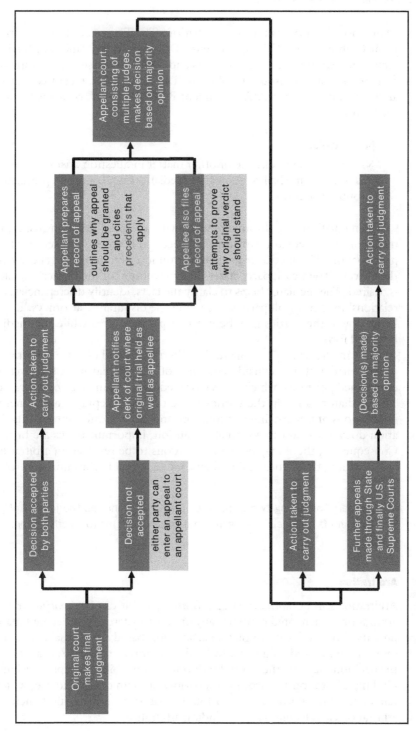

Figure 11.8 Post-trial phase of litigation

Many model contracts require mediation. The Engineers Joint Contract Development Committee (EJCDC), ConsensusDOCS, and the American Institute of Architects (AIA) contracts between owners and design professionals require mediation as a first step in formal dispute resolution. The AIA's most recent contract between an architect and an owner, AIA Document B101™ - 2007, contains the following contract clause:

§8.2 Mediation

§8.2.1 Any claim, dispute or other matter in question arising out of or related to this Agreement shall be subject to mediation as a condition precedent to binding dispute resolution . . .

In mediation, the disputants meet with a neutral decision-maker, in this case a mediator. The mediator explains the process and reviews the reasons for the parties' participation. The mediator goes over ground rules, decorum, and confidentiality of the process. Then each party states their perception of the matter, the facts, and what is desired. The mediator helps to clarify the facts, identify discrepancies, and assess the relationships among all parties, including the disputants' attorneys. If joint meetings break down, the parties may be put in separate rooms while the mediator shuttles back and forth.

The mediator may be able to find alternatives and creative solutions that the disputants had not yet considered. One of the mediator's objectives is to loosen entrenched positions, which is easier if mediation is pursued sooner rather than later. The mediator can help the parties move toward acceptable adjustments, stress the implications of not arriving at a solution, and put any offers into words. Even if mediation does not result in a formal resolution, important issues are brought to light. Consequently, there may be fewer questions to be resolved in arbitration or at trial. In mediation, the parties, rather than a judge, jury, or arbitrator, control the result.

> Joinder: Combining two or more elements into one, as the joinder of parties as coplaintiffs or codependents in litigation or as parties to an arbitration.

Arbitration

Arbitration can be *mandatory* or *voluntary*, and it can be *binding* or *nonbinding*. If arbitration is mandated contractually or court-ordered, it is deemed mandatory. Disputants also can decide to pursue arbitration based on mutual agreement, in which case it is considered voluntary. In binding arbitration, the disputants must adhere to the decision made by the neutral (arbitrator). Possibility for appeal is limited. In nonbinding arbitration the arbitrator provides an opinion and advice, which the disputants decide to follow or not to follow. Based on mutual agreement, nonbinding arbitration can be converted to binding arbitration.

In mandatory binding arbitration the disputants present their cases to one arbitrator or a panel of three arbitrators, depending on the complexity of the problem. Usually, Construction Industry Arbitration Rules of the American Arbitration Association (AAA) are followed. Discovery and the right of appeal are limited; and the arbitrator's decision is binding and enforced by the courts. An AAA-based arbitration procedure begins when one of the involved parties files a letter or AAA form, with an accompanying filing fee, requesting arbitration. Based on information about the claim outlined in this document, the AAA prepares a list of potential arbitrators.

Frequently the disputants and their attorneys convene with a senior arbitrator in a prehearing meeting. At this meeting the parties discuss: undisputed facts, the estimated number of hearings required, schedule, arbitrator compensation, claims and counterclaims, lists of witnesses, and arrangements for site inspection, if necessary. Sometimes this meeting results in a "meeting of the minds," and the parties arrive at a solution to their problem. In a sense, the senior arbitrator has served the purpose of a mediator.

Following the prehearing meeting, disputants select the arbitrator(s) from the AAA list of potential arbitrators. The arbitrator has the power to consider amendments to the claim or counterclaim, control the timing of hearings, and continue without the presence of a party, given appropriate notice. The arbitrator also can subpoena witnesses and documents, when permitted by law.

There are some drawbacks to this process. Discovery is limited, so the potential for surprise at hearings exists, perhaps to the detriment of one of the parties. Also, although arbitrators are familiar with technical issues arising in the construction industry, they may not be well-informed about recent legal precedents guiding the development of contract clauses. Additionally, arbitrators typically write brief decisions and are not required to include their rationale. Courts do not like to interfere with arbitrators' decisions. Seldom is appeal granted, usually only when the arbitrator has been shown to have a conflict of interest or when the rules of arbitration have been breached seriously. See Table 11.6 for a comparison of litigation, arbitration, and mediation.

Mini-Trial

Mini-trials may be best utilized in complex cases when disputants essentially disagree on the way precedents and the law should be applied to the facts. The term "mini-trial" is somewhat a misnomer because the process is more like solving a business problem than conducting a trial. Because the scope of preparation and presentation is limited, each party must formulate its best case. The parties directly involved in the problem, experts, and attorneys present their case to members of top company management, who have the authority to settle.

The American Arbitration Association has developed a mini-trial process, though the disputants are free to establish their own mutually agreed upon procedures. A neutral party is usually selected to guide the process and to advise the parties about the relative strengths and/or weaknesses of their cases. Without the well-established

Table 11.6 Litigation, Arbitration, and Mediation Compared
(Adapted from Howard Goldburg, Esq., "Dispute Resolution Methods," *The Architect's Handbook of Professional Practice*, pp. 396–404.)

Factor	Litigation	Arbitration	Mediation
Overall satisfaction	Generally poor	Generally high	Very high
Characterization	Adversarial	Adversarial	Conciliatory
Confidentiality	Highly public	Private	Private
Phase of dispute	Last resort	When problems first arise or following mediation	When problems first arise and when negotiation fails to work
Cost of resolution	Very expensive	Administrative fees high, but discovery/trial costs lower than litigation	Inexpensive
Speed of resolution	Very slow	Fast	Fastest
Sustainability of parties' relationship(s)	Least possible	Possible	Most possible
Discovery	Extensive	Limited	Limited
Possibility of appeal	Yes	No	No
Trier(s) of fact, neutral	Judge and/or jury	Arbitrator	Mediator
Selection of trier(s) of fact, neutral	Appointed by legal system and drawn from general public	Informed professional selected by involved parties	Informed professional selected by involved parties
Atmosphere of surroundings	Very formal—courtrooms symbolize the power of the law	More relaxed	More relaxed
Finality of decision	None, until appeals exhausted	Final	Final, if settlement reached
Fairness of decision	Varies—courts apply applicable laws; but jurors have little knowledge of issues	Fair because decision made by experienced arbitrator who is industry professional	Ultimate in that each party must agree to the resolution

rules and limitations imposed by litigation, the parties most knowledgeable about the problem can arrive at a settlement that works for them. The resulting solution is business-oriented and may not have the winner-loser characteristic of litigation.

Dispute Review Board

Dispute Review Boards (DRBs) are an ADR technique used mainly by public agencies. Once a construction contract has been signed and before any disputes arise, a DRB is formed. One member is selected by the owner, subject to contractor approval. Another is selected by the contractor, subject to owner approval. These two members select a third, who serves as board chairperson. Regular DRB meetings are held on-site to review disagreements. The DRB's decisions are not binding and do not preclude later mediation, arbitration, or litigation. However, they do provide an excellent record of real-time events, facts, and opinions of an impartial panel of experts.

AFFIRMATIVE ACTION, EQUAL OPPORTUNITY, AND DIVERSITY

Various laws affect the practice of civil engineering, including those governing affirmative action (AA), equal opportunity (EO), and diversity in the workplace. Several government organizations are responsible for enforcing federal laws that make discriminating against a job applicant or an employee because of the person's race, color, national origin, sex (including pregnancy), age (40 or older), religion, or disability illegal.

U.S. Anti-Discrimination Laws

The following are specific U.S. anti-discrimination laws:

- *Title VII of the Civil Rights Act of 1964 (Title VII)*
  This law makes discriminating against someone on the bases of race, color, religion, national origin, or sex illegal.
- *The Pregnancy Discrimination Act*
  This law amended Title VII to making discriminating against a woman because of pregnancy, childbirth, or a medical condition related to pregnancy or childbirth illegal.
- *The Equal Pay Act of 1963 (EPA)*
  This law makes paying different wages to men and women if they perform equal work in the same workplace illegal.
- *The Age Discrimination in Employment Act of 1967 (ADEA)*
  This law protects people who are 40 or older from discrimination because of age.
- *Title I of the Americans with Disabilities Act of 1990 (ADA)*
  This law makes discriminating against a qualified person with a disability in the private sector and in state and local governments illegal.
- *Sections 501 and 505 of the Rehabilitation Act of 1973*
  This law makes discriminating against a qualified person with a disability in the federal government illegal. The law also requires that employers reasonably accommodate the known physical or mental limitations of an otherwise qualified individual with a disability who is an applicant or employee, unless doing so would impose an undue hardship on the operation of the employer's business.
- *The Genetic Information Nondiscrimination Act of 2008 (GINA)*
  This law makes discriminating against employees or applicants because of genetic information illegal. Genetic information includes information about an

individual's genetic tests and the genetic tests of an individual's family members, as well as information about any disease, disorder, or condition of an individual's family members (i.e., an individual's family medical history).

In all of the above laws, retaliating against a person because the person complained about discrimination, filed a charge of discrimination, or participated in an employment discrimination investigation or lawsuit also is illegal.

Enforcement of Anti-Discrimination Laws

The U.S. Equal Employment Opportunity Commission (EEOC) has the authority to investigate charges of discrimination against employers who are covered by the law. Most employers with at least 15 employees are covered by EEOC laws (20 employees in age discrimination cases). Most labor unions and employment agencies are also covered. The laws apply to all types of work situations, including hiring, firing, promotions, harassment, training, wages, and benefits. If discrimination has occurred, EEOC will attempt to settle the charge. If unsuccessful, EEOC has the authority to file a lawsuit to protect the rights of individuals and the interests of the public.

Another organization is the U.S. Department of Labor's Office of Federal Contract Compliance Programs (OFCCP). Since 1965 the OFCCP has ensured that federal contractors comply with the equal employment opportunity and the affirmative action provisions of their contracts. OFCCP administers and enforces Executive Order 11246, as amended, which prohibits federal contractors and federally assisted construction contractors and subcontractors, who do over $10,000 in government business in one year, from discriminating in employment decisions on the bases of race, color, religion, sex, or national origin. The Executive Order also requires federal contractors to take affirmative action to ensure that equal opportunity is provided in all aspects of their employment.

Affirmative Action Requirements

Each federal contractor with 50 or more employees and $50,000 or more in government contracts is required to develop a written affirmative action program (AAP) for each business entity. The written AAP helps employers identify potential problems in the participation and utilization of women and minorities in their workforce. If there are problems, the employer can state in its AAP the specific procedures it will follow to provide equal employment opportunity. Companies can include outreach, recruitment, and training as affirmative steps in helping members of protected groups to compete for jobs on a level playing field with other applicants and employees.

SUMMARY

The legal aspects of professional practice are extensive. This chapter has covered topics of particular concern to civil engineers: the U.S. legal system; statutory and contract law; contracts used in project delivery; risk management; insurance and bonds; dispute resolution; alternative dispute resolution; and affirmative action, equal opportunity, and diversity. The references listed below provide copious, detailed information and are valuable resources for further study.

REFERENCES

The American Institute of Architects. (2003). *The Architect's Guide to Design-Build Services.* Ed. by G. William Quatman II and Ranjit (Randy) Dhar. John Wiley & Sons, Inc. Hoboken, NJ. ISBN 0-471-21842-1.

The American Institute of Architects. (2008). *The Architect's Handbook of Professional Practice, 14th edition* . Ed. by Joseph A. Demkin. John Wiley & Sons, Inc. Hoboken, NJ. ISBN 978-0-470-00957-4.

Bachner, John Philip. (1991). *Practice Management for Design Professionals: A Practical Guide to Avoiding Liability and Enhancing Profitability.* John Wiley & Sons, New York. ISBN 0-471-52205-8.

Beard, Jeffrey L., Michael C. Loulakis, and Edward C. Wundram. (2001). *Design Build: Planning through Development.* McGraw-Hill, Boston, MA. ISBN 0-070-06311-9.

Clough, Richard H., Glenn A. Sears, and S. Keoki Sears. (2005). *Construction Contracting: A Practical Guide to Company Management,* 7th edition . John Wiley & Sons, Inc. Hoboken, NJ. ISBN 0-471-44988-1.

Design Build Institute of America (DBIA). (2009). *Design-Build Contract & Risk Management.* Course Materials, Design-Build Institute of America, Washington, DC.

Design Build Institute of America (DBIA). (2009). *Fundamentals of Project Delivery.* Course Materials, Design-Build Institute of America, Washington, DC.

Jackson, Barbara. (2010). Design-Build Essentials. Delmar Cengage Learning. ISBN-10: 1-428-35303-8/ISBN-13: 978-1-428-35303-9.

O'Reilly, Michael. (1999). Civil Engineering Construction Contracts, 2d edition. Thomas Telford, London. ISBN: 0-727-72785-0.

Smith, Currie, & Hancock's Common Sense Construction Law: A Practical Guide for the Construction Professional, 3rd edition . (2005). Ed. by Thomas J. Kelleher, Jr. John Wiley & Sons, Inc. Hoboken, NJ. ISBN 0-471-66209-7.

Introduction
Ethics
Professional Engagement
History
What Engineers Deliver
Emerging Technology
1 2 3 4 5 6 7 8 9 10 11 12 13 14 15 16 17 A B C D E F
Engineer's Role in Project Development
Permitting
Leadership
Managing
Having a Life
Executing a Professional Commission
Client Relationship
Legal Aspects
Communicating
Globalization
Sustainability

Chapter **12**

Managing the Civil Engineering Enterprise

Big Idea

Profit and exceptional client service are essential for the continued existence of the consulting engineering enterprise.

> Our job as leaders—the alpha and the omega and everything in between—is abetting the sustained growth and success and engagement and enthusiasm and commitment to Excellence of those, one at a time, who directly or indirectly serve the ultimate customer.

—Tom Peters

Key Topics Covered

- Introduction
- The Influence of Economics on Project Development
- Financial Reporting
- Professional Human Resources Management
- Career Planning and Execution
- Specialization
- Certification and Registration
- Professional Services Marketing
- Professional Business Development
- Professional and Trade Organization Activities
- Summary

Related Chapters in This Book

- Chapter 3: Ethics
- Chapter 4: Professional Engagement
- Chapter 7: Executing a Professional Commission
- Chapter 14: Having a Life

(Continued)

Related to *ASCE Body of Knowledge 2* Outcomes

ASCE BOK2 outcomes covered in this chapter

Foundational
1. Mathematics
2. Natural sciences
3. Humanities
4. Social sciences

Technical
5. Materials science
6. Mechanics
7. Experiments
8. Problem recognition and solving
9. Design
10. Sustainability
11. Contemporary issues & historical perspectives
12. Risk and uncertainty
13. Project management
14. Breadth in civil engineering areas
15. Technical specialization

Professional
16. Communication
17. Public policy
● 18. Business and public administration
19. Globalization
20. Leadership
21. Teamwork
22. Attitudes
23. Lifelong learning
24. Professional and ethical responsibility

INTRODUCTION

Even as technological advances continue to accelerate the technical capabilities of the engineering practice, a fundamental precept that major engineering decisions are based on economic considerations remains unchanged. The basic economics of the engineering practice are essential to understanding their successful application in a larger business administration context.

The discussion of the economic considerations in this chapter is within the convention of the *successful delivery of a project*. Several terms need to be defined to within the following phrase:

A *project* is a temporary endeavor undertaken to create a unique product or service. "Temporary" means that every project has a distinct beginning and definite end. "Unique" means that the product or service provided is different in some way from all other similar products or services. (Duncan 1996)

Delivery is the presentation of the final product or service a client asked for at the genesis of the project. It is the reason the project came into existence, and upon its conclusion, the essential lifespan of the project is deemed complete.

Successful is defined in this instance as the submittal of the final product or service to the client that not only meets the agreed-to terms and stipulations of the arrangements made between the engineer and the client, but also provides a mutual sense of accomplishments and satisfaction to both the engineer and the client. In other words, the client feels that the product or service received was well worth the money paid to obtain it, and the engineer considers the fee paid to provide a reasonable profit for the labor and materials expended to produce it. These concepts are also related to the engineer's ethics as presented in greater detail in Chapter 3, Ethics.

The mechanics of executing a successful project were discussed at length in Chapter 7, Executing a Professional Commission. The focus of Chapter 7 is primarily the processes and procedures of project management. This chapter focuses on the business aspects of projects and discusses how to attain a profitable result. This portion of the handbook relates to the application of tools to assist in budgeting a project for a successful business outcome—work performed at a profit for the engineering organization.

THE INFLUENCE OF ECONOMICS ON PROJECT DEVELOPMENT

During the excitement of planning and initiating the development of a project, the question, "Can we do it?" seldom is followed with the question, "Should we do it?" The "Should we do it?" question addresses whether or not pursuing the project is in the best interest of the engineering firm. Many engineering firms have implemented a process, with varying degrees of formality, commonly referred to as a "Go/No-Go" decision process. In final consideration of a decision to proceed with a project (or not), the engineer should consider the material presented in Chapter 14, Having a Life, especially as it relates to Chapter 3, Ethics.

The Go/No-Go Decision Process

While it may seem obvious, frequently decisions are made to pursue the development of a project that are independent of the technical and economic merits of the project outcome. The rationale for these pursuits may be driven by a variety of factors, including relationships with the client, potential expansion into new lines of engineering, political factors, or even the vanity and hubris of a firm's senior leadership. Using a go/no-go process to quantify the economic impacts a project may have on a firm's business should be an essential first step in the project development process. By implementing this process, the engineer may avoid the prospect of performing a project that fails to deliver a desired result within any prescribed schedule, possibly at an enormous cost to the client and/or the firm.

The go/no-go process is essentially a series of questions asked of the potential project manager to consider *prior to* moving into the project initiation stage. While there is no prescribed approach for conducting this exercise, the typical questions that are almost universally asked are:

1. How well is the work understood?
2. How well-defined are the client's expectations?
3. What degree of profit can reasonably be expected?
4. What kind of risk factors are presented by this project? (For instance, is the project in a foreign country, are there unusual terms and conditions, is it a design-build project, is hazardous material/waste present, is there extensive public and/or regulatory agency involvement, and so forth.)

Assuming that the risk and technical factors can be addressed such that the go/no-go process answer to the question "Should we do the work?" is affirmative, the motivational factor of developing a project budget that reflects reasonable and accurate expenditures and commensurate profit becomes essentially the ultimate goal of the project, when viewed from a business economics (as opposed to technical) perspective. One of the first things to understand, then, is how the costs of a project, including indirect and direct overhead, and labor and nonlabor needs affect the creation of an appropriate budget.

Overhead and Direct Labor

Chapter 7 discusses the development of a project budget; key among that budget is the level of effort (hours) needed to perform a given task, and the mix of professional disciplines required to complete that task properly. Each discipline has a different hourly labor rate, but how are these rates determined?

When budgeting for a project, a labor rate that is "loaded" or "fully loaded" is what is typically used to establish budget cost, as opposed to that of a "raw" or "direct" labor rate. In this case, a fully loaded labor rate accounts for the overhead as well as the direct labor costs of a worker on the project. Thus, fully loaded labor rates more accurately reflect the actual costs involved in providing a service. "Overhead," in this instance, is defined as a cost or expense (such as for liability insurance, rent, and utility charges) that relates to an operation of the engineer's firm as a whole, and cannot be applied or traced to any specific unit of output. Overhead is considered an indirect cost.

A subset of the development of a fully loaded labor rate includes the fringe benefits costs associated with the firm's employment of engineering and other professional staff. "Fringe benefits" are compensation paid by the firm in addition to direct wages or salaries. Fringe benefits include items such as medical insurance, vision and dental benefits, paid holidays, 401(k) matches, stock options, and subsidized meals.

Additionally, in the calculation of a fully loaded labor rate, the general and administrative (G&A) costs must be factored into the labor rate calculation. These general and administrative costs reflect those expenditures necessary for the operations of the firm, but are not directly associated with developing a product or providing a service. Examples of G&A-related costs include account invoicing staff, office receptionists, and human resources personnel.

Finally, a calculation of profit may be considered in the development of the fully loaded labor rate, or it may be a calculated value based on the total value of the project. Profit determination varies among contract types, and the description of different contract types is included in Chapter 11. Several examples of the determination of a fully loaded rate, both including and excluding profit calculations, are shown below to better illustrate their development.

Example 1: A newly hired engineer has a base salary that pays $40 per hour. The firm's fringe benefits costs, an indirect cost, average 35 percent for its employees. The overhead costs on an hour of labor for this firm are 65 percent. Lastly, the firm has a G&A rate of 10 percent and an assumed profit of 9 percent for all of the firm's projects. To determine the fully loaded hourly labor rate for this engineer, perform the following calculations:

Multiply raw labor rate by fringe benefits, add to the raw rate: **($40.00 × 0.35) + $40.00 = $54.00**

Multiply the result by the overhead rate and add the result: **($54.00 × 0.65) + $54.00 = $89.10**

Next, multiply this result by the G&A rate and add the result: **($89.10 × 0.10) + $89.10 = $98.01**

Finally, assuming that profit is calculated in the fully loaded rate, multiply the results above by the profit to determine the final fully loaded labor rate: **($98.01 × 0.09) + $98.01 = $106.83**

In this example, the firm would realize this projected 9 percent profit (based on this employee's efforts) only if the employee managed to be 100 percent billable to the client's project work, and if the firm actually collected that invoice in full.

Example 2: Perform the same function for a senior project manager with a base salary of $50 per hour, but with profit being considered a separate determination from labor.

Multiply raw labor rate by fringe benefits, add to the raw rate: **($50.00 × 0.35) + $50.00 = $67.50**

Multiply the result by the overhead rate and add the result: **($67.50 × 0.65) + $67.50 = $111.38**

Next, multiply this result by the G&A rate, add the result: **($111.38 × 0.10) + $111.38 = $122.52**

Since profit is a separate calculation in this instance, the result above is the final fully loaded labor rate for this engineer.

Some cautionary words regarding loaded labor rate calculations: the description of the above determination of loaded labor rates is subject to a wide degree of variability. The engineer's firm may calculate their loaded rates in a different fashion, by defining whether certain administrative costs are G&A or true overhead costs instead. Likewise, the ratios shown in the above examples are typical, but by no means standard, for the engineering industry. The fringe benefit costs, overhead, and G&A are determined by the firm through a detailed review of expenditures, and if the firm is performing work for the federal government, the rates may be subject to extensive audits and revision by the auditing agency for which the engineer's firm is proposing to perform contract work. There may also be strategic factors in terms of providing favorable rates to clients in certain circumstances whereby a firm may choose to provide discounted labor rates by lowering their indirect or overhead costs.

Multipliers

In light of the above discussion, the use of multipliers as a means of developing loaded rates and comparing the performance of a project from a business perspective is a very effective tool to determine rapidly if a project is performing in accordance with its budgeted baseline, or if staff usage is consistent with the hours allocated in the original budget. Several distinct multipliers will be discussed: the labor multiplier, the budget multiplier, and the effective multiplier. All are closely related, but can be used to track or budget different aspects of a project.

Labor Multiplier: The labor multiplier is determined by dividing the fully loaded rate in a contract or a budget by the raw or direct labor rate of a given individual. So, for the two examples given above for determining labor rates, the labor multiplier of the new engineer, with profit included, is 2.67 ($106.80 ÷ $40.00). For the senior project manager, the labor rate without profit included is 2.45 ($122.52 ÷ $50.00). If profit were not included in the loaded labor rates for the new engineer in Example 1 above, the resulting labor multiplier would be 2.45 as well.

Labor multipliers vary widely for engineering firms, as there are numerous factors that influence their development. Typically in engineering firms, labor multipliers range from a low of 2.2 to a high of over 4.0. Higher multipliers are often found in higher risk projects, or for extremely technical work where the work product or knowledge required is rare and specialized. Conversely, lower multiplier work is found in projects where there may be a wide range of qualified engineering firms capable of providing the product or services, or the work is considered fairly standard and of fairly low technical complexity. But this is not universal; very complex projects may have low multipliers, and simple projects may have very high multipliers if the terms of the contract are structured in such a manner to track profit or control risks borne by the engineer for their work.

In this example, the manager of this engineering enterprise will rely on the firm's business plan that projects work and the Marketing Department that finds the opportunities and helps procure the work in this firm's business sector. Using three tools the Engineering Manager can effectively manage the business aspects of the firm in a four-step process. The first step calculates the billable versus nonbillable hours. The second step determines the salary of direct labor personnel for billable versus nonbillable/overhead hours. The third step establishes the overhead costs and rates to calculate the "breakeven multiplier." And finally, the Engineering Manager will back-calculate and verify the breakeven multiplier since this manager is business savvy. A detailed calculation of these calculations and labor multipliers is presented in Figure 12.1.

Budget Multiplier: The budget multiplier is a straightforward calculation. It is the value of the net contract amount of a project divided by the direct (or raw) labor costs. The net contract amount is the total contract amount less any raw subcontractor costs and less any other direct costs. For example, if a contract were signed for $16,000, with a subcontractor's raw costs being $5,000 of the amount, and with $1,000 of other direct costs such as report reproduction, travel costs, and equipment rental, the net contract amount would be $10,000. If the direct (or raw) labor costs for the contract totaled $4,000, then the budget multiplier for this project is $10,000 ÷ $4,000 = 2.5.

Effective Multiplier: The effective multiplier is a measure of the actual financial performance of the contract when compared with the budget multiplier. It is defined as the actual fully loaded costs for the work, subtracting out all raw nonlabor costs. This number is then divided by the raw direct labor costs to determine the effective multiplier. So, if the contract has had $15,000 of fully loaded costs placed against it, with $5,000 subcontractor raw costs, $500 spent to date for other raw nonlabor costs, and with $4,200 in direct (raw) labor costs being the total expenditures at a certain point in the project's lifecycle, the resulting effective multiplier is: ($15,000 − $5,000 − $500) ÷ $4,200 = $9,500 ÷ $4,200 = 2.26.

Having an effective multiplier lower than the budget multiplier is an indication that the actual costs for performing the work are exceeding what the budgeted costs were for the same work. This would suggest some type of corrective action may need to be taken, depending on the terms of the contract and the amount of work remaining to be performed to produce the final product or service. While an effective multiplier that is lower than a budget multiplier is an indicator that the profit potential for the project is decreasing, it does not mean necessarily that the project will lose money for the engineering company. That may be dependent on whether the firm has a breakeven multiplier for a project that can demonstrate the lowest effective multiplier threshold a project can reach and not result in a loss of net revenue for the firm.

Step 1: Billable versus Nonbillable Hours*

Staff Title	Payroll Burden Holidays	Paid Leave	Indirect Labor Hours Marketing	Admin	Other	Total O/H (Hours)	Total Billable (Hours)
PM	80	120	800	0	100	1100	980
Staff Eng.	80	120	400	0	100	700	1,380
Support	80	120	100	500	100	900	1,180

*Total Annual Hours: 2,080 = 40 hours/week × 52 weeks

****

Step 2: Salary of Direct Labor versus Nonbillable

Staff Title	Annual Salary, $	Raw Hourly Rate, $$	Billable Hours	Raw Salary for Billable Hrs, $	Overhead Hours	Raw Salary for Overhead Hrs, $
Senior PM	104,000	50	980	49,000	1100	55,000
Staff Eng	83,200	40	1,380	55,200	700	28,000
Support	52,000	25	1,180	29,500	900	22,500
Totals	239,200			133,700		105,500

****

Step 3: Overhead Cost and Rate Calculation

COSTS	Amount, $
Payroll Burden: (200 hr × each raw hourly salaries, $ (50 + 40 + 25))	23,000
Indirect Labor: (900 hr × $50) + (500 hr × $40) + (700 hr × $25)	82,500
G & A + Overhead Expenses: (10% + 65%) × 239,200	179,400
Total Overhead:	284,900
Overhead/Direct Labor (284,900/239,200, %)	119%
Break-Even Multiplier:	2.19

****

Step 4: Cross-Check for Break-Even Calculation

Staff Classification	Billable Hours	Hourly Rate, $/Hour	Total
Senior PM	980	109.50	107,310
Staff Eng	1,320	87.60	115,632
Support	1,180	54.75	64,605
TOTAL			**287,547**

The resulting calculation of $287,547 is within 1% of the Total Overhead figure $284,900 above and verifies the cross-check.

Figure 12.1 Example Calculations of Labor Multipliers

A few words about multipliers: It's in the best interest of the accounting manager and the engineer to keep comparative multipliers in perspective. Having an effective multiplier lower than a target budget multiplier doesn't mean the project is a failure, but some accountants/managers may make such a black-and-white comparison. A better comparison would be comparing effective multipliers against profit baseline or breakeven multipliers to see if the project is making money or actually costing the firm money to perform. In short, multipliers should not be used as the single "pass/fail" measure of a project's success; there are other factors, most notably client satisfaction and potential for follow-on work that also should be considered when evaluating project performance.

FINANCIAL REPORTING

Income Statement (Profit and Loss)

An income statement may also be called a profit and loss (P&L) statement. It displays the company's financial position for a specific period of time and shows how the company's gross revenue is transferred to net income. The gross revenue is the total revenue from the sales of products and services and is sometimes referred to as the "top line" on the P&L statement. The company's net income is the amount of revenue remaining after "expenses" is subtracted from the gross revenue and is sometimes referred to as the "bottom line." The income statement shows the company's management and investors how well (or not well) the company is performing during the specific reporting period (typically monthly, quarterly, or annually). (Schroeder, Richard, et al. 2010)

For purposes of this chapter, the income statement is calculated on the basis of the single step method which is the simple approach of totaling revenues and subtracting expenses to find the bottom line. There is a more complex method that involves multiple steps of calculations including inventory, equipment depreciation, and income yield from operations. (Makoujy, Jr., Rick J., 2010) Both accounting methods show income before taxes. Net income will show the final income after the applicable taxes are subtracted from the income. However, the income statement does have some limitations.

Example Income Statement

Kando Engineering submits a year-to-date income statement to Capital City Bank to obtain financing for their office expansion for a 12-month period, as of December 31, 2010. Using the figures below, Capital City Bank determines whether to fund Kando's planned office expansion.

- Project Revenue = $218,000
- Other Income = $1,000
- Direct Labor = $84,000
- Indirect Labor = $34,000

- Payroll Burden = $25,000
- Rent & Utilities = $12,000
- Equipment Leases = $5,000
- Maintenance & Repairs = $1,000
- Supplies = $2,000
- Insurance = $15,000
- Professional Services = $2,000
- Interest = $2,000
- Other = $8,000
- Federal Taxes = 30%

This information results in an income statement as follows:

Income Statement—December 31, 2010

REVENUE		
Project Revenue	$218,000	
Other Income	$1,000	
Total	$219,000	$219,000
EXPENSES		
Direct Labor	$84,000	
Indirect Labor	$34,000	
Payroll Burden	$25,000	
Rent & Utilities	$12,000	
Equipment Leases	$5,000	
Maintenance & Repairs	$1,000	
Supplies	$2,000	
Insurance	$15,000	
Professional Services	$2,000	
Interest	$2,000	
Other	$8,000	
Total	$190,000	$190,000
GROSS PROFIT		$29,000
FEDERAL TAXES	$8,700	−$8,700
NET PROFIT		$20,300

This income statement results in the following summary:

1. Kando's *Total Revenue* for 2010: $219,000
2. Kando's *Total Expenses* for 2010: $190,000
3. Kando's *Federal Taxes* for 2010: $8,700
4. Kando's *Gross Profit* for 2010: $29,000
5. Kando's *Net Profit* for 2010: $20,300

- It's important for the engineer to know the difference between revenue and profit. Revenue is recognized as the monies received (income) during an invoice period. Profit is defined as revenue minus expenses.

Limitations of Income Statements

Generally income statements help company managers, investors, and creditors to assess the past performance of the company. This statement also provides an indication of the future performance, and can provide information for generating future cash flows. (Schroeder, Richard, et al. (2010) However, income statements have limitations:

- Future results are not directly related to past performance.
- Some results on the income statement depend on the specific accounting methods, described above, and cannot measure inventories.
- Some results on the income statement are judgmental, such as final salvage values, useful life, depreciation, or market conditions. For example, the retail industry does more than 50 percent of their business and relies heavily on the fourth-quarter months of September through December. After a detailed evaluation of their income statement in June, a prospective investor may have a significantly different opinion of this company's performance to a similar evaluation of the January results.

Statement of Financial Position (Balance Sheet)

A statement of financial position, also referred to as a balance sheet, is a summary of a company's financial balance. Balance sheets are typically produced at the end of their financial year and reflect a company's assets, liabilities, and owners' equity. The balance sheet is sometimes referred to as a snapshot of a company's financial condition (Williams, et al., 2008).

The company balance sheet has three parts: assets, liabilities, and owners' equity. The main categories of assets are usually listed first and typically in order of liquidity (Daniels, 1980). Assets are followed by the liabilities. The difference between the assets and the liabilities is known as equity, or the net assets or the net worth or capital of the company and, according to the accounting equation, net worth must equal assets minus liabilities.

In summary, assets equal liabilities plus owners' equity. Looking at the equation in this way shows how assets were financed: either by borrowing money (liability) or by using the owners' money (owners' equity). Balance sheets are usually presented with assets in one section and liabilities and net worth in the other section with the two sections being in "balance" (Williams et al., 2008).

In general, individuals and small businesses tend to have simple balance sheets (Gitman, 2005). Larger businesses tend to have more complex balance sheets, which are typically presented in the organization's annual report. It may be useful to compare a balance sheet from one year to the next to look at a company's financial progress in time.

Balance Sheet Example

Kando Engineering also needs to submit a year-to-date statement of financial position to its Board of Directors:

As of December 31, 2010:

- Cash = $20,000
- Short-Term Investments = $7,000
- Work in Progress = $21,000
- Accounts Receivable = $45,000
- Property & Equipment = $29,000
- Accounts Payable & Accrued Expenses = $15,000
- Deferred Income Taxes = $45,000
- Long-Term Debt & Liabilities = $15,000
- Capital Stock = $25,000
- Retained Earnings = $22,000

Statement of Financial Position—December 31, 2010

ASSETS		LIABILITIES	
Current Assets		**Current Liabilities**	
Cash	$20,000	Accounts payable	$15,000
Short-term investments	$7,000	Deferred income taxes	$45,000
Work in progress	$21,000	Long-term debt and liabilities	$15,000
Accounts receivable	$45,000		
Fixed assets		**Owners' equity**	
Property & equipment	$29,000	Capital stock	$25,000
		Retained earnings	$22,000
TOTAL ASSETS	$122,000	**TOTAL LIABILITIES & OWNERS' EQUITY**	$122,000

Kando's total *Current Assets:* $93,000

Kando's total *Fixed Assets:* $29,000

Kando's *Current Liabilities:* $75,000

Kando's *Owners' Equity:* $47,000

Kando's *Total Liability and Owners' Equity:* $122,000

Cash Flow

The flow of revenue (cash flow) refers to the movement of money into or out of a project or a company. Cash flow is usually measured over a finite period of time. Evaluating cash flow is used to:

- Assess a project's or company's rate of return on the investment. The amount of money flow into and out of projects is used as inputs and outputs to calculate an internal rate of return, and net present value.

- Assess potential problems with a business's liquidity and net result in the operating checking account. Money is the fuel to keep the employees and vendors paid to continually provide the company's products and services. A profitable company can fail because of a shortage of cash, especially if it cannot pay the employees or vendor partners.

- Evaluate and verify the company's profits when it is believed that accrual accounting concepts aren't accurate. For example, it may appear that a company may be profitable but could be generating cash by issuing shares, selling equipment, or increasing debt.

- Evaluate the actual income generated by accrual accounting. If net income is composed of items other than actual cash, it is considered low quality and potentially suspicious.

- Evaluate risks such as default risk, bonuses, or reinvestment requirements, to mention a few.

Cash flow is considered a generic term and depends on context. Companies that offer professional services commonly wait long periods to receive payment for those services. Cash flow can refer to actual past flows or projected future flows. It can refer to the total of all the flows involved or to only a subset of those flows. Net cash flow is based upon operational, investment, and financing cash flows. To reconcile the ending cash balance, all cash flows should be considered.

PROFESSIONAL HUMAN RESOURCES MANAGEMENT

Every civil engineering firm has a human resources function; and in large firms there usually is a separate Department of Human Resources. The Human Resources Department typically is responsible for recruiting and training, dealing with performance issues, managing compensation and 401(k) retirement packages, and ensuring that the firm's practices conform to various laws and regulations.

However, within the context of successful project execution, the project manager also plays a critical role in managing human resources. The project manager must balance the limitations imposed by the project budget with the technical requirements and demands of the client for the quality end product. Staffing a project entirely with proven senior professional engineers may be desirable; but if the project budget is based on a preponderance of junior and entry-level labor to produce the work, the likelihood of the project being completed within the original budget is greatly reduced. Even if the senior staff being used are much more efficient than the junior-level staff, the budget likely will be in jeopardy.

Extending this example, if the mix of staff identified to perform the work is appropriate, various technical resource managers will need to be consulted to ensure the availability of staff at key points in the project's life. For example, if a particular junior engineer is the perfect staff member for the job, but is on a remote assignment in a foreign country for the duration of the project, alternative resources need to be allocated. With highly technical or specialized disciplines, staff may be in special demand. Consulting the resource manager early in the process helps to ensure the project manager is obtaining the right professional staff in a timely manner for a given phase of the project's schedule.

Another aspect in the development of skill mixes is the assignment of categories to specific staff. For a given client or task, is it appropriate to downcharge (charge at a lower billing rate or labor multiplier) an experienced engineer to meet a client's budget, or to upcharge (charge at a higher labor multiplier) a junior staff member to a more senior rate to reflect the actual activities the staff is undertaking? These are situational questions, and some clients, notably the federal government, may explicitly forbid the practice of downcharging or upcharging. Private sector clients may take a less formal approach to these nuances of human resourcing and project budget development, and so may afford the project manager more flexibility in developing profitable project budgets and schedules.

Therefore, in the project planning and budgeting stages, a project manager's key responsibility is to develop an appropriate mix of technical and professional staff (with cost-effective billing rates) to complete the tasks identified for the work. In the course of developing a competitive project budget, the project manager should prepare a (working) cost estimating spreadsheet that shows a representative sample of the labor categories and other direct costs associated with the project. But after the contract award or at the project kick-off meeting, specific names must be assigned to these labor categories, or the project's chances of ultimate success will be hampered.

CAREER PLANNING AND EXECUTION

The role of career development is an integral component in any discussion of business administration for projects. The critical role an engineering project manager may play in the development of the staff careers, particularly junior staff, cannot be overstated. The project manager engineer must be mindful of the potential tendency to slot personnel that excel in a given role for a project into that same role repeatedly. The project may benefit, but "pigeonholing" a given staff member into a particular activity for the sake of project efficiency may stunt the development of a staff member and frustrate the employee, creating the potential for lowered morale or even the loss of the staff member to another firm.

In the context of the engineer as project manager serving to advance the financial objectives of the firm, project staffing decisions must relate to the career arcs of the individuals participating in the project. While having an entry-level engineer collect samples in the field as a technician may be appropriate in order to help them understand the nuances of data processing and management, the same engineer should be given new opportunities to expand his or her sphere of experience to gain broader, overall engineering perspectives in professional practice.

The project manager role is not one of supervisor; "mentor" would be a more apt description. The project manager, often even more so than an engineer's supervisors, can gain an accurate feel of the performance strengths, weaknesses, and interests of project staff members. By informally checking in with staff to gauge where their interests lie, and where they seem to excel professionally, the engineer as project manager can assist in guiding staff into opportunities that will further their technical interests in their career, ideally within the structure of the engineer's firm.

SPECIALIZATION

Discussing the importance of providing a broad range of technical opportunities to engineering staff and then immediately talking about the merits of encouraging technical specialization may appear contradictory. However, if the engineer identifies, either for him/herself or a colleague, a particular area of interest, specialization in a particular technical subarea may lead to extensive opportunities.

Experience by itself is a necessary, but not sufficient, constituent of a successful career. The engineer needs to obtain training to supplement the experience gained through their specialization. There are many professional organizations (ASCE, ACEC, ASFE, NSPE, SAME, to name just a few) that offer training programs to their members. Additionally, many engineering firms have internal and/or external training programs. Training, coupled with specific relevant experience, can lead an engineer to being recognized within the firm as the resident expert on a particular technical matter, and as the "go-to" person to perform that work on a myriad of projects. From a professional development standpoint, there may be commensurate potential salary growth, if the specialized expertise gained is rare and highly sought after.

> There's an old adage that it's "better to be a mile wide and an inch deep than the other way 'round." It often seems that the shelf-life of an expert is shrinking as technology accelerates change and improves data accessibility. Technical specialization can be a lucrative career path, but technical or market developments may necessitate a wholesale mid-career change. Before deciding to go down the path of any particular specialization, the engineer is advised to research carefully whether the field is new, established, or undergoing transition. This due diligence may be vital to longevity with this career path.

But, there is a cautionary note to consider with regard to technical specialization. As technology accelerates the forces of change in the engineering profession, an area of expertise that may be in high demand presently, may be supplanted almost entirely by new technology. There are numerous examples; for example, hand-held GPS units now deliver vertical and horizontal data via remote downloads with accuracies that rival formal land surveys. The use of Geographic Information Systems (GIS) for design has transformed the development and presentation of design drawings. Calculations for sizing pipes and pumps that used to rely on careful analysis of multiple complex graphs are completed by computers with minimal input. Information modeling is becoming more and more prevalent in the engineer's workplace.

Therefore, there is a danger in becoming too specialized in a given technical field, as that field's technical advantage may be changed dramatically and quickly with an advance in technology. The expert whose specialized services are in tremendous demand one year may find the next year that his/her body of knowledge is now considered commonplace; suddenly, they must compete with other technical staff in order to stay gainfully employed.

CERTIFICATION AND REGISTRATION

As established in Chapters 1, 2 and 11, civil engineering is a recognized profession with a well-established set of standards of professional care. Licensing, or registration, for the engineer is vital in the civil engineering profession; it may be desirable, but is less important, in other more specialized engineering fields. For the civil engineering firm, employing registered engineers is valuable, for it provides the means to perform work in the public sphere may be denied otherwise. Like a law firm without any lawyers who have passed the bar exam, or a medical practice without any licensed physicians, an engineering firm without registered professional engineers will not last long in a competitive business environment.

Nevertheless, in a diverse, multidisciplinary environment, professional registration of every engineer in the firm is not absolutely essential for the firm to be successful, or for individual projects to be profitable. The engineer should not mistake professional registration with financial profitability; one does not guarantee the other. Professional registration offers opportunity, but with it comes professional responsibility, and the two should not be separated.

In a similar vein, certification of technical staff, such as for industrial hygiene, project management, or other aspects of professional performance, is a more specialized, focused type of recognition of professional competence. The liability issues pertaining to certification are not as great as with registration typically, and the corresponding opportunities and requirements for having certified staff performing tasks on a project not as crucial. Nonetheless, work performed on a project done by certified professionals is likely to make acceptance of the final products or services easier for the firm, because such certifications generally reassure clients.

In the role of project manager, the civil engineer should work with the supervisor of technical staff to support the development of staff in their desire to obtain various professional certificates and/or professional registration. This can be done by advocating to the supervisor on their behalf, expanding staff's work experience to comply with given certification program or registration requirements, and providing the schedule flexibility for staff to pursue these goals without affecting either project performance or their general health and well-being.

PROFESSIONAL SERVICES MARKETING

Assuming that the role of marketing and long-range business planning falls outside the purview of the engineer may be tempting. After all, these are administrative functions of the firm, not at all related to technical work, the true domain of the engineering staff. Such a viewpoint is both short-sighted and detrimental to the long-term well-being of the firm. The engineer has vital insights in the workings and relationships of clients and the products or services they demand. Consequently, the engineer absolutely must be included in marketing and the business of planning activities of the firm. It is a quaint, archaic notion for an engineer to assume that he or she can perform their work successfully without interacting and building relationships with their clients. Introversion may come naturally to engineers, but it has no place in developing a successful business model.

Marketing professional services is a formal way of describing and selling the firm and, by extension, the individual engineers employed by the firm. Marketing is not a haphazard, random series of glad-handing events, at least not for those who wish to market their services successfully. Successful firms use a strategic, focused method for selling and marketing. Many Fortune 500 firms use specific marketing processes, and frequently train their professional and technical staff in their application. Why do so many companies use a specific process for selling and marketing? Because they find that using a process works, and as a result, they are able to obtain work. To compete today, successful firms must use focused marketing.

Since the strongest lead for the next project or opportunity for work will always be an existing client, the engineer should continue to develop and ensure strong relationships with current clients. This may seem obvious to most, but it is very easy to take current key clients for granted and not spend the necessary time with them to maintain and continue the relationship. By being client-focused, the engineer will lower business development costs in the long run, while positioning the firm for follow-on business as the incumbent supplier.

Going after every opportunity that uses the word "engineering" or "environmental" or "transportation" (whatever the buzz word is for your group) is like chasing fish or rabbits . . . can you really catch one? Do you really know if the pursuit is worthwhile? Do you know the potential of that client? Does the client know you? The engineer must focus on systematically working to grow the best prospects for the firm.

So how do you gather the information you really need to know about your client? Many civil engineering firms have well-qualified marketing personnel that can help the engineer prepare prior to meeting a potential client. (See Chapter 9 and later in this chapter.) The firm may have identified a process such as the one depicted in Figure 4.1 and may have an established database for tracking leads. Knowing where the firm is in the process is extremely important because all firm representatives need to project a "united front" and a consistent message.

When meeting with the client regarding future opportunities, listening is far more important than selling. Through listening and asking appropriate questions of the client, the engineer may gain the necessary information to help the firm decide to pursue the opportunity (make the go/no-go decision) and develop a proposal.

The process of proposal development is somewhat subjective, and the mechanics of proposal assembly are often the purview of the firm's marketing professionals. However, the content (as opposed to the organization and formatting or appearance) of the proposal often relies extensively, or even exclusively, on the input of the engineer. The next two subsections briefly discuss some of the areas the engineer must be cognizant of when preparing technical and persuasive proposals for winning project work for the firm.

Resume Updates

Just as a civil engineer's personal resume is vital for obtaining employment, a corporate resume is equally vital for successfully obtaining work for the engineer's firm. The value and benefit of the formatting, emphasis, placement, and appearance of engineers' corporate resumes is well-established, but the actual appearance of such resumes varies from firm to firm. A corporate resume should contain the following components for the engineer, or any other technical staff, included in a proposal to a client: the name and title of the individual; their proposed role for the project or contract; their education and certifications/registrations; their general technical experience, including years of service in the field; the *relevant* project experience (defined as that which is pertinent to the scope of services being requested by the client); pertinent publications; and professional citations/memberships. The relative importance of these items is somewhat determined by the nature of a given solicitation for services by the client, but the elements listed are most frequently asked for by prospective clients.

A corporate professional resume should be developed as a comprehensive listing of all technical work performed during the course of the engineer's career, as well as the listing of professional training, certifications, registrations, and publications collected during one's professional tenure. It is not unusual for senior technical engineers with decades of experience to develop corporate resumes that are

60 pages in length or greater; it is their responsibility, however, to work with marketing staff to winnow these resumes down to the page limits specified, and to include only the relevant technical information requested by the client's request for a proposal or qualifications.

Demonstrating that the firm's proposed team members are qualified to meet the client's needs is a critical component of a winning proposal. The client does not want to pay engineers to be educated on how to solve their (the client's) problem; however, clients are willing to pay more for experienced engineers who have learned from their mistakes on someone else's project.

As a rule of thumb, the engineer should update his or her corporate resume every six months and should work with technical staff and supervisors to ensure that other key project personnel are providing these updates to their resumes at the same frequency.

Project Descriptions

Project descriptions, simply put, are the best way of concisely showing a prospective client that the engineer and the firm he/she represents possess all of the experience and skills necessary to perform the project work requested by the client. Project descriptions must be factual, but they also should paint a compelling picture of the work that was performed by the firm on previous, similar or related projects.

As with resumes, the format and organizational presentation of the project description can vary depending on the preference of either the firm or the prospective client, but the content of the project description should be developed by the project manager and pertinent staff most involved with the performance of the project. Project descriptions have essential components that should include: the dollar value of the project; its status (current or completed); the planned and actual duration of the project; the client's information regarding the project (key point of contact for the client and their contact data); a brief description of the scope of services or products provided; any key specialty subcontractors or vendors used to perform the work; where the work was performed; and the role the firm and members of the team proposed, if applicable, played in the performance of the project. The importance of a well-written and timely project description cannot be over-emphasized in convincing a prospective client to select the engineer and the firm for the work being solicited.

> In a proposal for engineering services, concise information combined with descriptive details, relevant financial information; and sharp photos provide the readers with a good idea of the firm's capabilities!

Business Planning

The role the engineer takes in planning the firm's future year budgets varies by firm. Business plans can vary in length and complexity but the input of company's operational and marketing staff is essential. In planning the future technical direction and markets the firm wishes to pursue, the accurate and well-thought-out perspectives of the engineer should be sought and incorporated into both short- and long-term planning documents.

The focus and goals of a firm often dictate the components of a business plan. The technical business development components that should incorporate the engineer's input include a description of a given market sector, and the short- and long-term outlook for work or project opportunities in that sector. This information should be derived from the professional relationships developed with clients by the engineer in order to gain as broad (and therefore reasonably accurate) perspective as possible. A discussion of market trends, growth drivers, and short- and long-term growth estimates or goals also should be developed. If possible, an analysis of the competition in the market sector and a comparison of the firm's capabilities and cost effectiveness against this competition should be utilized. Lastly, the goals for growth within the market should be stated clearly, the objectives and plans for achieving these goals clearly laid out, and the responsible parties for implementation identified.

A Great Business Plan Should Include Five Related Factors Critical to Every New Venture (Sahlman, 2008)

1. The Specific Venture: The business plan includes a detailed description of the business opportunity, the specific need, the competitors, the potential customers, the product details, the marketing plans, distribution and outlets.

2. The Business Team: The venture team should include specific managers, staff, subcontractors, suppliers, and vendors. Other support staff should be mentioned such as legal staff, human resources, and business support among others.

3. The Work Plan: This section includes specifics like financing options and investment opportunities, potential customer base and demographics, banking details such as interest rates and investment relationship, projected rates of return, schedules, product or service details and delivery plans, inventory control, accounting software, projections and basis of assumptions.

4. The Potential Rewards and Associated Risks: A complete plan will include the details on the total investment funding needed, the likely sources, rates of return, return periods, and product/service delivery schedules. This description will be accompanied by a statement of risks and a mitigation plan that include cash flow projections, sales and marketing, invoicing, cash receivable projections, market risks, and contingencies.

5. A Complete Financial Analysis: This analysis includes key elements of the four components above folded into a summary statement that includes the income statement, balance sheet, and cash flow details.

A business plan is less a blueprint and more of a roadmap for growing the size and profitability of a firm. It should be considered a living document, and subject to revision and adjustment throughout the course of its implementation phases. Course corrections to the plan are necessary to ensure its overall success, or else it will be nothing more than a stagnant comparative tool for basing performance data against subsequent business plans.

PROFESSIONAL BUSINESS DEVELOPMENT

The term "business development" may have a loaded connotation of either engineering staff sitting at a display booth in a large technical conference trying to attract attention to potential clients, or that of connected "players" smoking cigars and making deals while playing a round of golf. The truth surrounding business development for the civil engineering services is that it is simultaneously more prosaic, yet sophisticated. The most basic element of professional business development is contact—contacting potential clients is the essence of any business development program. If the firm wishes to perform work for a client in the future, contact must be made as soon as possible with that client. Staying at the forefront of a client's thoughts—to be "on the radar"—is critical, and to do that requires making frequent contact with that client.

There are numerous publications regarding the art of making sales and the psychology of buying and selling. A recurring theme that seems to be widely accepted is that a client will buy engineering services on emotion. The client will like the work the firm does, will like the staff used to perform the work, and/or will like how they feel about interacting with the engineering staff when project work is being done. While there are many structured and formal approaches to generating the requirements for official requests for proposals, and equally as many logical, objective procedures for evaluating the merits of competitive proposals, ultimately how the client feels about the firm and the staff will affect the selection process the most. A client cannot like a firm they don't know as much as one they do, assuming the established firm has either met or exceeded the client's expectations. The only way for a client to get to know a firm is for the firm's representative to make contact with that client.

"You can't listen your way out of a sale, but you can sure talk your way out of one."
—Zig Ziglar

The engineer has to overcome the often predominant trait of introversion to be successful in business development endeavors. The temptation is to let professional marketers handle these types of contacts, but many clients also have technical or engineering staff who may be introverted by nature. The client often feels intimidated by exuberant marketers, but contact with similar technical people can put them more at ease.

There are some fundamental aspects of contacting clients that the engineer should consider before moving down this path. First, keep a positive attitude, even if the initial responses are negative or neutral in tone. Very rarely does new work come from a single business development meeting with a client. The more realistic scenario is that multiple meetings with the same prospective client are required, so following-up is a crucial component. The engineer should be professional in these interactions with the client, so that even if work is not forthcoming in the short run, a favorable impression of the firm and its people is left in the mind of the client. Finally, relation-ships need to be nurtured; clients need to be thanked for their time, and as the initia-tor of the meeting, the engineer must be an active listener in these engagements. While the key selling points of the firm should be discussed, the business develop-ment meetings that are the most successful are the ones where the prospective client does the majority of the talking.

Every contact made with a prospective client should be designed. There are nu-merous mechanisms used to contact a client, including phone calls, meetings, formal presentations, e-mails, marketing brochures, or cut sheets. Every one of these mecha-nisms can be used when designing the contact. Meetings should not be haphazard; would the civil engineer ever construct a high-rise building without a design?

To design a client contact, the engineer should first define the objective of the con-tact. That is, what is the ultimate outcome desired from this meeting? Once this is done, the strategy to achieve this objective needs to be developed. What is the stated basis of the contact? Then, a game plan to implement this strategy should be scoped out by defining the analyses and methods to be used for the contact. Finally, the game plan should be checked against the objective, or in design terms, a quality assurance check should be performed to ensure that the game plan will get the firm what it wants. More information on forming these relationships is provided in Chapter 9.

Of similar importance is the mindset the engineer must have when performing the work that has been won. A project should be considered an audition for another project in the future. Successfully performing a project for a client, by demonstrating that their needs are understood and by keeping commitments regarding cost and schedule, is ar-guably the most powerful business development tool of all. It is a cliché, but there is truth in the old saying that "the reward for good work is more work." The converse is true as well—the surest way to lose a client permanently is to produce a poor quality product or service that is not what the client wanted or expected, and to do so at a cost far higher than originally estimated. While the actions of the engineer performing on a project are necessary to ensure follow-on work, the other key aspects of business devel-opment also must be in place to ensure future work with that client.

PROFESSIONAL AND TRADE ORGANIZATION ACTIVITIES

There are numerous benefits to the engineer for engaging in professional and trade organization activities outside of work. From a professional development standpoint, through technical presentations, seminars, and conferences, these organizations offer opportunities for continuing education. Life-long learning is an important aspect of professional development, and many organizations offer formal recognition of this education with continuing education units and/or professional development hours.

Additionally, professional organizations offer a forum for the engineer to interact with colleagues from other firms, to learn about various markets and clients, and to cross-feed technical information. Prospective clients are often members of the same organizations. The engineer should not feel limited only to professional organizations in the technical field; however, there are other organizations, either community- or market-centric where joining would be beneficial for the engineer. Examples include local chambers of commerce, real estate developer associations, and environmental outreach organizations.

Voluntary Activities and Sponsorship

Community involvement generates a number of significant benefits for the civil engineer's firm and its clients. Through volunteer activities, staff from the firm can develop a higher profile in the community and a stronger community connection to the firm, thereby contributing positively to staff recruitment, retention, and public image. Increasingly, clients appreciate firms that are recognized as strong contributors to the community. Such involvement often provides business development opportunities as an additional benefit.

There are numerous opportunities for volunteer activities and sponsorships through involvement in professional and trade organizations within the local community. One theory in marketing is that a potential client has to hear a name, phrase, or company name nine times before it takes root in the mind. When a potential client connects a firm's name and volunteer staff with a community project, it counts as one or possibly more of these connections.

SUMMARY

Financial management is the life blood of the civil engineering enterprise, whether public service or engineering consulting. "Profit and exceptional client service" are essential for the continued existence of the consulting engineering enterprise. The project manager is at the heart of the exceptional client service aspect and the engineering manager (or financial manager) has to manage the CE enterprise's cash flow to keep the operation viable.

REFERENCES

Daniels, Mortimer. (1980). *Corporation Financial Statements.* Arno Press, New York. ISBN 0-405-13514-9.

Duncan, William R. (1996). *A Guide to the Project Management Body of Knowledge.* PMI Standards Committee, Darby, PA. ISBN: 1-880-41013-3.

Gitman, Lawrence J. (2005). *Principles of Managerial Finance*, 11th ed . Addison Wesley, Boston, MA.

Makoujy, Jr., RickJ., (2010). *How to Read a Balance Sheet*. McGraw Hill Companies, New York. ISBN: 978-0-071-70033-7.

Sahlman, William A. (2008). *How to Write a Great Business Plan*, Harvard Business School Publishing Corporation, Harvard Business Review Classic, Watertown, MA. ISBN 978-1-422-12142-9.

Schroeder, Richard, et al. (2010). *Financial Accounting Theory and Analysis—Text and Cases*. John Wiley and Sons, Hoboken, NJ. ISBN: 978-0-470-64628-1.

Williams, Jan R., Susan F. Haka, Mark S. Bettner, and Joseph V. Carcello (2008). *Financial & Managerial Accounting*. McGraw-Hill Companies, Columbus, OH. ISBN 978-0-072-99650-0.

The Winning Proposals Method Handbook, Radian Corporation, 1998 Edition, Austin, TX.

Chapter **13**

Communicating as a Professional Engineer

Big Idea

Communication provides the conduit for the engineer to transmit extensive information, knowledge and experience for effective use by clients and others on the project team. Excellent advice for engineers relating to clients is, "no surprises"!

"The most important thing in communication is to hear what isn't being said."

—Peter F. Drucker

Key Topics Covered

- Introduction
- Communication Conduits
- E-mail Use and Limitations
- Conflict Resolution
- Behavioral Characteristics of Team Members, Friends, or Family
- Typical Report Format
- Useful Forms for the Engineer
- Sample PowerPoint Presentation
- Summary

Related Chapters in This Book

- Communication is related to every chapter in this book

(Continued)

Related to *ASCE Body of Knowledge 2* Outcomes

ASCE BOK2 outcomes covered in this chapter

Foundational
1. Mathematics
2. Natural sciences
3. Humanities
4. Social sciences

Technical
5. Materials science
6. Mechanics
7. Experiments
●8. Problem recognition and solving
9. Design
10. Sustainability
11. Contemporary issues & historical perspectives
●12. Risk and uncertainty
●13. Project management
14. Breadth in civil engineering areas
15. Technical specialization

Professional
●16. Communication
17. Public policy
18. Business and public administration
19. Globalization
●20. Leadership
●21. Teamwork
22. Attitudes
23. Lifelong learning
24. Professional and ethical responsibility

INTRODUCTION

Communication is a three-dimensional process where information is exchanged to meet a mutual objective in a relevant time period. Effective communication includes:

- A message sent from the sender to the receiving party, signifying the transfer of information
- Acknowledgment of receipt (or clarification, if needed) of this information
- The "time element" for the transfer of this information

> Communication is the process of exchanging ideas, information, feelings, or data from one party to another. The information transfer can occur verbally, nonverbally, electronically, physically (by touch), or in writing. Complete communication occurs when all modes of information transfer are engaged.

These points make the difference between effective and ineffective communication. For example, just because one party transferred information to another does not mean that the message was received, understood, or acknowledged. The acknowledgment is an essential component of the information transfer and signals to the sender that the message was understood.

With any two (or more) party exchange, a summary statement is an effective means of wrapping up and getting a feel for the level of understanding the other party has regarding the information presented. An engineer should end a meeting with this sort of statement, "So, in summary, we are offering to prepare a work plan that addresses the issues of bridge abutments on the river bank for the fixed fee of $150,000. Does that adequately address your concerns?" This gives the other party, in this case a prospective client, a chance to respond with an affirmation or to come back with their own understanding of the exchange.

This type of communication must take place during the initial conversations on a project or proposal, or both parties may spend countless hours working on something that could wind up being a waste of time. Engineering managers have all heard the excuses from staff and clients, "Sorry, I did not get the message, so I'm not responsible." In order to circumvent this example of failed communication, when an experienced manager does not receive a message confirmation he/she will often send another message to the receiving party and ask what the party perceived as the request and if they would acknowledge or possibly even repeat back what was said. This confirmation loop will demonstrate that all parties received the original transmission and that everyone is in agreement. A simple acknowledgment shows respect and saves time for all parties.

An example of this informational transfer is shown in Figure 13.1. The figure also illustrates how the sender should consider and vary the level of detail in communication, depending upon the relative comprehension of the subject matter by the

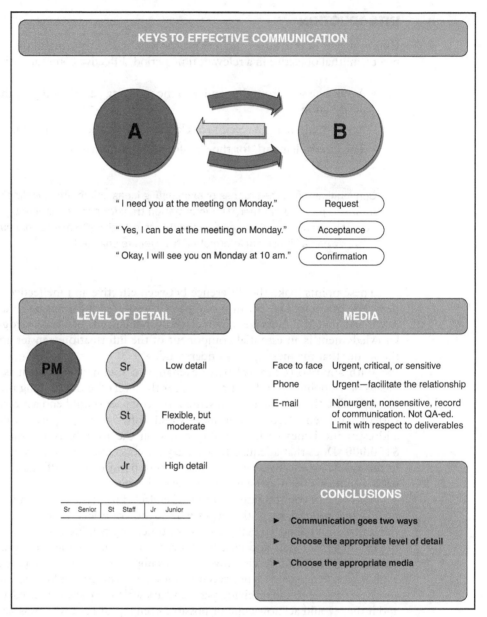

Figure 13.1 Keys to effective communication

listener. For example, a very experienced engineer may become offended if a new engineer were to go into great detail when requesting assistance from the senior.

Time relevance is also a critical factor for effective communication. Information transfer is related to time. Regardless of whether the information is transmitted via e-mail or face-to-face contact, the receiving party should regard an acknowledgment and/or instructions with a sense of urgency. This urgency may be minutes, hours, or days, but it is professional to respond "as soon as practical" or immediately if the task has become an urgent issue. An outdated acknowledgment or delayed action upon a request that was time critical shows a lack of attention to the job and a lack of consideration with regard to those who depend on a timely response.

Sidebars

Sometimes in the context of an active conversation or e-mail exchange, another urgent but related issue may come up that needs to be immediately addressed. One way to pursue this additional related conversation is referred to as a "sidebar." Engineers have borrowed this informational exchange from the legal profession. During an active trial the judge may call a sidebar where the attorneys approach the judge to have a separate private conversation related to the trial. This process is an acceptable way of calling a timeout from the original active conversation to immediately address a side issue related to the original informational exchange. However, the engineer should exercise judgment and care if this process is used because it may not be appreciated by the other party, and it can be disruptive. However, use of this sidebar tool can sometimes be advantageous if used wisely to break up a tense conversation or to introduce a time delay in the conversation.

COMMUNICATION CONDUITS

Information is exchanged through some type of conduit or medium. The most frequent conduit is the atmosphere, or air, by verbal communication. Another frequent medium is the electronic medium, meaning phone lines or cell towers or electronic messages such as e-mails. The auditory method, speaking, can be transmitted in air by simple dialogue or electronically by cell tower or land line.

Communication is a process by which we assign and convey meaning in an attempt to create shared understanding. This process requires a vast repertoire of skills in intrapersonal and interpersonal processing, listening, observing, speaking, questioning, analyzing, and evaluating. These processes are developmental and transfer to all areas of life: home, school, community, work, and beyond. It is through communication that collaboration, cooperation, and subsequent action occur.

Tailor communications to the knowledge and experience level of the listener!

Systems of signals, such as voice sounds (intonations or pitch), gestures, or written symbols communicate thoughts or feelings. There are three major parts in human face-to-face communication; body language, voice tonality, and words. According to communications research:

- 55 percent of impact is determined by body language—postures, gestures, and eye contact
- 38 percent by the tone of voice, and
- 7 percent by the content or the words used in the communication process.

The exact percentage of influence by factors other than spoken words may differ with regard to variables such as the listener and the speaker. There are many scientific studies dealing with "body talk," how to interpret motions and gestures, facial expressions, eye contact, touch, and even foot placement. Kevin Hogan, author of *The Secret Language of Business: How to Read Anyone in 3 Seconds or Less*, breaks down body language into eight key elements: eyes, face, gestures, touch, posture, movement, appearance, and voice. This example illustrates that body language is a very important component of communication and that "actions may speak louder than words."

Therefore, engineers should be aware of nonverbal communication as a way of interpreting language and also recognize that the importance of picking the proper conduit for transmitting information. For example, experienced engineers will assess the content and nature of a message and choose the most appropriate conduit for delivery of this information. Conveying a convenient meeting time or location is appropriate for e-mail delivery but, conveying an employee injury on the jobsite is an appropriate in-person or phone message delivery.

Communication for private sector consulting engineers is often concise and direct. The project team is very aware of the fixed relationship between scope of work, schedule, and budget. Often in private industry all three elements of this relationship are critical, so efficient communication is essential to stay viable and profitable. Frequently, there's little time for pleasant discussions or conversations.

Communication for public sector engineers is also concise and direct, but often more parties are involved in the communication loop for informational purposes (depending upon the actual agency or department). The public sector team is also very aware of the fixed relationship between scope of work, schedule, and budget. Many public agencies are trying to do more with less, so efficient communication is essential in order to be responsive to the public and governing political body, e.g., legislature, board of supervisors, city council, and so forth.

E-MAIL USAGE AND LIMITATIONS

What a difference a decade makes. Many older (40+) engineering professionals could never have imagined the day when entire reports, including graphics, could be sent virtually instantaneously via electronic media. Now electronic communication and data transmittal is routine and pervasive.

Most e-mail transmissions are informal and many use verbal shorthand; salutations are minimal or nonexistent and punctuation is often limited to emoticons (}:[= angry, frustrated, for example). Here are some of the many advantages of e-mail communication:

- You can communicate quickly with anyone on the Internet.
- You don't have to worry about interrupting someone when you send e-mail.
- You can deal with your e-mail at a convenient time.
- You don't have to be shy about using e-mail to communicate with anyone.
- The cost for e-mail has nothing to do with distance, and in many cases, the cost doesn't depend on the size of the message.

However, although essential, electronic communications do have their drawbacks. Many have great "oops" stories, such as sending messages to the wrong recipient, sending the wrong version of a report or other document, or sending an off-color joke to someone who was offended. There are many limitations of e-mail communication, including the lack of privacy. System administrators can read e-mails, others can bypass security, and still others can save and/or possibly send your message to unknown parties. Additionally, some e-mail systems can send or receive text files only. It's good to remember that:

- It's possible to forge e-mail (think viruses).
- It's difficult to express emotion using e-mail.
- You can receive too much, unwanted e-mail.
- You may not know about the person with whom you are communicating.

Writing (for print) developed over centuries, with an entire industry of writers, editors, proofreaders, publishers, printers, and book stores. E-mail has been in wide use for fewer than 20 years and has its own set of unofficial rules. Most of us use our own professional judgment when writing/responding to e-mail communication. Here are some suggestions to keep in mind when using e-mail for business and engineering applications.

Use Clear Subject Lines

State your desired response like "Please Review/Comment by a Date/Time," "For Your Records," "Please Approve," and you may receive a faster response.

Use a simple, clean format (no fancy fonts, cute symbols, or color). Other suggestions include:

- Group messaging: Restrict names in the copy (cc) line to those who need to know. Use the blind copy (bcc) feature when writing to a large list so that the recipient doesn't have to go through a page of names before getting to the message.
- Confidentiality: Don't write anything you wouldn't want your first-grade teacher to see.
- Monitoring: Try to check your business e-mail account every hour, if at all possible.
- Temper: Don't write or respond in anger. Try waiting for an hour before responding.
- Proofread: Use caution when employing spell check because it's not a replacement for actually reading the message and verifying the attachments are correct. Spell check will not catch all *miss steaks*, as illustrated with this example.
- Efficiency: Use your electronic address book, which will save time and energy when looking for a contact name.
- Develop a signature block with your name and contact information. Better yet, set up your e-mail system so that this is attached to all outgoing messages.

Some servers have limits on the size of attachments, which may restrict the addressee from receiving the intended message and files. It's a good idea to send a test e-mail to the receiving party and/or to verify they received the message and files after it has been sent.

If you would like more information on e-mail communication, visit this site from Microsoft:

http://office.microsoft.com/en-us/help/HA010429671033.aspx

CONFLICT RESOLUTION

Even when the proper conduit for delivery of information has been chosen, sometimes communication of a message will cause conflict. The conflict may be a result of miscommunication, the transmittal of unfortunate news (like a delay in the project), or receipt of information in an untimely manner (late transmission).

There are multiple methods to resolve conflict and volumes of publications and many management courses on handling conflict (Weeks, 1992) (Erickson, McKnight 2001) (Winslade, Monk, 2008). The following discussion of the 4 Cs of conflict resolution is not meant to be a treatise on those publications and courses. Instead, some general tools are offered for handling conflict with colleagues, clients, and stakeholders in engineering business practice.

The 4 Cs of Conflict Resolution

Conflict is inevitable; engineers will experience conflict with clients or stakeholders in their careers and with friends and family in their personal lives. When in a potential or real conflict situation, listening is essential. Empathizing and being cognizant of a client's (or other's) position is crucial, especially when the conflict is related to a project's scope of work or schedule and budget impacts. The engineer should consider alternative approaches that could achieve a winning scenario for all.

The 4 Cs of conflict resolution are four general approaches to dealing with conflict and resolving issues, not always with the best-case scenario result for both parties. As discussed below, the 4 Cs framework includes collaboration, compromise, co-existence, and capitulation.

Collaboration

Collaboration is a way of dealing with one another that moves beyond differences to find a "common solution," allowing differing views or desires to be valued, realized, and accomplished. It involves working together to create a win-win solution, or a solution that lets both parties feel validated. One additional tool that can be employed in this situation is referred to as mitigation. Mitigation can be employed when a common solution is available for each party but one party still suffers some damage. It may be possible to offset this damage if the advantaged party considers offering some type of compensation. This compensation may be in the form of additional work, time savings, money or some other appropriate form of value to be given to the disadvantaged party in order to complete the collaboration. This method of conflict resolution is often the most productive and rewarding for all parties concerned.

Compromise

The art of compromise is the second "C" of dealing with differences. Compromise, similar to co-promise, essentially refers to the effort of negotiating with each other about differences to create a "mutual way" that requires each person to give in to a degree. Ultimately a compromise is a "partial win—partial win" solution for each party. This method of conflict resolution has been employed successfully in the realm of politics and was used often by our founding forefathers when the original colonies agreed to form one nation.

Coexistence

This strategy refers to the decision simply to accommodate another person's desire or view and to simply coexist, in essence, agreeing to disagree. For example, this situation exists in many households where one partner in the home is affiliated with a specific political party and the other partner is affiliated with the opposite party. This doesn't mean that they can't live in the same home, it simply means they have agreed to respect one another's position and they agree to disagree on this particular issue. Managers, coworkers, clients, or the public rarely will all share the same opinion on a matter, but that's usually okay as long as there's mutual respect. However, sometimes each party's desires will be so different that the best solution is for each person to go their separate ways.

Capitulation

Capitulation involves giving in or acquiescing to the other person's wishes. If you decide to capitulate, you decide to handle a difference by agreeing to the other individual's view or desire. This method of conflict resolution is a "win—lose" solution for the parties involved.

BEHAVIORIAL CHARACTERISTICS OF TEAM MEMBERS, FRIENDS, OR FAMILY

Getting acquainted with general character types and traits may be helpful for the engineer to understand "why" some people do what they do and why they do it a particular way. Table 13.1 describes general character types and traits that people often display when communicating with one another. Knowing the personality traits of those you work with and for (and those you live with) can make both personal and professional life less stressful. For example, if you know a colleague is a "conflict

Table 13.1 Typical Behavioral Characteristics

Perfectionist *Advantage*: These individuals are detail-oriented and respect quality and accuracy in their work products. *Disadvantage*: These individuals may need to perform task assignments beyond perfection. There may be an inability to let go of the assignment possibly due to fear of being judged, fear of the next assignment, or other complex reasons.	Controller *Advantage:* These individuals have leadership qualities and can provide inspiration to the team members. *Disadvantage:* These individuals may need to perform or be overly involved in many task assignments which could result in negative impacts on the schedule or budgets for these tasks. This behavior may be due to a lack of self-confidence, the need to have an over-abundance of work, the need to be in constant control, or simply the inability to delegate tasks.
People Pleaser *Advantage:* These individuals want to please people and are generally liked by clients and the public because they appreciate the "can-do" attitude and immediate positive responses from the people pleaser. *Disadvantage:* These individuals have a need to please people and may possibly try to receive immediate gratification from pleasing people with overly positive statements. They may make promises that they cannot keep with a tendency to get the project or assignment deeper in the hole or further behind schedule.	Conflict Avoider *Advantage*: These individuals want to avoid conflict and are generally easy going, very likeable, and generally pleasant. They are team players and they want to perform well but may be internally conflicted by the need to avoid conflict. *Disadvantage:* These individuals will avoid direct conflict and sometimes avoid direct questions. They may have a tendency to follow and "go with the flow" and postpone inevitable conflicts that could be avoided earlier in their development. Often when conflicts are recognized early but go unresolved they can become larger and more difficult to resolve. There may also be unstated resentment with "general compliance."

avoider," when asked whether they will meet a deadline a likely response will be "sure, no problem." The more prudent question to this conflict-avoiding colleague may be, "What percentage of the task is currently complete and may I see the product?" Assuming the colleague complies with the request, the next questions could be, "How much effort is needed to complete the task?" And finally, "Do you believe there is sufficient time between now and the deadline to complete this task?"

This discussion and list of behavioral characteristics is a brief summary of years of management and supervision from a practical perspective. There are numerous published and unpublished references and management training series available on these topics.

By understanding the various types of people and the advantages and disadvantages of each character type and trait, directions and responses can be tailored to get the best performance from each individual.

TYPICAL REPORT FORMAT

Business, industry, and public service often demand short technical reports. They may be proposals, progress reports, trip reports, completion reports, investigation reports, feasibility studies, or evaluation reports. As the names indicate, these reports are diverse in focus and intent, and differ in structure. However, one goal of all reports is the same: to communicate to an audience. A typical example of an engineering Feasibility Study Report is shown in Appendix C - Example Feasibility Study Report and an example short technical report is shown in Appendix D - Example Short Technical Report.

The following section describes a general format for a short report, which can be adapted to the needs of specific reporting requirements. A format, however helpful, cannot replace clear thinking and strategic writing. Organizing ideas carefully and expressing them coherently is a must. Precision and conciseness also are essential.

Typical Report Sections or Chapters

1. **Title Page:** The essential information here is the company or agency name, the title of the project, authors or contributors, and the date. The title of a report can be a statement of the project and the specific subject of discussion. An effective title is informative but reasonably short. If an engineering report contains conclusions or recommendations it is generally required that the report be prepared under the direction of a registered professional engineer, signed, dated, and stamped by this engineer. The specific regulations of the state for the project and the location of the engineer's office should be checked.

2. **Executive Summary:** This section contains the salient components of the report in a paragraph for a very short report, or a page or two for a medium length report (10 to 50 pages), or maybe up to 2 to 10 pages for a volume. The whole report is summarized in this one section. Writing one sentence or a paragraph that summarizes each of the traditional report chapters is useful.

The problem should be emphasized, and a short narrative on the objective and approach helps set the framework for the reader to understand the scope of work that was derived to reach the conclusion. The data collection, summary analyses, and recommendations also should be addressed. The engineer should not copy a whole paragraph from various sections in the report for placement in the executive summary. The executive summary condenses and emphasizes the most important elements of the whole report and it cannot be written until after the report is complete. This summary should be concise and specific with summary details. A technical report is not a mystery novel and giving the report conclusion right away is recommended.

3. **Introduction:** Whereas the executive summary summarizes the whole report, the introduction of a technical report identifies the subject, the purpose (or objective), and the approach or the plan of development of the report. The subject is the "what," the purpose is the "why," and the plan is the "how." This section acquaints the reader with the problem and the overall approach to solving it. The introduction provides the reader with background information needed before launching into the body of the report. It may be necessary to define the terms used in stating the subject and provide background, such as theory or history of the subject.

4. **Background:** If the introduction requires a large amount of supporting information, such as a literature review or a description of a process, then the background material should form its own section. This section may include the previous history, a review of previous research, and regulations or formulas the reader needs to understand the problem.

5. **Discussion:** This section leads to the most important part of the report. It takes many forms and may have subheadings of its own. The basic components are methods, findings (or results), and evaluation (or analysis). In a progress report, the methods and findings may dominate. A final report should emphasize the evaluation. The discussion should answer the questions: who? when? where? what? why? how?

6. **Conclusion:** The conclusion should be explained in terms of the preceding discussion. It is common to see some repetition of the most important ideas presented in the discussion section, but duplication should be avoided.

7. **Recommendations:** The recommendations should be clearly connected to the results of the rest of the report. Those connections should be made explicit at this point so the reader does not have to guess at the real meaning. This section also may include a statement as to whether any further investigation or work tasks are required.

8. **Attachments:** Typical attachments may include references, appendices, or other relevant documents. Research or conclusions derived from other sources referred to in the report must also appear in a list of references at the end of the main report. Appendices may include raw data, calculations, graphs, and other

quantitative materials that were part of the research but would be distracting to the report itself. Each appendix should be referred to at the appropriate point (or points) in the text. In industry, a company profile and profile of the professionals involved in a project might also appear as appendices.

An excellent example of a short report titled, "The Benefits of Green Roofs" appears in Appendix D. The author did a good job of incorporating the photos and exhibits into the text to achieve an excellent flow and presentation.

Key Words

To communicate important concepts while reporting information, key words can be used that will relay the information more quickly.

Some words related to research functions follow:

Diagnosed	Evaluated	Examined
Extracted	Identified	Inspected
Interpreted	Interviewed	Investigated
Organized	Reviewed	Summarized
Researched	Surveyed	Systematized

Some words related to detail task functions follow:

Approved	Arranged	Catalogued
Classified	Collected	Compiled
Dispatched	Executed	Generated
Prepared	Recorded	

USEFUL FORMS FOR THE ENGINEER

In an effort to do more with less an engineer will often struggle with efficiency. Therefore, some useful forms are offered to the reader for carrying out daily or frequently performed tasks.

Much engineering work is conducted through "team efforts" with client interactions. Therefore, some of the most useful forms are "telephone record," Figure 13.2, and "meeting notes," Figure 13.3. The forms are designed so the engineer can capture the salient points of the discussion and specifically record the action items, the party responsible for performing these actions, and the targeted completion dates for the action items. These forms offer the basic record information for the engineer. Revisions and modifications to the forms are invited for custom use.

Another form often used by experienced engineers is the "transmittal form," Figure 13.4. This form includes pertinent project information, states the type of

Company or Personal Name Here	Telephone Conversation Record			
Phone:				
SAMPLE ONLY				
☐ TO: ☐ FROM:			Phone Number	
Name / Title				
Company / Agency			Date:	
Subject:			Time:	
			File / Project #	

Discussion

Item #	Action Item or Task	Responsible Person	Due Date

Figure 13.2   Telephone conversation record

deliverable product, delivery information, reason for transmittal (for review or approval), the present form of the deliverable product expressed as draft, final, or 90 percent drawings, and is offered as an alternate to a custom cover letter. Cover letters are a useful tool if time permits custom preparation for each deliverable

Company or Personal Name Here Phone: SAMPLE ONLY	**Meeting Notes**	
	Revision #	

Project / File:		Date	
Purpose:		Time	
Attendees:			

Discussion / Conclusions

Item #	Action Item or Task	Responsible Person	Due Date

Figure 13.3   Meeting notes

Company or Personal Name Here		Letter of Transmittal			
Phone:					
SAMPLE ONLY		Revision			
To:			Subject		
Company:			Phone Number		
Address:			Date:		
			Time:		
			File / Project #		

Item #	Deliverable	Version
	Remarks	

	Delivered Via		Transmitted
☐	Overnight Delivery	☒	As Requested
☒	Messenger	☐	For Approval
☐	US Mail	☒	For Review and Comment
☐	Email	☐	For Your Reference

Figure 13.4   Letter of transmittal

product. However, the transmittal form serves the same purpose and is easier and quicker to prepare in the event time is limited.

Finally, one additional form offered for consideration is referred to as the "Response to Comment" (RTC) table, Figure 13.5. After transmittal of a deliverable

Company or Personal Name Here Phone: **SAMPLE ONLY**	 **Revision #**

RESPONSE TO COMMENTS TABLE

Agency:		Project Location:	APN	Date:
Project Name:		Plan Check No.:		
Document Being Reviewed/Revision Number:				

COMMENT NO.	SHEET NO./ DWG NO	COMMENT	RESPONDENT	RESPONSE	CONCUR	NON-CUNCUR	FIO
REVIEWER:							
GENERAL AND CIVIL DRAWINGS							

Figure 13.5 Response to comment

product, such as 60 percent drawings or a draft report to a client, the engineer will inevitably receive comments for one or, more likely, several other team partners on the project. The RTC table is a valuable tool to record these comments from multiple sources for the record and includes key project data, summary comments, a space for the engineer's response to the comment, and a summary conclusion on how the comment was handled. Revisions and modifications to the forms are invited for custom use.

SAMPLE POWERPOINT PRESENTATION

PowerPoint has advantages and disadvantages for presentation of information to groups. However, it is still a popular tool for informational presentations. An example of a good PowerPoint presentation prepared by senior civil engineering students is presented on the support website at www.wiley.com/go/cehandbook, Example PowerPoint Presentation. This presentation was prepared in response to the sample "Request for Proposal" discussed in Chapter 4 and included in Appendix A, Example RFP. The presentation is the culmination of the Final Feasibility Study Report prepared by the student group that called themselves CVision Engineering. The Feasibility Study Report is also presented in Appendix C, Example Feasibility Study Report for consideration as a good example of a report.

SUMMARY

Communication is the process of exchanging ideas, information, feelings, or data from one party to another. The information transfer can occur verbally, nonverbally, electronically, physically (by touch), or in writing. Complete communication occurs when all modes of information transfer are engaged. An acknowledgment of information receipt and a brief reiteration of the message by the listener will provide the sender with an opportunity to confirm or deny the accuracy of the transmission. Examples of efficient communication tools are discussed and included for the engineer's use. Examples of typical short engineering reports and medium-length technical reports are also included for reference.

Valuable Lessons I Learned from My Clients

Bridget Crenshaw Mabunga, Adjunct Professor and Technical Writer

It's amazing how easy it is to forget our audience. Whether through written or oral communication, it is imperative to understand the needs of your audience. When I teach I consider what my students may or may not be familiar with, and I try to be as transparent as possible. The same transparency is necessary in my work with clients. When in doubt, I make sure to clarify my communication and check in with the client or student to verify that we have a shared understanding of the expectation or goal at hand.

In addition to knowing our audience, we must make sure that we develop what I call multicultural communication awareness. I teach a multitude of students with a variety of cultural backgrounds, and each cultural group has particular communication strategies or comfort levels. For example, I know that often my students of Asian descent (Chinese, Japanese, Filipino, Vietnamese, for example) struggle with asserting their points and opinions because their cultures are largely focused on community and less focused on the individual, as we are in American business culture. Thus, I have to spend a little extra time helping them navigate the rigors of academic writing, which requires asserting points and making arguments.

Another cultural consideration is the relationship with time. After having lived in Costa Rica, I learned that time is not as fixed as it is here in the States. I would make appointments to meet my Tico friends and I would show up on the dot, but they consistently arrived up to an hour late, and were unapologetic. Their relationship with time was more flexible than mine, and I had to adjust to that. In addition, in many Latin cultures, a business lunch will begin with friendly conversation and may take quite some time to get to the business at hand, as it is culturally valuable to develop a friendly relationship before getting into the business aspect of the meeting. Ultimately, when considering multicultural communication, we may have to set our cultural norms aside and do a little more negotiating to make sure our message gets through in the way we intend within the time frame necessary. The best way to navigate these cultural differences is to know your audience; if you are working with a client from another cultural background, take some time to familiarize yourself with the client's cultural norms in order to stave off communication struggles over the course of your relationship.

REFERENCES

Erickson, S, and M. McKnight. (2001). *The Practitioner's Guide to Mediation: A Client-Centered Approach.* John Wiley & Sons, Inc. New York. ISBN 0-471-35368-X.

Hogan, Kevin. (2008). *The Secret Language of Business: How to Read Anyone in 3 Seconds or Less.* John Wiley & Sons, Inc. Hoboken, NJ. ISBN 978-0-470-22289-8.

Weeks, D. (1992). *The Eight Essential Steps to Conflict Resolution.* Tarcher Putnam, New York. ISBN 0-874-77751-8.

Winslade, J., and G. Monk. (2008). *Practicing Narrative Mediation—Loosening the Grip of Conflict.* A Wiley Imprint, Jossey-Bass. San Francisco, CA. ISBN 978-0-787-99474-7.

Introduction

1   2   Ethics   Professional Engagement

History   3   4   5   What Engineers Deliver   Emerging Technology

Engineer's Role in Project   6   Permitting   Leadership   Managing   Having   A   B   C   D   E   F
Development   7   a Life
                         8   9   10 11 12 13 14 15 16 17
Executing a Professional   Client Relationship   Sustainability
Commission   Communicating
                           Legal Aspects   Globalization

C h a p t e r **14**

Having a Life

Big Idea

Successful engineers must learn how to integrate a busy career with community, family, and self. The balance among these elements is essential for "having a life."

> Life loves to be taken by the lapel and told: "I am with you kid. Let's go."

> —Maya Angelou

Key Topics Covered

- Introduction
- The Mind
- The Body
- The Spirit
- Laugh and Have Fun
- Self-Assessment Test
- Analysis of Self-Assessment Test
- Summary

Related Chapters in This Book

- Communication is related to every chapter in this book

(*Continued*)

Related to *ASCE Body of Knowledge 2* Outcomes

ASCE BOK2 outcomes covered in this chapter

Foundational
1. Mathematics
2. Natural sciences
3. Humanities
4. Social sciences

Technical
5. Materials science
6. Mechanics
7. Experiments
8. Problem recognition and solving
9. Design
10. Sustainability
11. Contemporary issues & historical perspectives
12. Risk and uncertainty
13. Project management
14. Breadth in civil engineering areas
15. Technical specialization

Professional
16. Communication
17. Public policy
18. Business and public administration
19. Globalization
20. Leadership
21. Teamwork
22. Attitudes
23. Lifelong learning
24. Professional and ethical responsibility

INTRODUCTION

Ironically, this chapter of the book is the shortest, but it is probably the most important. Why . . . ? Because ignoring our innermost desires and needs can extinguish a portion of ourselves. This does not mean that we should forgo the pursuit of a meaningful career, love, or family; but we should create a mechanism and time to pursue the things in life that we enjoy and "have fun."

Very experienced engineers have commented on this chapter and provided their input on balancing an engineering career and a personal life. And, that's just it: life should be balanced. The balance won't look like the scales of justice where the "lady of justice" carries scales that balance the strength of a legal case against the opposition to the case.

Life balancing is more like balancing a pizza on a pencil. The number and size of pizza slices represent the many components of our lives. One can imagine that when any pizza slice becomes too large and overwhelming, the pizza becomes unstable and out of balance. The slices are analogous to the components of our lives—career, family, personal time, fun, exercise, sleep, volunteer work, and more.

> Time is so precious we should treat it like it deserves to be treated: an extremely valuable investment both in our careers and in our happiness! A 2002 study conducted at the University of Illinois by Diener and Seligman found that the most salient characteristics shared by the 10 percent of students with the highest levels of happiness and the fewest signs of depression were their strong ties to friends and family and commitment to spending time with them. It is important to work on social skills, close interpersonal ties, and social support in order to be happy (Diener 2008).

> "Having a Life" in the professional world requires learning to balance the potential resentment of working too much (being burned out) against the guilt of taking personal time.

So, what does *having a life* really mean? Some would say, "having time to do what I love to do," "being around friends and loved ones," "having a meaningful relationship and being part of a family," "making a contribution," "having children," "being famous," or "being wealthy." Life is more than the tiny electrochemical charges that stimulate our mind and body. Life is laughing, crying, sharing your time or meals with someone, and seeing wonderful things like the colors of nature.

Life is many of these things and *taking the time* to make them happen, celebrating these events, reflecting upon the events, and planning and enjoying new ones. The important element is taking the time because our lives get cluttered, and we almost always have somewhere to go or something to do. So, it will be up to you to take the time to have a life, to feel fulfilled, engaged, happy, and enthusiastic. And, finding a balance among career, family, friends, and personal time in your life will be critical.

The key components to life are:

- Mind—The Command Center
- Body—Our Home
- Spirit—Our "inner self," sometimes referred to as our soul

Establishing a real life balance with mind, body, and spirit will lead toward achieving a fulfilling life.

THE MIND

The Command Center of the Body and Our Inner Self

The mind is like the "bridge" on a ship or the "cockpit" on a jet. It's the control center for the body and spirit. The mind is the center for all intellectual thought, memory, emotion, and free will. It's the control center for all the major systems in the body and without it the body cannot function.

If you maintain the ability to use your memory, concentrate on intellectual thought, and focus critical thinking toward solving a problem, your mind should operate near its highest potential. Studies have shown that frequent use of the brain as you age helps maintain the ability to function effectively. Keeping your mind intellectually stimulated may strengthen brain cell networks and help preserve mental functions (Katz L., Rubin M., 1999). If you can synchronize the optimum operation of the mind with the body and the spirit, you can operate at the highest level possible for you as an individual.

Memories

We can never relive the precious moments we have experienced!

The most important professional component that our minds control is the belief in what we do and how we view our job performance. A good self-image shines through to others, especially when combined with a confident, positive attitude. "How" we handle problems and challenges can be more important than the actual solutions we devise. Self-confidence balanced with professionalism and humility is a tremendous asset to an engineer. The art of presenting yourself at your best is simple when character and personality are genuine traits. Spend some time identifying your best "selling" points and build on those as you enhance your mind.

Believe It or Not, We Sell Ourselves More Than We Think. Think About It

- To enter college you had to sell yourself on your application and explain why you should be chosen above other applicants.

- To become employed you had to demonstrate how you could be an asset to your employer.

- To meet people (that are now your friends) you had to relax, be yourself, and show how you might be interesting to another.

- We are constantly selling ourselves as responsible citizens, dedicated employees, worthy partners, concerned parents, and good people.

What About Stress?

You just yelled at your life partner. Your children have taken to calling you the Commander. You start daydreaming about drinks after work. You find you're forgetting things, important things. Your family says you're never home. You can't seem to concentrate. What's your problem?

In a word: Stress. (Sternberg, 2001)

Welcome to the club. Almost half of all Americans are concerned about the level of stress in their lives, according to the American Psychological Association's 2009 and 2010 Stress Survey. Chalk it up to "our overscheduled, harried 21st-century lifestyle, which can wreak havoc with our relationships and our work," says Bruce S. McEwen, M.D., a coauthor of *The End of Stress as We Know It*.

Stress also plays a heavy role in our overall health. Chronic stress weakens the immune system and increases the risk for a range of illnesses, including heart disease and depression. Stress drives people to eat too much, sleep too little, skimp on exercise, and shortchange fun. It doesn't have to be toxic; a little stress can sharpen focus, improve memory, and heighten emotions. But sometimes good stress goes bad, and researchers have just begun to figure out how to deal with it. "By understanding what makes it go wrong," says McEwen, "we have the power to make it right."

The Stress Response

In its most basic form, the stress response is known as "fight or flight," and it swings into action whenever you're confronted with a novel or threatening situation.

"If you step off the curb in front of an oncoming bus, your body reacts automatically to protect you," says Esther M. Sternberg, M.D., author of *The Balance Within: The Science Connecting Health and Emotions*. In a matter of seconds and without even thinking, you begin pumping out brain chemicals and hormones, including adrenaline and cortisol. Your heart rate accelerates, oxygen-bearing red blood cells flood the bloodstream, the immune system gears up for the possibility of injury, and energy resources are diverted to your muscles, brain, heart, and lungs and away from functions, such as digestion and hunger, which can wait until the crisis has passed.

Meanwhile, the brain releases a cascade of endorphins, the body's natural opiates, to dull the pain of those potential injuries. You're ready for action, whether it's a full-out battle or a hasty retreat in this case, fleeing back onto the sidewalk to escape the speeding bus.

When the danger has passed, all these systems are restored to their normal resting state. "Your stress response makes you get out of danger," says Sternberg. "Without it, you'd be dead." Many of the physical changes that energize you to get out of the way of the bus are the same ones at work in more positive situations. Your heart races when you're falling in love.

Developing a practice of *mindful relaxation* can be a great help. Physiologically, relaxation is the opposite of stress. When you're relaxed, your breathing and heart rate slow and your mind clears. Mindfulness is a way to achieve this level of relaxation using a variety of techniques, including yoga, meditation, and simple relaxation exercises.

Mindfulness quiets your mind by teaching you how to observe your thoughts and feelings without seeing them as positive or negative. It trains you to use your breathing and an awareness of your body to focus on the here and now. The basic relaxation response was first described in 1975 by Harvard Medical School researcher Herbert Benson.

His approach has two steps: First, close your eyes and focus on your breath (that's the foundation). Second, choose a phrase, a word, or a prayer and repeat it to stay in the moment and be mindful. "I use two phrases," says Bernadette Johnson, director of integrative medicine at Greenwich Hospital, in Connecticut. "I'm breathing in 'relaxation and peace' when I inhale. I'm breathing out 'tension and anxiety' on the exhale."

Ideally, you'd begin and end your day with 10 to 20 minutes of regular relaxation exercise. But should you find your tension rising during the day, "take a deep breath, hold it for a count of four, and exhale for a count of four," Johnson says. "That's what we call a mini, and if it's built on a foundation of regular, longer relaxation exercises, you can tap into it whenever you need it."

If a thought or an emotion intrudes on your mindfulness and threatens to take you out of the moment, observe it but don't react to it. Think of it as a leaf floating by on a slow-moving stream.

> As a young engineering PM in a prominent engineering consulting firm, I observed that the firm's engineering directors and vice-presidents worked excessively long hours but received handsome bonuses for their efforts. I later realized that all but one of six of these senior managers were divorced or separated from their spouses and paid spousal and child support that exceeded half of their salaries. It was ironic that my "take-home" compensation as a PM exceeded the residual salary of a senior manager.
>
> Source: Anonymous Engineer

THE BODY

Consider your body as your home for your mind and inner self. In the engineering profession we consider the operation and maintenance (O & M) of any structures we design or build. Good O & M practices extend the usefulness, functionality, and viability of these structures. In our career as engineers, these structures may be buildings, roads, bridges, dams, treatment plants, airports, wetlands, landfills, or operational processes. As we carry out our engineering practice, we should recognize that we are operating out of our bodies, our home. And, maintaining our home will extend our own viability and usefulness.

A trip to your medical practitioner will provide detailed advice and instructions for each individual. There is no substitute for this professional counsel. There are literally volumes and millions of pages of references on health and the human body. In general, here are some relatively simple rules from the personal experience of the authors to increase our effectiveness:

- Eat Well
- Sleep Well
- Exercise Well

Eat Well—The Balanced Diet

Food provides the nutrients to run an extremely complicated machine—our bodies. As learned in elementary school and virtually every health magazine and other prestigious references, a balanced diet is required to help maximize performance. Maximum performance will be required to excel in an engineering career.

Sleep Well—Save Space for Dreams

Our bodies require rest to accommodate for daily stresses and to relax the muscular, circulatory, and nervous systems. Most references state that adults should strive for about eight hours of sleep each night. In general, they also state that sleep deprivation often seriously affects memory and cognitive abilities and that deficit hours cannot be made up on the weekends. In an effort to excel at our careers and pack the most into our lives, sleep is often sacrificed. There is a price to be paid for sleep deprivation and it varies from slower response to higher accident rates. Sleep is very important for adults to excel in their careers. Sleep disturbance can be a sure sign of stress or other impacts on your life, and it should be addressed with your medical practitioner.

If sleep is disturbed with thoughts of "things to do" or the "next day's activities," keeping a pad and pen next to the bed may help to purge these thoughts, to allow sleep to come. Other tips include allowing time for some pleasant reading material (no, not technical journals or homework), soft music, or favorite pleasant sounds like a running stream to induce rest. Some additional advice includes thinking pleasant thoughts as one drifts off to sleep.

Exercise Well—A Healthy Body Will Facilitate a Healthy Mind

Our bodies also require exercise to maintain the muscle groups, the heart, and circulatory system. Most references cite the need for at least 30 minutes of strenuous exercise per day, at least 3 to 4 times per week. The exercise could be as simple as a brisk walk for 30 minutes or mixing exercise and fun like tennis, biking, swimming, or other active sports.

In general, health magazines and references state that the lack of exercise can induce weight gain and early onset of diseases. Like sleep deprivation, the lack of adequate exercise cannot be made up on the weekends. This condition is referred to as a "weekend warrior" and is often accompanied by soft tissue, muscle, or other injuries. Exercise is very important for adults to excel in their careers.

THE SPIRIT

Spirit is our inner self where we store our feelings and thoughts. Spirit is referred to by some as our soul or inner essence and is therefore metaphysical in nature. In this regard the spirit is closely related to our mind and body. It can be thought of as the spark of life that actually displays our character, emotion, personality, and consciousness that connects our mind and body to form unique individuals different from any other in the universe. Many people connect spirit with a religion or their own belief or concept of a god, deity, or higher power. Other people may not believe or accept a god, deity, or higher power but that doesn't mean they don't have spirit.

Favorite Places

I have favorite places that I like to go to where I let my mind wander, just stare in amazement, relax, and breathe. These are kind of safe havens and there's one that I can walk to, one that I can to drive to within an hour, and a favorite vacation place. For me all three favorite places have water—two have lakes and one has the ocean, and all of them have rolling terrain, lush trees and natural beauty. I find that I get restless if I don't go there enough, as if my job or life stresses consume my inner peace.

It took me a while to realize this fact and even more time to actually articulate it. But now that I know this, I try to go to a favorite place to re-group, to re-kindle that inner flame of peace and re-fuel my spirit at least once, twice a week or more. Sometimes, if I'm having an especially difficult time (like in the dentist's chair or a stressful moment at work) I will close my eyes and go there for a few moments in my mind. I find this brief retreat settles me and allows me to regain energy, proceed, and engage.

Our intellect, learned information, and innate information (instinct) are housed in our mind. Our spirit seems to be primarily in our mind and hearts, contained within our bodies.

For an example of innate information and pure instinct, one can observe baby sea turtles. After hatching from the sandy beach, they automatically move toward the ocean on their own without any guidance from their parents. Instinct can be thought of as our innate behavior. These instinctual actions are contrasted to learned information, which is taught to us and stored in our memory (minds) for future retrieval and use. Sometimes instincts are hard-wired without our knowledge, ready for implementation after a maturation process. An example of this behavior appears with horses and colts. For example, in one hour a baby colt can learn to stand, walk, glide, skip, hop, and run. After much training they may be able to run one and one-quarter miles in about two minutes, pull a wagon, or come after the owner's whistle.

THE EFFECTIVE COMBINATION OF MIND, BODY, AND SPIRIT

Okay, so how does this all relate to *having a life* and being an engineer? It's very difficult to articulate, but it seems that if we can integrate the learned and innate information in our minds while maintaining a well-tuned and healthy body and synergistically combining our individual spirit and personality, we have the ability to solve (or answer) extremely complex and sophisticated problems. In this manner, our conscious minds can draw upon megabytes of learned information from all conscious and subconscious levels (innate thoughts) integrated with our spirit (and personality) operating multi-dimensionally to reveal these answers.

High Performance Careers

In my career I have seen individuals with a "great mind" excel (chess champion), individuals with a "great body" excel (college quarterback), and individuals with a "great spirit" (senior class president) excel in their careers. But, it's okay if you don't happen to have a "great" mind, body, or spirit. It's also possible to excel with a "great balance" of mind, body, and spirit!

The Integration of Mind, Body, and Spirit

A true, integrated connection of mind, body, and spirit can provide the engineer with distinct advantages in his or her career. Recognition and comprehension of this experience can provide the engineer with the self-confidence and capability to excel!

LAUGH AND HAVE FUN

Laughing—Don't Be Too Serious

What's the value of a laugh? Laughing is fun, it's healthy, and contagious. Humor can make your job enjoyable and rewarding. It can lighten stress and tense moments and be one of your secret weapons. If you have some tense moments consider whether a lighthearted statement and a brief laugh would lighten the mood, then, move on to address the situation.

Laughter reduces pain and allows us to tolerate discomfort. It reduces blood sugar levels, increasing glucose tolerance in diabetics and nondiabetics alike. It improves your job performance, especially if your work depends on creativity and solving complex problems.

Its role in intimate relationships is vastly underestimated and it really is the glue of many good marriages. It synchronizes the brains of speaker and listener so that they are emotionally attuned. Laughter establishes—or restores—a positive emotional climate and a sense of connection between two people. In fact, some researchers believe that the major function of laughter is to bring people together. And all the health benefits of laughter may simply result from the social support that laughter stimulates.

Recently, there has been hard evidence that laughter helps your blood vessels function better. It acts on the inner lining of blood vessels, called the endothelium, causing vessels to relax and expand, increasing blood flow. In other words, it's good for your heart and brain, two organs that require the steady flow of oxygen carried in the blood.

The research doesn't say for sure exactly how laughter delivers its heart benefit. It could come from the vigorous movement of the diaphragm muscles as you chuckle or guffaw. Alternatively, or additionally, laughter might trigger the release in the brain of such hormones as endorphins that have an effect on arteries.

It's also possible that laughter boosts levels of nitric oxide in artery walls. Nitric oxide is known to play a role in the dilation of the endothelium. Thirty minutes of exercise three times a week, and 15 minutes of laughter on a daily basis is probably good for the vascular system (McGhee, 1999).

Personalize Your Fun Time

Personalized fun time can be a real treat. Think of something you're good at and something you really like to do. Maybe this activity will be swimming, fishing, running, biking, volunteering, snowboarding, or skiing. Now, think about how you might personalize this activity. Personalize it with your own style and grace, something unique

After completing my Bachelor's Degree in Civil Engineering, I celebrated by touring the western states. I visited the Grand Canyon and remember coming to the first overlook and looking over the North Rim. I was speechless and I can still recall that feeling!

about you that applies your personal attributes to the activity. Then, embrace it, increase your passion for it, learn from others and share your knowledge and experiences.

SELF-ASSESSMENT TEST—PLEASE CHALLENGE YOURSELF

Mind: What Do You Think, Really?

It's often a challenge for analytical people like engineers to assess their own thoughts and feelings and express them, even to ourselves. It's important to get in touch with your inner thoughts and to capture these feelings for yourself. It may help to go to a very quiet place where you feel safe and very comfortable.

1. Concentrate on your life components and "having a life."
2. Think about how you feel emotionally.
3. Focus and come up with some things that are going right for you.
4. Now consider some things that may need attention.

Body: How Does Your Body Feel, Really?

Many of us don't like to look too closely in the mirror. Make an appointment with your closest glossy surface and take a close look. Check for signs of stress, your range of motion, try touching your toes, or just stand up straight. When was the last time you ran or walked or even noticed the weather? If the weather question stumped you, walk to a window in your office at least once a day for a month. You'll probably feel more like going outside if you see daylight on a regular basis. Give yourself permission to take a 15-minute walk during your lunch break.

1. What is your honest assessment after the mirror exercise?
2. Think about how you feel physically.
3. Focus and come up with some things that are going right for you.
4. Now consider some things that may need attention.

Spirit: How Do You Feel, Really?

Check in with yourself regularly. Ask yourself how things are going, where they are going, where you would like them to go. Include all "the slices of the pizza" such as your career, your hobbies or favorite sport, your family and friends.

1. What is your honest assessment of your feelings?
2. Think about your relationships with family, friends, coworkers, and community.
3. Focus and come up with some things that are going right for you.
4. Now consider some things that may need attention.

ANALYSIS OF THE ASSESSMENT TEST

On a Scale of 1 to 10 . . .

How happy are you today? How happy are you compared to last week, last month, or last year? What could you do to be even happier? Is it time to make a plan?

- So, how do you feel—mad, sad, glad, or scared? Do you need to make any adjustments or revisions? Are any life components out of balance? Is it time to make a plan? It may help to take some brief notes to capture your thoughts. Now, place these feelings or notes in a little box in your mind (or someplace safe if it's your personal notes) and temporarily put them away for the night or for a day or two.

- Take out the box and re-examine the thoughts you had last time you put them away. How do you feel now? Are any revisions necessary? Do you need to take any action at all?

Life is about balance among mind, body, and spirit. When things are given equal importance and one thing does not overshadow the other, then you have the perfect "pizza on a pencil" balance. If you are out of balance, identify the area of weakness and plan and fortify this asset. For example, if you need to sharpen your mind, take some college courses, learn a foreign language, or take a cooking class. If your body needs attention, find a personalized sport or activity where you can combine fun with exercise. If your spirit is winsome, reconnect with an old friend, make new connections with colleagues or neighbors, or spend a day with your child and see the world through their eyes.

Some Additional Thoughts

Here are some important concepts that may help you in your challenge to find balance in your life and career:

- Respond to life and its changes . . . Don't just react to them.
- Being proactive is much easier and rewarding than being reactive.
- Seek growth, be inspired, and display good character.
- Avoid arrogance—it can sink your boat . . . and career.
- Respect for others and humility for oneself are great character traits.
- Remember that attitude makes all the difference . . . especially a winning, positive attitude.

(Continued)

- If you think you can, you can, and if you think you can't . . . you probably won't.
- Have a sense of urgency and appreciate the "value of time."

Take Time . . .

- Take time to enjoy the little things. Make a conscious effort to notice things you enjoy, focus on them during the moments you have, and remember them as part of your day. Okay, you might say what things are you talking about? Let's think about your commute to work. Do you see anything interesting like your neighbors, flowers in someone's yard, school children laughing while waiting for a bus, a smile on a stranger's face, a creek or a river, a really large old tree or an interesting building? Can you remember this interesting mind-picture, reframe it and compare it to the previous day or save it for a brief discussion with a colleague? Don't just dismiss it and forget it. Savor the treat and consider it as an asset to your day.
- Take time to enjoy your family and friends. When visiting family or friends you haven't seen for a while, have you ever said, "Wow, the kids have grown," or "he/she looks older or different than I remember," or something similar? People are busy, we're all busy, but time continues on and we can't get it back . . . ever. The question to ask yourself is, "Are your family and friends a priority?" Be honest with yourself and act accordingly.
- Take time during your career to learn something unique from every employer. Make an effort to use this learned skill to become a more qualified engineer, a smarter and better person.

And, Here are Some Other Thoughts from Wise Men and Women

Kites rise highest against the wind—not with it. (Winston Churchill)

We cannot become what we need to be by remaining what we are. (Max DePree)

(Continued)

Ideas without action are worthless. (Harvey Mackay)

A goal is a dream with a plan of action, schedule and budget. (Zenobia)

Don't postpone joy! (Pamela M. Peeke, MD, MPH)

Since you get more joy out of giving joy to others you should put a great deal of thought into the happiness you are able to give. (Eleanor Roosevelt)

Happiness is a by-product of a well-lived life, and it is achieved through the pursuit of endeavors that are meaningful and sometimes painful. (Mark O'Connell)

Happiness and moral duty are inseparably connected. (George Washington)

It's just over the next rise . . . (Sacajawea spoken in Native American language)

SUMMARY

Life is about "balance" between many competing factors relating to career, community, family, and personal time. Balancing these components of our lives is a lot like trying to balance a pizza on the tip of a pencil. Integrating mind, body, and spirit is a means to "having a life."

REFERENCES

Diener, Ed, and Robert Biswas-Diener. (2008). *Happiness: Unlocking the Mysteries of Psychological Wealth.* Blackwell Publishing, Malden, MA. ISBN 978-1-405-14661-6.

Katz, Lawrence, and Rubin, Manning. (1999). *Keep Your Brain Alive.* Workman Publishing Company. New York. ISBN 13: 978-0-761-11052-1.

McGhee, Paul. (1999). *Health, Healing, and the Amuse System: Humor as Survival Training.* Kendall/Hunt Publishers. ISBN-13: 978-0-787-25797-2.

SternbergM.D., EstherM. (2001). *The Balance Within: The Science Connecting Health and Emotions.* W. H. Freeman and Company. New York. ISBN-13: 978-0-716-74445-0.

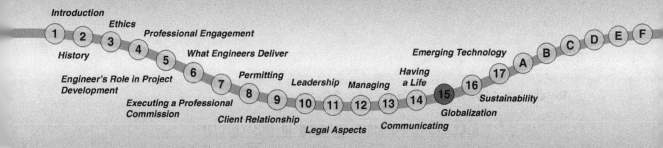

Introduction
Ethics
Professional Engagement
1 2 3 4
History
5
What Engineers Deliver
6
Engineer's Role in Project
Development
7
Permitting
8
Executing a Professional
Commission
9
Leadership
10
Client Relationship
11
Managing
12
Legal Aspects
13
Having
a Life
14
Communicating
15
Globalization
16
Sustainability
17
Emerging Technology
A B C D E F

Chapter **15**

Globalization

Big Idea

Civil engineers need to understand the issues and concepts surrounding globalization—rapid changes in the planet's economic order and climate are ushering in large-scale opportunities and threats that will leave few unaffected.

glob•al•i•za•tion: the development of an increasingly integrated global economy marked especially by free trade, free flow of capital, and the tapping of cheaper foreign labor markets.

—Merrriam Webster OnLine, www.merriam-webster.com

Key Topics Covered

- Introduction
- The Globalization Process
- Global Climate Change—From a World View and a State Perspective
- Outcomes of Globalization and Climate Change
- Learning to Project Manage a Mega-Project—The Case of BAA and Heathrow Terminal 5
- Civil Engineering Practice—A Wider Community Viewpoint
- Summary

Related Chapters in This Book

- Chapter 3: Ethics
- Chapter 4: Professional Engagement
- Chapter 5: The Civil Engineer's Role in Project Development
- Chapter 6: What Engineers Deliver
- Chapter 13: Communicating as a Professional Engineer
- Chapter 16: Sustainability
- Chapter 17: Emerging Technologies

(*Continued*)

Related to *ASCE Body of Knowledge 2* Outcome

ASCE BOK2 outcomes covered in this chapter

Foundational
1. Mathematics
2. Natural sciences
3. Humanities
4. Social sciences

Technical
5. Materials science
6. Mechanics
7. Experiments
8. Problem recognition and solving
9. Design
● 10. Sustainability
11. Contemporary issues & historical perspectives
12. Risk and uncertainty
13. Project management
14. Breadth in civil engineering areas
15. Technical specialization

Professional
16. Communication
17. Public policy
18. Business and public administration
● 19. Globalization
20. Leadership
21. Teamwork
22. Attitudes
23. Lifelong learning
24. Professional and ethical responsibility

INTRODUCTION

This chapter gives civil engineers a basic understanding of how globalization affects competition, utility of engineering services, design, cost-effectiveness, construction, and decision-making. The effects of globalization are being felt now and just are beginning to be understood. Civil engineers can expect additional, needed attention on infrastructure, increased requirements for ethical standards as the global community gets smaller, more complex problem solving, and increased competition for financial resources.

The term "globalization" became widely used and accepted by economists and other social scientists in the 1960s. Its use was widespread in the world news in the late 1980s. Since its inception, the concept of globalization has inspired numerous competing definitions and interpretations (Steger 2009). For the purpose of this handbook, globalization is the continuous evolution of separate economies, nations, and cultures toward homogenization and integration through a worldwide network of communication, travel, and trade.

In an economic context, globalization refers to the reduction and removal of barriers between national borders in order to facilitate the flow of goods, capital, services, and labor. Although considerable barriers remain on the flow of labor, globalization is not a new phenomenon. It began in the late 19th century, but its spread slowed during the period from the start of WWI until the early 1970s. This slowdown can be attributed to the inward-looking policies pursued by a number of countries in order to protect their respective industries. However, the pace of globalization picked up rapidly during the last 25 years of the 20th century (Ritzer 2010).

THE GLOBALIZATION PROCESS

According to Rosebeth Moss Kanter of Harvard University, globalization is a process of change stemming from a combination of increasing cross-border activity and information technology (IT) enabling virtually simultaneous communication worldwide. Its promise is to make the world's best accessible to everyone. Four broad processes (shown in Figure 15.1) put more choices in the hands of individual consumers and organizational customers, generating a "globalization cascade" with reinforcing feedback loops that strengthen and accelerate globalizing forces. More information on these processes follows.

Process 1: Mobility—Capital, People, Ideas. The key business ingredients of capital, people, and ideas are increasingly mobile. Capital mobility is noted often, but migrant professionals and managers are now joining more traditional migrant workers in an international labor force. Ideas move around the world through academic research and through global media, such as CNN, the Internet, and popular websites. High-speed information transfer makes location irrelevant: American Airlines' data entry point for tickets is in Barbados (at the time of this writing, anyway); British credit bureaus, European patent offices, and

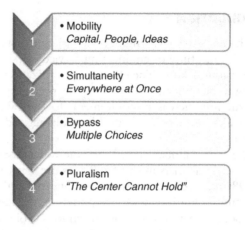

Figure 15.1 Four processes contribute to globalization

switching networks are located in the Philippines; and a Swedish fire department reaches its databases of street routes through a computer in Ohio.

Process 2: Simultaneity—Everywhere at Once. Globalization means that goods and services are increasingly available in many places at the same time. The time lag between the introduction of a product or service in one place and its adoption in other places is decreasing radically. The slow rollout from local test to home country launch to adjacent country availability is becoming less common. The newer the technology or application, the more likely it is to be designed with the whole world in mind.

Process 3: Bypass—Multiple Choices. Cross-border competition supported by easier international travel, deregulation, and privatization of government monopolies aids globalization and increases alternatives. Innovators can use alternative channels and new technology to go around established players rather than competing with them head-to-head. "Bypass" first referred to the rise of private switching networks that went around American regional telephone operating companies' wires; now wireless networks such as cellular and satellite systems bypass even more easily. Bypass creates numerous alternative routes to reach and serve customers. Japanese mail-order companies save 20 to 30 percent of postal costs by sending catalogues to Hong Kong for mailing back to Japan, thereby bypassing Japan's expensive postal service monopoly.

Process 4: Pluralism—"The Center Cannot Hold". There is a relative decline of monopolist centers of expertise and influence; activities concentrated in a few places are being decentralized. Traditional centers often still direct action and are main beneficiaries, but their automatic dominance or power to shape events declines when expertise and influence spread. "National champions," especially government-owned enterprises, are being reorganized and opened up to competition. Corporate headquarters functions are being dispersed and "centers of excellence" are being created in many parts of the world. Hewlett-Packard's corporate headquarters is in Palo Alto, California, but its world center for medical equipment is in Boston; for personal computers it's in Grenoble, France; for fiber-optic research it's Germany; for computer-aided-engineering software it's Australia; and for laser printers it is in Singapore.

An idea left over from the industrial economy now being discredited is that power comes from control over the means of production. In the global information economy, power comes from influence over consumption. Globalization of markets increases customers' choices, requiring producers to think more like customers. For example:

- Producers think they are making products; customers think they are buying services.
- Producers want to maximize return on the resources they own; customers care about whether resources are applied for their benefit, not who owns them.
- Producers worry about visible mistakes; customers are lost because of invisible mistakes.
- Producers think their technologies create products; customers think their needs create products.
- Producers organize for internal managerial convenience; customers want their convenience to come first.

Companies positioned to be successful in global markets put an emphasis on innovation, learning, and collaboration. They:

- Organize around customer logic
- Set high goals
- Select people who are broad, creative thinkers
- Encourage enterprise (on the part of employees)
- Support constant learning
- Collaborate with partners

GLOBAL CLIMATE CHANGE—A WORLD VIEW AND A STATE PERSPECTIVE

Though there are those who will debate its causes, global climate change appears to be real. The implications for civil engineers regarding global climate change are significant, perhaps more so than for other engineering disciplines. This section reports on the finding of three organizations: 1) the US Environmental Protection Agency (EPA); 2) the Intergovernmental Panel on Climate Change (IPCC), a scientific body established by the United Nations Environment Program (UNEP) and the World Meteorological Organization (WMO); and 3) the State of California Natural Resources Agency, in conjunction with numerous other state agencies.

Potential Global Impacts

According to the U.S. EPA,

> "Many elements of human society and the environment are sensitive to climate variability and change. Human health, agriculture, natural ecosystems, coastal areas, and heating and cooling requirements are examples of climate-sensitive systems.
>
> Rising average temperatures are already affecting the environment. Some observed changes include shrinking of glaciers, thawing of permafrost, later freezing and earlier break-up of ice on rivers and lakes, lengthening of growing seasons, shifts in plant and animal ranges and earlier flowering of trees." (www.epa.gov/climatechange/effects/index.html)

There are inter-relationships of climate impacts to conditions affecting public health that will have direct and indirect impacts on civil engineering design requirements (see Figure 15.2).

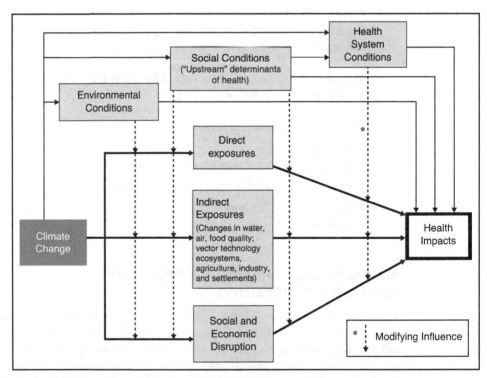

Figure 15.2 Flow diagram of effect of climate change on health

The U.S. EPA has additional relevant information on their website and also has directed readers to the IPCC website. (www.ipcc.ch/) In 2007, IPCC shared the Nobel Peace Prize with former Vice-President Al Gore for their work on climate change, which the Nobel Prize Committee recognized as having a connection with peace and war. The IPCC website contains much useful information, including an IPCC report titled, *Climate Change 2007: Impacts, Adaptation and Vulnerability*. In addition to this report, further information is available from recent IPCC meetings and publications. The reader is encouraged to visit the website to access these materials.

IPCCs 2007 report is rather detailed, so a brief summary is presented here:

"Observational evidence from all continents and most oceans shows that many natural systems are being affected by regional climate changes, particularly temperature increases (very high confidence). A global assessment of data since 1970 has shown it is likely that anthropogenic warming has had a discernible influence on many physical and biological systems . . ."

"For physical systems, climate change is affecting natural and human systems in regions of snow, ice and frozen ground, and there is now evidence of effects on hydrology and water resources, coastal zones and oceans . . ."

"There is more evidence, from a wider range of species and communities in terrestrial ecosystems than reported in the Third Assessment, that recent warming is already strongly affecting natural biological systems. There is substantial new evidence relating changes in marine and freshwater systems to warming. The evidence suggests that both terrestrial and marine biological systems are now being strongly influenced by observed recent warming . . ."

"The number of people living in severely stressed river basins is projected to increase significantly from 1.4−1.6 billion in 1995 to 4.3−6.9 billion in 2050 . . . (medium confidence)."

"The resilience of many ecosystems (their ability to adapt naturally) is likely to be exceeded by 2100 by an unprecedented combination of change in climate, associated disturbances (e.g., flooding, drought, wildfire, insects, ocean acidification), and other global change drivers (e.g., land-use change, pollution, over-exploitation of resources) (high confidence) . . ."

This information is accompanied by detailed descriptions, graphics, and references from around the globe. Some of the more interesting graphics are presented in Figure 15.3 to illustrate the potential severity and gravity of climate change.

Figure 15.3 illustrates the potential impacts to systems/resources due to an average temperature rise of from 1 to 5 degrees Celsius. The impacts on fresh water systems, ecosystems, and food production are reflected in stress on public health, pressure on water resources, increased species extinction, a greater number of wildfires, decreased production in some crops, and changing coastlines. The impacts from 0 to 2 degrees Celsius are significant but become very dramatic and extremely problematic from 3 to 5 degrees Celsius.

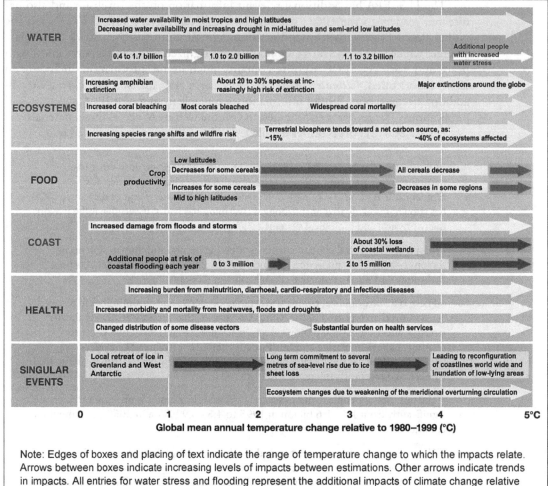

Figure 15.3   Example of impacts to systems/resources related to temperature rise

Figure 15.4 illustrates the potential impacts to the major continents, polar regions, and small islands related to rises in average temperatures from 1 to 5 degrees Celsius. The specific systems and resources impacted in these land masses are highlighted. The potential impacts to North America will be an increased need for cooling systems within buildings, increased frequency of high ozone pollution days, an increase in crop yield where reliable water supplies are available, and increased wildfires. The impact for Europe will be an increase of water resources in northern Europe with an accompanied decrease of water resources in southern Europe, and variable increase

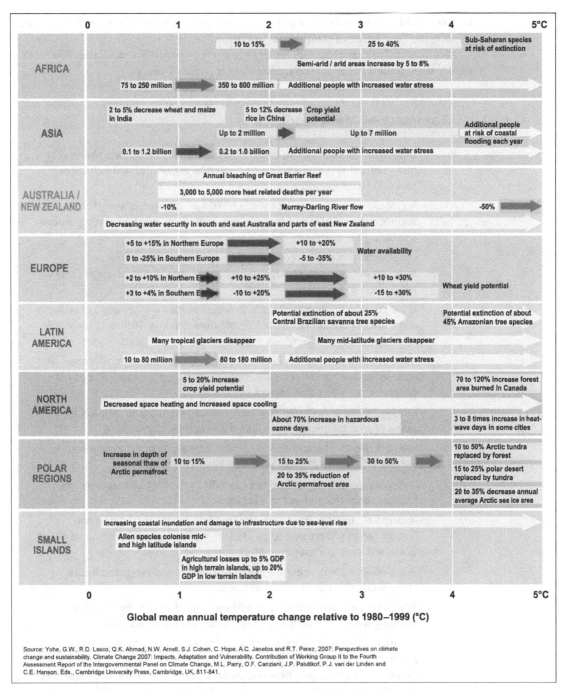

Source: Yohe, G.W., R.D. Lasco, Q.K. Ahmad, N.W. Arnell, S.J. Cohen, C. Hope, A.C. Janetos and R.T. Perez, 2007: Perspectives on climate change and sustainability. Climate Change 2007: Impacts, Adaptation and Vulnerability. Contribution of Working Group II to the Fourth Assessment Report of the Intergovernmental Panel on Climate Change, M.L. Parry, O.F. Canziani, J.P. Palutikof, P.J. van der Linden and C.E. Hanson, Eds., Cambridge University Press, Cambridge, UK, 811-841.

Figure 15.4   Projected global temperature rise and impacts

in crop yield where reliable water resources exist. Other continents have varied impacts over the range of temperature increases.

Some conclusions that can be drawn from Figure 15.4 and the 2007 IPCC report are included in the textbox, *Summary of Main Findings.*

Summary of Main Findings

- Observational evidence from all continents and most oceans shows that many natural systems are being affected by regional climate changes, particularly temperature increases.

- A global assessment of data since 1970 has shown it is likely that anthropogenic warming has had a discernible influence on many physical and biological systems.

- Other effects of regional climate changes on natural and human environments are emerging, although many are difficult to discern due to adaptation and nonclimatic drivers.

- More specific information is now available across a wide range of systems and sectors concerning the nature of future impacts, including for some fields not covered in previous assessments.

- More specific information is now available across the regions of the world concerning the nature of future impacts, including for some places not covered in previous assessments.

- Magnitudes of impact can now be estimated more systematically for a range of possible increases in global average temperature.

- Impacts due to altered frequencies and intensities of extreme weather, climate and sea-level events are very likely to change.

- Some large-scale climate events have the potential to cause very large impacts, especially after the 21st century.

- Impacts of climate change will vary regionally but, aggregated and discounted to the present, they are very likely to impose net annual costs which will increase over time as global temperatures increase.

- Some adaptation is occurring now, to observed and projected future climate change, but on a limited basis.

- Adaptation will be necessary to address impacts resulting from the warming which is already unavoidable due to past emissions.

- A wide array of adaptation options is available, but more extensive adaptation than is currently occurring is required to reduce vulnerability to future climate change. There are barriers, limits and costs, but these are not fully understood.

- Vulnerability to climate change can be exacerbated by the presence of other stresses.

- Future vulnerability depends not only on climate change but also on development pathway.

- Sustainable development can reduce vulnerability to climate change, and climate change could impede nations' abilities to achieve sustainable development pathways.

- Many impacts can be avoided, reduced or delayed by mitigation.

- A portfolio of adaptation and mitigation measures can diminish the risks associated with climate.

Source: IPCC Technical Report Titled: Climate Change 2007: Impacts, Adaptation and Vulnerability www.ipcc.ch/publications_and_data/publications_ipcc_fourth_assessment_report_wg2_report_impacts_adaptation_and_vulnerability.htm

Potential Impacts on California and the Western States

A similar report from the State of California also exists. The report, titled *2009 California Climate Adaptation Strategy*, was chosen because it presents some global climate impacts that will affect the State, as well as a strategy for how the State will handle climate change. Specifically, the report summarizes the best known science on climate change impacts in seven specific sectors. Here are some excerpts from the report to the Governor of the State of California in response to the Governor's Executive Order S-13-2008:

"Climate change is already affecting California. Sea levels have risen by as much as seven inches along the California coast over the last century, increasing erosion and pressure on the state's infrastructure, water supplies, and natural resources. The state has also seen increased average temperatures, more extreme hot days, fewer cold nights, a lengthening of the growing season, shifts in the water cycle with less winter precipitation falling as snow, and both snowmelt and rainwater running off sooner in the year . . ."

"These climate driven changes affect resources critical to the health and prosperity of California. For example, forest wildland fires are becoming more frequent and intense due to dry seasons that start earlier and end later. The state's water supply, already stressed under current demands and expected population growth, will shrink under even the most conservative climate change scenario . . ."

"If the state were to take no action to reduce or minimize expected impacts from future climate change, the costs could be severe. A 2008 report by the University of California, Berkeley and the non-profit organization Next 10 estimates that if no such action is taken in California, damages across sectors would result in 'tens of billions of dollars per year in direct costs' and 'expose

trillions of dollars of assets to collateral risk.' More specifically, the report suggests that of the state's $4 trillion in real estate assets '$2.5 trillion is at risk from extreme weather events, sea level rise, and wildfires' with a projected annual price tag of up to $3.9 billion over this century depending on climate scenarios (www.next10.org/research/research_ccrr.html) . . .'"

"California's ability to manage its climate risks through adaptation depends on a number of critical factors including its baseline and projected economic resources, technologies, infrastructure, institutional support and effective governance, public awareness, access to the best available scientific information, sustainably-managed natural resources, and equity in access to these resources . . ."

"To effectively address the challenges that a changing climate will bring, **climate adaptation and mitigation (i.e., reducing state greenhouse gas (GHG) emissions) policies must complement each other** and efforts within and across sectors must be coordinated. For years, the two approaches have been viewed as alternatives, rather than as complementary and equally necessary approaches. Adaptation is a relatively new concept in California policy. The term generally refers to efforts that respond to the impacts of climate change—adjustments in natural or human systems to actual or expected climate changes to minimize harm or take advantage of beneficial opportunities . . ."

A Collaborative Approach to the Adaptation Strategy

"The development of the adaptation strategies was spearheaded by the state's resource management agencies, CNRA staff who worked with seven sector-based Climate Adaptation Working Groups (CAWGs) focused on the following areas:

- Public health
- Ocean and coastal resources
- Water supply and flood protection
- Agriculture
- Forestry
- Biodiversity and habitat
- Transportation and energy infrastructure . . .

This adaptation strategy could not have been developed without the involvement of numerous stakeholders. Converging missions, common interests, inherent needs for cooperation, and the fact that climate change impacts cut across jurisdictional boundaries will require governments, businesses, non-governmental organizations, and individuals to minimize risks and take advantage of potential planning opportunities . . ." (www.climatechange.ca.gov/adaptation/)

The report, *2009 California Climate Adaptation Strategy*, went on to conclude potential impacts to public health and biodiversity. These conclusions are highlighted below:

Public Health and Environmental Impacts Due to Warming

- Higher rates of mortality and morbidity
- Increased air pollution
- Seasonal changes and increases in allergens
- Changes in prevalence and spread of disease vectors
- Possible decrease in food quality and Security
- Reduction in water availability
- Increased pesticide use

Public Health Impacts Due to Sea-Level Rise

- Wastewater issues with flooding of septic systems near coastline
- Salt water intrusion—risks to drinking water
- Threats of injury and even death during coastal storms
- Emotional and mental health impacts related to more coastal flooding and erosion
- Emotional and mental health impacts related to internal displacement and migration of coastal residents

Biodiversity and Habitat Impacts Due To Warming

- Higher Barriers to species migration and movement
- Temperature Rise—lakes, streams, and oceans
- Increase in invasive species potential
- Changes in natural community structure
- Threats to rare, threatened, or endangered species
- Altered timing of phenological events
- Timing disruptions between predators and prey and pollinators and plants
- Loss of ecosystem goods and services

Biodiversity and Habitat Impacts Due to Precipitation Changes

- Stream flows—Impact to fish passage
- Distribution/longevity of surface water, impact to wildlife
- Changes in riparian communities and structure
- Decreased water availability—fish, wildlife, and plants
- Water temperature, pollution, and sediment load changes
- Impacts to water-dependent species
- Surface water allocations—impact all water users (Humans & Wildlife)
- Increased susceptibility to pests, disease, wildfires and invasive species
- Habitat conversions—changes in biodiversity

Biodiversity and Habitat Impacts Due to Sea-Level Rise

- Inundation of permanent coastal habitat
- Alteration of dune habitat and coastal wetlands
- Coastal habitat loss of migratory birds, shellfish and endangered plants
- Reduction of fresh water resources due to salt water intrusion for coastal land areas
- Sedimentation increases may increase pollution and run off
- Degradation of aquatic ecosystem
- Increase in invasive species
- Competition shifts in urban growth and development
- Agricultural relocation
- Alterations of ecological reserves, wildlife areas, undesignated lands, mitigations sites and easements
- Groundwater recharge & overdrafting
- Water management and water transfer conflicts
- Reduction in wetland habitat on commercial and sport fisheries

After presenting the background, potential impacts to the State, and conclusions, the report presents Preliminary Recommendations as highlighted below.

Preliminary Recommendations

"The preliminary recommendations outlined in the adaptation strategy were developed by CA Natural Resources Agency staff, other CA Agencies and from public comments . . . . Stakeholder comments covered many topics, with the most common being the need for more coordination and guidance, funding, and outreach. All public input on the CAS Discussion Draft can be viewed on the web at: www .climatechange.ca.gov/adaptation/.

It is recognized that implementation of the following strategies will require significant collaboration among multiple stakeholders to ensure they are carried out in a rational, yet progressive manner over the long term. These strategies distinguish between near-term actions that will be completed by the end of 2010 and long-term actions to be developed over time . . . Key recommendations include:

1. A Climate Adaptation Advisory Panel (CAAP) will be appointed to assess the greatest risks to California from climate change and recommend strategies to reduce those risks building on California's Climate Adaptation Strategy . . .

2. California must change its water management and uses because climate change will likely create greater competition for limited water supplies needed by the environment, agriculture, and cities . . .

3. Consider project alternatives that avoid significant new development in areas that cannot be adequately protected (planning, permitting, development, and building) from flooding, wildfire and erosion due to climate change. The most risk-averse approach for minimizing the adverse effects of sea level rise and storm activities is to carefully consider new development within areas vulnerable to inundation and erosion. State agencies should generally not plan, develop, or build any new significant structure in a place where that structure will require significant protection from sea level rise, storm surges, or coastal erosion during the expected life of the structure . . .

4. All state agencies responsible for the management and regulation of public health, infrastructure or habitat subject to significant climate change should prepare as appropriate agency-specific adaptation plans, guidance, or criteria by September 2010 . . .

5. To the extent required by CEQA Guidelines Section 15126.2, all significant state projects, including infrastructure projects, must consider the potential impacts of locating such projects in areas susceptible to hazards resulting from climate change . . .

6. The California Emergency Management Agency (Cal EMA) will collaborate with other State Agencies to assess California's vulnerability to climate change, identify impacts to state assets, and promote climate adaptation/ mitigation awareness through the Hazard Mitigation Web Portal and My Hazards Website as well as other appropriate sites. Special attention will be

paid to the most vulnerable communities impacted by climate change in all studies . . .

7. Using existing research the state should identify key California land and aquatic habitats that could change significantly during this century due to climate change. Based on this identification, the state should develop a plan for expanding existing protected areas or altering land and water management practices to minimize adverse effects from climate change induced phenomena . . .

8. The best long-term strategy to avoid increased health impacts associated with climate change is to ensure communities are healthy to build resilience to increased spread of disease and temperature increases. The California Department of Public Health will develop guidance by September 2010 for use by local health departments and other agencies to assess mitigation and adaptation strategies, which include impacts on vulnerable populations and communities and assessment of cumulative health impacts. This includes assessments of land use, housing and transportation proposals that could impact health, GHG emissions, and community resilience for climate change, such as in the 2008 Senate Bill 375 regarding Sustainable Communities . . .

9. The most effective adaptation strategies relate to short and long-term decisions. Most of these decisions are the responsibility of local community planning entities. As a result, communities with General Plans and Local Coastal Plans should begin, when possible, to amend their plans to assess climate change impacts, identify areas most vulnerable to these impacts, and develop reasonable and rational risk reduction strategies using the CAS as guidance. Every effort will be made to provide tools, such as interactive climate impact maps, to assist in these efforts . . .

10. State fire fighting agencies should begin immediately to include climate change impact information into fire program planning to inform future planning efforts. Enhanced wildfire risk from climate change will likely increase public health and safety risks, property damage, fire suppression and emergency response costs to government, watershed and water quality impacts, and vegetation conversions and habitat fragmentation . . .

11. State agencies should meet projected population growth and increased energy demand with greater energy conservation and an increased use of renewable energy. Renewable energy supplies should be enhanced through the Desert Renewable Energy Conservation Plan that will protect sensitive habitat that will while helping to reach the state goal of having 33 percent of California's energy supply from renewable sources by 2020 . . .

12. Existing and planned climate change research can and should be used for state planning and public outreach purposes; new climate change impact research should be broadened and funded. By September 2010, the California Energy Commission will develop the CalAdapt Web site that will synthesize existing California climate change scenarios and climate impact research and to encourage its use in a way that is beneficial for local decision-makers . . ."

OUTCOMES OF GLOBALIZATION AND CLIMATE CHANGE

Analysis of the information contained in these reports and background literature leads to the conclusion that globalization and climate change may influence civil engineering applications, education, and business on a very large scale. Several possible outcomes globalization and climate change include:

Outcome 1: The competition for resources and financial capital will become greater.

There appears to be a growing backlog of deferred maintenance in the United States and other large economies that has caused or may result in failures. The "American Society of Civil Engineers Infrastructure Report Card" documents this problem and gives the nation's infrastructure a near failing grade. (http://www.infrastructurereportcard.org/)

Searching the Report Card yields the most recent "engineering failures." The root cause of many of these failures is disregard for the engineer's recommendations, including often deferred periodic maintenance. Civil engineers recognize that capital budgeting is necessary and critical to the maintenance of civil structure and/or improvements. However, once the improvement is placed in service, the maintenance budgets are often drawn from general or operating funds. Finance and budgeting staff often ignore Operation and Maintenance (O & M) requirements for infrastructure and remove capital budgeting from the hands of engineers. Frequently, the O & M budgets are lumped into the general, local, or municipal funds to compete with all other general expenditures. There are many competing segments of the public exercising their influence over the budgeting of these general funds, and maintenance often falls to the bottom to satisfy the voting constituency. This situation will likely worsen before it gets better, or we may only see spotty improvements across. Therefore, we believe that competition for resources and financial capital will become greater.

Outcome 2. Civil engineers will see increased emphasis on cost and low maintenance requirements for project goals.

One way to combat maintenance is in the selection, preparation, or original assembly of the civil structure. However, low maintenance options in civil design are often initially more expensive than other options. This competing scenario (higher initial cost versus lower maintenance and overall costs) will require the engineer to skillfully communicate the advantages of various alternatives presented to the client or owner.

The Rise of China and India—Tectonic Economics

"The two countries have one thing in common: their transformations—and the way they will transform the globe—are as stunning as any the world has seen

(Continued)

since America itself emerged onto the world economic stage. The impact can be seen from the falling prices on Wal-Mart's shelves, the rising prices at local gas stations, the shrinking size of many American paychecks, even in the air we breathe. It can be heard in the voices on the end of tech-support phone calls. It is noticeable from the way freighters float low in the waters of the South China Sea because they are so heavily loaded with goods flowing out of new Chinese factories. Most plainly, it can be seen in the raw numbers: India and China have become the fastest-growing big economies on the planet. They look set to stay that way for decades and are on their way to becoming economic giants within a generation . . .

Yet the rise of India and China is about much more than jobs moving overseas: it is about a major shift in post–Cold War geopolitics, about quenching a growing thirst for oil, and about massive environmental change. This is tectonic economics: the rise of India and China has caused the entire earth's economic and political landscape to shift before our eyes."

—Robyn Meredith. (2008). *The Elephant and the Dragon*, pp. 11–13, Norton & Company, New York. ISBN 978-0-393-33193-6 pbk.

Outcome 3. American civil engineers will need to consider potential economic impacts from India and China when serving global clients and pricing competitively.

But India and China not only consume engineering services, they educate young engineers. Both countries' cultures value education and view engineering as a well-respected profession. The sheer size of the populations of these countries and the growing supply of engineers and other labor will be factor in the global economy. The key question is whether these countries will graduate more engineers and other skilled labor in greater numbers that their own countries require.

Outcome 4. Civil engineers will need to consider global climate change and more severe weather in their civil designs.

The U.S. Army Corps of Engineers has a publication that predicts the height of rising oceans in the next 30 years. Many parts of the nation and the world seem to be experiencing greater fluctuations in temperature and rainfall. Civil engineers would be prudent to clearly identify project design criteria that consider these trends in the "basis of design." Having the client or owner approve these climate related design criteria, assumptions, and details as part of the project records also would be prudent.

Outcome 5. Civil engineering and many other technical disciplines will continue to move toward a higher degree of specialization.

A higher degree of complexity in projects seems to be a prevalent condition. This complexity is leading to increased specialization within engineering disciplines. ASCE recognizes this trend, which is evidenced in the *Body of Knowledge 2 (BOK2)* and discussed in Chapter 1. An engineer equipped with a B.S. degree in civil engineering may not be sufficiently prepared for the projected complexity of engineering tasks on many projects in the future.

Accompanying the trend toward increased specialization will be a need for civil engineers to be conversant in advanced computer tools involving simulation and data modeling. Clients, such as the federal government, already are requiring designers to supply building information models (BIM) as part of a project's "as-builts." Contracting methods, like design build and integrated project delivery, and new techniques, like lean construction, are converging with 3-D computer aided-design (CAD), 4-D CAD (3-D plus project scheduling), and 5-D CAD (4-D plus budget/cash flow). (See Chapter 17, Emerging Technologies.)

Outcome 6. Project Management Excellence—Excelling at project management will be a great asset in the 21st century.

If projects seem to be more complex and the competition for resources and financial capital is increasing, then it seems logical that the need to remain on schedule and within budget will become even more critical. As discussed in Chapter 4, Professional Engagement, many clients believe in the "open window" concept where the longer the window (their project) remains open and active the more dollars (their money/budget) fly out. And, in some ways, these clients are correct because most schedule slippages are accompanied by budget increases. Many experienced civil engineers will agree that, depending upon the specific project, labor costs are a very significant component of the overall budget. Each hour the project is extended generally is accompanied by additional labor hours to manage, direct, and construct the project. Therefore, the demand for effective project managers will likely increase and we may also see an increase in innovative project delivery systems such as design build (DB) and integrated project development (IPD), as clients struggle to keep their projects on schedule and within budget.

Outcome 7. Civil engineering must respond proactively to increasingly complex challenges related to public health, safety, and welfare.

The 2001 American Society of Civil Engineers (ASCE) report titled *Engineering the Future of Civil Engineering*, acknowledged that civil engineers will need to respond proactively to increasingly complex challenges related to public health, safety, and welfare. If we review the six outcomes presented above in addition to the data in the previous section titled, Global Climate Change—From a World View and a State Perspective, it is apparent that civil engineers will have a challenging future responding to increased needs for infrastructure; water resources; building design and construction; response to disasters from high winds, floods, and wildfires, among other public health and safety demands.

The following case study looks at ways that a project team responded to scarce resources and financial capital accompanied by an increased emphasis on low cost and very low maintenance. The example considered is the recent construction of Terminal 5 at London's Heathrow airport.

Learning to Project Manage a Mega-Project—The Case of BAA Airports Ltd. and Heathrow Terminal 5

Abstract

This case study examines how BAA implemented a strategic programme of capability building to improve the management of its construction projects, ranging from routine capital projects to a one-off mega-project—Heathrow's Terminal 5 (T5). It focuses on the learning gained from previous projects, individuals and organisations that contributed to the innovative approach used to manage T5 Europe's largest and most complex project. The project is an example of a 'mega-project' because of its scale, complexity and high cost, and its potential to transform the project management practices of the UK construction industry.

   BAA Airports Ltd., (BAA) owner and operator of seven UK airports is one of the largest transport companies in the world. BAA stems originally from British Airports Authority and was purchased by Airport Development and Investment Limited, a company formed and owned by a consortium led by Grupo Ferrovial, a Spanish firm specializing in infrastructure.

Introduction

This chapter examines how BAA—a major construction client— implemented a strategic programme of capability building to improve the management of projects at its airports. These range from routine capital projects to a one-off 'mega-project'—Heathrow's Terminal 5 (T5). The focus is on the learning gained from previous projects, individuals and organisations that contributed to the innovative approach used to manage the T5 project. The T5 project used 'integrated team working' to ensure that safety, time, budget and quality constraints were met. It was completed in March, 2008, on time, within budget and with a high safety record. BAA developed an innovative form of cost-reimbursable contract—the 'T5 Agreement'—under which BAA held all the risks associated with the project, rather than transferring them to external suppliers, and guaranteed a level of profit for suppliers.

Background

BAA is the world's largest commercial operator of airports, responsible for seven UK airports and managing a range of activities in various other airports around the world. It undertakes hundreds of capital projects as part of its ongoing operations, as well one-off mega-projects, such as the design and construction of Stansted Airport and Heathrow's T5. In 1994 BAA's capital projects programme was running at £500 million but it was also planning a series of major projects, including the development of T5 (the largest project BAA has undertaken). This ambitious capital projects programme had to be undertaken against a background of steadily rising costs of building, while the charges they could levy on their building occupants were rising at a lower rate since they were subject to regulation. The T5 project was one of Europe's largest and most complex projects. It is an example of a mega-project (Flyvbjerg et al., 2003[i]) because of its scale, complexity and high cost and its potential to transform the project management practices of the UK construction industry. It was broken down into 16 major projects and 147 sub-projects. At any one time the project employed up to 6000 workers, and as many as 60,000 people were involved in the project over its lifetime. The goal of the project was to increase the airport's capacity of 67 million passengers a year to 95 million.

Towards Developing Repeatability and Predictability in Construction

Based on his previous experience in the car industry, Sir John Egan, BAA's Chairman during the early 1990s recognised that BAA could make radical improvements to the way it traditionally delivered projects. Egan wanted to emulate the continuous improvements in performance achieved in car manufacturing, by creating an orderly, predictable and replicable approach to project delivery. Egan brought in experienced people from outside BAA to spearhead this strategy—people who had worked on massive projects for demanding and sophisticated clients in other sectors.

This new thinking was embodied in a UK government-sponsored report called Rethinking Construction (1998) authored by Sir John Egan, which became a manifesto for the transformation of the UK construction industry. The report argued that the client could play a role in this transformation by abandoning competitive tendering and embracing long-term partnerships with suppliers, based on clear measures of performance. Partnerships would provide a continuous stream of work over an extended period and the stable environment that contractors needed to achieve systematic improvements in the quality and efficiency of their processes. The report recommended that firms focus more strongly on customer needs, integrating processes and teams, and on quality rather than cost.

(Continued)

Under Egan's guidance, BAA's senior managers began applying the principles laid out in the Egan Report to improve BAA's project processes and relationships with suppliers, and by the late 1990s BAA had made significant improvements to its project execution capabilities, reflected in a greater degree of predictability in terms of time, cost and quality.

BAA's first phase of moving towards being able to deliver T5 centred around trying to get more predictability in terms of time and cost in its projects. There were four key strands to this:

- Developing new and improved project processes
- Managing the supply chain
- Standardisation and prefabrication
- Integrated team working

Project Process Improvement

BAA's desire to improve its project processes was led by its CIPP (continuous improvement of the project process) programme—to develop a new process for organising projects in its capital investment programme—which was introduced in 1995. It aimed to establish a consistent best practice process applied to all projects with a value of over £250,000. The process was designed around a typical £15 million building project but it was expected that it could be used regardless of the size of the project right across BAA's business. A taskforce was set up, with project representatives from all parts of the group, with the aim of capturing all the best parts of existing practice and creating a single system. More than 300 people were consulted over a period of 18 months, both from within BAA and from other companies and industries. The issuing of the CIPP handbook was accompanied by a long-term training programme to provide a firm basis of understanding for implementation.

The CIPP handbook laid down a set of key policies or principles—safe projects, a consistent process, design standards, standard components, framework agreements (see below), concurrent engineering and pre-planning—which all capital projects had to adhere to. It provided a template for the organisation of BAA projects and outlined a seven-stage process covering the project life cycle, from inception through to operation and maintenance. Each stage included a series of checkpoints which had to be completed and a series of evaluation gateways, where the project was assessed by an evaluation team before going to approval gateways for sign-off from local and/or group capital project committees. To successfully pass through a gateway, eight key sub-processes needed to be managed and the outputs from each co-ordinated: development management, evaluation and approval, design management, cost management, procurement management, health and safety, implementation and control, and commission and handover. BAA developed a process map showing each stage along the top and the sub-processes within each stage, outputs and gateways.

Managing the Supply Chain: The Framework Programme

At the same time as it was developing the CIPP process, BAA started the development of what it called the Framework Programme to work with a number of preferred suppliers on an ongoing basis. Up to that point, every time BAA embarked on a capital project it went to tender and through a whole process of qualifying, with the result that everyone had to go up the learning curve every time. The five-year framework agreement provided suppliers with an opportunity to learn and to make continuous improvements year by year that would benefit both BAA and the supplier.

This was first attempted in 1993–1994 and subsequently became widely used. Framework agreements were not restricted to first-tier suppliers—they encompassed a wide range of services, including specialist services, consultancy services (design and engineering), construction services, etc. Each agreement was structured slightly differently, depending on the nature of the service, but the concept was applied consistently. Standardisation of components helps to reduce unit costs, but the framework agreements also incentivise the suppliers to improve the products and performance. For example, in the case of lifts and escalators, the installation contractors went to BAA and would analyse the process, saying that if they did it like this they could get the lift installed quicker and this would save time and money.

In 1998 BAA still had as many as 23,000 suppliers working throughout the company, and each of its seven UK airports had developed its own unique approach to supply chain management. BAA recruited Tony Douglas as Group Supply Chain Director and tasked him with profiling and reducing BAA's number of suppliers to a much more manageable level (Douglas, 2002[ii]). He was able to draw upon his own experience in the car industry and many studies of electronics, aerospace and other industries that had already developed sophisticated approaches to supply chain management. Strong internal capability had to be developed so that BAA could better manage its external suppliers.

By 2002 BAA had developed a second generation of framework agreements to achieve more accurate project costs, to implement best practice and to work with suppliers in longer-term partnerships. The key difference between first- and second-generation framework agreements is that the latter were devised to source the best-in-class capability and were valid for a period of 10 years rather

(Continued)

than five. BAA developed strict criteria to select the best partner for each project. All aspects of its business are examined, including all its systems and processes covering quality, people, its supply chain, finance, R&D, and business development.

Under the second-generation agreements, suppliers worked with BAA in integrated project teams to cultivate close co-operation, to leverage the right expertise needed for specific projects and to reduce costs. BAA injected commercial rigour into its second-generation framework agreements by creating an annual review to measure the supplier's performance against projects that it delivers. If the supplier failed to perform, as measured against an agreed set of performance criteria, BAA set an improvement plan. A supplier would remain in BAA's family of framework suppliers if it achieved the targets set in the improvement plan. If it failed, it was deselected and replaced by another firm from the list of framework suppliers.

Standardisation and Prefabrication

Several projects carried out at BAA prior to the T5 project were experiments in predictability and repeatability. CIPP was applied to four different clusters of BAA products: car parks, pavements, baggage handling and buildings. For example, an early attempt to create predictability in BAA's project delivery was the design of standardised BAA office products. Standardised design was first developed for three offices at BAA's World Business Centres in Heathrow, which were built in succession. Previously each office would have been dealt with as a separate project, starting from scratch each time. Under the new approach BAA designed and built the first one and, rather than starting from scratch again with a new design, decided to replicate the first on a site next door but with a target of reducing the cost by 10% and the time by 15%. BAA then built a third to the same design with similar targets for time and cost reduction over the second one.

The design was then used at Gatwick Airport and reused at lower cost for Stansted Airport's second satellite office buildings. For the Stansted project 90% of the team from Gatwick was retained. The frame that had been used for the Gatwick project had had some problems related to component fit and watertightness, but these were overcome at Stansted, where the frame utilised pre-cast columns with cast-in slab/column connectors (patented by Laing O'Rourke plc). Prefabrication of the external stairs in entire floor height modules was another feature. BAA reaped significant learning-curve cost advantages from this approach to product standardisation. The above-ground construction time was reduced by five weeks, saving around £250,000.[iii]

Involvement in framework agreements on a long-term basis has enabled other suppliers to develop repeatable processes and improvements in productivity over time, as lessons have been learned and shared. For example, AMEC—which has framework agreements covering building services, airfield pavements, aircraft stand services and general consultancy to support the development and upgrade of airport infrastructure across the UK—claims that in excess of 100 projects have been delivered on time with savings of 30% achieved overall.[iv]

Integrated Team Working

One of the central features of the BAA approach was integrated team working. The Heathrow Express tunnel collapse in 1994 played a key role in the development of this form of partnering. BAA had contracted Balfour Beatty as the main contractor for the construction of a fast passenger link between Heathrow and London, the Heathrow Express. Balfour Beatty had been chosen because they had just completed the Channel Tunnel and a crucial engineering concern was the tunnels needed within the airport.

The collapse of the Heathrow Express tunnel put the entire £440 million project in jeopardy. The normal construction industry response would have been to go down the litigation route and place the blame on the main contractor, Balfour Beatty. Instead BAA chose to work together with Balfour Beatty and the other main parties as partners to resolve the situation. A new contract was drawn up between BAA and Balfour Beatty, which established what became known as 'the single team', in which all the parties worked as one team instead of a collection of separate project groups. This integrated team, which also included Mott MacDonald, succeeded in delivering the new project in June 1998—only six months behind the original schedule.

BAA appointed a new Construction Director for the recovery project, who demonstrated and promoted a culture of co-operation instead of rivalry. He employed:

> 'a team of specialist change facilitators and behavioural coaches to work with the team on changing the tradition of adversarial working in order to make the single team "statement of intent" a reality. This was demonstrated in the everyday behaviour of the construction team.'[v]

The success of the Heathrow Express recovery project showed how partnering and trust could be made to work.[vi]

At one point Heathrow Express was 24 months behind programme, but eventually became operational only nine months after the original projected date. It was a huge success story within BAA. Once finished, the Construction Director recruited to work for BAA on that Heathrow Express recovery project was transferred to the central BAA capital projects team as Group Construction Director.

(Continued)

Much of the learning was channelled into developing the CIPP and applying the integrated team working concept to different clusters: the pavement team (for runways, taxiways, links and resurfacing); the baggage handling team (clusters of suppliers); and a cluster of suppliers for buildings (shell and core, fit-out, etc.). Teamwork was mentioned as a major success factor in the Terminal 1 International Arrivals concourse refurbishment project, which we studied in 1999. There it was claimed that teamwork had been excellent, both at the Heathrow Airport Limited level and through to construction activities where the co-location of the team provided huge benefits. It was also noted that the team members 'left their companies at the door' when they came to work on the project.

Preparing for T5: Creating the T5 Approach

The steps that BAA had taken up to this point—the development of the CIPP, the first- and second-generation framework agreements, and integrated team working—had improved project predictability and repeatability. However, a more radical approach was needed to deliver T5, where there was a much higher level of uncertainty involved. T5 had to be constructed while causing minimal disruption to the operations of Heathrow Airport.

Between 2000 and 2002 BAA carried out an in-depth analysis of every major UK construction project in the last 10 years (valued at over £1 billion) and every international airport that had been opened over the last 15 years. This analysis showed that:

- No single UK construction project of that size had been delivered on time, on budget, safely and to the quality standards that had originally been determined, and
- Not a single international airport had worked properly on day one.

Based on this analysis BAA predicted that T5 would be 18 to 24 months late, over budget by a £1 billion, and six people would be killed during its construction.[vii] The airport case studies showed that it would take three years to build up to moving 30 million passengers a year through the new terminal. BAA expects T5 to do that in its first year of operation, since the passengers are already there in Terminals 1 and 4.

Following the public inquiry, severe restrictions were placed on traffic volumes and routes outside the site and there was only one viable entrance to the site. According to Tony Douglas, Managing Director of the T5 project:

'. . . if we were to build it conventionally, not only would we require about 7000 additional workers on site, but there would be a 40ft vehicle load passing through the gate every 40 seconds or so for the next four years. . . . The only possible solution is to develop standard solutions that include pre-assembled products, manufactured off site, then assembling at the airport with the minimum of human intervention. This also brings environmental benefits by reducing the impact of thousands of construction workers on local infrastructures and, as everybody recognises, safety performance is infinitely better in factory conditions than on a construction site.'[viii]

There was recognition in BAA that the only way to deliver T5 was to change 'the rules of the game' by creating a set of behaviours that allow people to be constructive and come up with innovative solutions to problems. The success of the Heathrow Express recovery project demonstrated that such an approach was viable. The T5 approach combined two main principles: the client always bears the risk; and partners are worth more than suppliers. These principles were embodied in the T5 Agreement—a new form of contract developed by BAA which guaranteed suppliers' profits while the client retains the risk.

The T5 Agreement provided an appropriate environment for integrated team working. Rather than transfer risks to its suppliers, BAA assumed responsibility for all project risks all the time and worked with its integrated team members to solve problems encountered during the project. The agreement included an incentive payment if a supplier achieved exceptional performance. This was designed to enable suppliers to work effectively as a part of an integrated team and focus on meeting the project's objectives not only in relation to the traditional time, budget and quality measures, but also in relation to safety and environmental targets.

BAA decided to adopt this approach since traditional liabilities—such as negligence, defective workmanship and the like—are extremely difficult to prove in an integrated team environment. BAA recognised that if suppliers were made jointly responsible for running significantly over budget then this would probably put them out of business. It decided to reimburse the costs of delivery and to reward exceptional performance and penalise inferior performance only in terms of profitability.

BAA's approach to the delivery of T5 has resulted in a number of innovations and examples of best practice in a number of areas (NAO, 2005).[ix]

- The contract form—the T5 Agreement is a cost-reimbursable contract, in contrast to the fixed-price contracts generally seen on large projects.
- Risk management—the client takes on the risk so that project management becomes a tool of risk management rather than vice versa.

(Continued)

- Sponsorship and leadership—the T5 project managing director is an executive main board member and receives regular and overt support from other board members.
- Logistics management—the constraints on access to the site have led to:
  - a focus on pre-assembly and off-site fabrication and testing wherever possible
  - the creation of two consolidation centres: the Colnbrook Logistics Centre—consisting of a railhead for the supply of all bulk materials, a factory for the prefabrication and assembly of rebar, and a laydown area—and the Heathrow South Logistics Centre for the preassembly of pile cages and later as an area for assembly of materials into work packages ready to deliver on site
  - the use of 'demand fulfilment software' designed to pull materials on site on a just-in-time basis.
- Insurance—BAA took out insurance for the whole project, lowering costs and avoiding unnecessary duplication.
- The approval process—a five-stage approval process based on the changing level of risk, which enables the project to move forward to the next stage without completing production design.
- Teamwork—the T5 Agreement incorporates integrated teams working to a common set of objectives and based on a capability approach, so that the team for each sub-project is assembled with the best possible expertise within the partner firms.

T5 opened on 27 March 2008, it was delivered on schedule and within budget. It was a major breakthrough in project management as it avoided the trend of all similar mega-projects being delivered late, over budget and often below quality expectations.

References

i. Flyvbjerg, B., Bruzelius, N. and Rothengatter, W. (2003), *Megaprojects and Risk: An Anatomy of Ambition*, Cambridge: Cambridge University Press.

ii. Douglas, T. (2002), 'Talking about supply chains', *Solutions: Projects and News*, WSP Group plc, Spring Issue, 5.

iii. Source: 'BAA Lynton roll out Mark 3', May 2000, www.m4i.org.uk.

iv. Source: 'Upgrading airports in partnership with BAA', www.amec.com.

v. Lownds, S. (1998), 'Management of change: building the Heathrow Express. Leveraging team skills to get a railway business rolling—the story of the change of culture on the construction of the Heathrow Express Railway', presented to Transport Economists' Group, University of Westminster, 25 November.

vi. 'Case Study 2 in Project Recovery: breaking the cycle of failure', Major Projects Association seminar, London, June 2000.

vii. National Audit Office (2005), 'Case studies: improving public services through better construction', 15 March.

viii. Douglas, T. (2002), 'Talking about supply chains', *Solutions: Projects and News*, WSP Group plc, Spring Issue, 5.

ix. National Audit Office (2005), 'Case studies: improving public services through better construction', 15 March.

—Dr. Tim Brady, Principal Research Fellow, CENTRIM,
University of Brighton, UK

Additional insights on practicing civil engineering on an international basis are offered in the article titled, "Civil Engineering Practice-A Wider Community View." Several approaches are presented that may spark the imagination of those civil engineers wishing to become involved with civil engineering beyond the borders of their own discipline or nation.

Civil Engineering Practice—A Wider Community Viewpoint

Introduction

When qualified with their degree, civil engineers tend to concentrate on one particular branch of the many within the discipline (at a time, perhaps), learning more about it and then practicing within that arena. The wider vision can be exciting, however, as the practicing civil engineer develops and becomes more involved in the wider built environment community as well as the wider community as a whole. This can include clients and the public who use that infrastructure, for example, as that essential broader scope develops.

These growing interactions and experiences can include many enriching aspects in a developing professional life. Some thoughts on examples of international and pan-discipline working can be considered in groupings such as those listed below:

- Forensic Engineering—A Worldwide Community
- Engineering Disciplines—The Similarity of Focus
- Working with a Range of Other Disciplines
- Risk Management—How Safe?
- Professional Communications—Bridging Cultural and National Practices
- Qualifying to Practice—Some Different Approaches
- Codes of Practice—*Sans Frontière*
- Conclusion—Continuing Professional Development: A State of Mind for Conscious Thinking

Forensic Engineering—A Worldwide Community

There is plenty of evidence to support the saying that travel broadens the mind and this is confirmed by the personal experience of many people. As an example, a visit to the American Society of Civil Engineers[1] (ASCE) Convention in Minneapolis in 1997 proved to be an inspiration because, among other things, the first congress on Forensic Engineering[2] was being held in parallel. Although not known at the time, this event stimulated the growth of a wider community of people interested in promulgating the understanding of both the fundamental reasons and also the wider reasons (or context) for our built environment sometimes not behaving as it was thought it would—or indeed as it was thought that it should. In other words, why did performance sometimes fall short of expectations to varying degrees, including those of the client and also of the designer? Occasionally that poor performance has proved to be catastrophic and caused loss of life.

The following year, a conference on Forensic Engineering[3] was held in London. This was organized by the U.K.-based Institution of Civil Engineers (ICE) in association with other relevant professional disciplines. It attracted wide interest, including from the BBC's World Service where it was featured in a radio program. An international series of conferences was thus born where, with the support of ASCE, it has helped to widen the worldwide family of civil engineers and other disciplines. In the more recent conferences, papers have been received

---

[1] www.asce.org/

[2] Rens, Kevin L., Editor. *Forensic Engineering: Proceedings of the First Congress.* ASCE. Reston, USA, 1997

[3] Neale, B.S., Editor. *Forensic Engineering—a professional approach to investigation*, Proceedings of the Institution of Civil Engineers First International Conference, 28–29 September 1998. Thomas Telford Ltd, London. 1999.

from all five continents of the world. More Forensic Engineering conferences have been held elsewhere in the world, for example, in Taiwan, India, Italy, and Spain. Conference proceedings[4,5,6] capture this knowledge base and thus the performance of facilities in ways that were as not intended. Building on the 1997 congress, a second one[7] was held in 2000 which established them as series, also, which has continued on a three-yearly basis since. Moreover, ASCE produces a publication—"The Journal of Performance of Constructed Facilities"—that builds on the success of that venture by their Technical Council for Forensic Engineering between congresses to continue the flow and exchange of information with regular publication dates.

The two series of Forensic Engineering conferences mentioned above have enjoyed continued support as they have continued to develop both in the USA and in the U.K., where ASCE and ICE, respectively, have continued to organize them with significant work and support from their members and many others. There is an international culture to "give something back" to the profession and thus the community as a whole.

Engineering Disciplines—The Similarity of Focus

When one examines catastrophic events that receive worldwide publicity and thus attention, original perceptions may change. Those events can include both "natural" events and "man made" ones. Here it is possible to think of examples of earthquakes and perhaps incidents in chemical works, respectively, where each type of event can (and has) led to significant loss of life. The international community pulls together to help where they feel they can—and learn for the future and thus future generations. Many engineering disciplines come into play including a number of specialist elements where civil and structural engineering are to the fore in respect of the all-important residual structural stability during and after rescue operations, for example.

Investigators in this context tend to be professional engineers, usually civil or structural, although chemical, electrical, fire, control systems and

---

[4] Neale, Brian S., Editor. *Forensic Engineering—from failure to understanding.* Proceedings of the Institution of Civil Engineers Fourth International Conference, 2–4 December 2008. Thomas Telford Ltd, London. 2009.

[5] Neale, B.S., Editor. *Forensic engineering—diagnosing failures and solving problems.* Proceedings of the Institution of Civil Engineers Third International Conference, 10–11 November 2005. Taylor and Francis, London. 2005.

[6] Neale, B.S., Editor. *Forensic engineering—learning from failures,* Proceedings of the Institution of Civil Engineers Second International Conference, 12–13 November 2001. Thomas Telford Ltd, London. 2001.

[7] Rens, Kevin L.; Rendon-Herrero, Oswald; Bosela, Paul A.; Editors. *Forensic Engineering: Proceedings of the Second Congress.* ASCE, Reston, USA. 2000.

(Continued)

mechanical engineers share similar interests, for example, as well as professions such as metallurgists and materials specialists. The fascinating behavioral study of people and why we humans do things or omit to do things in a particular way, or in a particular order or time, becomes more understandable when listening to the input from ergonomists and psychologists, for example, who can and do contribute.

We thus have community which can be considered rich in diversity in ways such as:

- Many engineering disciplines
- Disciplines other than engineering
- National and international variations in culture—and thus possible approaches

Working with a Range of Other Disciplines

Working in Forensic Engineering, for example, often means interacting with a range of other disciplines. Hence, the experiences of investigators, clients, law enforcers, lawyers, code writers, loss adjusters, researchers, and teachers—to name just some—have been shared to help disseminate these global experiences of performance and ''knock-on'' activities. The rich mix of disciplines can help inspire by cross-fertilization of ideas with pan-discipline cultures. This compliments the, albeit, expected inspiration from experiencing the variety of national cultures with, perhaps, some different approaches, as mentioned above.

A further example of this nexus can be seen in an organization based in Europe that was set up following major incidents worldwide with the objective of acting as a forum, primarily for engineers, to exchange views and discuss significant issues at a senior level to help contribute to prevention of future occurrences of such incidents. A major strength of this body, the Hazards Forum,[8] is its rich diversity of members and contributors. The focus is not Forensic Engineering, as such, as it focuses on future developments and management, including the avoidance of—or appropriate mitigation or amelioration of hazards—and thus potential risks, particularly pan-discipline. To look at the future, however, there is a need to examine the catalogue and pathology of past events and these aspects are included as appropriate.

For example, the Hazards Forum has looked at the next generation of new nuclear power station design issues; risks with some alternate fuels; and carbon capture in a particular series. To show the variety of topics presented and discussed, however, the event immediately preceding that ''Energy series'' was

---

[8] www.hazardsforum.org.uk

from the medical sector and from an ergonomics point of view. Whereas this might have been perceived as near the outer bounds of the remit of the forum and of little interest to engineers, discussions centered on the ease with which prescriptions could be (and are) sometimes incorrectly administered through people error. A solution offered that significantly helped was better labeling and packaging in more than one way—perhaps not surprising. One speaker presented the way forward in the form of a hospital direction sign to the various internal departments where the strap line for a better way forward could simply be—Safety by better design. Does this sound familiar to the construction industry and to civil engineers? If not, perhaps every civil engineer should consider "wearing this on their sleeve" as one fundamental maxim with which to undertake their personal practice. The sign, for further thought, is reproduced in the following figure.

A Wider Community Viewpoint
(Acknowledgment: Professor Peter Buckle, Robens Centre for Public Health, University of Surrey from Hazards Forum Newsletter No. 64, Report of Seminar How ergonomics improves patient safety. [Peter Buckle, p.buckle@surrey.ac.uk and www.hazardsforum.org.uk/publications/publications_newsletters.asp])

Risk Management—How Safe?

This brings us into the realms of a discussion on hazards, risk, and how to approach this and which approaches for the assessment of hazards and risk would be better for which situations. For example, international variations include:

- The principles of the risk hierarchy[9]
- The precautionary principal[10]

---

[9] www.hse.gov.uk/
[10] Precautionary Principal reference - http://europa.eu/legislation_summaries/consumers/consumer_safety/l32042_en.htm and http://www.jncc.gov.uk/page-1575

(Continued)

- As low as reasonably practical (ALARP)
- So far as is reasonably practical (SOFARP)
- The traffic signals system (red, amber, green)
- Others . . .

There will be various tools that can be used within assessments, some of which will include tools for analysis, a word sometimes used where "assessment" would be more appropriate and therefore better advised. Where analysis takes place the users need to be very aware of the accuracy and significance of any data used—and of course of the significance of their outputs.

National enforcers can, of course, contribute to the discussion on optimum approaches to ameliorate hazards and risks, although international influences are often incorporated in various ways. National regulations may also affect the approach, such as Building Regulations which may have requirements to resist disproportionate collapse, such as in the U.K.[11] Other legislations, such as for occupational health and safety and the safety of the public as a result of those activities, can affect the construction sector, including designers where they may become subject to criminal law—again, such as in the U.K.[12] The risk creators have much guidance available to them, as do other duty holders.

Professional Communications—Bridging Cultural and National Practices

We can learn so much from others, therefore, if we open up our minds to other cultures and ways of doing things. This can include, for example, the international civil engineering community and still further in the wider engineering professional community—and others. When doing this, however, we need to ensure that our communications are effective, and thus accurate. Do we know how successful we are at this?

Communications is an essential part of a professional's job and one that sometimes achieves less than 100 percent success. We know from studies outside the engineering community the verbal communication can be less than 10 percent of that which happens in face-to-face encounters, partly depending on whether it is a social interaction or a business one—apparently. There are, of course, many tools that engineers use for communication, but how often do we know that those with whom we communicate have exactly the idea of what we

---

[11] www.communities.gov.uk/planningandbuilding/buildingregulations/
[12] www.hse.gov.uk

expect to happen, that is, the "correct" idea? Many examples of poor performance and other failures have been shown to be the result of "misunderstandings."

In one's professional practice, how often do we ensure that ideas have been communicated accurately, bearing in mind that there are two (or maybe more) parties to any communication. We may be satisfied with what is transmitted, but what assurance do we have that the receiver has it correctly and if not, what margin of error would be acceptable? Perhaps none! Well documented fatalities have occurred around the world because of inaccurate communications—or misunderstandings. Part of communication relies on particular competencies and perhaps assumptions about competencies of people and organizations, although it has been said that nothing should be assumed. Within parts of occupational health and safety legislation in Europe there are requirements for the competence of those involved, both organizations and individuals, which are in criminal legislation. This is generally seen as not unreasonable. How does one define competence, however, and even more difficult, how does one measure appropriate competence for the numerous tasks that are undertaken in creating and maintaining the built environment and not forgetting the removing, reusing, and recycling stages also.

Defining competence has been seen described as problematic, but tends to include a combination of relevant and appropriate education, training, and experience as a start—although there are variants. Each of those elements then needs to be expanded for effective application by both those attempting to demonstrate their competence and also by those seeking to be convinced of those relevant competencies, of course.

Mentioning Europe above brings to mind another communications thought—how many continents are there in the world? When next with a group of people, perhaps a group that you may not normally be with, it can be interesting to ask them—and to note the variations in the answer. We may be used to America being divided into two—north and south, with the Arctic and Antarctic adding a further two. These would give more than the basic five. We may be used to subdivisions also, such as the Indian sub-continent. One example that might take many people by surprise is the combining of two continents rather than splitting down into smaller sets as mentioned above, such as the word "Eurasia," which is sometimes used! This is another communications example of one person's knowledge being outside the scope of another's—or is that competence? How careful we need to be!

Qualifying to Practice—Some Different Approaches for Competency

Approaches to professional practice vary across the globe where, for example, in the USA engineers are licensed to practice on a state by state basis and on

(Continued)

defined or particular aspects. This can be seen as a qualification of competence—for the period of the license which will need to be renewed and perhaps altered to suit developments, new requirements (perhaps in legislation), or interests of the applicant, for example. The need for continuing professional development is seen as an essential component for practicing engineers. This is seen as applicable for professional engineers across the world, although the basis and way in which the state gets involved varies. As an example there is not a renewable type of PE qualification in the U.K., but a CEng award. CEng is the recognized shortened form for Chartered Engineer which is a qualification awarded by the Engineering Council UK[13] (EC_{UK}) under a Royal Charter which recognizes a level of competence gained—at a particular time. That level is tested by a professional institution such as the U.K.-based Institution of Civil Engineers for corporate membership of that body.

To achieve success a candidate must have appropriate academic qualifications from an accredited university together with appropriate post-graduate experience, where the latter is the particular focus of the examination for corporate membership. This can result in some people with extensive experience being tempted to apply to a number of institutions which can result in a multitude of letters after a person's name—a system which some cultures find bemusing! To return to CEng, however, that is a one-off award and is usually retained for the whole of a career. To put this into context, however, these engineers are expected to keep up with current relevant developments appropriate to their personal work area through recognized Continuing Professional Development (CPD). To add another overlay of a difference in practices, the Engineer of Record system in the USA is not a system practiced in the U.K.,[14] as such.

Hence, although there are many similarities between, for example the ASCE and the ICE, one fundamental difference is that ICE is a qualifying body for its members, with most engineers who qualify choosing to apply for the CEng award that corporate membership allows them.

To go to another part of the world, in Taiwan, the Taipei Professional Civil Engineers Association[15] states in an introduction to the Association that "The Association is an assembly of professionals and is responsible for upholding the legal rights and interests of all its members and is on hand at all times to provide various engineering services for the government, the engineering community and

---

[13] www.engc.org.uk/

[14] Kardon, Joshua B. *The responsible engineer: the concept of the "Engineer of record"*. Proceedings of the Institution of Civil Engineers Fourth International Conference, 2–4 December 2008; Neale, Brian S., Editor. *Forensic Engineering—from failure to understanding*. Thomas Telford Ltd, London. 2009.

[15] www.tpcea.tw

society at large.'' This is perhaps another variation in the model for professional civil engineers.

Codes of Practice—*Sans Frontière*

Codes of practice, or standards, can be an interesting area for example of sharing experiences and approaches (especially across national boundaries) where many options will be explored before deciding on a common, or agreed path to follow and adopt. They are therefore a uniting force across cultures for both users and drafters. As national standards organizations look more and more to unified codes on a global or world region basis, others look more to having their own codes adopted more widely among their member-ship communities. The existence of ISO[16] (International Organization for Standardization) codes tends to be well known, with many being adopted as both world region and also as national standards. As an example, in Europe and the U.K., there are now standards with the nomenclature BS EN ISO XXXX, which can be seen as a British Standard EuroNorm International Stan-dard! There may also be another version,[17] for example.

How are these achieved? An example close to civil engineers is the Structural Eurocodes produced for CEN[18] (European Committee for Standard-ization), the European standards-making body. The number of states involved is twenty plus, with almost as many languages and of course, many diverse cultures with a very wide variation in weather conditions. How does one move forward to establish agreement on codes when often a member state will have thought that it had produced the ideal code for it own purposes, for example. Perhaps above all, how does one move forward with the plethora of languages in use?

There are just three official languages—German, French, and English. The official versions of each code are produced in those three languages with other national versions produced by member states in other languages as required from that base. To prepare these codes, however, they are drafted, discussed, devel-oped, and finalized in one language. That working language was English and was used even outside the meeting caucus where knowledge of further languages may still not offer a compatible mode of communication. An interesting side effect of being a natural English speaker in a community where English is the working language but not the home language of most, there was, on returning to the U.K., a short period of readjustment that was required to help to re-order words as usually spoken! Another positive spin-off was that the English used in

---

[16] www.iso.org/
[17] BS ISO 13824:2009 General principles on risk assessment of systems involving structures.
[18] www.cen.eu/cenorm/homepage.htm

(Continued)

those CEN meeting tends to be more precise than used at home and thus aided better communications, it is thought. The following example is given from a "Loads" code[19]:

- Climatic—instead of environment for weather conditions
- Actions—instead of loadings
- Execution—instead of construction which can include demolition, for example

The wider context for this code is the head code[20] for the actions series. There are codes, however, that do not as yet come under the ambit of CEN and whereas member states are required to withdraw codes that are covered by equivalent CEN codes, they are free to develop or maintain those national codes that are not. An example[21] in the U.K. is one that includes sustainability issues with respect to environmental consideration and the broad issue of demolition, partial demolition, decommissioning, and structural refurbishment where a number of options can be considered on a broad risk management approach. A performance-based code for demolition activities has proved advantageous and is one used in other countries too. The code also includes planning ahead for ways of dealing with waste streams where considerations such as re-use and re-cycling are helpful in establishing the materials that are to be removed before planning the works. The Demolition Protocol[22] can help. Minimizing waste during construction is also helped by planning ahead, of course.

Hazard assessments and risk assessments are part of a way of civil engineering life with assessments for accidental actions becoming more relevant. There is a Structural Eurocode[23] on the topic that may be of interest to those considering accidental actions.

---

[19] BS EN 1991-1-6:2005 Actions on structures—Actions during execution (A part of Eurocode 1). BSI, London. 2005. To use standard the appropriate National Annexe is required, such as for the UK - BS EN 1991-1-6:2008 National Annex for Actions on structures—Actions during execution.

[20] BS EN 1990:2002 Eurocode—Basis of structural design. Eurocode 1. BSI, London. 2002.

[21] BS 6187:2000 Code of practice for demolition. BSI, London. 2000. To be updated by 201? version.

[22] ICE Demolition Protocol 2nd Edition. ICE London 2008.

[23] BS EN 1991-1-7:2008 Actions on structures—Accidental actions during execution (A part of Eurocode 1). BSI, London. 2008. To use standard the appropriate National Annex is required, such as for the UK - BS EN 1991-1-7:2008 National Annex for Actions on structures—Accidental actions.

Conclusion

Are we alone as civil engineers? The answer is a resounding No!, of course. It is good to pause occasionally and think of the wider (worldwide) community in which we live, practice, and develop. It could be said that we need to keep developing and expanding our menu of experiences for the greater good of our profession and the global community—as well as ourselves, perhaps. This may lead us to a performance-based approach to engineering solutions that can both stimulate thought and inspire innovation—when considered with an open mind and with forward thinking which anticipates future demands and needs.

Hence, if continuing professional development (or continuing education and training) becomes a way of thinking for personal development there are many optimistic avenues to pursue, some of which are mentioned above.

—Brian S. Neale CEng, FICE, FIStructE, Hon FIDE
Cheshire, UK
brian.s.neale@virgin.net

SUMMARY

There has been a "global explosion" of technical achievements in the last 200 years since the initial age of industrialization and civil engineers have been involved in many of these achievements. The growth and maturity of the profession has been astounding and complementary.

Judging from the potential impacts of the four broad processes (depicted in Figure 15.1) that place more choices in the hands of individual consumers and organizational customers, we can expect a "globalization cascade" with reinforcing feedback loops that strengthen and accelerate globalizing forces. These forces, including global climate change, will have dramatic effects on the civil engineering profession with anticipated demand for engineers, innovation, client service, and decentralized centers of expertise and influence.

Hot, Flat, and Crowded

"We are the first generation of Americans in the Energy-Climate Era. And what we do about the challenges of energy and climate, conservation and preservation, will tell our kids who we really are. After all is said and done, I am still an optimist that we will rise to this challenge. I am certain that my children and grandchildren will live in a cleaner world and a safer world and a more sustainable world. Why? Because technology today is allowing us to connect and leverage more and more

(Continued)

brainpower than ever before. Whole swaths of the world that really could not collaborate in solving problems are being brought into the discussion. That is hugely important and the reason that I believe we will work this out – we will learn as nations and individuals that we cannot afford to grow the old-fashioned way – by just mining the global commons and by thinking that the universe and nature revolve around us, and not the other way around.''

—Thomas L. Friedman. (2009). *Hot, Flat, and Crowded*, p. 474.

In addition, seven possible outcomes of globalization and climate change are presented for the reader's consideration, including:

Outcome 1: The competition for resources and financial capital will become greater.

Outcome 2: Civil engineers will see increased emphasis on cost and low maintenance requirements as project goals.

Outcome 3: American civil engineers will need to consider potential economic impacts of India and China when serving global clients and pricing competitively.

Outcome 4: Civil engineers will need to consider global climate change and more severe weather in their civil designs.

Outcome 5: Engineering and many other technical disciplines will continue to move toward a higher degree of specialization.

Outcome 6: Project management excellence—excelling at project management—will be a great asset in the 21st century.

Outcome 7: Civil engineering must respond proactively to increasingly complex challenges related to public health, safety, and welfare.

REFERENCES

Friedman, Thomas L. (2009). *Hot, Flat, and Crowded: Why We Need a Green Revolution—and How It Can Renew America*. Picador/Farrar, Straus and Giroux, New York. ISBN-13: 978-0-312-42892-1, ISBN-10: 978-0-312-42892-2.

Kanter, Rosebeth Moss. (1995). *World Class: Thriving Locally in the Global Economy*. Simon & Schuster. New York.

Meredith, Robyn. (2008). *The Elephant and the Dragon*. Norton & Company, New York. ISBN: 978-0-393-33193-6.

Ritzer, George. (2010). *Globalization: A Basic Text*. Wiley-Blackwell. West Sussex, United Kingdom. ISBN: 978-1-4051-3271-8.

Steger, Manfred B. (2009). *Globalization—A Very Short Introduction*. Oxford University Press, New York. ISBN: 978-0-199-55226-9.

Introduction
Ethics
Professional Engagement
1 2 3 4 5 6 7 8 9 10 11 12 13 14 15 16 17 A B C D E F
History
What Engineers Deliver
Emerging Technology
Engineer's Role in Project
Development
Permitting
Leadership Managing
Having
a Life
Executing a Professional
Commission
Client Relationship
Legal Aspects
Communicating
Sustainability
Globalization

Chapter **16**

Sustainability

Big Idea

As the global population grows and standards of living improve, there will be increasing stress on the world's limited resources. Thus, engineers of the future will be asked to use the Earth's resources more efficiently and produce less waste, while at the same time satisfying an ever-increasing demand for goods and services.

> "Knowledge of the principles of sustainability, and their expression in engineering practice, is required of all civil engineers."
>
> —ASCE's *Civil Engineering Body of Knowledge for the 21st Century*, 2008, page 128.

Key Topics Covered

- Introduction
- Sustainability Defined
- Sustainable Engineering
- Ecodesign
- Toward New Values and Processes
- Sustainable Design and Materials Strategies
- Lifecycle Cost Analysis
- Leadership in Energy and Environmental Design (LEED)
- Future Directions
- Summary

Related Chapters in This Book

- Chapter 5: The Engineer's Role in Project Development
- Chapter 15: Globalization
- Chapter 17: Emerging Technologies

(Continued)

439

Related to *ASCE Body of Knowledge 2* Outcomes

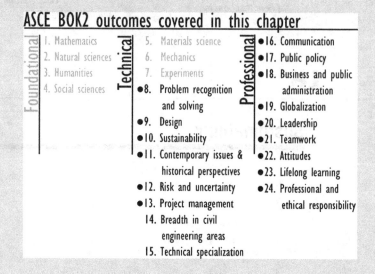

ASCE BOK2 outcomes covered in this chapter

Foundational
1. Mathematics
2. Natural sciences
3. Humanities
4. Social sciences

Technical
5. Materials science
6. Mechanics
7. Experiments
●8. Problem recognition and solving
●9. Design
●10. Sustainability
●11. Contemporary issues & historical perspectives
●12. Risk and uncertainty
●13. Project management
14. Breadth in civil engineering areas
15. Technical specialization

Professional
●16. Communication
●17. Public policy
●18. Business and public administration
●19. Globalization
●20. Leadership
●21. Teamwork
●22. Attitudes
●23. Lifelong learning
●24. Professional and ethical responsibility

INTRODUCTION

For the past 200 years, humanity gradually has been polluting and modifying the natural environment at an increasing rate. Added to the pollution is the quantum depletion of nonrenewable energy in fossil fuels. Why should civil engineers, architects, and builders care? As Kenneth Yeang, author of *Ecodesign*, states:

> Our health as human beings, and as one of the millions of species in nature, depends upon the air that we breathe and the water that we drink, as well as on the uncontaminated quality of the soil from which our food is produced.

The built environment is constructed from renewable and nonrenewable energy and material resources. Built systems are dependent upon the earth as a supplier of energy and material resources. Understanding the basis of sustainability enables us to comprehend the interconnections and processes that make up the environment and can help us to recognize the causes of degradation due to humankind's development and progress.

This chapter explores the background of ASCE's call for renewed professional commitment to stewardship of natural resources and the environment in 1996. It also provides some principles of sustainability, and their expression in engineering practice, as required of all civil engineers by the ASCE. The chapter concludes with what civil engineers can do to promote sustainability and how sustainability can be incorporated into the contemporary design process.

SUSTAINABILITY DEFINED

"Sustainability" has numerous meanings in the English language. The most widely quoted definition is from the UN's 1987 Brundtland Commission Report, *Our Common Future:* "meeting the needs of the present without compromising the ability of future generations to meet their own needs." The Brundtland Report also pointed out the importance of evaluating actions in terms of what has been called the Three Es:

- Environment
- Economy
- Equity

The Three Es force us to examine the cause and effects of our actions in relationship to the systems of the Earth and also to examine issues of justice—both economically and socially—for our fellow humans. This is a very challenging concept!

Our Common Future laid the foundation for the "Earth Summit" at Rio de Janeiro, Brazil, in 1992, which marked the real beginning of international environmental protection.

Sustainability Definitions

In 1983, Gro Harlem Brundtland, a female physician who later became Prime Minister of Norway, was invited by then United Nations Secretary-General Javier Pérez de Cuéllar to establish and chair the World Commission on Environment and Development (WCED), also known as the Brundtland Commission. Through extensive public hearings, the Commission published a report in April 1987 called *Our Common Future*.

The complete quote from the Brundtland Commission Report is:

"Sustainable development is development that meets the needs of the present without compromising the ability of future generations to meet their own needs. It contains within it two key concepts:

1. the concept of 'needs,' in particular the essential needs of the world's poor, to which overriding priority should be given; and

2. the idea of limitations imposed by the state of technology and social organization on the environment's ability to meet present and future needs."

Additional definitions of sustainability abound, and they are many and varied. However, as does the Brundtland definition, most definitions of sustainability include three basic components: 1) environmental protection; 2) economic development; and 3) equity (social).

SUSTAINABLE ENGINEERING

As the global population grows and standards of living improve, there will be increasing stress on the world's limited resources. Thus, engineers of the future will be asked to use the Earth's resources more efficiently and produce less waste, while at the same time satisfying an ever-increasing demand for goods and services. To prepare for such challenges, engineers will need to understand the impact of their decisions on built and natural systems, and must be adept at working closely with planners, decision-makers, and the general public. Figure 16.1 illustrates how engineers should seek opportunities in planning and designing to improve:

- Biodiversity
- Ecological connectivity
- Biointegration with local habitants

In 2005, Carnegie Mellon University, the University of Texas at Austin, and Arizona State University established the *Center for Sustainable Engineering*, supported

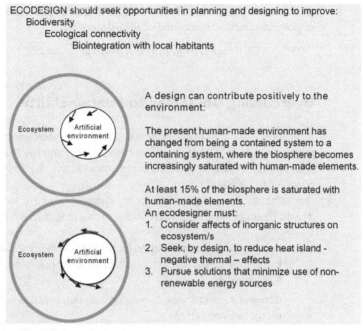

ECODESIGN should seek opportunities in planning and designing to improve:
Biodiversity
Ecological connectivity
Biointegration with local habitants

A design can contribute positively to the environment:

The present human-made environment has changed from being a contained system to a containing system, where the biosphere becomes increasingly saturated with human-made elements.

At least 15% of the biosphere is saturated with human-made elements.
An ecodesigner must:
1. Consider affects of inorganic structures on ecosystem/s
2. Seek, by design, to reduce heat island - negative thermal – effects
3. Pursue solutions that minimize use of non-renewable energy sources

Figure 16.1   Ecodesign
(Adapted from Kenneth Yeang, *Ecodesign*)

by the National Science Foundation (NSF) and the Environmental Protection Agency (EPA). The Center for Sustainable Engineering uses the Brundtland definition of sustainable development and cites the following examples of sustainable engineering:

- Using methods that minimize environmental damage to provide sufficient food, water, shelter, and mobility for a growing world population
- Designing products and processes so that wastes from one are used as inputs to another
- Incorporating environmental and social constraints as well as economic considerations into engineering decisions

The American Society of Civil Engineers has taken a strong position in support of sustainability:

ASCE embraced sustainability as an ethical obligation in 1996, and policy statements 418 and 517 point to the leadership role that civil engineers must play in sustainable development. The 2006 ASCE Summit on the Future of Civil Engineering called for renewed professional commitment to stewardship of natural resources and the environment. Knowledge of the principles of sustainability, and their expression in engineering practice, is required of all civil engineers.

Civil Engineering Body of Knowledge for the 21st Century, 2008, page 128.

Overcoming Obstacles to Sustainability

At a 2009 workshop in Sacramento, California, members of ASCE, ASCE's Environmental & Water Resources Institute (EWRI), and the Floodplain Management Association (FMA) came together to discuss sustainability. Participants observed the following:

Although many California communities are eager to incorporate sustainability into their planning efforts, a number of challenges make this difficult:

- In many cases, communities lack a sustainability vision, state leadership, tools, incentives, and indicators to effectively ensure their own sustainability.
- Community sustainability needs financing, political will, and appropriate protective regulations.
- Communities are not linked with State actions and policies on sustainability.
- Addressing the connection between land-use planning, flood management, water supply, energy consumption, and natural resource protection is still in the early stages of implementation.
- Water policy does not sufficiently incorporate the economics of water, or the significant connection between water use and greenhouse gas production.
- Planning efforts at local, regional, and State levels lack sufficient coordination and integration.

These challenges are not unique to California. What is encouraging is that civil engineers are developing plans to overcome these obstacles.

Systems Thinking

Sustainability encompasses a set of diverse concerns such as global climate change, environmental degradation, pressures on food and water supplies, loss of species, consumerism, and pollution. In essence, sustainability includes many systems and subsystems. These systems include very complex issues, such as climate, demographics, and global economics. The reductionist analytical approach, which breaks a whole into its parts to be studied independently, fails to enable understanding of systems-level questions. As the system gets larger, more complex, and more dynamic, the analytical approach becomes even less effective in dealing with the complexities of sustainability. Thus, sustainable engineering begins with study and analysis that requires a systems perspective.

Systems thinking provides a framework based on the theory that the component parts of a system can be understood best in the context of relationships with other components and with other systems, rather than in isolation. Understanding why a problem occurs and persists requires understanding the part in relation to the whole.

Since built systems can be composites containing both natural and artificial components, establishing the boundaries between these components is crucial. *System theory* defines a system boundary as:

> A physical or conceptual boundary that contains all the system's essentials and effectively isolates it from its external environment, except inputs and outputs that can move across the system's boundary. Establish the designed system's boundaries in relationship to the ecosystem.

ECODESIGN

Much can be learned about the systems approach from exploring the work of noted architect, Kenneth Yeang. He has developed the concept of *Ecodesign*, which he uses in both theory and practice. Ecodesign integrates artificial systems with natural systems. *Integrate* is the key word. In Ecodesign, design of the built environment should reflect the relationship between our human-made environment and the natural environment. However, Ecodesign is not an assembly of ecological-technological systems and gadgets in a single project or product. *The ultimate objective is environmental integration by design.*

Since Ecodesign involves the integration of artificial systems with natural systems, determining the level of environmental integration that can be achieved is essential:

- Theoretically, there are limitless ways of making a design ecologically responsive.
- The critical issue is where to stop biointegration.
- There is a dynamic balance between the standard of living and environmental consequences.

Ecodesign incorporates the entire ecosystem concept where maintaining the web of species within an ecosystem and the web connecting all four systems—the biosphere, the ecosystem, the community, and the population—is essential. When the web is damaged,

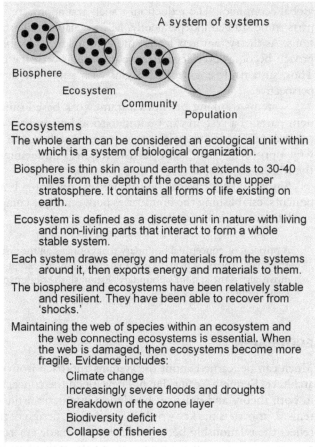

A system of systems

Biosphere

Ecosystem

Community

Population

Ecosystems

The whole earth can be considered an ecological unit within which is a system of biological organization.

Biosphere is thin skin around earth that extends to 30-40 miles from the depth of the oceans to the upper stratosphere. It contains all forms of life existing on earth.

Ecosystem is defined as a discrete unit in nature with living and non-living parts that interact to form a whole stable system.

Each system draws energy and materials from the systems around it, then exports energy and materials to them.

The biosphere and ecosystems have been relatively stable and resilient. They have been able to recover from 'shocks.'

Maintaining the web of species within an ecosystem and the web connecting ecosystems is essential. When the web is damaged, then ecosystems become more fragile. Evidence includes:

    Climate change
    Increasingly severe floods and droughts
    Breakdown of the ozone layer
    Biodiversity deficit
    Collapse of fisheries

Figure 16.2 Ecosystems
(Adapted from Kenneth Yeang, *Ecodesign*)

then the overall health of the ecosystem becomes more fragile. Figure 16.2 illustrates this concept.

Ecodesign planning principles provide guidance to the design engineer and facilitate the integration of ecosystem concepts into our civil structures and our communities. Figure 16.3 describes the key principles to accomplish this goal.

Effective planning begins with an initial assessment in the early design phase of the input and output of materials and their environmental consequences, followed by integrating efficient and eco-friendly transportation into the community, preserving landforms for ecological corridors, encouraging human inhabitants' perception of the natural beauty of the environment by enhancing distinctive site features, limiting waste output, enhancing resource conservation, controlling noise, and incorporating energy efficiency.

Key planning principles can include:

◊ Plan, manage, and integrate human vehicular and urban transportation system:

- Minimize routes traversing ecologically sensitive sites
- Avoid fragmentation of existing ecosystems that would create isolated island habitats
- Shape routes and landscape to assure continuity of vegetation patterns
- Encourage patterns that minimize the use of fossil fuels and emissions that add to greenhouse gases

◊ Preserve what exists or use opportunities in landforms to create new ecological corridors

◊ Encourage human inhabitants' perception of the natural environment by enhancing distinctive site features like unusual rock formations, topographic configurations, vistas, etc.

◊ Limit the amount of sewage emissions and other outputs, such as waste generated for disposal off-site

◊ Increase water conservation

- Filter run-off from impervious surfaces and return back to the ground
- Harvest rainwater

◊ Control noise emissions

◊ Assess the designed energy system in terms of its use and management

◊ At beginning of design, assess the input and output of materials and their environmental consequences

◊ Manage all demolition and construction activities to minimize their effects on ecological systems

Figure 16.3   Ecodesign planning principles
(Adapted from Kenneth Yeang, *Ecodesign*)

An award-winning project for sustainability is depicted in Figure 16.4. This project followed the Ecodesign planning principles and won accolades for this accomplishment.

TOWARD NEW VALUES AND PROCESSES

In the traditional business environment within the architectural/engineering/construction (AEC) industry, projects typically are defined and delivered as a *linear* process that begins with response to a problem, a need, an opportunity, or a desire and ends with the delivery of a completed facility or civil infrastructure system. (See Chapter 5, The Engineer's Role in Project Delivery.) The focus of this linear definition and delivery process is on the efficiency and productivity of: (1) the management of the planning, the design, the procurement and construction, and the commissioning and start-up processes; and

Figure 16.4 Singapore national library
(Kenneth Yeang, Wikimedia Commons)

(2) the management and use of the resource base, which can include economic and financial resources; physical resources such as materials, equipment, and tools; human resources such as technical, nontechnical, and administrative personnel; technological resources such as computing, communication, collaboration, and management of information technologies; and miscellaneous other resources such as data/information, knowledge/experience, abilities/skills, and technological proficiency.

Fortunately, there are alternatives available to these potentially environmentally wasteful processes. They begin with considering the design, construction, and operations over the entire life time of the building. An *integrated design* process helps to establish goals for the design, foresee the impacts of construction, and plan the operations and maintenance of a building.

It's worth remembering that:

Human welfare is related to the
  material standard of living, which depends on the
    provision of manufactured goods, which requires the
      consumption of natural capital, which means the
        extraction of natural resources, which involves
          discharges of waste, which result in
            impacts on human welfare.

Ecomimicry

Properties of ecosystems that can be applied to design and construction, including:

- Optimization of energy and materials consumption
  - Minimization of waste generation
    - Use of effluents (discharge/waste) of one process serve as raw materials for another

Specific human ecomimicry objectives:

1. Reduce dependency on nonrenewable energy in a system's lifecycle
2. Change from nonrenewable to renewable sources of energy
3. Increase efficiency in energy use
4. Reduce wasteful use of nonrenewable energy resources
5. Recycle materials and outputs
6. Balance producers, manufacturers, and services with consumers
7. Increase diversity to stabilize system
8. Increase compact spatial efficiency
9. Have high community organization (many networks to provide information)
10. Provide global protection from environmental perturbations
11. Conserve resources, use sustainably in order to buffer and cope with changes
12. Adopt self-correcting systems for environmental stability

Sources: Kenneth Yeang, *Ecodesign*, and Janis Birkland, *Design for Sustainability: A Sourcebook of Integrated Eco-logical Solutions*.

Sustainability Principles

Specific education and research opportunities exist within the broad body of general knowledge on sustainability. For example, the International Institute for Sustainable Development provides an extensive compilation of sustainable development principles from numerous sources that address three major aspects: environment, economy, and community.

There is also an extensive body of specific knowledge on built environment sustainability. The key is to investigate how to adapt and customize this knowledge to the specific reality of an architectural/engineering/construction (AEC) project. For example, some selected examples of principles that can be used to implement and achieve best practice include:

- The Precautionary Principle, which guides human activities to prevent harm to the environment and to human health

- The Earth Charter Principles, which promote respect and care for the community of life, ecological integrity, social and economic justice, and democracy, nonviolence, and peace

- The Natural Step System Conditions, which define basic principles for maintaining essential ecological processes, and recognizing the importance of meeting human needs worldwide, as integral and essential elements of sustainability

- The Daly Principles, which address the regenerative and assimilative capacities of natural capital, and the rate of depletion of nonrenewable resources

- The Ceres Principles, which provide a code of environmental conduct for environmental, investor, and advocacy groups working together for a sustainable future

- The Bellagio Principles, which serve as guidelines for starting and improving the sustainability assessment process and activities of community groups, nongovernment organizations, corporations, national governments, and international institutions, including the choice and design of indicators, their interpretation, and communication of the result

- The Ahwahnee Principles, which guide the planning and development of urban and suburban communities in a way that they will more successfully serve the needs of those who live and work within them

- The Interface Steps to Sustainability, which were created to guide the interface company in addressing the needs of society and the environment by developing a system of industrial production that decreases their costs and dramatically reduces the burdens placed upon living systems

- The Hannover Principles, which assist planners, government officials, designers, and all involved in setting priorities for the built environment, and promoting an approach to design which may meet the needs and aspirations of the present without compromising the ability of the planet to sustain an equally supportive future

- Design through the 12 Principles of Green Engineering, which provide a framework for scientists and engineers to engage in when designing new materials, products, processes, and systems that are benign to human health and the environment

Source: Karen Lee Hansen and Jorge A. Vanegas. (2006). "A Guiding Road Map, Principles, and Vision for Researching and Teaching Sustainable Design and Construction." *American Society for Engineering Education (ASEE), Chicago, IL, Conference Proceedings.*

Expanded Project Delivery Process

Construction costs can represent 5 to 15 percent of a facility's lifecycle cost; design costs are typically less than 1 percent. Operations and renovations constitute most of the remaining costs. For the least expense, design can have the greatest impact on long-term sustainability. Therefore, planning and designing facilities and civil infrastructure systems sustainably is critical.

For built environment sustainability, the project delivery process needs to:

- Address AEC projects from their complete lifecycle perspective, including operations and maintenance, and also the end-of-service life decision

- Emphasize the use of sustainable resources

- Monitor and document the outcomes resulting from the use of the facility or civil infrastructure system delivered

- Provide a post-occupancy evaluation and feedback to the project originator that, depending on what the project driver was, answers the question: Did the delivered project solve the problem? Satisfy the need? Capitalize on the opportunity? Realize the desire?

The fundamental approach for enabling and achieving sustainable facilities or civil infrastructure systems at a *strategic level* is shown in Figure 16.5. The basis for implementation is to frame an AEC project within a contextual envelope, which (1) is defined by the *requirements and characteristics* of the specific facility or civil infrastructure system, the *processes followed in its delivery and use*, and the *resources*

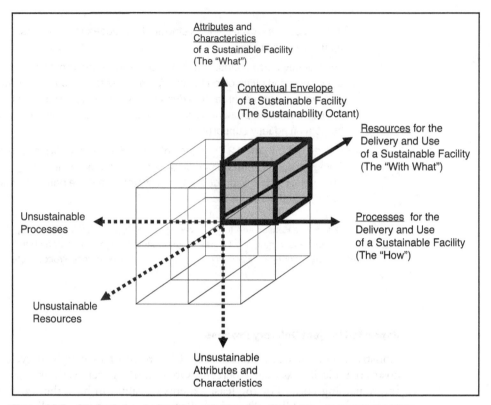

Figure 16.5 Strategic level of implementation of built environment sustainability
(Adapted from Dr. Jorge A. Vanegas, Dean, College of Architecture, Texas A&M University)

consumed in its delivery and use; and (2) uses sustainability as a fundamental criterion in making decisions, choosing among various options, or taking actions.

The contextual envelope for the project is represented as an (x,y,z) diagram defined by the specific requirements and characteristics of the specific facility or civil infrastructure system *x*, its processes *y*, and its resources *z*, expressed as relative points on an axis, with a scale that spans from what is unsustainable to what is sustainable in each one, and with the thresholds that separate the two extremes within each axis as point (0,0,0). The strategy then, is to maintain project decisions, choices, and actions within the *octant* where all three (*x,y,z*) points are sustainable. While this may be conceptually simple, in reality it is quite complex, and much research is yet to be done to provide clear and absolute definitions of what is sustainable, what is unsustainable, and what is the threshold that separates them.

Integrative Approaches

If the model of sustainable engineering uses a systems approach that integrates diverse natural and man-made elements into an ecological system, then the project delivery process also should be integrated and balanced. The design, construction, and operation of facilities and civil infrastructure systems should be part of a restorative, integrated process, rather than a wasteful, linear one.

Collaborative project delivery processes facilitate integration of design strategies among all disciplines and players. Integrating design requires an inclusive project team that involves users, contractors and subcontractors, and future maintenance staff. Using this approach, the design team has more opportunities to discover creative solutions, to save time, and to minimize costs. For example, contractors included in the design phase understand project goals and can work toward suitable solutions with the designers. Integrated design also can lead to better facility performance, thereby saving costs. For example, overall energy consumption can be optimized if site design, facility "envelope," and electrical and mechanical systems are thought of together.

These synergies and savings require collaboration among engineers, architects, and others. Synergies can provide multiple benefits, much like the collaboration of civil engineers, mechanical engineers, traffic engineers, and maintenance staff can lead to low-impact development, preserving open space, decreasing stormwater treatment, and reducing radiant heat and the heat island effect from the landscape.

Impact of the Construction and Operation of Facilities on the Environment

The construction and operation of facilities have both direct and indirect impacts on the environment. Direct impacts include the energy used for electricity and heating, ventilating, and air conditioning (HVAC) systems, materials and resources required for construction, the water used in facilities/buildings, and the stormwater and open space impacts of displacing greenfield sites.

In the United States, buildings consume 39 percent of primary energy use (including fuel input for production) and contribute 38 percent of CO_2 emissions. Buildings represent 72 percent of U.S electricity consumption and use 13.6 percent of all potable water, or 15 trillion gallons per year. The EPA estimates that 170 million tons of building-related construction and demolition debris was generated in the United States in 2003, with 61 percent coming from nonresidential and 39 percent from residential sources.

Indirect impacts include the inputs to building material manufacturing processes, the fossil fuels used to transport the materials, and the CO_2 emissions from both. Additionally, the emissions of volatile organic compounds (VOCs) from many building materials adversely impact people's well-being.

Source: U. S. Green Building Council Website (www.usgbc.org)

Solving multiple environmental challenges requires an integrated, multidisciplinary design team. Using integrated design practices to maximize sustainability, rather than focusing on minimum requirements, opens up synergies and solutions that otherwise might remain undiscovered. The project team needs to set sustainability goals early, design to achieve those goals, and verify that performance meets those goals. Through this process, sustainability becomes integral to design and construction, not an "add-on."

SUSTAINABLE DESIGN AND MATERIALS STRATEGIES

Realization of a sustainable facility or civil infrastructure system starts long before the actual design process begins. The owner, perhaps in conjunction with a professional consultant, needs to establish sustainable design goals early, embed a multidisciplinary effort into the design process, and budget for the necessary meetings and analyses critical to ensuring sustainability from design through construction through postoccupancy.

Sustainable Design Strategy

Sustainable design and materials strategies are defined by *requirements and characteristics* of the specific facility or civil infrastructure system, the *processes* followed in its delivery and use, and the *resources* consumed in its delivery and use. These strategies include, but are not limited to, the following:

- Siting and design considerations that optimize local geographic features to improve sustainability, such as proximity to public transportation and maximizing use of vistas, microclimate, and prevailing winds
- Durable systems and finishes with long lifecycles that minimize maintenance and replacement needs
- Layouts and designed spaces that can be reconfigured for adaptive reuse (versus demolition);
- Systems designed for optimization of energy, water, and other natural resources
- Optimization of indoor environmental quality
- Utilization of environmentally preferable products and processes, such as recycled-content and recyclable materials
- Procedures that monitor, trend, and report operational performance as compared to the optimal design and operating parameters
- Durable systems and finishes with long lifecycles that minimize maintenance and replacement

In order to operationalize these strategies, the owner must allocate adequate budget and schedule. Enovity and HOK, two firms with expertise in sustainability and alternative energy solutions, suggest that the following meetings, analyses, and actions be incorporated into the project:

- Sustainable design goals meeting
- Design and strategy charrette with entire multidisciplinary project team
- Energy and daylighting analysis
- Lifecycle cost analysis
- Scope and extent of commissioning
- Comprehensive maintenance plan
- Postoccupancy evaluation, including measurement and verification
- Regular design sessions to develop and implement integrated design

See Table 16.1 for more information regarding actions that support a sustainable design.

Table 16.1   Useful Meetings, Analyses, and Actions That Support Sustainable Design (Adapted from Enovity—HOK)

	Meeting, Analysis, or Action	Description
1	Sustainable design goals meeting	During predesign phases (Scoping, Feasibility, Programming), conduct a sustainable design goals meeting. Owner personnel and the project team should commit to a sustainable design vision, measurable goals, and methods of measuring and verifying the achievement of the goals. Prioritize according to project-specific opportunities and constraints.
		The results of the sustainable design goals meeting should inform the equivalent of the Owner's Project Requirements (OPR) document, as required for Leadership in Energy and Environmental Design (LEED) certification. The OPR should explain the owner's goals and expectations of the facility's program, operation, energy efficiency, and sustainability.
2	Design a strategy *charrette* with entire multidisciplinary project team (*Charrette: intense effort to complete a project within a limited period of time*)	Hold workshop intended to establish specific ways to achieve the goals and intentions identified in the sustainable design goals meeting. Define a design methodology based on analysis, integrated design, and optimized sustainability, rather than simply meeting minimum standards.
		Discuss the cost implications of sustainability strategies, such as a set-aside budget for technologies with a good lifecycle value and cost-effective energy-efficiency upgrades that may help supplement construction budget.
		Consider design strategies' costs and benefits. Discuss the schedule implications of sustainability measures, such as commissioning and testing.
		(Continued)

Table 16.1   (*Continued*)

	Meeting, Analysis, or Action	Description
3	Energy and daylighting analysis	Optimize designs with the assistance of energy and daylighting analysis tools and software that can help evaluate performance. Computer programs to assess energy efficiency include: DOE-2, Energy-10™, and eQuest®. Daylighting evaluation programs include Radience, Lumen Micro, and the Lightscape visualization system. A Helidon, a device that uses a light source and physical model, can also simulate the effects of sunlight.
4	Lifecycle cost analysis	Require designers to incorporate lifecycle cost analysis (LCCA) into the design to predict long-term building costs and minimize energy and replacement costs. LCCA compares the total costs and benefits over the entire lifecycle of a system, component, or material. It allows for future costs and benefits to be incorporated into present-day decision-making.
5	Scope and extent of commissioning	Select commissioning authority, determine scope and extent of commissioning. This information is needed early so that it can be incorporated into the owner's budget and the project schedule.
6	Comprehensive maintenance plan	After the design and strategy charrette, develop a maintenance plan that helps foresee operations challenges and ensures that original design intentions are met for the lifetime of the facility. Newer facilities will have advanced technologies which may require training or different operations practices. For these reasons, those responsible for building operations and maintenance should be included in early goals and strategy meetings.
7	Postoccupancy evaluation, including measurement and verification	Plan and schedule a postoccupancy evaluation from the outset. Set regular reviews of energy and water end use, air quality and heating, ventilation, and air-conditioning (HVAC) performance, and waste metrics. Implement the measurement and verification plan to ensure that energy and HVAC design intentions are being met. Use the measurement and compliance strategies appropriate to the project.
8	Regular design sessions to develop and implement integrated design	Hold regular project meetings, including designers, stakeholders, and—if possible—key contractor(s) and subcontractors. Collaborative, multidisciplinary, integrated design teams need to communicate well and regularly in order to be effective. Some team members may be required to perform tasks outside their typical roles. For example, civil engineers may be asked to design biologically based stormwater treatment systems.
9	Miscellaneous actions	Include sustainability in RFPs and RFQs. Build a design team with members who are experienced and committed to sustainable design and working collaboratively. Appoint a Sustainability Champion. Design for flexibility and adaptability. Give priority to building materials and systems of durable and repairable assemblies (replacement or repair of isolated areas possible, without the need to replace the entire system).

Sustainable Materials Strategies

- **Limit the Use of Materials with Hazardous Content**

Materials with hazardous content can increase health risks of facility users, generate hazardous construction waste, or require a toxic manufacturing process. Avoiding these materials whenever possible is recommended.

- **Look for Salvaged Materials**

Used bricks and timbers, asphalt paving, and concrete all present opportunities for reuse or refurbishment. Alternatively, upon disassembly or tear-down of old facilities, owners can retain reusable materials for future use or sale to a salvage yard.

- **Seek Products with Recycled and Renewable Content**

The EPA promotes the recovery of materials from solid waste streams through its Comprehensive Procurement Guidelines. Consult the guidelines at www.epa. gov/cpg/products.htm.

- **Identify Local Manufacturers, Regionally Appropriate and Locally Available Materials**

Locally produced goods reduce the costs and environmental impact of transportation. Local goods are also more likely to be adapted to the regional climate and building requirements. At the same time they support the local economy. Regional and local materials should account for at least 20 percent of material costs in new construction. A list of local manufacturers and materials should be maintained and incorporated into standard designs and future projects.

- **Use Low-Emitting Materials**

Adhesives, sealants, carpeting, paint, coatings, and composite wood products that emit low amounts of volatile organic compounds (VOCs) are all available on the market and should be considered for use. As demand for sustainable buildings increase, more options are becoming available.

- **Create a Policy of Diverting 95 Percent of Construction Waste**

Instituting the 95 percent goal as a policy will provide design teams with an official and identifiable goal to strive for, facilitating design optimization and creative reuse.

Source: Enovity—HOK

LIFECYCLE COST ANALYSIS

Lifecycle Cost Analysis (LCCA) can inform decisions about a multitude of factors, thereby helping to control the initial and the future costs of facility ownership and maintenance. LCCA is a measure of "cradle to grave" costs that can be performed on a full range of projects, from entire site complexes to specific system components, both large and small. For example, civil engineers can use LCCA to choose between concrete or asphalt paving and between a reinforced concrete or steel structures. As contrasted with many decisions made in conventional project development processes, initial cost is a factor, but not the only factor.

Lifecycle Cost (LCC) is the total discounted (present value) dollar cost of owning, operating, maintaining, and disposing of a facility or system over a period of time. The LCC equation contains three variables: the pertinent *costs* of ownership, the period of *time* over which these costs are incurred, and the *discount rate* that is applied to future costs to equate them with present-day costs.

The Basic Lifecycle Cost Analysis Process

- Estimate costs and benefits
- Estimate timing
- Discount future costs and benefits
- Compare net present values

Following are the key concepts used in LCCA.

Costs

The two major cost categories by which projects are evaluated in an LCCA are initial expenses and future expenses. *Initial expenses* are all costs incurred prior to occupation of the facility. *Future expenses* are all costs incurred after occupation of the facility. Defining exact costs of each expense category tends to be difficult since few costs are known at the time of the LCCA. Careful, well-documented assumptions are necessary for preparation of a realistic LCCA.

Residual Value

One future expense that warrants further explanation is residual value. *Residual value* is the net worth of a facility or system at the end of the LCCA study period. Unlike other future expenses, an alternative's residual value can be positive or negative. LCC is a summation of costs, so a negative residual value indicates that there is value associated with the facility at the end of the study period. The value could be in a component that was replaced recently or in the facility's superstructure that could function for another 30 years; or the costs could be related to abatement of hazardous material or demolition of the structure. A positive residual value indicates disposal costs associated with the facility at the end of the study period. The residual value is a tangible asset that should be included in the LCCA.

Zero residual value indicates that there is no value or cost associated with the facility at the end of the study period. This situation is rare and occurs, for example, if the intended use of the facility terminates concurrently with the end of the study period, the owner is unable to sell the facility, or the owner is able to abandon the facility at no expense.

Study Period

Time is the second component of the LCC equation. The *study period* is the time over which ownership and operations expenses are to be evaluated. A typical study period ranges from 20 to 40 years. The length of the study period depends on the owner's preferences, the stability of the user's program, and the designed overall life of the facility, though the study period usually is shorter than the intended life of the facility.

Some LCCA approaches, such as that defined by the National Institute of Standards and Technology (NIST), break the study period into two phases: the planning/construction period and the service period. The planning/construction period is the time period from the start of the study to the date the building becomes operational (the service date). The service period is the time period from the date the building becomes operational to the end of the study.

Discount Rate

The third component in the LCC equation is the discount rate. The *discount rate* is the rate of interest reflecting the investor's time value of money—either the minimum acceptable rate of return for investment purposes (frequently used by owners in private industry) or the current rate of interest for borrowing (frequently used by public owners). As world economics change, so does the discount rate.

Constant versus Current Dollars

Constant dollars can be defined as dollars of uniform purchasing power tied to a reference year, exclusive of general price inflation or deflation. *Current dollars* can be defined as dollars of nonuniform purchasing power that include general price inflation or deflation.

The use of constant dollars simplifies LCCA. For example, suppose one wants to evaluate a product over a 30-year period. However, one product must be replaced after 20 years. How much will the replacement of the product cost in 20 years? By using constant dollars, estimating the escalation of labor and material costs is eliminated. The future constant dollar cost (excluding demolition) to install a new product in 20 years is the same as the initial cost to install it. Any change in the value of money over time will be accounted for by the discount rate.

Escalation must be considered when future costs differ from inflation. For example, energy costs may rise more quickly, perhaps 1 to 2 percent, than inflation.

Present Value

To combine initial expenses with future expenses accurately, the present value of all expenses must be determined. *Present value* can be defined as the time equivalent value of past, present, or future cash flows as of the beginning of the base year. The present value calculation uses the discount rate and the time a cost was or will be incurred to establish the present value of the cost in the base year of the study period.

Initial expenses are considered to occur during the base year of the study period since most of these expenses occur at approximately the same time. Consequently, the present value of these initial expenses does not need to be calculated because their present value is equal to their actual cost. The present value of future costs is time dependent. The time period is the difference between the time of initial costs and the time of future costs. Initial costs are incurred at the beginning of the study period at Year 0, the base year. Future costs can be incurred anytime between Year 1 and Year n. The present value calculation is the summation of initial and future costs. Along with time, the discount rate also dictates the present value of future costs. Because the current discount rate is a positive value, future expenses will have a present value less than their cost at the time they are incurred.

Future Costs

Future costs can be broken down into two categories: one-time costs and recurring costs. *Recurring costs* are costs that occur ever year over the span of the study period. Most operating and maintenance costs are recurring costs. *One-time costs* are costs that do not occur every year over the span of the study period. Most replacement costs are one-time costs. To simplify the LCCA, all recurring costs are expressed as annual expenses incurred at the end of each year and one-time costs are incurred at the end of the year in which they occur. If costs in a particular cost category are equal in all project alternatives, they can be documented as such and removed from consideration in the LCC.

Present value calculations can be made using the formulas included in Table 16.2.

Table 16.2 Present Value Formulas

Type of Cost	Application	Constant Dollars	Current Dollars
First	Initial capital investments	$P = P$ (no conversion)	$P = P$ (no conversion)
Future	Replacements or alterations	$P = F \times \dfrac{1}{(1+i)^n}$	$P = F \times \dfrac{(1+e)^n}{(1+i)^n}$
Future Series	Energy or maintenance	$P = A \times \dfrac{\left(\frac{(1+e)^n}{(1+i)^n}\right) - 1}{1 - \left(\frac{(1+e)^n}{(1+i)^n}\right)}$	$P = A \times \dfrac{\left(\frac{(1+e)^n}{(1+i)^n}\right) - 1}{1 - \left(\frac{(1+e)^n}{(1+i)^n}\right)}$

P = present value
F = future value
i = real interest rate
i = total interest rate (real interest rate plus inflation)
e = escalation
n = time (expressed as number of years)
A = annual amount

Alternatives

Prior to beginning an LCCA, several (usually three) distinctly different and viable solutions—*alternatives*—should be established. The LCCA describes each project alternative as well as the rationale for its inclusion.

Some possible options that can be considered while selecting the most viable, reasonable, and cost-effective alternatives include:

- Renovation and addition to an existing facility
- Rental and remodeling of an existing facility
- Purchase and remodeling of an existing facility
- Demolition of existing facility and construction of a new facility on the same site
- Sale of existing facility and construction of a new facility on a new site
- The use of double labor shifts or some other solution not involving construction that increases facility capacity
- Alteration of the facility's function in some way, also a nonconstruction approach

A "No Action" alternative also frequently must be considered.

Lifecycle Cost Analysis Terminology

Lifecycle Cost: the total discounted (present value) dollar cost of owning, operating, maintaining, and disposing of a facility or system over a period of time

Alternative: distinctly different and viable solution

Constant dollars: dollars of uniform purchasing power tied to a reference year, exclusive of general price inflation or deflation

Current dollars: dollars of nonuniform purchasing power that include general price inflation or deflation

Discount rate: rate of interest reflecting the investor's time value of money

Future expenses: costs incurred after occupation of the facility

Initial expenses: costs incurred prior to occupation of the facility

One-time costs: costs that do not occur every year over the span of the study period

Present value: time equivalent value of past, present, or future cash flows as of the beginning of the base year

(Continued)

> **Recurring costs:** costs that occur every year over the span of the study period
>
> **Residual value:** net worth of a building at the end of the LCCA study period

Limitations of LCCA

Defining accurate costs and discount rates necessary for realistic LCCA is difficult at best. Where to establish cost boundaries often is not obvious. For example, fuel costs are considered in transportation input, but what about the costs associated with manufacturing vehicles and maintaining highways? Naturally, manufacturers have incentives for limiting boundaries in their studies. Uninformed users may employ LCCA data without realizing its limitations.

Lifecycle Assessment is a next step beyond LCCA. LCA translates the flow of resources and waste into overall environmental impact. The lifecycle of any given product may result in many emissions, each affecting the environment to varying degrees. Because the impact of each emission varies, the overall effect of each emission on greenhouse gases is expressed in units of carbon dioxide (CO_2) equivalents. The Environmental Protection Agency (EPA) uses a similar process to estimate environmental impacts in other categories. EPA's Tool for the Reduction and Assessment of Environmental Impacts (TRACI) includes variables such as ozone depletion, global warming, acidification, photochemical smog, human carcinogenic effects, eco-toxity, fossil fuel use, land use, and water use, among others. (See Figure 16.6.)

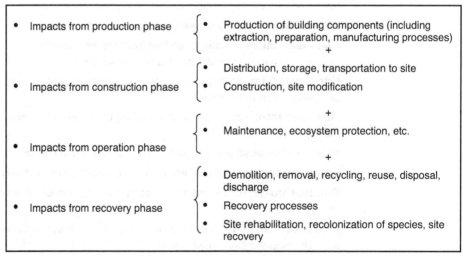

Figure 16.6   Lifecycle impacts of a designed system
(Adapted from Kenneth Yeang, *Ecodesign*)

Cradle to Cradle (2002) by architect William McDonough and chemist Michael Braungart proposes that an industrial system that "takes, makes and wastes" can transform itself into a creator of goods and services that generates ecological, social, and economic value. The authors maintain that products and manufacturing practices stemming from the Industrial Revolution have resulted in a series of unintended, disastrous consequences. But with today's greater understanding of Earth as a living system, the authors imagine that nature and commerce can coexist in new products, industrial systems, buildings, and regional plans. The book itself is an example of "cradle to cradle" principles: It is printed on a synthetic, treeless "paper," made from plastic resins and inorganic fillers, designed to be recycled and used again.

LEADERSHIP IN ENERGY AND ENVIRONMENTAL DESIGN

Leadership in Energy and Environmental Design (LEED), developed by the U.S. Green Building Council (USGBC), is a voluntary certification program that measures how well a project or community compares to others in achieving predetermined sustainability goals. The LEED framework provides metrics involving energy savings, water efficiency, CO_2 emissions reduction, improved indoor environmental quality, and stewardship of resources and sensitivity to their impacts. The framework can be applied throughout the facility lifecycle—design and construction, operations and maintenance, tenant improvement, and significant retrofit. A category called Neighborhood Development extends the LEED framework beyond the building footprint into the surrounding neighborhood.

As shown in Figure 16.7, LEED evaluates sustainability performance in several areas:

- Sustainable Sites
- Water Efficiency
- Energy & Atmosphere
- Materials & Resources
- Indoor Environmental Quality
- Locations & Linkages
- Awareness & Education
- Innovation in Design
- Regional Priority

Other certification programs exist—Green Globes is used in Canada and BREEAM (Building Research Establishment Environment Assessment Method) in the United Kingdom. However, LEED currently is the most widely adopted third-party sustainability certification system, perhaps because many public owners have

Sustainable Sites
Choosing a building's site and managing that site during construction are important considerations for a project's sustainability. The Sustainable Sites category discourages development on previously undeveloped land; minimizes a building's impact on ecosystems and waterways; encourages regionally appropriate landscaping; rewards smart transportation choices; controls stormwater runoff; and reduces erosion, light pollution, heat island effect and construction-related pollution.

Water Efficiency
Buildings are major users of our potable water supply. The goal of the Water Efficiency credit category is to encourage smarter use of water, inside and out. Water reduction is typically achieved through more efficient appliances, fixtures and fittings inside and water-wise landscaping outside.

Energy & Atmosphere
According to the U.S. Department of Energy, buildings use 39% of the energy and 74% of the electricity produced each year in the United States. The Energy & Atmosphere category encourages a wide variety of energy strategies: commissioning; energy use monitoring; efficient design and construction; efficient appliances, systems and lighting; the use of renewable and clean sources of energy, generated on-site or off-site; and other innovative strategies.

Materials & Resources
During both the construction and operations phases, buildings generate a lot of waste and use a lot of materials and resources. This credit category encourages the selection of sustainably grown, harvested, produced and transported products and materials. It promotes the reduction of waste as well as reuse and recycling, and it takes into account the reduction of waste at a product's source.

Indoor Environmental Quality
The U.S. Environmental Protection Agency estimates that Americans spend about 90% of their day indoors, where the air quality can be significantly worse than outside. The Indoor Environmental Quality credit category promotes strategies that can improve indoor air as well as providing access to natural daylight and views and improving acoustics.

Locations & Linkages
The LEED for Homes rating system recognizes that much of a home's impact on the environment comes from where it is located and how it fits into its community. The Locations & Linkages credits encourage homes being built away from environmentally sensitive places and instead being built in infill, previously developed and other preferable sites. It rewards homes that are built near already-existing infrastructure, community resources and transit, and it encourages access to open space for walking, physical activity and time spent outdoors.

Awareness & Education
The LEED for Homes rating system acknowledges that a green home is only truly green if the people who live in it use the green features to maximum effect. The Awareness & Education credits encourage home builders and real estate professionals to provide homeowners, tenants and building managers with the education and tools they need to understand what makes their home green and how to make the most of those features.

Innovation in Design
The Innovation in Design credit category provides bonus points for projects that use new and innovative technologies and strategies to improve a building's performance well beyond what is required by other LEED credits or in green building considerations that are not specifically addressed elsewhere in LEED. This credit category also rewards projects for including a LEED Accredited Professional on the team to ensure a holistic, integrated approach to the design and construction phase.

Regional Priority
USGBC's regional councils, chapters and affiliates have identified the environmental concerns that are locally most important for every region of the country, and six LEED credits that address those local priorities were selected for each region. A project that earns a regional priority credit will earn one bonus point in addition to any points awarded for that credit. Up to four extra points can be earned in this way.

Figure 16.7   Categories used in various LEED certification programs

put the requirement for LEED certification in their requests for proposals (RFPs) and requests for qualifications (RFQs).

The LEED rating scorecards are available for the following project types:

- New Construction
- Existing Buildings: Operations & Maintenance

- Commercial Interiors
- Core & Shell
- Schools
- Retail
- Healthcare
- Homes
- Neighborhood Development

At this point, LEED is oriented more toward buildings than civil infrastructure. However, several rating categories involve civil engineers, such as stormwater provisions in New Construction and some of the land use and transportation categories in Neighborhood Development. Even if LEED is not a part of project requirements, the LEED approach is useful in helping to identify areas where sustainable principles can be applied. See Figures 16.8 and 16.9.

Importance of Evaluation, Measurement, and Verification

"Studies have shown that simply by operating most existing buildings as they were originally designed, a 20 percent energy savings could be achieved. Unfortunately, this does not happen often enough because those operating the building were not part of the original design, or designers do not stay in touch with the operations of the building. Ensuring buildings perform as well as they are designed is critical to sustainability . . . Design and LEED certification are therefore only the beginning of sustainability. And many LEED credits acknowledge the need for follow-up evaluation, measurement and verification."

Source: Enovity—HOK

FUTURE DIRECTIONS

A serious movement began in 1969 with the U.S. National Environmental Policy Act (NEPA). NEPA declared as its goal a national policy to "create and maintain conditions under which [humans] and nature can exist in productive harmony, and fulfill the social, economic and other requirements of present and future generations of Americans." Most Americans recognize that a balance between nature and humans must be achieved but progress has been slow. In the short run, ignoring the concepts expressed by Ecodesign is less expensive than dealing with the

California Academy of Sciences
Project # 10000723
Certification Level: PLATINUM
October 8, 2008

U.S. Green Building Council — LEED — USGBC

LEED for New Construction v2.0/2.1

54	Points Achieved		Possible Points: 69
Certified 26 to 32 points	Silver 33 to 38 points	Gold 39 to 51 points	Platinum 52 or more points

14 | Sustainable Sites — Possible Points: 14

Y	Prereq 1	Erosion & Sedimentation Control
1	Credit 1	Site Selection
1	Credit 2	Development Density
1	Credit 3	Brownfield Redevelopment
1	Credit 4.1	Alternative Transportation, Public Transportation Access
1	Credit 4.2	Alternative Transportation, Bicycle Storage & Changing Rooms
1	Credit 4.3	Alternative Transportation, Alternative Fuel Vehicles
1	Credit 4.4	Alternative Transportation, Parking Capacity & Carpooling
1	Credit 5.1	Reduced Site Disturbance, Protect or Restore Open Space
1	Credit 5.2	Reduced Site Disturbance, Development Footprint
1	Credit 6.1	Stormwater Management, Rate & Quantity
1	Credit 6.2	Stormwater Management, Treatment
1	Credit 7.1	Landscape & Exterior Design to Reduce Heat Islands, Non-Roof
1	Credit 7.2	Landscape & Exterior Design to Reduce Heat Islands
1	Credit 8	Light Pollution Reduction

5 | Water Efficiency — Possible Points: 5

1	Credit 1.1	Water Efficient Landscaping, Reduce by 50%
1	Credit 1.2	Water Efficient Landscaping, No Potable Use or No Irrigation
1	Credit 2	Innovative Wastewater Technologies
1	Credit 3.1	Water Use Reduction, 20% Reduction
1	Credit 3.2	Water Use Reduction, 30% Reduction

9 | Energy & Atmosphere — Possible Points: 17

Y	Prereq 1	Fundamental Building Systems Commissioning
Y	Prereq 2	Minimum Energy Performance
Y	Prereq 3	CFC Reduction in HVAC&R Equipment
1	Credit 1.1	Optimize Energy Performance, 15% New / 5% Existing
1	Credit 1.2	Optimize Energy Performance, 20% New / 10% Existing
1	Credit 1.3	Optimize Energy Performance, 25% New / 15% Existing
1	Credit 1.4	Optimize Energy Performance, 30% New / 20% Existing
1	Credit 1.5	Optimize Energy Performance, 35% New / 25% Existing
1	Credit 1.6	Optimize Energy Performance, 40% New / 30% Existing
1	Credit 1.7	Optimize Energy Performance, 45% New / 35% Existing
1	Credit 1.8	Optimize Energy Performance, 50% New / 40% Existing
1	Credit 1.9	Optimize Energy Performance, 55% New / 45% Existing
1	Credit 1.10	Optimize Energy Performance, 60% New / 50% Existing
1	Credit 2.1	Renewable Energy, 5%
1	Credit 2.2	Renewable Energy, 10%
1	Credit 2.3	Renewable Energy, 15%
1	Credit 3	Additional Commissioning
1	Credit 4	Ozone Depletion
1	Credit 5	Measurement & Verification
1	Credit 6	Green Power

7 | Materials & Resources — Possible Points: 13

Y	Prereq 1	Storage & Collection of Recyclables
1	Credit 1.1	Building Reuse, Maintain 75% of Existing Shell
1	Credit 1.2	Building Reuse, Maintain 100% of Shell
1	Credit 1.3	Building Reuse, Maintain 100% Shell & 50% Non-Shell
1	Credit 2.1	Construction Waste Management, Divert 50%
1	Credit 2.2	Construction Waste Management, Divert 75%
1	Credit 3.1	Resource Reuse, Specify 5%
1	Credit 3.2	Resource Reuse, Specify 10%
1	Credit 4.1	Recycled Content, Specify 5%
1	Credit 4.2	Recycled Content, Specify 10%
1	Credit 5.1	Local/Regional Materials, 20% Manufactured Locally
1	Credit 5.2	Local/Regional Materials, of 20% Above, 50% Harvested Locally
1	Credit 6	Rapidly Renewable Materials
1	Credit 7	Certified Wood

14 | Indoor Environmental Quality — Possible Points: 15

Y	Prereq 1	Minimum IAQ Performance
Y	Prereq 2	Environmental Tobacco Smoke (ETS) Control
1	Credit 1	Carbon Dioxide Monitoring
1	Credit 2	Ventilation Effectiveness
1	Credit 3.1	Construction IAQ Management Plan, During Construction
1	Credit 3.2	Construction IAQ Management Plan, Before Occupancy
1	Credit 4.1	Low-Emitting Materials, Adhesives & Sealants
1	Credit 4.2	Low-Emitting Materials, Paints
1	Credit 4.3	Low-Emitting Materials, Carpet
1	Credit 4.4	Low-Emitting Materials, Composite Wood & Agrifiber Products
1	Credit 5	Indoor Chemical & Pollutant Source Control
1	Credit 6.1	Controllability of Systems, Perimeter
1	Credit 6.2	Controllability of Systems, Non-Perimeter
1	Credit 7.1	Thermal Comfort, Comply with ASHRAE 55-1992
1	Credit 7.2	Thermal Comfort, Permanent Monitoring System
1	Credit 8.1	Daylight & Views, Daylight 75% of Spaces
1	Credit 8.2	Daylight & Views, Views for 90% of Spaces

5 | Innovation & Design Process — Possible Points: 5

Y	Credit 1.1	Innovation in Design: Transportation Incentives
1	Credit 1.2	Innovation in Design: Green Building Education
1	Credit 1.3	Innovation in Design: Pest Management
1	Credit 1.4	Innovation in Design: Flexible Exhibition System
1	Credit 2	LEED® Accredited Professional

Figure 16.8 LEED project scorecard

Figure 16.9   Natural Academy of Sciences Building—San Francisco, California

impacts of nonsustainable development on public health and the environment. Ultimately the cost of those impacts has to be paid, usually at a substantially higher price.

Five-hundred years ago, ignorance of the environment led to mass deaths from polluted water and dirty air. Even in the late 20th century civil engineers struggled to quantify and control waste discharges. In December of 2009, on the eve of the United Nations Climate Change Conference Copenhagen, the EPA Administrator signed two distinct findings regarding greenhouse gases under Section 202(a) of the Clean Air Act:

> **Endangerment Finding:** The Administrator finds that the current and projected concentrations of the six key well-mixed greenhouse gases—carbon dioxide (CO_2), methane (CH_4), nitrous oxide (N_2O), hydrofluorocarbons (HFCs), perfluorocarbons (PFCs), and sulfur hexafluoride (SF_6) —in the atmosphere threaten the public health and welfare of current and future generations.

> **Cause or Contribute Finding:** The Administrator finds that the combined emissions of these well-mixed greenhouse gases from new motor vehicles and new motor vehicle engines contribute to the greenhouse gas pollution, which threatens public health and welfare.

This action means that the EPA is now authorized and obligated to take reasonable efforts to reduce greenhouse pollutants under the Clean Air Act.

Another organization, Architecture 2030, was established by architect Edward Mazria in 2002. Edward Mazria, author of *The Passive Solar Energy Book* in 1979, has been a long-term supporter of sustainability principles. Architecture 2030's mission is to transform the U.S. and global building sector from the major contributor of greenhouse gas emissions to a central part of the solution to the global warming crisis. The organization has established specific targets for the AEC industry.

These examples are indicative of a movement from pollution control to pollution prevention and now to sustainability.

The 2030 Challenge Targets

All new buildings, developments, and major renovations shall be designed to meet a fossil fuel, greenhouse gas (GHG)-emitting, energy consumption perform- ance standard of 50 percent of the regional (or country) average for that building type.

At a minimum, **an equal amount of existing building area shall be reno- vated annually** to meet a fossil fuel, GHG-emitting, energy consumption per- formance standard of 50 percent of the regional (or country) average for that building type.

The **fossil fuel reduction standard** for all new buildings and major renovations shall be increased to 60 percent in 2010, 70 percent in 2015, 80 percent in 2020, and 90 percent in 2025.

And, finally, to meet the 2030 challenge targets new buildings shall be **Carbon-neutral in 2030,** meaning that they will use no fossil fuel GHG-emitting energy to operate.

Source: 2030 Challenge,/www.architecture2030.org/2030_challenge/index.html

SUMMARY

The Architecture, Engineering, and Construction (AEC) industry plays a critical role in delivering a diverse range of facilities and civil infrastructure systems, including res- idential, building, industrial facilities, and transportation, energy, water supply, waste management, and communications systems. It also plays a critical role in maintaining their quality, integrity, and longevity. At the same time, the AEC industry contributes to natural resource depletion, waste generation and accumulation, and environmental impact and degradation. As a result, a range of constituencies have been attempting to define the attributes and characteristics, the processes for the delivery and use, and the resources consumed in the delivery and use of facilities and civil infrastructure systems as possible mechanisms to slow, reduce, and eliminate these impacts.

Traditional approaches of environmental regulatory compliance or reactive cor- rective actions have proven to be consistently costly, inefficient, and often ineffective. In a sustainable approach to design and construction, decision-makers integrate sus- tainability at all stages of the project lifecycle, particularly the early funding allocation, planning, and conceptual design phases.

Existing standards, such as LEED (Leadership in Energy and Environmental De- sign) and BREEAM (Building Research Establishment Environment Assessment Method), are helpful but may give designers the perception that meeting these pre- scribed targets will result in satisfactory environmental performance. *Sustainable engi- neering* must address issues that go beyond checklists.

REFERENCES

The American Institute of Architects. (2007). *The Architects Handbook of Professional Practice*, Joseph A. Demkin, ed. John Wiley & Sons, Inc. New York.

American Society of Civil Engineers. (2008). *Civil Engineering Body of Knowledge for the 21st Century, 2d edition*. ASCE Report, Reston, VA.

Birkland, Janis. (2002). *Design for Sustainability: A Sourcebook of Integrated Ecological Solutions.* Earthscan, London.

Enovity—HOK. (2008). *Sustainable and Strategies Report for California State University, Sacramento*, July 2008.

Hansen, Karen Lee, and Jorge A. Vanegas. (2006). "A Guiding Road Map, Principles, and Vision for Researching and Teaching Sustainable Design and Construction." *American Society for Engineering Education (ASEE), Chicago, IL, Conference Proceedings.*

Mazria, Edward. (2002). *2030 Challenge.* http://www.architecture2030.org/2030_challenge/index.html

McDonough, William, and Michael Braungart. (2002). *Cradle to Cradle.* North Point Press. New York. ISBN 0-865- 47587-3.

Vanegas, Jorge A., ed. (2004). *Sustainable Engineering Practice: An Introduction.* ASCE Press, Reston, Virginia. ISBN 10: 0-784-40750-9, ISBN 13: 978-0-784-40750-9.

Yeang, Kenneth. (2006). *Ecodesign: A Manual for Ecological Design.* Wiley Academy, London.

Introduction

Ethics

Professional Engagement

1 2 3 4 5

History

What Engineers Deliver

Emerging Technology

A B C D E F

Engineer's Role in Project Development

6 7 8

Permitting

Leadership *Managing*

Having a Life

17

Executing a Professional Commission

9 10 11 12 13 14 15 16

Sustainability

Client Relationship

Legal Aspects

Communicating

Globalization

Chapter **17**

Emerging Technologies

Big Idea

Changing technologies, as well as change in general, are facts of life. In order to use change to their advantage, civil engineers must develop ways of maximizing the application of new technologies.

> Time is a sort of river of passing events, and strong is its current; no sooner is a thing brought to sight than it is swept by and another takes its place, and this too will be swept away.

> —Marcus Aurelius, 2d century A.D.

Key Topics Covered

- Introduction
- The Nature of Change
- Information Technology—Enabled Process Change
- Engineering Thinking
- Summary

Related Chapters in This Book

- Chapters 4: Professional Engagement
- Chapter 5: Engineer's Role in Project Development
- Chapter 6: What Engineers Deliver
- Chapter 7: Executing a Professional Commission
- Chapter 11: Legal Aspects of Professional Practice
- Chapter 12: Managing the Civil Engineering Enterprise
- Chapter 15: Globalization
- Chapter 16: Sustainability

(Continued)

Related to *ASCE Body of Knowledge 2* Outcomes

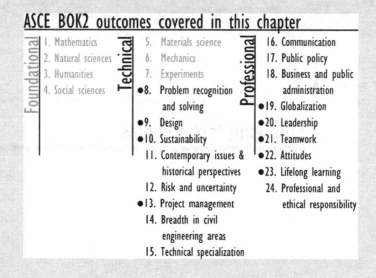

ASCE BOK2 outcomes covered in this chapter

Foundational
1. Mathematics
2. Natural sciences
3. Humanities
4. Social sciences

Technical
5. Materials science
6. Mechanics
7. Experiments
●8. Problem recognition and solving
●9. Design
●10. Sustainability
11. Contemporary issues & historical perspectives
12. Risk and uncertainty
●13. Project management
14. Breadth in civil engineering areas
15. Technical specialization

Professional
16. Communication
17. Public policy
18. Business and public administration
●19. Globalization
●20. Leadership
●21. Teamwork
●22. Attitudes
●23. Lifelong learning
24. Professional and ethical responsibility

INTRODUCTION

Technologies have helped form both our physical environment and world socio-economic systems. In his book *Guns, Germs, and Steel: The Fates of Human Societies*, Jared Diamond argues that advances in military technologies have been a key factor in shaping world history and development. By enabling the use of armored cavalry, something as simple as the adoption of the stirrup has determined who were the vanquished and the victors in more than one historic battle.

Today civil engineers are exposed to a heady array of new technologies. With such variety, decisions about which technology to adopt can be baffling. Civil engineers may ask:

- How can this technology create value for our clients/customers?
- What problems can this technology solve?
- What processes can this technology improve?
- How have some organizations enhanced their effectiveness/profitability with this technology?

This chapter addresses the use of emerging information technologies and processes. It highlights several new developments in engineering materials and methods and outlines some of the world's key engineering challenges. The chapter concludes with suggestions on how the development of *engineering thinking* can assist civil engineers to make the most of change.

> tech•nol•o•gy: the practical application of knowledge, especially in a
> particular area
>
> emer•ging: newly formed or prominent
>
> —Merriam Webster OnLine,/www.merriam-webster.com

THE NATURE OF CHANGE

For several decades, most organizations have found themselves in the midst of rapid business, technological, and process change. Now environmental change has been thrown into the mix. (See Chapter 15, Globalization and Chapter 16, Sustainability.) Some reasons for rapid change in the business environment include:

- Globalization of competition
- Strengthened role of powerful clients
- Increased regulation
- Internationalization of technologies and tools
- Growing distance between world-class firms and local "backbone"

Table 17.1 A Newer Perspective on Complex Issues
(Adapted from Waldrop, *Complexity: The Emerging Science at the Edge of Order and Chaos*)

Old Economics	New Economics
Based on 19th-century physics—equilibrium, stability, deterministic dynamics	Based on biology—structure, pattern, self-organization, lifecycle
People identical	Focus on individual life; people separate and different
If only there were no externalities and all had equal abilities, we'd reach Nirvana	Externalities and differences become driving force; no Nirvana, system constantly unfolding
Elements are quantities and prices	Elements are patterns and possibilities
No real dynamics in the sense that everything is at equilibrium	Economy rushes forward—structures constantly coalescing, decaying, changing
See subject as structurally simple	See subject as inherently complex
Economics as soft physics	Economics as high-complexity science

These changes have lead to a fundamentally different way of considering business and economics. Table 17.1 contrasts the old and new views of economics.

Intertwined with these changes are rapid, continuous changes in information technology hardware and software. Some of these developments include:

- Ever increasing processor speeds
- Miniaturization
- Exponentially expanded storage capacity
- "Hardened" hardware, i.e., hardware that functions on dirty construction sites
- Global positioning systems (GPS) and geographic information systems (GIS)
- Radio frequency (RF) tracking
- Integration of systems
- Interoperability standards
- The Internet
- Social networking
- Mobilization

Responses to these changes have been varied. Many involve efforts to improve processes with the help of new technologies and tools. Process improvement efforts include:

- Business Process Reengineering (BPR)
- Lean Value Chains (e.g., Lean Construction)
- Concurrent Engineering
- Continuous Improvement

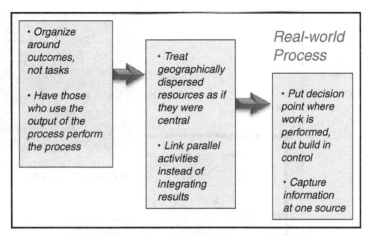

Figure 17.1 Fundamental principles of business process reengineering

- Total Quality Management (TQM)
- Total Enterprise Management

These approaches have ranged from the BPR perspective to highly quantitative methods. BPR essentially is an integration mechanism for holding together organizational forms and linking the operations of an organization to the requirements of customers. One dramatic result of BPR has been the pervasive outsourcing of all but the most essential functions in many U.S. corporations. Toward the other end of the process improvement spectrum are methods used in manufacturing. In this approach, a list of activities is coupled with the way in which these interact with resources. The resulting data can then be used to generate a mathematical model that can be solved to arrive at a production plan or distribution plan or plant design. See Figure 17.1 for the principles of BPR and Figure 17.2 for some process definition models used in process improvement.

INFORMATION TECHNOLOGY—ENABLED PROCESS CHANGE

For many years, architectural, engineering, and construction industry (AEC) practitioners and academics have believed that appropriate implementation of information technology (IT) design, engineering, and management support systems could help significantly to improve performance in project delivery. Yet the experience often was one of implementation in "islands of automation," such as in the use of computer-aided design (CAD) by design firms, or cost-estimating and scheduling systems by contractors. This, at best, resulted in small improvements in performance. The potential benefits, which may flow from the use of integrated systems, was largely imagined, rather than realized.

An approach has been needed in which firms learn how to deploy technology to move from the limited benefits achievable through the substitution of information

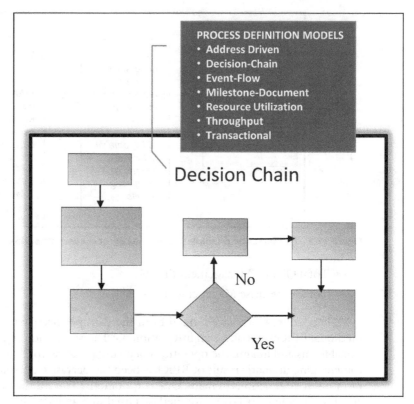

Figure 17.2   Process definition models, decision chain example

technologies for existing technologies, resulting in the automation of individual work steps, to a wider transformation of the entire process. A three-part model of the information technology adoption process illustrates the possibilities for firms to improve performance incrementally by moving through each phase, and more radically by pursuing strategies to transform processes with the assistance of information technology systems. (See Figure 17.3.)

Early Developments

The seeds for this transformation were sown many years ago. AEC firms have responded to change through the adoption of new tools such as:

- 3D CAD
- Simulation
- VR (Virtual Reality)
- Rapid Prototyping
- Modeling and Workflow
- Automated Process Improvement

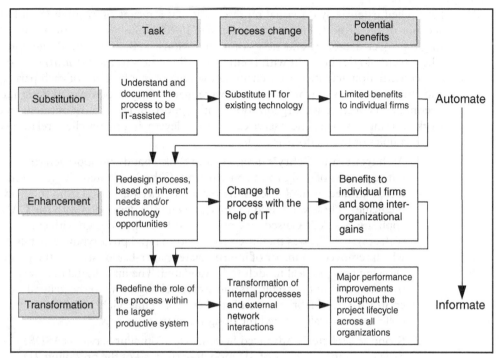

Figure 17.3 IT adoption and business process change
(Adapted from Gann & Hansen, *IT Decision Support and Business Press Change in the United States*)

Perhaps for historical reasons, many of these tools have focused on design. Over the last 40 years or so, there have been numerous attempts to devise explicit models of the design process, including approaches which assist in finding ways of construing a rational model for architecture, for example, exploring the total industrial process of construction, sometimes linked to an even larger project process including land development; or developing a generic approach to design, including product design, manufacturing design, chemical engineering, and so forth.

The most obvious thrust of design method work was toward CAD drawing systems and attempts to link these either to prospective component or building systems or to form-finding methods that exploited classic theories of typology, proportion, and composition. There also were attempts to link such programs to databases of components used as part of a maintenance strategy for buildings. In the 1970s and 1980s, architectural researchers were unable to devise the types of coherent computer-aided design—computer-aided manufacturing (CAD–CAM) systems developed in some manufacturing engineering industries, neither were they able to develop a total computerized architecture in the vision of Negroponte's 1970 book, *The Architecture Machine*. (Nicolas Negroponte was a pioneer in the field of computer-aided design and has been a member of the MIT faculty since 1966.)

One of the difficulties was that, by contrast, until relatively recently there have been poor methods for representing construction processes. Consequently, designers have relied extensively on visual representation as a means of maintaining a holistic account of the project. However, in the mid-1990s some firms sought to introduce single project databases and new communications media into the design process. These technologies brought with them a new dimension to the *integration* of design and construction activities; they changed the type of involvement of each participant and altered the ways in which decisions were made. They also provided design, engineering, and construction organizations with opportunities to carry out new types of work, offering customers new services and thus developing better client relationships.

Examples of these efforts include:

1. Architect **Frank Gehry's** design studio took a hands-on approach to the entire process of design and construction. Gehry's sculpted physical models were digitized using 3D scanning technologies (transferred from the field of medicine). These were used to drive a 3D model developed with the CAD application, CATIA (used by Chrysler and Boeing to design with curved surfaces). CATIA was then used to provide a rapid prototyping capability which removed a number of intermediate (paper-based) steps in the process, allowing new physical models to be validated. The final digital data was transferred from the architects and structural engineers to the general contractor, steel fabricator, and steel erector. Thus, Gehry and his colleagues radically altered the design-construct process.

2. **Stone & Webster's** Advanced Systems Development Services (ASDS) was an early user and developer of 3D modeling (instead of the prevailing 2D systems). It was specialized in project-specific systems integration, developing and customizing applications software to link CAD with databases and knowledge-based systems. The result was an "as needed" approach to integrating systems, creating a "bricolage model." ASDS solutions have been used effectively in mechanical engineering (by firms such as Chrysler and Mercedes-Benz). This was a particular approach to middleware development in which the project appears to induce the selection of data, knowledge, and applications, which are then transformed into practical tools.

3. **Parsons** adopted an approach to sharing project information aimed specifically at extending the market for their services from engineering, procurement, and construction, forward into early project decision-making and downstream into facility management. To achieve this, they developed their existing Computer-Integrated-Engineering IT support systems to form a new Computer-Integrated-Project system. This was supported by a variety of technologies such as GIS in early project decision-making. The adoption of this approach resulted in the need for internal business process changes and new relationships with suppliers and other design and construction organizations. It created opportunities for the company to provide its clients with new value-added services, extending Parsons' markets.

4. **Bechtel** used Virtual Reality systems to share information with their clients in order to reduce risk and uncertainty and improve predictability in design

decisions. The technology provided the client with a decision-making tool, in which transnational customer links facilitated visualizations of prospective facilities enabling clients to modify design decisions at little expense. The system also helped to reduce overall project times and saved on travel costs. Other developments at Bechtel included simulations of heavy lifting processes and digital data collection tools for site work.

As the new millennium has come into focus and then moved on, these earlier approaches have become less leading-edge and more the norm.

Building Information Modeling

Advances in *interoperability*, the ability to exchange data between computer programs, have hastened the development of new technologies. Among these is building information modeling (BIM). One of the promises that BIM holds out is the potential to bridge the gap between conceptual and technical problem solving. (See Figure 17.4.)

Design, engineering, and construction work involves integrating and assembling different subsystems and components. Many of the challenges facing the AEC industry relate to problems encountered at the interfaces between the work of different professional disciplines. It is here that BIM may be of particular assistance in helping to improve information flows between different experts, professionals, technicians, and trades, and across building lifecycles.

No single computer program has yet been able to support all tasks associated with design and construction; but interoperability allows data to flow from one application to another. Thus, many experts and applications can be included in the

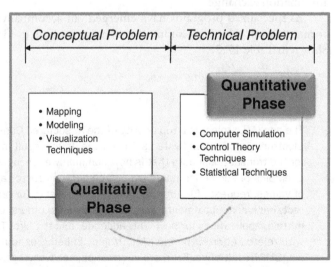

Figure 17.4   Bridging the gap between conceptual and technical problems

Table 17.2   Methods of Exchanging Data
(Adapted from Eastman, Teicholz, Sacks, and Liston, *BIM Handbook: A Guide to Building Information Modeling for Owners, Managers, Designers, Engineers, and Contractors*. p. 67.)

	Medium	Description
1	Direct, proprietary link between specific BIM tools	Provides an integrated connection between two applications. Relies on middleware software interfacing capabilities like ODBC or COM or proprietary interfacing capabilities like ArchiCad's GDL or Bentley's MDL, all of which use C, C++, or C# languages.
2	Proprietary file exchange format, primarily dealing with geometry	Interfaces within specific company's own applications. In the AEC industry, well-known formats include: DXF (Data eXchange Format), developed by Autodesk; SAT by Spatial Technology; STL for stereo lithography; and 3DS for 3D-Studio.
3	Public product data exchange format	Involves an open-standard building model. IFC (Industry Foundation Class) and CIS/2 for steel are the principal options. Carries information regarding object and material properties and relations between objects as well as geometry.
4	XML-based exchange format	Supports exchange of many types of data between applications. XML is eXtensible Markup Language, an extension of HTML, the base language of the Web. Especially good at exchanging small amounts of business data between two applications set up to do so.

design-construct process. Still, two applications can export or import different information for describing the same object. In the United States, there is an effort to standardize the data required for particular workflow exchanges. The main endeavor is called the National BIM Standards (NBIMS) and is being conducted by the National Institute of Building Sciences. Table 17.2 summarizes the primary methods used in information exchange.

As specialized programs have emerged, an accompanying plethora of exchange formats has developed. Common exchange formats used in the AEC industry are shown in Table 17.3.

agcXML

The agcXML project is a top priority of the Associated General Contractors (AGC) Electronic Information Systems Committee. It will result in a set of XML schemas for the transactional data that is now commonly exchanged in paper documents. Examples of such documents include owner/contractor agreements, schedules of values, requests for information (RFIs), requests for proposals (RFPs), architect/engineer supplemental instructions, change orders, change directives, submittals, applications for payment, addenda, and the like. To ensure compatibility with related efforts, the agcXML Project is being executed as part of the buildingSMART Initiative. For more information, see www.agcXML.org.

Table 17.3 Common Exchange Formats Used in the AEC Industry
(Adapted from Eastman, Teicholz, Sacks, and Liston, *BIM Handbook: A Guide to Building Information Modeling for Owners, Managers, Designers, Engineers, and Contractors.* **p. 69.)**

Format Type	Variability	File Extensions
Image (raster)	Compactness, number of possible colors per pixel, data loss with compression	JPG, GIF, TIF, BMP, PIC, PNG, RAW, TGA, RLE
2D Vector	Compactness, line widths and pattern control, color, layering, and types of curves supported	DXF, DWG, AI, CGM, EMF, IGS, WMF, DGN
3D Surface and Shape	Types of surfaces and edges represented, whether surfaces and/or solids are represented, material properties of shapes (color, bitmap, texture map), and viewpoint information	3DS, WRL, STL, IGS, SAT, DXF, DWG, OBJ, DGN, PDF(3D), XGL, DWF, U3D, IPT, PTS
3D Object Exchange	Geometry according to the 2D or 3D types represented, object properties, and relations between objects	STP, EXP, CIS/2
Game	Types of surfaces, whether they carry hierarchical structure, types of material properties, texture and bump map parameters, animation, and skinning	RWQ, X, GOF, FACT
GIS	Geographical information system	SHP, SHX, DBF, DEM, NED
XML	Information exchanged and workflows supported	AecXML, Obix, AEX, bcXML, AGCxml

Communication among project participants occurs formally and informally, and on a variety of levels. Eastman, Teicholz, Sacks, and Liston have identified four different types of communication exchange (as opposed to data transfer) that transpire in a BIM process. These are shown in Table 17.4.

Table 17.4 Types of Communication in BIM Processes
(Adapted from Eastman, Teicholz, Sacks, and Liston, *BIM Handbook: A Guide to Building Information Modeling for Owners, Managers, Designers, Engineers, and Contractors.* **pp. 123—124.)**

	Communication	Description
1	Published snapshots	One-directional, static views that provide the receiving party with access only to visual or filtered meta-data, such as bitmap images.
2	Published BIM views and meta-data	Viewing access to the model with limited ability to edit or modify data, such as PDF or DWF. Receiving party can perform query functions on the model, comment, mark-up, and change certain view parameters.
3	Published BIM files	Access to the native data through proprietary and standard file formats such as DWG, RVT, and IFC.
4	Direct database access	Access to the project database through a dedicated or distributed project server. Model data controlled through access privileges or more sophisticated edit and change capabilities.

Based on interviews with hundreds of owners; architects; civil, structural, and MEP (mechanical, electrical, plumbing) engineers; construction managers; and general contractors and subcontractors currently using BIM, a McGraw-Hill report (*Smart Market Report: Building Information Modeling*, p. 27) found that the most valuable aspects of BIM are:

- Easier coordination of different software and project personnel
- Improved efficiency, production, and time savings
- Lifecycle analysis, including modeling energy usage
- Better communication
- Improved quality control/accuracy
- Visualization (ability to keep owners informed and to clarify construction tasks to workers)
- 3D modeling and coordination, including interference checking/clash detection
- Keeping pace with advances by competition and others in the marketplace

The report also found that the U.S. Army Corps of Engineers is requiring BIM-based deliverables as part of its Centers for Standardization program, an effort involving 43 standard facility types.

Hurdles on the Path to BIM Adoption

Adequate Training

Training is often the biggest challenge with the adoption of any new technology. Because few users have expert backgrounds, there is a shortage of training resources. As more expertise develops in universities, within firms, and from consultants/trainers, the challenge of training should be reduced. Among architects, engineers, contractors, and owners, engineers are most concerned about training.

Costs of Software and Hardware Upgrades

Issues related to cost also are common with the adoption of new technologies. Increased costs of software and hardware upgrades are significant concerns in BIM adoption. These costs are of greater importance to architects and engineers than contractors and owners.

Senior Management Buy-In

Higher-level management is less likely than any group to embrace BIM adoption. This could be because they have to justify the costs of adoption (in terms of training, software, and hardware) or because they are more comfortable with "tried-and-true" methods. Junior-level staff buy-in is considered least challenging, possibly because they may have been exposed to BIM as part of their education, are more open to change, and/or are less aware of associated risks.

Other Factors

Among architects, engineers, contractors, and owners, engineers are most likely to see a lack of external incentives or directives moving them to use BIM.

Both architects and engineers are challenged by the potential loss of intellectual property and increased liability associated with BIM.

—McGraw-Hill Construction, *Smart Market Report: Building Information Modeling*, p. 9.

More recently, developments in computer information technologies have converged with new project delivery methods and contract structures, such as Design-Build, Design-Assist, and Integrated Project Delivery. (See Chapter 11, Legal Aspects of Professional Practice.)

Integrated Project Delivery

The American Institute of Architects (AIA) has become active in promoting integrated project delivery (IPD). From the AIA's perspective [AIA National and AIA California Council, *Integrated Project Delivery: a Guide*]:

Integrated Project Delivery (IPD) is a project delivery approach that integrates people, systems, business structures and practices into a process that collaboratively harnesses the talents and insights of all participants to optimize project results, increase value to the owner, reduce waste, and maximize efficiency through all phases of the design, fabrication and construction.

IPD principles can be applied to variety of contractual arrangements and IPD teams can include members well beyond the basic triad of owner, architect, and contractor. In all cases, integrated projects are uniquely distinguished by highly effective collaboration among the owner, the prime designer, and the prime constructor, commencing at early design and continuing through to project handover.

Table 17.5 Traditional versus Integrated Project Delivery
(Adapted from AIA National and AIA California Council, *Integrated Project Delivery: A Guide*, p. 1.)

Element	Traditional Project Delivery	Integrated Project Delivery
Teams	Fragmented group of prime designer and subconsultant representatives, assembled on "just-as-needed" or "minimum necessary" basis; strongly hierarchical, controlled	Integrated team entity composed of key project stakeholders (owner, architect, engineers, contractor, subcontractors, others) assembled early in the process; open, collaborative
Process	Linear, distinct, segregated; knowledge gathered "just-as-needed"; information hoarded; silos of knowledge and expertise	Concurrent and multilevel; early contributions of knowledge and expertise; information openly shared; stakeholder trust and respect
Risk	Individually managed by each entity; transferred to the greatest extent possible	Collectively managed; appropriately shared
Compensation/ reward	Individually pursued by each entity; minimum effort for maximum return; (usually) first cost basis	Team success tied to project success; value based
Communications/ technology	Paper-based; analog; 2 dimensional	Digitally based; Building Information Modeling (3-, 4-, 5-dimensional BIM)
Agreements	Unilateral effort encouraged; risk allocated and transferred; no risk sharing	Multilateral open sharing and collaboration encouraged, fostered, promoted, and supported; risk sharing

IPD changes contract structures, the way project teams are formed, the manner in which the teams interact, and the technologies used in project delivery. See Table 17.5 and Figure 17.5.

IPD is a new process enabled by software, but it is not the software itself. BIM and its accompanying interoperability aid the project team in accomplishing the primary goal: satisfying (or more than satisfying) a client's need within a specific time period and for a given budget.

The basic principles of IPD include:

1. *Mutual respect and trust:* All members of the integrated team—owner, designers, consultants, contractor, subcontractors, and suppliers—value collaboration and are committed to working as a team in the best interests of the project.

2. *Mutual benefit and reward:* All participants and/or team members benefit from IPD. Compensation structures recognize the need for and reward early involvement. Compensation is based on value added, such as incentives tied to achieving project objectives.

3. *Collaborative innovation and decision-making:* Freely exchanged ideas among all participants stimulates innovation. Ideas are judged on merits, not on their

author's role or status. To the greatest extent possible, decisions are made unanimously.

4. *Early involvement of key participants:* Participants are involved at the earliest practical moment, thereby improving decision-making through the influx of knowledge and expertise of all key participants. Decisions that are made early have the greatest effect. (See Chapter 4, Engineer's Role in Project Development.)

5. *Early goal definition:* Project goals are developed early, agreed upon, and respected by all participants. Project outcomes are held at the center of a framework of individual objectives and values.

6. *Intensified planning:* Increased efforts in planning result in increased efficiency and savings during execution. The thrust of IPD is not to reduce the design effort but to improve the design results and thereby streamline and shorten the construction period.

7. *Open communication:* Team performance is based on open, direct, and honest communication. A no-blame culture leads to early identification and resolution of problems. Disputes are recognized as they occur and are resolved promptly.

8. *Appropriate technology:* Proper technology is specified at project initiation to maximize functionality, generality, and interoperability. Technology that complies with open standards is used whenever possible because it best enables communications among all project participants.

9. *Organization and leadership:* Project team members are committed to the team's goals and values. Leadership is taken by the team member most capable with regard to specific work and services—often design professionals and contractors lead in their areas of traditional expertise. Roles are defined clearly but do not create artificial barriers.

Dimensions Defined

2D—2-dimensional project representation; x and y coordinates only; often paper-based

3D—3 dimensions; x, y, and z coordinates included in a geometric digital model, sometimes with additional "intelligence" attached to drawing objects

4D—dimension of time incorporated into the 3D digital model so that the construction schedule can be visualized

5D—dimension of cost incorporated into the 3D digital model in order to automate quantity take-offs. When used with the 4D feature, 5D also can predict cash flow.

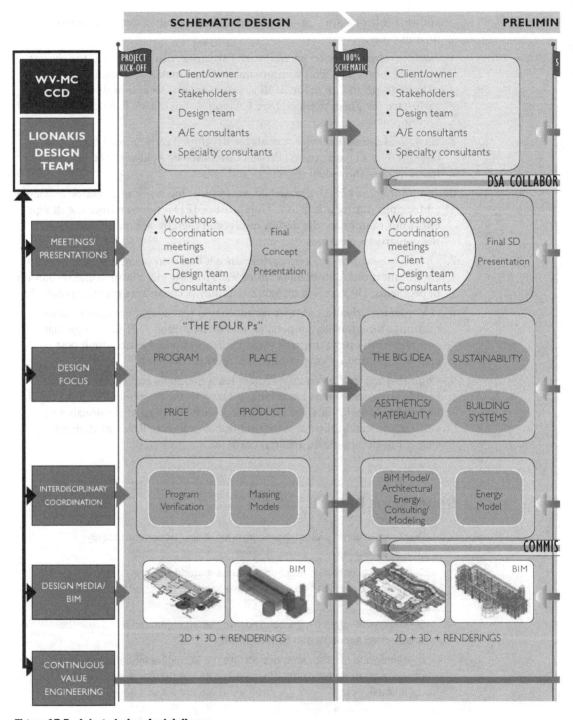

Figure 17.5 Integrated project delivery

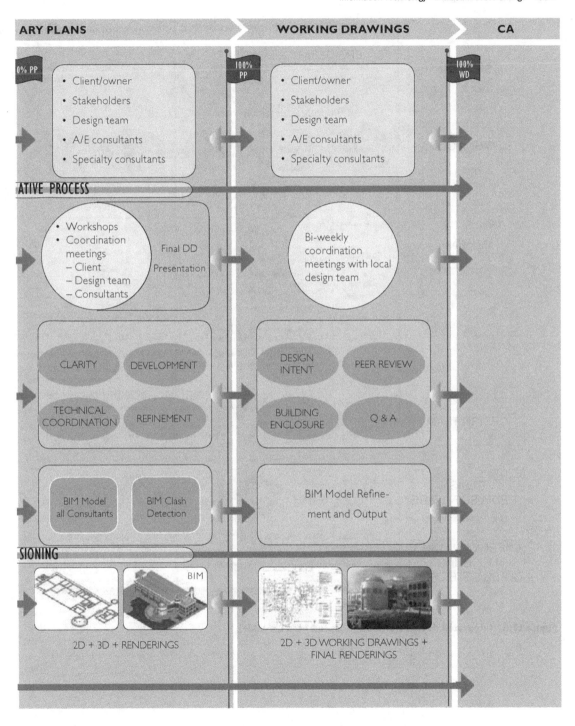

ARY PLANS | WORKING DRAWINGS | CA

0% PP
- Client/owner
- Stakeholders
- Design team
- A/E consultants
- Specialty consultants

100% PP
- Client/owner
- Stakeholders
- Design team
- A/E consultants
- Specialty consultants

100% WD

ATIVE PROCESS

- Workshops
- Coordination meetings
  – Client
  – Design team
  – Consultants

Final DD Presentation

Bi-weekly coordination meetings with local design team

CLARITY   DEVELOPMENT

TECHNICAL COORDINATION   REFINEMENT

DESIGN INTENT   PEER REVIEW

BUILDING ENCLOSURE   Q & A

BIM Model all Consultants   BIM Clash Detection

BIM Model Refinement and Output

SIONING

BIM

2D + 3D + RENDERINGS

2D + 3D WORKING DRAWINGS + FINAL RENDERINGS

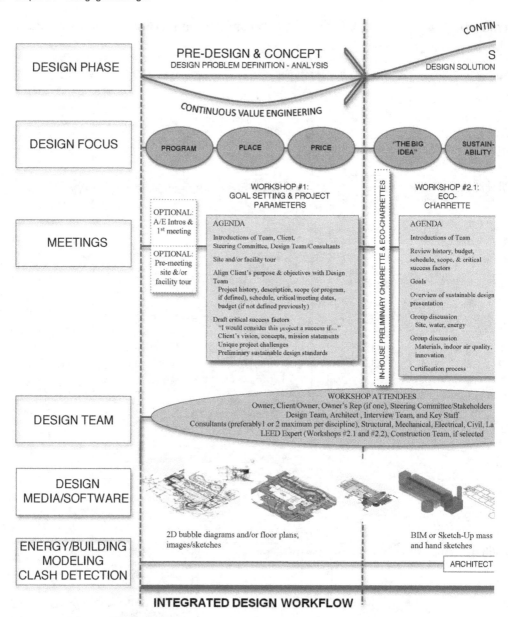

Figure 17.6 Integrated design workflow (adapted from Lionakis)

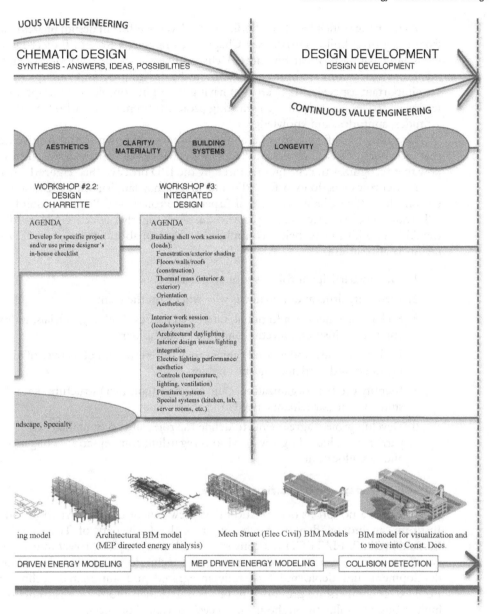

UOUS VALUE ENGINEERING

CHEMATIC DESIGN
SYNTHESIS - ANSWERS, IDEAS, POSSIBILITIES

DESIGN DEVELOPMENT
DESIGN DEVELOPMENT

CONTINUOUS VALUE ENGINEERING

AESTHETICS   CLARITY/
MATERIALITY   BUILDING
SYSTEMS   LONGEVITY

WORKSHOP #2.2:
DESIGN
CHARRETTE

WORKSHOP #3:
INTEGRATED
DESIGN

AGENDA

Develop for specific project
and/or use prime designer's
in-house checklist

AGENDA

Building shell work session
(loads):
    Fenestration/exterior shading
    Floors/walls/roofs
    (construction)
    Thermal mass (interior &
    exterior)
    Orientation
    Aesthetics

Interior work session
(loads/systems):
    Architectural daylighting
    Interior design issues/lighting
    integration
    Electric lighting performance/
    aesthetics
    Controls (temperature,
    lighting, ventilation)
    Furniture systems
    Special systems (kitchen, lab,
    server rooms, etc.)

ndscape, Specialty

ing model   Architectural BIM model
(MEP directed energy analysis)   Mech Struct (Elec Civil) BIM Models   BIM model for visualization and
to move into Const. Docs.

DRIVEN ENERGY MODELING   MEP DRIVEN ENERGY MODELING   COLLISION DETECTION

As the basic principles of IPD indicate, the level of effort in design phases changes from design-bid-build (DBB). (See Chapter 5, Engineer's Role in Project Development and Chapter 6, What Engineers Deliver.) Because of the collaborative environment, many decisions are brought forward in the design process. Consequently, two very important aspects of IPD are: (1) having the right contract for the specific professional services needed; and (2) selecting project team members who have the right attitudes, aptitudes, and knowledge.

Lionakis, a California-based architectural practice known for innovative design that incorporates sustainability and technology, has developed an approach to workflow that recognizes the changes required by the IPD process. (See Figure 17.6.)

Model contracts do exist for IPD, and an attorney familiar with IPD also should be consulted. (See Chapter 11, Legal Aspects of Professional Practice.) Building the IPD teams involves selecting people who can work together effectively. According to the AIA, in addition to being committed to the collaborative process, IPD team members also should:

1. Identify participant roles as soon as possible

2. Prequalify firms and individuals who will be on the team

3. Seek involvement of additional, interested parties (building officials, utility companies, insurers, sureties, and other stakeholders)

4. Define mutually understood values, goals, interests, and objectives of team members and participating stakeholders

5. Identify the IPD organizational and business (contract) structure that is best suited to the participants' needs and constraints

6. Develop project agreements to define the roles and accountability of participants, including key provisions regarding compensation, obligation, and risk allocation

FIATECH Roadmap—An Organizing Principle

FIATECH is a nonprofit organization that grew out of the Construction Industry Institute, an independent research center at the University of Texas at Austin. Formed in 2000, FIATECH is a consortium of leading capital project industry owners, engineering construction contractors, and technology suppliers that advocates development and deployment of fully integrated and automated technologies. FIATECH members are grounded in business and their motivation is to deliver the highest business value throughout the lifecycle of capital projects.

FIATECH's *Capital Projects Technology Roadmap* (see Figure 17.7) presents a vision for the capital projects industry (i.e., the industry that executes the planning, engineering, procurement, construction, and operation of predominantly large-scale buildings, plants, facilities, and infrastructure). According to FIATECH and many others, the capital projects industry greatly lags other sectors in exploiting technological advances:

> It is characterized by vast disparities in business practices and levels of technology application. It is fragmented, with great divergence in tools and technologies from

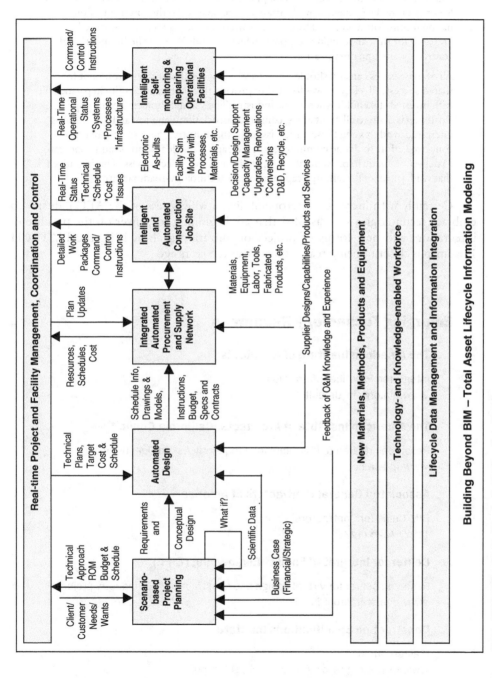

Figure 17.7 FIATECH's Capital Projects Technology Roadmap

company to company and across its supply chains. New pressures, such as Homeland Security in the U.S., have moved infrastructure security to the forefront of our national consciousness . . . The capital projects industry generally is not well prepared for this far-reaching response, which extends beyond the boundaries of control for any one organization.

All of these issues can and should be addressed in a collaborative environment for shared success. FIATECH was formed to provide that integrating entity in partnership with invested stakeholders across the industry. The Capital Projects Technology Roadmap is open to all companies, consortia, associations, and research institutions interested in addressing these critical issues to the industry. Presently, there is no concerted effort to define common goals, leverage available resources, and cooperate to deliver dramatic improvements in capability and cost-effectiveness. This initiative fills that void. [http://fiatech.org/capital-projects-technology-roadmap.htm]

Certainly with increased interoperability, a willingness of the AEC industry to embrace technologies such as BIM that are beginning to deliver on the promise of integration, and new forms of project organization such as IPD, the FIATECH Roadmap is closer to being reality than it was ten years ago.

Emerging Technology Resources

The American Institute of Architects

Integrated Practice Information
www.aia.org/ip_default

The American Institute of Architects, California Council

Resources related to IPD including Frequently Asked Questions
www.ipd-ca.net

Associated General Contractors of America

BIM Guide for Contractors
http://agc.org/

Center for Integrated Facility Engineering (CIFE)

Research center for virtual design and construction AEC industry projects
www.cife.stanford.edu

Construction Specifications Institute

MasterFormat
www.csinet.org/s_csi/docs/9400/9361.pdf

Construction Users Roundtable (CURT)

Owners' views on the need for integrated project delivery
www.curt.org/

Design Build Institute of America (DBIA)

Library of information and case studies related to design-build
www.dbia.org

FIATECH

Consortium of leading capital project industry owners, engineering
construction contractors, and technology suppliers that provides
global leadership in development and deployment of fully integrated and
automated technologies
http://fiatech.org/

International Alliance for Interoperability (IAI)/buildingSMART Alliance

International organization working to facilitate software interoperability and
information exchange in the AEC/FM industry
www.iai-na.org/

LEAN Construction Institute

Nonprofit corporation dedicated to conducting research to develop
knowledge regarding project-based production management in the design,
engineering, and construction of capital facilities
www.leanconstruction.org/

McGraw-Hill Construction

Source for design and construction industry information regarding IPD
www.construction.com/NewsCenter/TechnologyCenter/Headlines/
archive/2006/ENR_1009.asp

National Institute of Building Sciences, National BIM Standards (NBIMS) Committee

Many related articles on IPD and BIM
www.facilityinformationcouncil.org/bim/publications.php

Cost Analysis of Inadequate Interoperability in the U.S. Capital Facilities
Industry
www.bfrl.nist.gov/oae/publications/gcrs/04867.pdf

(*Continued*)

UNIFORMAT II Elemental Classification for Building Specifications, Cost Estimating, and Cost Analysis

www.bfrl.nist.gov/oae/publications/nistirs/6389.pdf

OmniClass

Classification structure for electronic databases
www.omniclass.org/

Open Geospatial Consortium

International, voluntary consensus standards organization that is leading the development of standards for geospatial and location-based services
www.opengeospatial.org/

Open Standards Consortium for Real Estate

Standards related to information sharing—BIM
http://oscre.org/

U.S. General Services Administration

Nation's largest facility owner and manager's program to use innovative 3D, 4D, and BIM technologies to complement, leverage, and improve existing technologies to achieve major quality and productivity improvements
www.gsa.gov/bim

Source: AIA National and AIA California Council, *Integrated Project Delivery: A Guide*

Some Technologies on the Horizon for Civil Engineering Projects

1. **Transdisciplinary, Transinstitutional, and Transnational:** Eliminating the artificial boundaries among disciplines and knowledge domains, institutions (public and private), and nations, in the pursuit of solutions

2. **Ubiquitous Computing:** Making many computers available to a user throughout the physical environment, while making them effectively invisible to the user, enabling the user to remotely interact with people and the natural, built, and virtual environments; remotely

monitor, collect, and access data, information, knowledge, experience, and wisdom; and remotely control devices.

3. **Ubiquitous Positioning Technologies:** Enabling the location of people, objects, or both, anytime, whether they are indoors or outdoors or moving between the two, at predefined location accuracies, with the support of one or more location-sensing devices and associated infrastructure

4. **Cloud Computing:** A style of computing in which capabilities related to Information Technologies (IT) are provided to users "as a service" allowing them to access technology-enabled services from the Internet ("in the cloud") without requiring knowledge of, expertise with, or control over the technology infrastructure that supports the services

5. **Augmented Reality:** A term for a live direct or indirect view of a physical real-world environment whose elements are merged with, or augmented by virtual computer-generated imagery, creating a mixed reality

6. **Collective Intelligence:** A shared or group intelligence that emerges through collaboration, innovation, and competition, from the capacity of human communities to evolve toward higher-order complexity and integration, which (1) appears in a wide variety of forms of consensus decision-making in bacteria, animals, humans, and computer networks; and (2) is studied as a subfield of sociology, of business, of computer science, of mass communications, and of mass behavior— from the level of quarks to the level of bacterial, plant, animal, and human societies.

7. **Automation and Robotics:** The application of science, engineering, and technology (particularly electronics, mechanics, control systems, computer-aided technologies, hardware and software, and artificial intelligence), in the design, manufacture, and application of autonomous devices and robots for industrial, consumer, or entertainment use, which reduce the need for human sensory and mental requirements, and which perform tasks that are too dirty, dangerous, repetitive, or dull for humans

8. **Nano-Bio-Info-Cogno Convergence:** The synergistic combination of four major provinces of science and technology, each of which is currently progressing at a rapid rate: (1) nanoscience and nanotechnology; (2) biotechnology and biomedicine, including genetic engineering; (3) information technology, including advanced computing and communications; and (4) cognitive science, including cognitive neuroscience

ENGINEERING THINKING

With so many changes unfolding simultaneously, how can civil engineers be prepared to answer the questions posed in the introduction to this chapter?

- How can this technology create value for our clients/customers?
- What problems can this technology solve?
- What processes can this technology improve?
- How have some organizations enhanced their effectiveness/profitability with this technology?

In truth, ancient, Renaissance, 19th-century engineers, basically all engineers who have preceded today's civil engineers, have pushed the limits of the technologies known to them. One way to thrive in an environment that involves continuous and rapid change is to develop *engineering thinking*.

Engineering Thinking

A desirable attribute of a professional engineer is to be a clear-thinking, innovative problem solver. The following section explores how such competence may be developed.

1. Knowledge

For the purposes of the discussion which follows, the following definitions are inferred:

Knowledge— that which is contained in the brain. What a person knows.

Information—a representation of knowledge outside the brain in the form of text, speech, graphics, mathematical models, and so forth.

Two types of knowledge are:

Explicit knowledge—can be represented as information

Tacit knowledge—cannot or has not yet been represented as information

1.1 Features of Tacit Knowledge

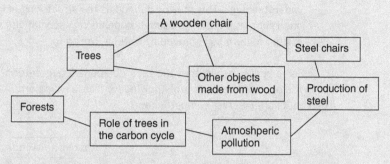

Figure 1   Example of knowledge associations

Since tacit knowledge, by definition, cannot be identified directly, it can only be inferred from outcomes which required its use. Features of tacit knowledge include:

Associativity—Items of knowledge can be deeply interconnected in the brain. For example, consider a simple object such as a specific wooden chair. The accompanying organizational chart shows a small set of the issues and entities that can be linked to a wooden chair. This does not form a simple hierarchy for which there are limits to the directions of the interconnections but is a distributed network for which there can be links between any node. While we can present as information small sets of such knowledge, the totality of the associations among items of knowledge in the brain is very, very large. Computers, as yet, do not come near to simulating the structure of such interconnectivity in the human brain. Therefore, the associations among the items of knowledge are mainly tacit. Such associativity is a major feature of the power of the human brain.

Intuition—We often know things without knowing why we know them. The brain is a phenomenally complex engine which can process knowledge subconsciously. Some believe that we do not take adequate advantage of our intuition (Gigerenzer, 2004).

Judgment—An important feature in the use of nondeterminate processes (Section 2) is that decisions can seldom be based on logic alone. Use of the word "judgment" tends to relate to such contexts.

(*Continued*)

Computers, probably because of their low level of associativity among items of information, cannot match the power of the brain in making judgments.

Understanding—This may be defined as the structuring of knowledge in the brain such that it can be used. It depends on associativity and, as discussed in what follows, is improved by working the brain.

Some people assert that the tacit component is greater and more important than the explicit component of knowledge. This is an important issue in the development of engineering competence.

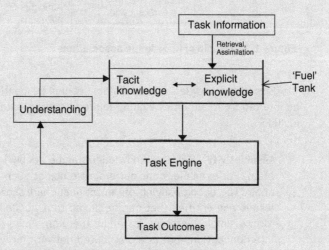

Figure 2   An engine model for information tasks

Figure 2 shows an engine model of competence. One can think of knowledge as the "fuel" needed to drive professional engineering tasks. Input information becomes explicit knowledge in the fuel tank. In the task process, explicit and tacit knowledge are mixed to achieve task outcomes. The processes of the task engine have another important outcome: They develop understanding which feeds back into the tank as new tacit knowledge. Therefore the mental task engines, unlike a combustion engine, enhance the quality of the fuel rather than consume it. The quality of the combination of explicit and tacit knowledge in the brain is fundamental to competence. Education tends to focus on explicit knowledge—one cannot transmit tacit knowledge. It develops by thinking.

It is clear that one cannot have too much explicit knowledge but its value is significantly less if it has not been used in tasks so as to develop corresponding tacit components. Therefore, to be a good engineering thinker one has to get the brain working hard.

2. Process Thinking

A *determinate process* has a unique outcome whereas a nondeterminate process may have more than one valid outcome. For example, when using a mathematical model the decision about which model to use will be nondeterminate but the calculation process will be determinate. With determinate processes the difference between the outcome and the correct result is error whereas with nondeterminate processes, acceptance of outcomes is in the realm of uncertainty. Most professional engineering processes are nondeterminate, often with determinate subprocesses.

All processes can be viewed as having three basic components:

- *Inception*—the requirements are established and information is gathered.
- *Conception*—the process is defined.
- *Production*—the process is implemented.

These components need to be controlled by asking relevant questions such as:

Inception: Are the requirements complete and clear? Is the input information adequate? (The assessment question)

Conception: Is the process capable of satisfying the requirements? Is the process the best in the context? (The validation question)

Production: Has the process been correctly implemented? (The verification question)

Table 1 shows review/control activities relevant to the three process components.

Table 1   Basic Process Model

Stage	Activity	Review/Control
Inception	Define the requirements, acquire information, investigate	Assess requirements, assess other input information
Conception	Identify options, evaluate, decide	Validate (ensure that the process can satisfy the requirements), optimize (seek to identify the best process)
Production	Implement	Verify (ensure that the process has been correctly implemented), interpret outcomes, revalidate

(Continued)

While successful engineers might not be explicit in expressing that they use the control strategies listed in Table 1, simple logic shows that they must do this. Under what circumstances can you achieve good outcomes if you do not have a clear idea of your objectives (requirements assessment) or if your process is not capable of satisfying the requirements (validation) or if the process has not been correctly implemented (verification)? The answer to this question is: "Only by luck!"

2.1 Process Model for Design

The process model of Table 1 applied to engineering design is shown in Table 2.

Table 2   Basic Process Model for Engineering Design

Stage	Activity	Review/Control
Inception	Define the requirements, acquire information, investigate the context, equip (in terms of staff competence, software, hardware, etc.)	Assess requirements, assess input information, assess equipment
Conception	Identify design options, evaluate, decide on the design solution	Validate the options, optimize the solution
Production	Technical design, produce drawings and specifications	Verify the outcomes against the requirements

2.2 Process Model for Analysis Modelling

The process model of Table 1 applied to analysis modelling (i.e., use of mathematical models to predict behaviour of engineering entities) is shown in Table 3.

Table 3   Basic Process Model for Analysis Modeling

Stage	Activity	Review/Control
Inception	Define the requirements, equip	Assess requirements
Conception	Establish the model	Validate the model (ensure that it can satisfy the requirements), optimize the model, validate, and verify the software
Production	Prepare data and carry out calculations	Verify the results (ensure that the model has been correctly implemented); interpret the results to identify behavior of the system being modeled. Carry out sensitivity analysis to gain understanding of behavior of the system

2.3 How Should the Process Model Be Used?

It is important to be constantly asking control questions and challenging outcomes. This is a main component of good engineering thinking.

The process model is used recursively, that is, it is used for the overall context and for detailed issues. For example, it is used for an overall design and for a detailed part of the system being designed. It can even be used on itself. For example, if you have to produce requirements, it may be worthwhile to establish requirements and a process for doing that.

2.4 Example of Challenge to Outcomes

This is an example from structural analysis. Figure 3 shows the deflected shape from a plane frame analysis model of a bridge truss. It is supported vertically at nodes 1 and 4 and has a single vertical point load at the central node 11. Even people who are not engineers are likely to suggest that there is something wrong with this shape. They would expect the deflection to be in the form of a smooth curve rather than being more of a "V" shape, as in the diagram.

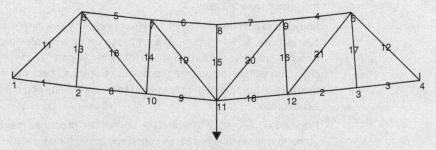

Figure 3   Deflected shape of a bridge truss

If you make such an observation and go on to rationalize the situation you can be in a win-win situation. If there is an error then you can identify the reason and you have discovered something very important by making the challenge. If the challenge is unfounded and you explain why this is so, then you learn about the behavior of the system. It is the latter situation that pertains in relation to Figure 3. (Both bending and shear mode deformation contribute to the deflection of the truss. In the case of the truss in Figure 3 the shear mode component, which gives linear displacement with constant shear force, dominates the behavior.)

(Continued)

3. Thinking for Innovation

One can define *creativity* as the production of new ideas/concepts whereas *innovation* is the development of creative ideas into useful outcomes. (MacLeod et al., 1996)

3.1 Producing Creative Ideas—Free Thinking

Albert Einstein said that "combinatory play seems to be the essential feature in productive thought" (Einstein 1954). That is, new ideas tend to result from the bringing together of concepts that were previously unrelated.

The combining of ideas may be affected by the following factors (MacLeod et al. 1998):

- **Innate Ability**—This is the ability that is not dependent on practice and received knowledge.
- **Knowledge**—Two types of knowledge are important. First, knowledge within the domain being considered is essential. Second, general knowledge is important because it may be advantageous to combine ideas from diverse areas. Being a specialist and a generalist is therefore useful for innovative engineers.
- **Knowledge Distance**—At high levels of creativity one is making connections between items of knowledge which were previously unconnected and the further these ideas are "apart" then the more difficult it will be to combine them. The "distance" between items is a measure of how close they appeared to be before any combining action takes place.
- **Practice**—The degree to which the person has practiced making combinations is likely to be an important factor for success in creative work.
- **Effort**—The amount and intensity of "effort" spent on the situation for which combination is required may be of significant importance in creative situations. Good creative ideas may need very intense mental effort.

There are techniques for developing ideas in groups such a brainstorming. When developing ideas with others it is a good strategy to arrange that one does not feel bad about being wrong. It can be worthwhile to agree that a session is for "free thinking" where ideas can pour out without being necessarily well thought out. Sometimes ideas that seem crazy at first turn out to have substance.

3.2 Making Creative Ideas Work—Focused Thinking

Producing creative ideas may be the easiest part of the innovative process. Converting them into useful outcomes can be the more difficult task. Since risk when innovating is normally increased, one has to pay closer attention to the process control strategies discussed in Section 2. Challenging outcomes is especially important. "Focused thinking" is needed where it is important not to be wrong.

3.3 Subconscious Thinking

A strategy used by people in innovative situations is to hold back from making decisions for as long as is practical. The subconscious mind can shape ideas. To get the subconscious to work requires hard conscious thinking followed by incubation periods where you move your thinking elsewhere—then come back to the problem. In the conscious thinking there should be a focus back to the requirements. Decision time looms up. It is important to leave enough time for implementation but not to rush to an early decision.

3.4 Knowing When to Innovate

It is important to know when innovation is needed. If you are innovating when there are standard ways of achieving a better result, then you may be considered to be incompetent.

4. Conclusion

We know that ability is a combination of what we were born with and lifetime experience. While nothing can be done to change the former attribute, the structuring of the latter is of fundamental importance. We are strongly influenced by principles that evolved in the culture in which we are raised. Some of these principles may not stand up to rational analysis but people still cling to them. The engineering approach is to analyze the way ideas are approached and to cut away the components of thinking which can be shown to be based on false logic, or no logic.

The way that we think, and, hence, behave also is deeply dependent on our interaction with others—our parents, siblings, friends, colleagues, managers. A very useful strategy is to seek to identify those whose mode of thinking is good, to identify principles that contribute to this competence, and to try to use these principles. Such principles can come from all walks of life. For example, the Duke of Wellington, one of the most successful battle commanders of all time, always did his own reconnaissance of a battleground rather than rely on reports from his subordinates. A good engineer also must satisfy herself or himself of the reliability of information that has been given.

(Continued)

Muscles need to be worked on to develop strength especially when one is young. It seems likely that the human brain has the same attribute. It needs to be worked hard in order to achieve optimum performance. The engine model of Figure 2 reflects this principle as does the discussion in Section 3.1.

Finally, engineering thinking involves ethical thinking. A main feature of a successful society is that resources are shared by the populace to an acceptable level. To achieve this, the society must be, in general, free from corruption. The behavior of the professionals in general, and of professional engineers in particular, has a very important role in setting the ethical standards for a country. For example, giving or taking a bribe is totally negative to good professional behavior. The work of professional engineers often affects the safety of the public. This issue should be at the forefront of their thoughts and actions.

References

MacLeod, I.A., B. Kumar, and J. McCullough. (1998). "Innovative Design in the Construction Industry." *Proc Inst of Civil Engrs*, Vol 126, No1, February, pp. 31–36.

—Iain A. McLeod, Ph.D., Chartered Engineer Professor Emeritus, Department of Civil Engineering, Strathclyde University, Glasgow, Scotland

SUMMARY

Recent developments in emerging information technologies will have a major impact on the way civil engineers work. As always, critical thinking—*engineering thinking*—is a key component in developing a successful and productive career in civil engineering.

"It is said that the present is pregnant with the future."

—Voltaire

"It is not the strongest of the species that survives, nor the most intelligent, but the one most responsive to change."

—Charles Darwin

"If you don't like change, you're going to like irrelevance even less."

—General Eric Shinseki, Chief of Staff, U.S. Army

REFERENCES

AIA National and AIA California Council. (2007). *Integrated Project Delivery: A Guide*. American Institute of Architects, Washington, D.C.

Eastman, Chuck, Paul Teicholz, Rafael Sacks, and Kathleen Liston. (2008). *BIM Handbook: A Guide to Building Information Modeling for Owners, Managers, Designers, Engineers, and Contractors*. John Wiley & Sons, Hoboken, New Jersey. IBSN: 978-0-470- 18528-5.

Gann, David M., and Karen L. Hansen (1996). *IT Decision Support and Business Process Change in the U.S.* Final Report to the U.K. Department of Trade and Industry (DTI) for Overseas Science and Technology Expert Mission.

http://fiatech.org/capital-projects-technology-roadmap.htm, accessed December 15, 2009.

McGraw-Hill Construction. (2008). *Smart Market Report: Building Information Modeling (BIM)—Transforming Design and Construction to Achieve Greater Industry Productivity*. The McGraw-Hill Companies, New York. ISBN: 978-1- 934- 92625-3.

Vanegas, Jorge A. (2009). *Is the Capital Projects Industry Observant? Is It prepared?* Invited Speaker within the Breakout Forum on a Futurist View: What's on the Horizon? at the 41st Annual ECC Conference: The Perfect Storm: Navigating through the Turbulence of Risk and Change, Engineering & Construction Contracting Association (ECC), Bastrop, Texas, September 2009.

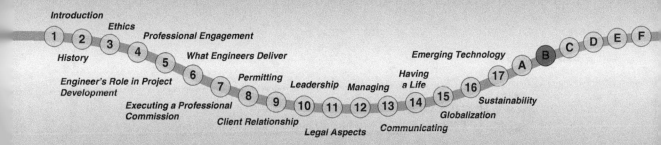

Introduction
1
2
Ethics
3
Professional Engagement
History
4
5
What Engineers Deliver
Engineer's Role in Project
Development
6
7
Permitting
Leadership
Managing
Emerging Technology
A B C D E F
Executing a Professional
Commission
8
9
10
11
12
13
14
15
16
17
Having
a Life
Client Relationship
Legal Aspects
Communicating
Globalization
Sustainability

Appendix B

Example Proposal

Placer County
California

Proposal for Applegate to North Auburn Wastewater Conveyance Pipeline Routing Study

October 3, 20XX

Mr. Bob Jones, P.E.
Mr. John Smith, P.E.

**RE: Request for Proposal to Provide Professional Engineering Services for the
Applegate to North Auburn Wastewater Conveyance Pipeline Routing Study**

Dear Mr. Jones and Mr. Smith,

Global Hydraulic Engineers, Inc. is pleased to submit our proposal for the Applegate to North Auburn Wastewater Conveyance Pipeline Routing Study. Our team's experience in planning, designing, and managing public works improvement projects and our thorough understanding of Placer County's requirements for this project will allow our team to develop a pipeline routing study that is fully accepted by Placer County and other major project stakeholders.

Global Hydraulic Engineers, Inc. (GHEI) understands the need for a new pipeline to connect the community of Applegate's wastewater collection system to the larger Placer County Sewer Maintenance District 1 (SMD1) system. The new connection will allow for decommission of the obsolete Applegate Wastewater Treatment Plant and will provide the opportunity for more of Placer County's residents to connect to the SMD1 collection system.

GHEI is aware of the challenges that this project will encounter. Crossing of both Interstate-80 and the Union Pacific Rail Road will play major roles in the development of this project. However, GHEI sees this as an opportunity for Placer County to commission a project that serves as a model for the type of coordination that is required between local and government agencies to provide the public with a well thought out and useful project. Our team's project manager has over three years experience working with Caltrans on various joint and oversight projects including the Sheldon Rd./SR 99 Interchange Reconstruction Project. His experience will prove to be a unique and valuable asset in the development of well thought out and feasible designs ideas.

GHEI is committed to designing projects that are both economical and environmentally sound. We are a well known and respected Sacramento-based firm that takes pride in employing engineers who see possibilities in the challenges we face. We are especially excited and look forward to working with you and your staff on this important project. If you have any questions or require additional information, please do not hesitate to contact our project manager at (555) 555-1212 or email@globohydro.com

Sincerely,
Global Hydraulic Engineers, Inc.

_____   _____   _____
Project Manager   Deputy Manager   Project Engineer

_____   _____
Project Engineer   Project Engineer

6000 J Street, Sacramento, CA 95819

Placer County
California

Proposal for Applegate to North Auburn Wastewater Conveyance Pipeline Routing Study

Proposal to Provide Professional Engineering Services for the Applegate to North Auburn Wastewater Conveyance Pipeline Routing Study

Presented to:

Placer County
California

Bob Jones, P.E.
John Smith, P.E.

Presented by:

Global HYDRAULIC
ENGINEERS INC.

6000 J Street, Sacramento, CA 95819
(555) 555-1212
email@globohydro.com

Global HYDRAULIC
ENGINEERS INC.

i

6000 J Street, Sacramento, CA 95819

Placer County
California

Proposal for Applegate to North Auburn Wastewater Conveyance Pipeline Routing Study

TABLE OF CONTENTS

Placer County
California

Section 1 - Project Description

1.1   Background

Placer County has initialized an effort to regionalize several small wastewater collection systems in the Auburn area. As a separable element of this large project, construction of a pipeline is needed to convey flows from the existing, obsolete wastewater treatment plant (WWTP) that serves the small unincorporated community of Applegate. This new pipeline would then need to connect to the existing North Auburn Sewer Maintenance District 1 (SMD1) collection system, allowing for decommission of the Applegate WWTP. The connection point that has been identified is near the intersection of Winchester Club Drive and Sugar Pine Road. Wastewater captured by the new pipeline will ultimately be conveyed to the SMD1 WWTP on Joeger Road in North Auburn. For this project to materialize, a pipeline routing study is needed. The study will identify and analyze possible alternatives to accomplish the task of rerouting Applegate's wastewater to the SMD1 collection system.

1.2   Project Details

GHEI's preliminary investigation of the project has revealed that routing of a pipeline from the Applegate WWTP to the connection point at Winchester Club Drive and Sugar Pine Road will cross both Interstate-80 (I-80) and the Union Pacific Rail Road (See Figure 1 – Project Area Map). Crossing of these facilities will require close involvement with representatives from both Caltrans and the Union Pacific Rail Road (UPRR) in the acquisition of encroachment permits, adherence to design standards, and discussion of design alternatives. Furthermore, the regional topography presents a unique challenge to the routing of the pipeline and may present the need for force main technology and pump stations in order for the wastewater to reach its destination (See Figure 2 – Project Area Topography and Figure 3 – Elevation Profile). Each of these major issues will be considered in the development of pipeline routing alternatives and study.

1.3   Site Description

The project area includes the Applegate WWTP located approximately 8 miles northeast of Auburn and the route of the pipeline to the connection point at Winchester Club Drive and Sugar Pine Road (approximately 3.2 miles southeast of the Applegate WWTP). The area surrounding the WWTP consists of the rail road tracks directly to the west (which run east parallel to Interstate-80), and several rural, single family homes. Also present is a natural gas pipeline that runs parallel to the UPRR tracks and directly adjacent to the WWTP.

Placer County
California

Figure 1 – Project Area Map

Figure 3 – Elevation Profile:

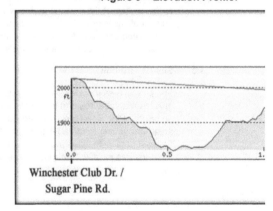

Winchester Club Dr. /
Sugar Pine Rd.

Global HYDRAULIC
ENGINEERS INC.

Figure 2 – Project Area Topography

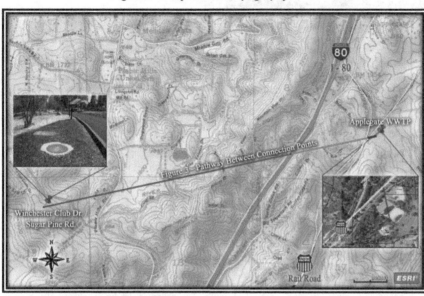

Applegate WWTP to Connection Point

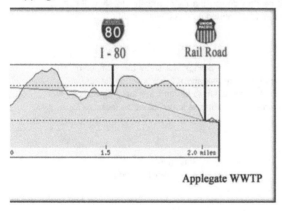

Section 2 - Work Plan

2.1   Task 1 – Project Management

Our team has selected a Project Manager. He will provide ongoing project management throughout the life of the project. In addition to coordinating the team's activities, he will serve as a readily available point of contact for the client and project team.

Task 1 Subtasks:

2.1.1   Kick Off & PDT Meetings

GHEI will organize and conduct a project kickoff meeting. The kickoff meeting will provide an opportunity for the client to be introduced to GHEI's project development team (PDT) members and to get an idea of the roles and responsibilities each will play. At the kickoff meeting GHEI will determine the preferred lines of communication for the client and will allow for input on streamlining of the project scope, schedule, safely issues and relevant information. GHEI will hold additional PDT meetings at least weekly depending on the nature of the project development and the needs of the client.

2.1.2   Project Schedule

GHEI will utilize Microsoft Project to establish and maintain a critical path project schedule to meet the client's project schedule. The schedule will be distributed at the kickoff meeting to allow the client to review, provide feedback, and endorse the project's schedule. Updates to the schedule will be provided to the client monthly to show the progress of each task and any revisions due to continued project development.

2.1.3   Client Billings

GHEI will prepare regular billings for the client. The billings will show the number of hours expended by project team members for each task and a breakdown of the hours remaining for those tasks. The billings will allow the client to further track the progress of the project and ensure that GHEI has allocated sufficient time to complete the tasks.

Task 1 Client Deliverables:

➢ Kickoff Meeting Agenda and
   Meeting Minutes

➢ Billing Statements
➢ Updates to Project Schedule

Section 2 - Work Plan

2.2   Task 2 – Project Research

Project research will consist of the following subtasks:

Task 2 Subtasks:

2.2.1   *Preliminary Research*

GHEI will gather pertinent existing information for the project locations to aid in choosing routing alternatives. This information includes, but is not limited to:

- Placer County Design Standards
- Aerial Site Mapping
- Available Roadway and Sewer As-Builts
- Caltrans Design Standards

- Available Utility Information/Maps
- Topographic Site Mapping
- Available Prior Studies
- Required Key Permits

2.2.2   *Project Stakeholders*

Through its ongoing project research GHEI will develop and maintain a list of stakeholders for the project. Some of the major project stakeholders have been identified as follows:

- Placer County
- UPRR

- Caltrans

2.2.3   *Site Investigation*

GHEI will perform up to three site visits to verify existing information obtained from preliminary research in addition to determining the physical constraints to the preliminary routing alternatives. Onsite exploration will help establish potential constraints and possible innovative ways to solve the design issues related to the project.

2.2.4   *Constraints*

GHEI has identified the following constraints during our initial investigation:

- Crossing of UPRR
- Crossing of I-80
Cross

- Regional Topography

GHEI will continue to determine project constraints which will be identified in a Project Constraints Table that includes information on how the constraint might impact the routing alternatives and possible mitigation for the each constraint.

Placer County
California

Section 2 - Work Plan

Task 2 Client Deliverables:

- ➢ Stakeholders List
- ➢ Project Constraints Table

Task 2 Assumptions:

- ➢ Pipeline is only for wastewater only; storm water drainage is not included.

2.3 Task 3 – Development of Alternatives

GHEI will develop a minimum of three (3) design alternatives for the Applegate/North Auburn Wastewater Conveyance Pipeline Routing Study to the extent necessary to establish the basic scope, feasibility, and cost for each. Each developed alternative will at a minimum address the costs, benefits, constructability, stakeholder interests and impacts to the environment.

Task 3 Subtasks:

2.3.1 Drawings and Specifications

GHEI will compose conceptual plan and profile drawings for each alternative in accordance with applicable design standards. Along with the pipeline layout drawings there will be detailed drawings and specifications of required pump stations including horizontal and vertical geometries gravity and force main technology and specific pump selection specifications and sizing. Other necessary specification for each alternative will be developed to the extent required for the feasibility study.

2.3.2 Material Lists

GHEI will provide a materials list for each alternative developed. The lists will later be used for estimating the cost of each alternative.

2.3.3 Available Construction Technology

GHEI will research available construction technology and or techniques that may help avoid or minimize impacts to physical, environmental, or other constraints in the project area.

Task 3 Client Deliverables:

- ➢ Pipeline Layout Drawings
- ➢ Specifications
- ➢ Pump Drawings and Specifications
- ➢ Materials Lists

Placer County
California

Proposal for Applegate to North Auburn Wastewater Conveyance Pipeline Routing Study

Section 2 - Work Plan

2.4 Task 4 – Prepare Cost Estimates

GHEI will prepare planning-level cost estimates for each developed alternative. Each cost estimate will include:

➤ Construction Cost
➤ Costs for operations and maintenance
➤ Land acquisition and Right of Way (R/W) services costs
➤ Cost for engineering fees

Cost estimates prepared by GHEI will be provided in both current year dollars and future year dollars. Current year cost estimates will allow for direct comparison of alternatives. Future year cost estimates are provided to allow for client budget programming. Engineering costs will be calculated as a percentage of the total construction cost.

Task 4 Client Deliverables:

➤ Cost Estimates for Each Alternative

2.5 Task 5 – Evaluate Alternatives

GHEI, with input from (client) and major project stakeholders, will develop a set of ranking criteria for the purpose of evaluating each alternative. The developed criteria will potentially be divided into the following 5 major categories:

➤ Benefits
➤ Costs
➤ Constructability

➤ Environmental Impacts
➤ Stakeholder Concerns/Benefits

Each alternative will be evaluated and ranked according to the final developed ranking weights and criteria.

Task 5 Subtasks:

2.5.1 Summary Table for Recommendation of Alternatives

GHEI will provide a summary table highlighting the evaluated alternatives and describing the positives and negatives for each. A recommendation for implementation of the highest ranked alternative will also be included. Viable evaluation information will be presented in the summary table in order to assure the client that a thorough and systematic evaluation process was followed.

2.8   Project Schedule

ID	ⓘ	Task Name	Start	Finish	Sep 21, '08 S M T W T F S	Sep 28, '08 S M T W T F S	Oct 5, S M T
1		Placer County/Applegate WWTP Pipeline Routing Study	Mon 10/6/08	Fri 12/12/08			▬
2		**Task 1 – Project Management (115 hrs.)**	**Mon 10/6/08**	**Fri 12/12/08**			▬
3	○	**Team Meeting (50 hrs.)**	**Wed 10/8/08**	**Wed 12/10/08**			
14	📅	Discuss Proposal with Client (10 hrs.)	Fri 10/10/08	Tue 10/14/08			
15	📅	Kickoff Meeting (2-3 hrs.)	Mon 10/13/08	Fri 10/17/08			
16	📅	Outline Submission	Fri 10/17/08	Fri 10/17/08			
17	📅	Annotated Outline Submission	Fri 10/31/08	Fri 10/31/08			
18	📅	Buyoff Meeting (2-3 hrs.)	Mon 11/3/08	Fri 11/7/08			
19	📅	90% Draft Report Due	Fri 11/14/08	Fri 11/14/08			
20	✓	Discuss 90% Report with Clients (10 hrs.)	Mon 11/17/08	Fri 11/21/08			
21	📅	Quality Control (30 hrs.)	Mon 10/6/08	Fri 12/12/08			▬
22	○	**Billing Statements (10 hrs.)**	**Mon 10/13/08**	**Mon 11/24/08**			
26	📅	Final Billing Statement	Mon 12/1/08	Mon 12/1/08			
27							
28		**Task 2 – Project Research (115 hrs.)**	**Mon 10/6/08**	**Fri 10/24/08**			▬
29	📅	Preliminary Research (70 hrs.)	Mon 10/6/08	Fri 10/17/08			▬
30	📅	Site Investigation (20 hrs.)	Mon 10/13/08	Fri 10/24/08			
31	📅	Develop Constrains Table (15 hrs.)	Wed 10/15/08	Wed 10/22/08			
32	📅	Develop Stakeholder List (10 hrs.)	Wed 10/15/08	Fri 10/17/08			
33							
34		**Task 3 – Development of Alternatives (160 hrs.)**	**Mon 10/13/00**	**Fri 10/31/08**			
35	📅	Brainstorming Alternative Ideas (10 hrs.)	Mon 10/13/08	Thu 10/16/08			
36	📅	Develop Minimum 3 Alternatives (50 hrs.)	Wed 10/15/08	Fri 10/24/08			
37	📅	Identify Constraints (20 hrs.)	Mon 10/20/08	Fri 10/24/08			
38	📅	Develop Drawings for Each Alternative (50 hrs.)	Mon 10/20/08	Fri 10/31/08			
39	📅	Specifications for Each Alternative (20 hrs.)	Mon 10/27/08	Fri 10/31/08			
40	📅	Materials Lists (10 hrs.)	Mon 10/27/08	Fri 10/31/08			
41							
42		**Task 4 – Preparation of Cost Estimates (25 hrs.)**	**Mon 10/27/08**	**Wed 10/29/08**			
43	📅	Cost Estmates for Each Alternative (25 hrs.)	Mon 10/27/08	Wed 10/29/08			
44							
45		**Task 5 – Evaluation of Alternatives (40 hrs.)**	**Wed 10/29/08**	**Fri 11/7/08**			
46	📅	Develop Ranking Criteria (20 hrs.)	Wed 10/29/08	Tue 11/4/08			
47	📅	Develop Summary of Alternative Evaluation (20 hrs.)	Wed 11/5/08	Fri 11/7/08			
48							
49		**Task 6 – Prepare Routing Study Report (185 hrs.)**	**Mon 10/27/08**	**Fri 12/12/08**			
50	📅	Prepare 90% Report (110 hrs.)	Mon 10/27/08	Fri 11/7/08			
51	📅	Revise 90% Report (40 hrs.)	Sat 11/8/08	Tue 11/11/08			
52		Print 90% Report	Tue 11/11/08	Tue 11/11/08			
53	📅	Submit 90% Report	Fri 11/14/08	Fri 11/14/08			
54	📅	Response to Comments (5 hrs.)	Mon 11/17/08	Mon 11/17/08			
55	📅	Prepare Final Report (30 hrs.)	Tue 11/18/08	Tue 12/9/08			
56	📅	Print Final Report	Tue 12/9/08	Tue 12/9/08			
57	📅	Deliver Final Report	Fri 12/12/08	Fri 12/12/08			
58							
59		**Task 7 – Oral Presentation to Clients (100 hrs.)**	**Mon 11/17/08**	**Fri 12/12/08**			
60	📅	Prepare Oral Presentation (80 hrs.)	Mon 11/17/08	Fri 12/5/08			
61	📅	Rehearse Oral Presentation (20 hrs.)	Mon 12/8/08	Thu 12/11/08			
62	📅	Oral Presentation	Fri 12/12/08	Fri 12/12/08			

Task	▬▬▬	Progress	▬▬▬	Summary	▬▬▬	External Tasks	▬▬▬
Split	....	Milestone	◆	Project Summary	▼▼▼	External Milestone	◇

Proposal for Applegate to North Auburn Wastewater Conveyance Pipeline Routing Study

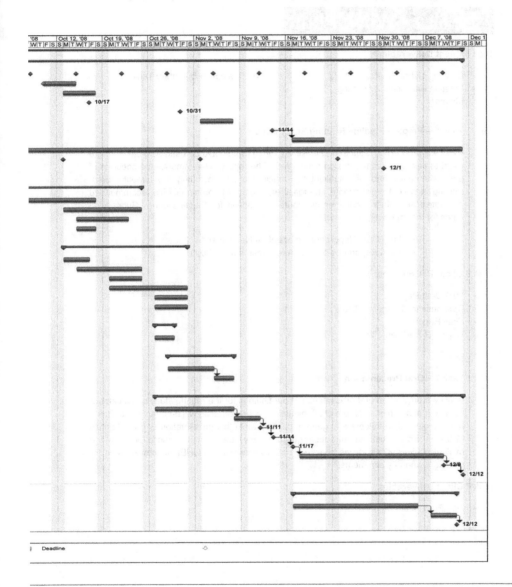

6000 J Street, Sacramento, CA 95819

Section 2 - Work Plan

Task 5 Client Deliverables:

- Detailed List of Ranking Criteria
- Meeting with Client & Meeting Minutes
- Evaluation Summary Table

2.6   Task 6 – Prepare Pipeline Routing Study Report

GHEI will prepare an engineering feasibility report that should be used as a basis for the client to select one of the design alternatives. The report will summarize the specifics of each alternative and will highlight their main features including the benefits, risks, ranking, and cost. Furthermore, the feasibility report will contain GHEI's recommended alternative and other recommended courses of action for Placer County. Submission dates for the report are as follows:

- 90% Draft Report due on or before November 14, 2008
- Final Report due on or before December 12, 2008

Task 6 Client Deliverables:

- 90% Draft Report
- Response to Comments Table
- Final Report
- Copies of Pertinent Files

2.7   Task 7 – Oral Presentation

GHEI will provide an oral presentation of our findings for the routing study on December 12, 2008; at a minimum length of 30 minutes. The presentation will be open to the Client, major project stakeholders and general public. Within this presentation GHEI will utilize Microsoft PowerPoint to introduce our company, the project description and the evaluated alternatives. At the conclusion of this presentation GHEI will provide time for questions and any comments.

Task 7 Client Deliverables:

- Oral Presentation

Section 3 - Consultant Assets and Qualifications

3.1 Assets

GHEI has offices conveniently located within 30 miles of the Applegate WWTP. GHEI is equipped with the necessary tools and software to design and analyze pipeline routing options; including: AutoCAD, Microsoft Office, Google Earth Plus, SewerCAD, ArcGIS and more.

For site visits GHEI will furnish its team with cell phones, vehicles, laptop computers, digital cameras, video recorders and all other equipment required for performing site visits. GHEI assures that each employee can effectively use these tools and is competent to operate them. GHEI provides each employee with extensive training to ensure safe and effective field operation procedures are followed.

In addition to the tools and capabilities mentioned, GHEI has working relationships with many agencies in the Sacramento and Placer regions including Caltrans and UPRR.

3.2 Team Organization & Level of Effort Summary

Figure 4 – Project Organization Chart

Table 1 – Project Availability

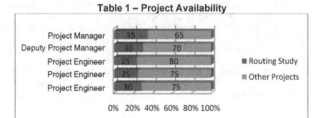

Placer County
California

Proposal for Applegate to North Auburn Wastewater Conveyance Pipeline Routing Study

Table 2 - Level of Effort

Project Tasks	Project Manager	Deputy Project Manager	Project Engineer	Project Engineer	Project Engineer	Budget Hours	Percentage of Budget Hours
Global Hydraulic Engineers, Inc.							
Project Management	35	35	15	15	15	115	16%
Project Research	15	15	35	15	35	115	16%
Development of Alternatives	25	25	25	60	25	160	22%
Prepare Cost Estimates	5	5	5	5	5	25	3%
Evaluate Alternatives	8	8	8	8	8	40	5%
Prepare Routing Study	40	40	40	25	40	185	25%
Oral Presentation	20	20	20	20	20	100	14%
Total Hours	148	148	148	148	148	740	100%

Section 4 – Supportive Information

Project Manager
Project Availability   35%
Budget Hours   148

Education	BS, Civil Engineering, California State University, Sacramento, December 2008
Experience Summary	He has more than five years experience working in the civil engineering industry. His experience comes from his work as a student intern for Sacramento County's CSD1 and more recently with Interwest Consulting Group, Inc at the City of Elk Grove. He excels as a project manager, ensuring projects stay on track and under budget.
Project Experience	**2008 – Sheldon Road/State Route 99 Interchange Reconstruction Project:** Assistant Project Manager. Prepared and secured the "Request to Proceed With Construction" (Caltrans: E-76) documents for the authorization and release of federal funding. **2007– Sheldon Interchange Demolition Project:** Project Manager. Prepared PS&E and monitored construction operations through completion.

Section 4 – Supportive Information

Deputy Project Manager
Project Availability   30%
Budget Hours   148

Education	BS, Civil Engineering, California State University, Sacramento, December 2008 Certification, Web Publishing, American River College, December 2008
Experience Summary	He has five years experience as a construction project manager for Ureta Construction. He managed and supervised all construction activities, coordinated client and company meetings, and communicated extensively with client to ensure customer satisfaction.
Project Experience	**2007 – National Cooperative Highway Research Program 12-74:** Played a key role in the development, fabrication, testing, and analysis of the experimental pre cast bent-cap system.

Project Engineer
Project Availability   25%
Budget Hours   148

Education	BS, Civil Engineering, California State University, Sacramento, December 2008
Experience Summary	He has worked in the civil engineering industry for over five years working for Caltrans and Nolte Engineering. His experience has come in the transportation, water, and wastewater engineering fields. He is adept in designing projects using AutoCAD.
Project Experience	**2006 to 2008 – Esparto Community Services District (CSD) Water/Wastewater System Improvements:** Assisted in the re-design and construction of major repairs to the Esparto, CA CSD Domestic Water and Wastewater Distribution System. **2006 to 2008 – City of Wheatland Wastewater System Improvements:** Assisted in the design and construction of major repairs to the City of Wheatland wastewater collection system.

Proposal for Applegate to North Auburn Wastewater Conveyance Pipeline Routing Study

Section 4 – Supportive Information

Project Engineer
Project Availability 30%
Budget Hours 148

Education	BS, Civil Engineering, California State University, Sacramento, December 2008
Experience Summary	She has three years experience in the civil engineering field working for David Evans and Associates. She has experience in all aspects of land development and transportation engineering. Tasks performed include, designing subdivision improvement plans, preparing cost estimations, assisting in interchange analysis, highway design, and assisting in project coordination.
Project Experience	**2008 - Hwy 65-70 Interchanges:** Assisted in the development of the feasibility studies as required by Caltrans. **2005 to 2007 Bickford Ranch:** Assisted in the design of trunk sewer and land development for 2000 plus homes.

Project Engineer
Project Availability 25%
Budget Hours 148

Education	BS, Civil Engineering, California State University, Sacramento, December 2008
Experience Summary	He has two years civil engineering experience working for Sacramento County in the fields of water resources and land development. His work involved collecting meter data, designing grading and drainage plans, and preparation of cost estimates.
Project Experience	**2006 - PFE Road Water Transmission Pipeline Project:** Performed design calculations and site visits **2006 – Lincoln Bypass Project:** Performed design calculations

Introduction
1 2 Ethics
3 Professional Engagement
History
4 5 What Engineers Deliver
Engineer's Role in Project Development
6 7 Permitting
8 9 Leadership Managing Emerging Technology
A B C D E F
Having a Life
17 16 15 14 13 12 11 10 Executing a Professional Commission
Client Relationship
Legal Aspects Communicating Globalization Sustainability

A p p e n d i x **C**

Example Feasibility Study Report

Engineering

December 12, 20XX

Mrs. Amanda, P.E.

Mr. Ryan, P.E.

Subject: Pipeline Routing Study for the Applegate to North Auburn Wastewater Conveyance Pipeline

Dear Clients,

CVision Engineering (CVE) is pleased to submit this Pipeline Routing Study for the connection of the existing wastewater collection system of Applegate to Sewer Maintenance District 1 (SMD-1) in Auburn at the intersection of Dry Creek Road and Blue Grass Drive. The study presents a summary of the findings of CVE's project research, based on which three feasible alternatives were identified and developed. The report also discusses in detail the various physical, environmental and design constraints for the project, as well as the multiplicity of stakeholders and their potential impact on the project. Included for each of the three alternatives are preliminary design drawings, a pump selection, a cost estimate and a detailed evaluation against a ranking criterion. Lastly, based on the ranking of the proposed feasible alternatives, CVE provides a recommended course of action for Placer County.

Should you have any questions, please feel free to contact CVE's Project Manager, at (555) 555-1212 or email@cvisionengineering.com Thank you.

Sincerely,

CVision Engineering

_____   _____
Project Manager   *Assistant Project Manager*

_____   _____   _____
Project Engineer   *Project Engineer*   *Project Engineer*

CVision Engineering · CSU Sacramento · 6000 J Street, Sacramento, CA 95819

December 12, 2008

Applegate to North Auburn
Wastewater Conveyance Pipeline
Pipeline Routing Study

December 12, 2008

PREPARED FOR:
COUNTY OF PLACER, CALIFORNIA

PREPARED BY:
CVISION ENGINEERING

California State University of Sacramento
6000 J Street, Sacramento, CA 95819

TABLE OF CONTENTS

VISION
Engineering

December 12, 2008

December 12, 2008

LIST OF TABLES

LIST OF FIGURES

1.0   INTRODUCTION

1.1   Project Overview

The Applegate to North Auburn sewer pipeline extension project is a part of Placer County's effort to regionalize the sanitary sewer system in the Auburn area. The need for the regionalization had been triggered by high costs of maintaining detention pond treatment facilities, such as the existing Wastewater Treatment Plant (WWTP) in Applegate, and lack of reliable performance as documented by the Central Valley Regional Water Quality Control Board (CVRWQCB). CVRWQCB has issued a number of Notice of Violations due to the consistent violations of the WWTP's permit. Furthermore, the current water treatment standards and effluent limitations are expected to become more stringent in the near future. To satisfy the terms and conditions of settlement as established by CVRWQCB, Placer County has agreed to decommission the Applegate WWTP and convey the existing wastewater flows to Sewer Maintenance District 1 (SMD-1) in North Auburn.

1.1.1   Project Location

Applegate is located in the foothills of Placer County, approximately 10 miles north-east of Auburn. The project area includes the proposed pipeline and the Applegate's WWTP with its treatment ponds and dechlorination facilities (Figure 1.1). The WWTP is located on the south side of Interstate-80 (I-80) and immediately east of the Union Pacific Railroad tracks at the location where Merry Lane ends. The north part of the project, from the WWTP to Placer Hills Road, is parallel to I-80. The south part of the project, from Placer Hills Road to the connection point with SMD-1 at Blue Grass Drive, follows Lake Arthur Road, which becomes Dry Creek Road at the intersection with Christian Valley Road. The entire project falls within the American River Watershed (Appendix A).

1.1.2   Existing Wastewater Collection and Treatment System

The existing wastewater collection system of Applegate was constructed in 1975 (Reference [1]) and consists of a combination of force main and gravity sewer leading to detention ponds and on-site dechlorination facilities (Figure 1.2). The capacity of the ponds is inadequate for the wastewater flows both because of inflow of rainwater during wet weather and also inflow of groundwater under artesian conditions. Per CVRWQB (Reference [2]), the groundwater inflow during the winter months is sufficient to overflow one of the detention ponds even if no wastewater is discharged into it. To avoid these violations, the ponds are temporarily decommissioned between the months of September and March and the wastewater is pumped into storage tanks and subsequently hauled away to SMD-1 by large tanker trucks.

1.1.2.1 Location and Service Area

The existing wastewater collection system is mostly within the Placer County right-of-way. Throughout much of the community, the pipe network is a gravity sanitary sewer conveyance system consisting of a 6-inch-diameter trunk with approximately 28 service connections. Within a section of Applegate Road, from Brick Road to Apple Court, the system is a 4-inch force main for approximately 2,000 feet (Figure 1.2). An existing lift station at the intersection of

---

Applegate Road and Brick Road, pumps some of the existing flows over a hill between Apple Court and Brick Road. After overcoming the elevation head between Apple Court and Brick Road, the pipe is again a 6-inch gravity flow system for the remainder of its way to the detention ponds. At the intersection of Applegate Road and Merry Lane, the pipe system leaves the County's right-of-way and turns due south. It follows Merry Lane, which is a private road. Before the sewer pipeline reaches the wastewater treatment plant, it crosses under the Union Pacific Railroad tracks (Figure 1.2). According to the Placer County sanitary sewer record drawings, the last service connection to the existing sewer pipeline is before the pipeline leaves the county right-of-way at the intersection of Applegate Road and Merry Lane.

1.1.2.2 Wastewater Treatment and Disposal Operations

The Applegate WWTP consists of three evaporation and percolation ponds, designed to operate in series. The ponds are only six feet deep and cannot meet the current freeboard requirements, which mandate the pond surface to be a minimum of two feet below the top of the embankment at all times. The anticipated overflow from the ponds into Clipper Creek led Placer County to the decision of implementing a disinfection and chlorination system to achieve the required at the time effluent water characteristics. These discharges are violations of the WWTP permit. Further improvement of the WWTP is unfeasible and hauling away the wastewater during the winter months is only a temporary solution.

1.2 Purpose and Limitations of the Study

The objective of the study is to identify and evaluate a minimum of three feasible pipeline routes to convey the existing wastewater flows from Applegate to the designated connection point with SMD-1 at Dry Creek Road and Blue Grass Drive in Auburn. Alternative routes were considered based on their potential to maximize project benefits and minimize cost and negative impacts to the project area.

The following is a list of the limitations of this study:

- Record drawings of existing utilities were not available at the time of the study. The locations of existing utilities as shown on the design drawings (Appendix B) are approximate and based on field observations. The horizontal and vertical controls of existing facilities in the project area need to be verified by a licensed surveyor prior to further design considerations. (*Appendixes to this report not included.*)

- The profile of the terrain along the selected alternative routes was first approximated using United States Geological Survey (USGS) 20-foot contour data for Placer County. However, field observations showed a discrepancy between the existing terrain conditions and the digitally produced vertical profiles. This was a result of the inability of the 20-foot contour GIS data to account for the cut and fill for the existing roads. To better represent the existing conditions and provide a more realistic pipeline design, adjustments were made to the vertical profiles using a GPS unit. The project site needs to be surveyed prior to further design considerations. Accordingly, adjustments in the alignments and profiles may be further required.

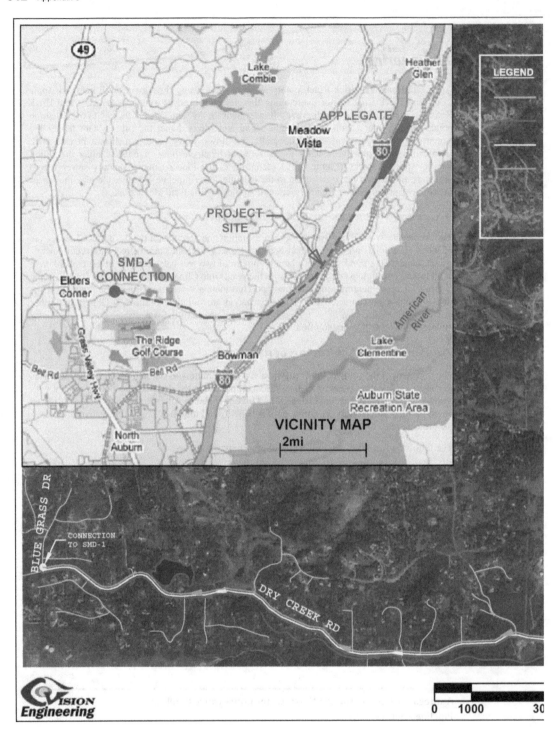

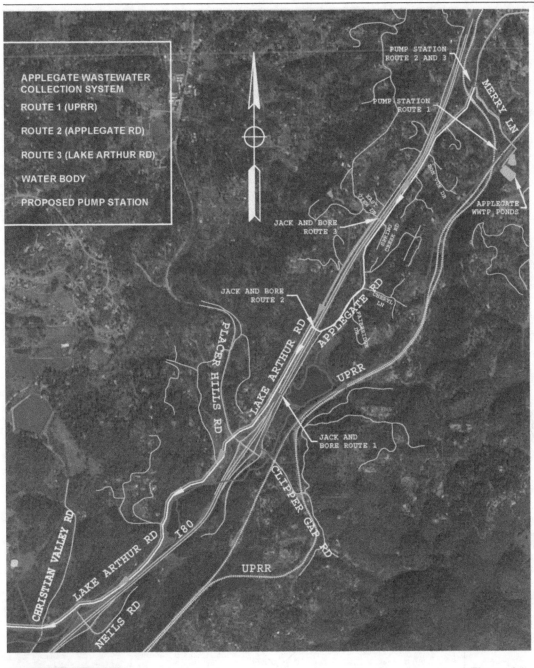

APPLEGATE WASTEWATER
COLLECTION SYSTEM

ROUTE 1 (UPRR)

ROUTE 2 (APPLEGATE RD)

ROUTE 3 (LAKE ARTHUR RD)

WATER BODY

PROPOSED PUMP STATION

PUMP STATION
ROUTE 2 AND 3

PUMP STATION
ROUTE 1

MERRY LN

APPLEGATE
WWTP PONDS

JACK AND BORE
ROUTE 3

JACK AND BORE
ROUTE 2

APPLEGATE RD

CHERYL LN

LAKE ARTHUR RD

PLACER HILLS RD

UPRR

JACK AND
BORE ROUTE 1

CLIPPER GAP RD

CHRISTIAN VALLEY RD

LAKE ARTHUR RD

I80

UPRR

NEILS RD

6000 ft

Figure 1.1
Project Location

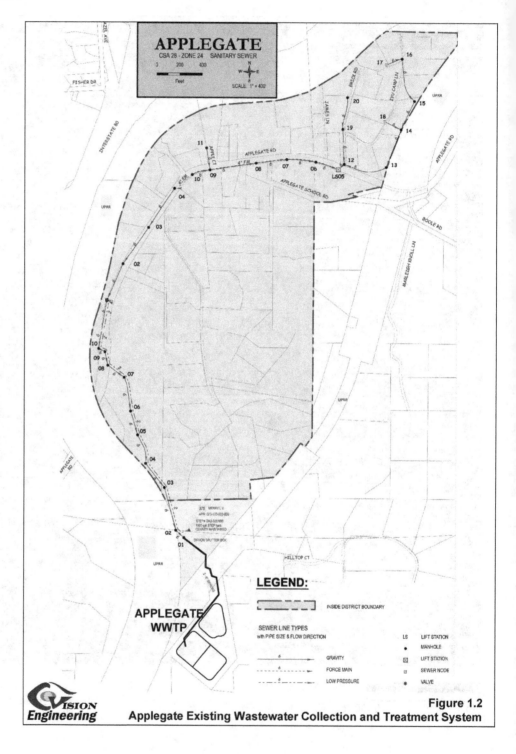

Figure 1.2

Applegate Existing Wastewater Collection and Treatment System

- The research of potential environmental impacts along the proposed alternatives was based on the Placer County Master Conservation Plan and existing Environmental Impact Reports (EIRs) filed in and around the project area. Based on the available data, no significant environmental concerns have been identified at this time. However, the findings of the Final EIR for the project, due in the summer of 2009, may require appropriate mitigation measures.

- This study considers existing flows only. Future connections will not be included in the development of the proposed alternatives.

- The North Auburn SMD-1 system is a Septic Tank Effluent Pumping system (STEP) at the designated for the project connection point. Although Applegate's wastewater flows are relatively low compared to the ones at the tie-in point, backup of the new pipeline at the connection to SMD-1 may occur. This issue could be evaluated through sewer modeling or measurements and observations after the pipeline is operational. Based on the results of this evaluation, the construction of a pretreatment facility for Applegate's wastewater flows, before they enter SMD-1, may be required. Further studies for the treatment of the existing wastewater flows from Applegate are beyond the scope of this routing study.

1.3   *Summary of Recommended Alternative*

The study rendered Route 3 (Lake Arthur Road) as the recommended alternative (Figure 4.3). The new pipeline would be a force main throughout. Since there are no service connections along the descent between the intersection of Applegate Road and Merry Lane and the detention ponds, the existing sewer pipeline within this stretch would be decommissioned (Appendix B, L3-6). A new lift-station would be constructed at the intersection of Applegate Road and Merry Lane (Appendix B, L3-6). This location avoids the additional 75 feet of elevation head between the wastewater treatment ponds and Applegate Road, which also reduces the size of the required pump. From the lift-station on towards the connection point with SMD-1, the recommended alternative follows Applegate Road and Lake Arthur Road (Figure 4.3) – the only two viable county roads leading from Applegate to Auburn. Both of these county roads are parallel to I-80, with Applegate Road to the East and Lake Arthur Road to the West of I-80. The jack-and-bore crossing of I-80 for Route 3 (Appendix B, L3-5 and X-2) was identified based on an overlay of the vertical profiles of Applegate Road, I-80 and Lake Arthur Road. The selected jack-and-bore location is favorable because of the short distance from Applegate Road to Lake Arthur Road and also because Applegate Road is elevated above Lake Arthur Road, thus eliminating additional elevation head and higher pumping requirements. The only reasonable continuation of the preferred alternative, from Lake Arthur Road to the connection point with SMD-1, is for the route to follow Dry Creek Road. In addition to this, Route 3 has less private residences and existing utilities along its way.

2.0 PROJECT CONSTRAINTS

An extensive research of the possible constraints in the project area was made prior to considering alternative routes. The research included review of relevant standards and regulations, available record drawings, and environmental studies for the project area. Additional information was obtained through contacting regulatory agencies and conducting field visits. The project research determined the following major groups of project constraints:

- Regulatory Compliance
- Physical Constraints
- Environmental Constraints
- Private Stakeholders' Interests

2.1 Design Standards and Permitting Requirements

The project faces regulatory constraints associated with the following agencies:

- Placer County
- California Department of Transportation (Caltrans)
- Union Pacific Railroad (UPRR)
- California Department of Public Health (CDPH)
- Central Valley Regional Water Quality Control Board (CVRWCB)
- Placer County Air Pollution Control District (APCD)
- U.S. Fish and Wildlife Service
- California Department of Fish and Game
- U.S. Army Corps of Engineers
- U.S. Environmental Protection Agency (EPA)
- California Department of Parks and Recreation, Office of Historic Preservation (OHP)

These regulatory constraints include both design standards and required permits that may be either very stringent and/or time consuming to obtain or satisfy. The design of the proposed alternatives is preliminary. Possible delays and design changes to the project may arise because of the complexity of the routes and their close proximity to some environmentally sensitive areas.

2.1.1 Design Standards

The proposed sewer pipeline design must follow the relevant engineering design standards of the regulatory agencies listed above. The following Table 2.1 is a summary of the main design considerations which were accounted for in the present routing study:

Table 2.1- Routing Study Design Considerations

No.	PLACER COUNTY (REFERENCE [3])	
	SPECIFICATION NO.	DESCRIPTION
1	71-1.02D	Force mains shall only be PVC pipes.
2	71-1.03	Excavation and backfill; trench excavation specifications.
3	71.1.05	Minimum cover of 3 feet; pipe depth may not exceed 20 feet for horizontal directional drilling method of construction.
4	71.1-1.06A	Bracing and shoring requirements.
No.	CALTRANS (REFERENCE [4])	
	SPECIFICATION NO.	DESCRIPTION
1	606.3	Transverse encroachment is permissible.
2	606.4	Longitudinal encroachments would not be allowed.
3	623.1	No open trenching or overhead sewer crossing is allowed.
		Only jack-and-bore and directional drilling would be allowed.
		Jack-and-bore installation for a 6-inch diameter force main would require a steel casing pipe with a minimum wall thickness of ¼ inch.
		Receiving pits for either jack-and-bore or directional drilling must be outside of the Caltrans right-of-way.
		No pump stations are allowed in the Caltrans right-of-way.
4	623.2D	Minimum 4 feet depth of pipe cover under I-80.
No.	UPRR (REFERENCE [5])	
	DESCRIPTION	
1	Longitudinal utility encroachment within UPRR right-of-way must be a minimum of 35 feet from centerline of the nearest railroad track.	
2	Utilities crossing railroad must be encased in a steel pipe; minimum thickness of a casing pipe with less than a 12-inch diameter is ¼ inches.	
3	Utility crossings must have a minimum cover of 4.5 feet and a maximum cover of 20 feet.	
No.	CDPH (REFERENCE [6])	
	DESCRIPTION	
1	Sewer line must have a minimum horizontal clearance of 10 feet from any drinking. water	
2	Sewer line must cross underneath potable water piping.	

2.1.2 Permitting Requirements

The majority of the construction work for the project would be in the Placer County right-of-way, which requires an Encroachment Permit (Reference [7] & [8]). The Improvement Plans for the new pipeline require the County's approval. Additional permits would need to be obtained from the regulatory agencies discussed above. Table 2.2 below lists the permitting requirements for the project:

Table 2.2- Permitting Requirements

No.	PLACER COUNTY (REFERENCE [7])
1	Improvement Plans approval/Encroachment permit – including approval of stormwater quality, noise pollution, traffic control, tree preservation, road restoration, grading, erosion and sediment control plans and measures.
No.	CALTRANS (REFERENCE [4])
1	Transverse encroachment permit.
2	Jack-and-bore permit.
3	Performing relocation work for utilities (if necessary).
No.	UPRR (REFERENCE [9])
1	Permit to be on railroad property for utility survey.
2	Longitudinal utility encroachment permit.
3	Encased non-flammable pipeline crossing permit.
No.	CVRWQCB (REFERENCE [10])
1	National Pollutant Discharge Elimination System (NPDES) – for construction projects, which encompass more than 5 acres of disturbed soil (Reference [11]).
2	Clean Water Act, Section 401 Certification – regulation of fill and dredged material.
No.	PLACER COUNTY APCD (REFERENCE [12])
1	Authority to construct permit (Reference [13]).
No.	U.S. FISH AND WILDLIFE SERVICE
1	Consultation may be required since the project has a potential to harm protected wildlife and plant species.
No.	CALIFORNIA DEPARTMENT OF FISH AND GAME
1	Consultation may be required since the project has a potential to harm protected wildlife and plant species.
No.	U.S. ARMY CORPS OF ENGINEERS (REFERENCE [14])
1	Any work within the waters of the State, including dredging or discharging sediment laden runoff into a creek requires a CWA Section 404 permit.
No.	CALIFORNIA DEPARTMENT OF PARKS AND RECREATION, OHP (REFERENCE [15])
1	Project will be required to ensure compliance with the National Historic Preservation Act and other regulations pertinent to the protection of cultural resources.

2.2 Physical Constraints

The proposed alternatives would face physical constraints that range from easements from various agencies to unforeseen conditions. Easements from agencies such as Caltrans and Union Pacific Railroad would have to be acquired in order for the sewer line to pass through their right-of-way. The proposed alternatives would require minor land acquisition from private landowners. Additional physical constraints such as Placer County's hilly terrain, deposits of hard rocks, creeks, and lakes, also pose some difficulties. Although the majority of the proposed routes follow existing roads, there is a possibility of uncovering archeological remains.

2.2.1 County Roads

- The narrow County roads along the proposed alternatives serve as the only access to private residences, which would require careful planning, phasing, re-routing and traffic control measures.
- Lane widths may be too narrow to work safely next to live traffic, which may require temporary road closures and/or traffic diversion.
- A proactive approach to public outreach would ensure that the concerns of the interested parties are addressed and possible conflicts that may arise are resolved in a timely manner.

2.2.2 I-80

- Installation of the sewer line would have to be through jack-and-bore or horizontal directional drilling (Table 2.1).
- The vertical and horizontal delineation of Eastbound and Westbound I-80 (Reference [16]) with respect to Applegate Road and Lake Arthur Road limit the reasonable locations for the jack-and-bore crossings. Crossing I-80 at different locations than those in the proposed alternatives would be more expensive due to additional elevation head and/or increased length of the jack-and-bore operation.
- Caltrans may impose time restrictions for construction along the freeway during peak traffic periods such as holidays or commute hours.

2.2.3 Union Pacific Railroad

- Installation of a sewer line along the UPRR right-of-way would require both a minimum of one transverse crossing (through jack-and-bore or directional drilling) and open trenching (Reference [17]).
- Some stretches of UPRR's right-of-way are currently surrounded by dense woods that provide a rich habitat to many species. Utilizing this land would require environmental mitigation.
- The difficult terrain and live train traffic would be major accessibility and safety constraints during construction (Figure 2.1).

Figure 2.1- UPRR Tracks 200 ft South of the Detention Ponds

- Special safety precautions per UPRR and the California Occupational Safety and Health Administration (Cal OSHA) should be taken during construction near the UPRR tracks.
- The lengthy processing time for permits (a minimum of 3 to 6 months) is also a constraint (Reference [9]).
- Depending on the proximity of construction, UPRR may require Railroad Protective Liability Insurance in addition to general liability insurance (Reference [9]).

2.2.4 Existing Utilities

- If properly submitted, the required Notice to Adjacent Utility Owners may take up to 30 calendar days before the County takes a course of action (Reference [7]).
- It may be unfeasible to relocate certain utility installations, such as high voltage PG&E lines, which would require design changes.
- Unmarked utilities pose a great concern for damage during construction.
- Limited space around existing utilities may be an issue during construction.
- Relocation of existing utilities may be required.

Figure 2.2 - Fiber Optic Line and Propane Gas Pipeline along UPRR

2.2.5 Creeks / Lakes / Culverts

- Due to the close proximity of creeks and lakes, care should be taken to prevent chemical contamination and sewer spills that may lead to public health issues, costly fines and time-consuming cleaning (Figure 2.3 – 2.4).
- Culverts that cross the roadway and are near the road surface would require an increased trenching depth (Figure 2.5).
- The installation of the sewer line must provide a minimum of 10 feet clearance to any water bodies or potable water supply lines.

Figure 2.3- Creek Under Applegate Road

Figure 2.4- View of Lake Arthur from Lake Arthur Road

Figure 2.5- Water Supply Pipe to the Helsey Power House on Dry Creek Road

2.2.6 Terrain

- The elevation changes along the alternatives, limit the opportunities to use gravity main and increase the need for force main technology.
- The rocky terrain of the project site (Appendix C), which is part of the Sierra Nevada Foothills, may make excavating trenches to the desired depth extremely difficult and time consuming if hard rock is encountered (Reference [18])
- There is a limited space to for the construction work due to cliffs and dense trees along the County roads.

2.2.7 Archeological

The area of the proposed project is considered to have a high possibility of artifacts from the California Gold Rush era and a moderate possibility of Native American artifacts. Even though the proposed routes mainly follow existing County roads, the depth of excavation still allows for the possibility of uncovering of artifacts with archeological significance. Uncovering such artifacts would affect the project schedule or may cause changes to the proposed routes (Reference [15]).

2.2 *Environmental Constraints*

Environmental impacts may also be a major constraint in the project area and need to be properly addressed. The following is a list of potential environmental issues:

- Air Pollution
- Sensitive Animal Species
- Sensitive Plant Species
- Naturally Occurring Asbestos
- Sediment Runoff

The Placer County Chapter of the Sierra Club is the most prominent environmental group in the area. The Sierra Club confirmed that they are not aware of any environmental concerns related to the proposed project. However, this might change based on the findings of the Final EIR and the proposed mitigation measures.

2.3.1 Air Pollution

Construction related activities have the potential to conflict with the Placer County APCD regulations, which follow the air quality standards set by the EPA and the California Air Resources Board (Reference [12]). The use of a generator at the proposed pump station would also require adherence to the air quality standards.

2.3.2   Sensitive Animal Species

The proposed alternatives could potentially affect the habitats of protected species (Reference [19]). Possible protected species in this area may include Elderberry Longhorn Beetle, California Red-Legged Frog, Foothill Yellow-Legged Frog and Western Pond Turtle (Figure 2.6 – 2.8). The possibility of these species to be encountered in the area would be addressed in the EIR. Harming of any threatened or endangered species during construction may pose serious consequences to the project, both in time delays and heavy fines.

Figure 2.6- Elderberry Longhorn Beetle

Figure 2.7- California Red-Legged Frog

Figure 2.8- Western Pond Turtle

2.3.3   Sensitive Plant Species

Possible sensitive plant species, such as Blue Oak Tree, Big Scale Balsamroot, Butte County Fritillary, Brandegee's Clarkia, Oval-Leaved Viburnum and Jepson's Onion, could also be affected by the proposed project (Figure 2.9 – 2.12). It is known that Blue Oak Trees are present along the proposed alternatives. The removal of any Blue Oak Tree during construction requires that three Blue Oak Trees are planted for every one that is removed (Reference [20]). The possibility that the rest of the sensitive plant species could be in the project area would also be addressed in the Final EIR. Disturbance or loss of these plants' habitats may pose time delays and heavy fines.

Figure 2.9- Blue Oak Tree

Figure 2.10- Butte County Fritillary

Figure 2.11- Brandegee's Clarkia

Figure 2.12- Oval-Leaved Viburnum

2.3.4 Naturally Occurring Asbestos

There is a high probability that a naturally occurring asbestos (NOA) will be encountered during construction because of past project experiences in the Placer County area. A preliminary geotechnical report for the project site has identified rocks, which are likely to contain NOA (Reference [18]). NOA has been previously encountered within metavolcanic and ultramific rock units in the project vicinity (Appendix D). The proximity of the site to fault fractures also suggests the existence of NOA in the project area. The presence of NOA may lead to the implementation of stringent mitigation measures that could delay the project.

2.3.5 Sediment Runoff

The proximity of the proposed alternatives to lakes and the crossing of a number of creeks along each the routes increases the possibility of sediment runoff entering these waterways. This can potentially violate Placer County's erosion control requirements (Reference [7]). Seasonal construction restrictions may be imposed due to high risk of sediment runoff, which could possibly lead to changes in the construction schedule. Construction during the wet season would require trench dewatering operations, thereby increasing project construction costs.

2.3 Private Stakeholders' Interests

- Nuisance complaints concerning noise pollution caused by construction equipment may be filed by residents. Noise from construction may not violate Placer County's noise ordinance of maximum 5 decibels (dB) of exterior ambient sound level or the standards (Reference [7]), whichever is greater.
- Nuisance complaints may also be filed due to loss of business (Figure 2.13) caused by construction activities and re-routing of traffic (Reference [7]). This may cause construction delays or time restrictions may be imposed.
- The minor land acquisition and temporary easements through private property may cause delays. It is desirable to start open communication with the affected property owners as early as possible.

Figure 2.13- Gas Station on Lake Arthur Road

December 12, 2008

3.0 DEVELOPMENT OF ALTERNATIVES

The flow chart in Figure 3.1 illustrates the general approach to identifying of the proposed feasible alternatives and the selection of the recommended alternative. The specific strategy used in the alternative selection process is discussed in this section.

Figure 3.1- Development of Alternatives Flow Chart

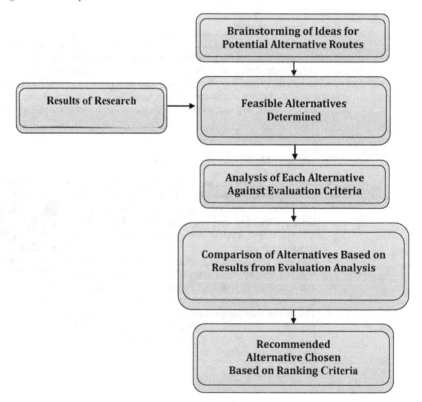

3.1   Strategy for Identifying of Alternatives

In developing a strategy for identifying feasible routes, the factors outlined below were considered. If these factors were not obviously present in an alternative, the route was not pursued as a feasible alternative. The alternatives presented in this study were identified based on the criteria listed below. This section also describes the reasons why the study proposes only three feasible routes.

3.1.1   Delineation Considerations

The path of the alternative is a major factor in the pipeline study. The routing of the alternatives was focused on minimizing the elevation head, which needs to be overcome. Furthermore, the goal was to minimize excavation, pavement replacement, length of pipe, and other high cost items for the new pipeline. Restrictions due to regulatory compliance require a wastewater pipeline route to adhere to standards set forth by Placer County. This limited the number of alternatives to a few possibilities that conform to sanitary sewer design criteria. Once this selection filter was applied, the following delineation considerations were made:

- Public right-of-way
- Existing easements
- Minimal distance
- Obstacles (existing structures, land features and utilities)
- Infrastructure (canals and storm drain)

Each criterion comes with certain challenges that were addressed in the development of each of the proposed alternatives and later evaluated according to weighted criteria.

3.1.2   Required Force Main Technology / Pump Stations

The Applegate Community and the SMD-1 tie-in point are separated by challenging land features characterized by foothill areas, spurs, draws, streams and elevations changes that do not allow a gravity flow system to be considered as an acceptable alternative. A method of minimizing excessive excavation or trenching applications for a gravity pipeline is to use force main technology. At the designated point of connection, the existing Auburn wastewater system is a force main, which requires the flows conveyed by the new pipeline to enter SMD-1 at least at the same pressure. If a partial gravity design / partial force main design is used, the gravity section of the pipeline would require numerous timed-valves and holding tanks in order to achieve the minimum gravity flow velocity of 2 feet per second as specified in the Placer County Design Standards. Also, the initial and operational costs would be increased by a partial gravity design because several pump stations and holding tanks would need to be placed along the pipeline length. Due to the above reasons, each of the proposed alternatives was designed as a force main for the entire length of the pipeline.

Force mains require periodic maintenance which was considered in the study as part of the weighted criteria to suggest a course of action to the County. Pumps, energy consumption, and maintenance personnel, associated with force mains were included as part of the alternatives' evaluation.

The existing Applegate lift station will remain in place with no changes. The project research showed that making upgrades to utilize it as an interception point of Applegate's wastewater flows would require overall improvements to the entire existing Applegate collection system. The reason for this is that only a portion of the service connections are pumped via the 4-inch force main which then outfalls to the 6-inch gravity sewer picking up the rest of the existing service connections (Figure 1.2). Utilizing the existing pump station would bypass a number of service connections, thus requiring their upgrade, which is beyond the scope of this study.

Additionally, each proposed lift station will have two identical pumps, one for normal operation and one for back-up in case of failure of the first pump. Emergency gasoline operated generator will also standby in case of power failure to keep the lift station operational until normal power is restored. Structural enclosures will house the electrical components and pump controls for weather protection and security purposes.

3.1.3   Conflict with Existing Utilities

The alternatives may cross or run next to existing utilities from Applegate to the point of connection to SMD-1. Regulations and improvement standards may provide guidance on the proper construction of wastewater pipeline to mitigate any potential interruption of services or damage to existing utilities. Since record drawings of existing utilities were not available at the time of this study, a consideration for necessary relocation of utilities could only be made after appropriate survey of existing utilities. Relocation of existing utilities may be necessary if it can significantly reduce the conflict with the preferred alternative route.

3.1.4   Land Acquisition / Easements / Right-of-Way

In order to minimize project costs, no land acquisitions are proposed along the pipeline delineation of either one of Route 2 or Route 3 except at the proposed pump station location at the intersection of Applegate Road and Merry Lane. The objective of the proposed alternative routes was to stay as much as possible within the Placer County right-of-way because the County would ultimately own, maintain and operate the proposed facilities. Route 1 would require limited traversing across private property.

3.1.5   Environmental Impacts

Environmental considerations are an important aspect of this project. The proposed alternatives attempt to minimize impact on wildlife or habitat in the project area. Noise, air quality, and other impacts on the environment as identified by regulations and improvement standards associated with the pipeline should be addressed and implemented by the contractor during the planning and construction phases of the project. The study identifies potential environmental impacts along each of the alternative routes. The proposed alternatives attempt to stay along existing right-of-ways to minimize the effects on undeveloped areas.

3.1.6   Cost

For each of the proposed alternatives, the study considered the estimated cost to construct the pipeline. The cost of force mains, pump stations, construction method, and other unique features were estimated to provide a comprehensive comparison between the alternatives presented in the

study. This also includes proposed land acquisition or utility relocation that may be necessary to pursue a feasible alternative.

3.2   Reasons for Disqualification

The following may be potential reasons to disqualify potential alternatives:

- Expected excessive costs due to constraints or limitations, whether regulatory or geographical, were not developed;
- Anticipated schedule delays for any portion or process of a route by external sources, i.e. regulatory agencies, stakeholders, or site conditions;
- Major stakeholder opposition to land use, location, or acquisition necessary for success of an alternative;
- Significant environmental impact due to alternative or construction thereof;
- Significant utility conflicts where a solution is not feasible or acceptable (such as the Kinder Morgan pipeline along the UPRR tracks or PG&E high voltage lines);
- Significant liability to the health or safety of the community;
- Unreasonable maintenance or operations of the proposed pipeline, whether it be access to the facility, dangerous for personnel, or hazardous for equipment (including vehicles).

Examples of routes, which were considered, but not further pursued, are:

- A route intercepting Applegate's sewer flows at the detention ponds and pumping it towards Applegate Road along Bon Vue Drive (Figure 1.1). This route would have required a larger pump due to the elevation head to be overcome. Moreover, Bon Vue Drive serves as the only access road to a cluster of private residences. Closure of this road due to construction would be, if not impossible, than extremely difficult to accommodate.
- Following County roads east of the detention ponds would have increased the length of the pipeline, and ultimately, the construction, operation and maintenance costs.
- Over-ground sewer pipeline route was considered along the UPRR tracks. Such route would have had a lower construction cost. However, safety precautions and maintenance complications deemed this design approach unfeasible.
- The unwillingness of Caltrans to allow the installation of the pipeline along the outside of the overpass at Placer Hills Road mandated the use of jack-and-bore.

4.0   PRIMARY ALTERNATIVES

The present routing study identified three feasible alternatives. As noted in the previous section, the research determined that other routes do not meet the expectation of the Client in terms of cost and schedule, and pose a significant potential for opposition by stakeholders. This section provides details and design information on each one of the proposed alternatives. Each alternative has elements that make it unique and different from the rest. Due to the constraints of the project site, the alternatives share common elements as an attempt to minimize costs related to construction and mitigation of significant impacts. The specific major groups of constraints for the alternative routes are:

- Regulatory Compliance
- Physical Constraints
- Environmental Constraints
- Private Stakeholders' Interests

The focus of this section is the design information related to each of the three alternatives.

4.1 Common Elements / Design Considerations

In all cases, the Applegate Treatment Plant would be decommissioned with the implementation of a new pipeline to convey wastewater flows to SMD-1. Therefore the study finds the following common elements for each alternative (Figures 4.1, 4.2 and 4.3):

- Cut off existing 6-inch gravity sewer to WWTP
- Construct a new pump station to connect the existing pipe to the proposed alternative. For alternative Route 2 (Figure 4.2) and Route 3 (Figure 4.3), this will occur before the UPRR overcrossing at Applegate Road and Merry Lane. The proposed pump station location for Route 1 (Figure 4.1) is at the detention ponds.
- From the point of connection in Applegate to SMD-1 the wastewater flows will be pumped via force main the entire distance.

The approach for construction of the pipeline shall be as follows whenever applicable:

- Open trench installation except for locations where directional drilling or jack-and-bore is required, i.e. crossing at creeks, canals, or the I-80 corridor.
- Minimum design depth of 3 feet per Placer County Standards. The design of the alternatives is based on the more conservative 5 feet minimum depth of cover, which accounts for the known and unknown locations where the pipeline needs to pass under shallow utilities or infrastructure, i.e. dry trench, water supply, drainage culverts, underground communication lines, etc. This approach also accounts for adequate clearance between the pipeline and such shallow utilities.
- There are five distinct creek crossings that all of the alternatives are going to encounter. Crossing underneath these channels will require a fairly deep directional boring operation. The footprint of each of these operations is proposed within the Placer County right-of way and will not require any easements over private land. Should the further field verification of the creek crossings uncover the lack of feasibility of directional

December 12, 2008

drilling, creek diversions and work in the stream bed will need to be pursued via special permits from the Army Corps of Engineers, the U.S. Fish and Wildlife Service, the California Department of Fish and Game and CVRWCB.

The final section of the pipeline along Dry Creek Road will be the same for all of the proposed Alternatives (Figures 4.1, 4.2 and 4.3):

- Installation of 6-inch force main in the public right-of-way along Dry Creek Road to Bluegrass Drive for a total of 5.1 miles.
- The designated tie-in point is into an existing 10-inch force main at Dry Creek Road and Bluegrass Drive.

For all of the alternatives there are three different construction methods: trenching, directional drilling and jack-and-bore. Trenching would require 15 to 20 feet of construction area centered around the proposed trench delineation in order to accommodate construction equipment and personnel. The depth of trench would dictate the shoring requirements and would not affect the width of the construction area. Directional drilling does not require any surface disturbance along the route except at the locations of the receiving pits. The size of the receiving pits would vary depending on the depth of the pipeline and would not exceed a 20 foot wide by 40 foot long footprint. Jack-and-bore is a method similar to directional drilling. The difference is that jack-and-bore can only be done in a straight line and the steel casing pipe remains in the ground. The jack-and-bore method needs the same construction footprint as the directional drilling method but would require a greater number of receiving pits.

Pump selection for each of the alternatives is challenging due to the low flows produced by the Applegate community and the high total head that needs to be pumped. Another challenge for the pump selection is the solids-handling during low flows. Since Auburn's SMD-1 wastewater system is a STEP system at the designated tie-in point (Reference [21]), Applegate's flows may require a pretreatment facility to avoid backup of the pipeline at the tie-in point. This potential issue would be handled by SMD-1. The specifics of the treatment of the Applegate's sewer flows are beyond the scope of this routing study.

The total operational time for the proposed pumps was selected to be 12 hours per day, which was based on an evaluation of head losses. The effective pumping rate, as shown in the design calculations per Appendix E, is double the assumed peak hour design flow rate, which was assumed to be 78,000 gallons of effluent wastewater per day. The existing effluent wastewater flows from the Applegate community were obtained from Placer County (Appendix E-1).

4.2   Route 1 (UPRR)

The pipeline for Route 1 follows a path along the Union Pacific Railroad (Figure 4.1). The Applegate's wastewater flows will be intercepted at the existing detention ponds. A lift station located at the existing detention ponds will pump the wastewater into the proposed pipeline under the westbound UPRR tracks through an existing tunnel (Reference [1]). From this point, the pipeline will head due south along the westbound UPRR tracks and cross underneath the eastbound UPRR tracks. For the length of the crossing, the proposed sewer pipeline must be incased in a steel pipe (Appendix B, X-1). After the second railroad crossing, the pipeline

---

Applegate to North Auburn Wastewater Conveyance Pipeline

Pipeline Routing Study

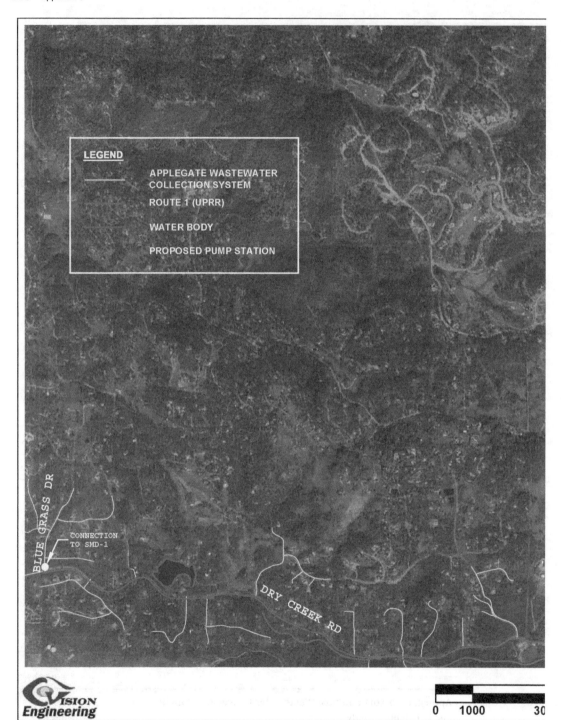

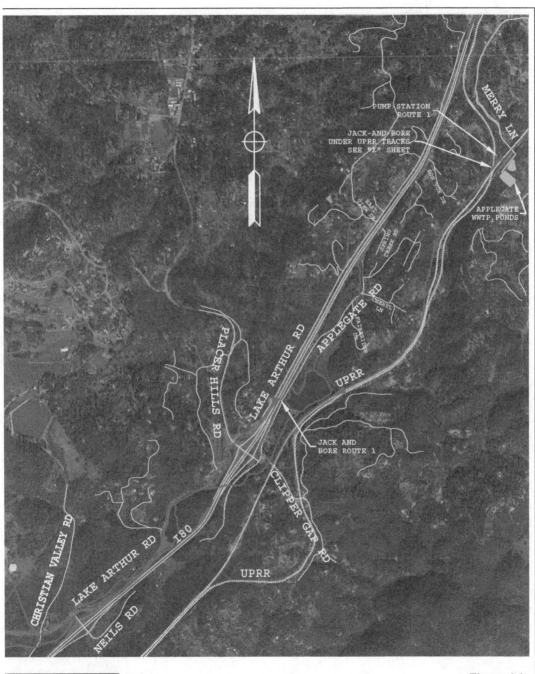

PUMP STATION
ROUTE 1

JACK-AND-BORE
UNDER UPRR TRACKS
SEE "X" SHEET

MERRY LN

APPLEGATE
WWTP PONDS

PLACER HILLS RD

LAKE ARTHUR RD

APPLEGATE RD

UPRR

JACK AND
BORE ROUTE 1

CLIPPER GAP RD

CHRISTIAN VALLEY RD

LAKE ARTHUR RD

I 80

NEILS RD

UPRR

6000 ft

Figure 4.1
Route 1 (UPRR)

follows a maintenance road along the West side of the UPRR right-of-way. The route turns west on Clipper Gap Road until it meets Applegate Road. There, the pipeline heads north to the proposed jack-and-bore crossing of I-80 (Figure 4.1). The Route 1 force main will convey the wastewater flows across the freeway to Lake Arthur Road and then south to Dry Creek Road, which leads to the designated point of connection with SMD-1. The total length of force main is approximately 8.1 miles.

A sewer easement would need to be obtained from UPRR for maintenance and construction of the pipeline, which would be secondary to the UPRR right-of-way easement. Because the pipeline is intended to follow the railroad, the entire construction footprint will remain in the UPRR right-of-way (Reference [21]). Thus, no easements over private land would be required except at the intersection of the UPRR tracks with Clipped Gap Road. There, a small sewer easement would be needed (Appendix B, PM-3).

The remainder of the pipeline, from the intersection of Clipper Gap Road with the UPRR tracks to SMD-1, remains within Placer County's right-of-way. A sewer easement would not be necessary because Placer County is to become the pipeline owner. Crossing underneath I-80 would require either a jack-and-bore or a directional drilling construction and a sewer easement would need to be obtained from Caltrans for this stretch of the pipeline delineation. The receiving pits must be outside of the Caltrans right-of-way.

4.2.1 Design Details

The design drawings for this alternative can be found in Appendix B. This alternative provides the following unique elements to meet the scope of the project:

- Force main route following the maintenance road on the West side of the UPRR tracks up to Clipper Gap Road.
- Force main turns west onto Clipper Gap Road and heads north along Applegate Road to avoid traversing private property before crossing I-80.
- Jack-and-bore crossing of I-80 between Applegate Road and Lake Arthur Road where the pipeline merges with the common for all of the proposed alternatives final stretch of the project (Appendix B, L1-4 and X-1)
- This alternative requires a single pump station at the point of flow interception at the existing detention ponds. The total elevation head that needs to be overcome by the pump is roughly 140 feet. The total dynamic head, accounting for friction losses, minor losses and elevation head, is 196 feet at a pumping rate of 110 gallons per minute identifying the pump operational point. The pump was sized utilizing a published catalog from an industry standard pump manufacturer – Goulds Pumps. Total head losses, system curve and pump selection can be found in Appendix E-2 and Appendix F.
- Route 1 requires one instance of land acquisition. The proposed land acquisition in form of an easement along the pipe delineation is located on parcel 077-130-026-000 which belongs to the Trustee. The easement is projected to be 67 feet wide by 71 feet long, totaling to 290 square feet of the aforementioned property (Appendix B, PM-3).
  No land acquisition for the pump station and facilities' footprint is needed for Route 1.

- In order to minimize project costs, Route 1 utilizes parcel number 073-120-013-000, which belongs to Placer County, for the construction of the required pump station facilities. The said parcel is the current site of Applegate's detention ponds and already has the necessary access roads in place.

4.2.2 Constraints

Table G.1 in Appendix G show the locations along Route 1 where certain physical, environmental or regulatory constraints may be an issue. Permits that would have to be obtained and their areas of concern along the route are also included. This table was used as an aid for comparing the constraints for each of the alternatives. The following is a summary of the constraints for Route 1:

- Maintenance road along the UPRR tracks

  - Hilly terrain may cause difficulties in installing the force main;
  - Encountering hard rock may have impact on the construction methods;
  - Live train traffic poses construction, accessibility and maintenance challenges;
  - Limited space for construction in some areas;

- Existing infrastructure

  - Construction nearby railroad tunnels;
  - Crossing under the railroad tracks;
  - Crossing drinking water supply lines such as the Bypass Canal near Clipper Gap Road;
  - Creek and canal crossings along Dry Creek Road;
  - Lack of public right-of-way or developed roads for construction crews and equipment along the UPRR easement – only dirt trails and narrow paths;

- Stakeholders

  - Possible delays due to opposition by UPRR to construct in their right-of-way;
  - Possible conflict with the existing high pressure liquefied propane gas pipeline by Kinder Morgan Inc.

- Environmental

  - Existing oak trees along the route that may need to be removed and replaced.

4.2.3 Cost Estimate

Among the proposed alternatives, Route 1 has the longest length of force main off of the existing paved roads. This means a lower cost for surface roadwork repair. Repaving will be the least for this alternative. However, the pump for this alternative (Appendix F) has to overcome nearly twice the total dynamic head and more than twice the elevation head alone. Table 4.1 presents a summary of the preliminary cost estimate for this alternative. A detailed preliminary cost estimate can be found in Appendix H.

December 12, 2008

Table 4.1- Summary of Preliminary Cost Estimate for Route 1

SUBTOTALS		
FORCE MAIN	$	2,556,060
PUMP STATION	$	475,000
SITE WORK	$	2,762,523
LAND ACQUISITION	$	18,850
ANNUAL MAINTENANCE	$	15,650
TOTAL	$	5,828,084

4.2.4   Alternative Evaluation

This alternative was originally intended to follow immediately next to the UPRR tracks, utilizing the existing, consistently downhill, elevation gradient. However, cobbles, boulders, narrow canyons, train tunnels and passing trains pose safety concerns and construction problems. After a close investigation of these conditions, this alternative had to be routed along the maintenance and access roads parallel to the UPRR tracks. These roads follow the terrain along the West boundary of the UPRR right-of-way, and often drastically change elevation. The rest of the route is along county roads with the exception of the jack-and-bore location to cross under the I-80 corridor. Back-tracking a short distance along Applegate Road was necessary to find a convenient crossing location and to avoid traversing private property. Almost the entire Applegate community is avoided by the delineation of this alternative.

Route 1 has the following benefits:

- Less than significant impact to most of Applegate's residents during construction;
- Minimized road repair since a long stretch of the pipeline is not following paved roads in the public right-of-way;

This alternative was pursued in attempt to mitigate the following:

- Avoid Creek crossings on Applegate Road;
- Avoid utility conflicts on Applegate Road;
- Minimize disruption of transportation and access disruption due to construction on public roads.

4.3 Route 2 (Applegate Road)

The pipeline for Route 2 (Figure 4.2) follows the Placer County's roads of Applegate, Lake Arthur and Dry Creek. It will pick up flows from the existing Applegate sewer system at the intersection of Applegate Road and Merry Lane. A lift station will pump the wastewater south, away from the community, along Applegate Road for approximately 1.8 miles to the location of the jack-and-bore crossing of I-80 (Figure 4.2). The force main will continue to convey flows across I-80 to Lake Arthur Road. Once across, the force main will continue south along Lake Arthur Road for 2.5 miles and west along Dry Creek Road for 2.6 miles before it injects flows into SMD-1. The total length of force main is approximately 6.9 miles.

From the tie-in point at the intersection of Applegate Road and Merry Lane the pipeline is proposed to stay within the Placer County right-of way. A sewer easement would not be necessary because Placer County is to become the pipeline's owner. Crossing underneath I-80 would require either a jack-and-bore or a directional drilling method of construction. A sewer easement would need to be obtained from Caltrans for this part of the pipeline delineation as well as both of the receiving pits. The receiving pits must be outside the Caltrans right-of-way. No surface work is permitted in the Caltrans right-of-way.

4.3.1 Design Details

The design drawings for this alternative can be found in Appendix B. This alternative provides the following unique elements to meet the scope of the project:

- Jack-and-bore crossing of I-80 between Applegate Road and Lake Arthur Road at a location further north than the one proposed in Route 1, north of Lake Theodore and where Applegate Roads veers away from I-80 (Appendix B, L2-4 and X-2)
- This alternative requires a single pump station at the point of Applegate's wastewater flow interception located at the intersection of Applegate Road and Merry Lane. The total elevation head that needs to be overcame by the pump is roughly 68 feet. The total dynamic head is 118 feet at a pumping rate of 110 gallons per minute, which identifies the pump's operational point. The pump was sized utilizing a published catalog from an industry standard pump manufacturer – Goulds Pumps. Total head losses, system curve and pump selection can be found in Appendix E-3 and Appendix F.
- Route 2 requires land acquisition in order to accommodate the footprint of the pump station and its facilities. Parcel 073-120-013-000 owned by XXX is a suitable location for the proposed pump station and facilities. The proposed land acquisition is estimated to be 100 feet by 35 feet and totaling to 350 square feet (Appendix B, PM-4).

4.3.2 Constraints

Table G.2 in Appendix G show the locations along Route 2 where certain physical, environmental or regulatory constraints may be an issue. Permits that would have to be obtained and their areas of concern along the route are also included. This table was used as an aid for comparing the constraints for each of the alternatives. The following is a summary of the constraints for Route 2:

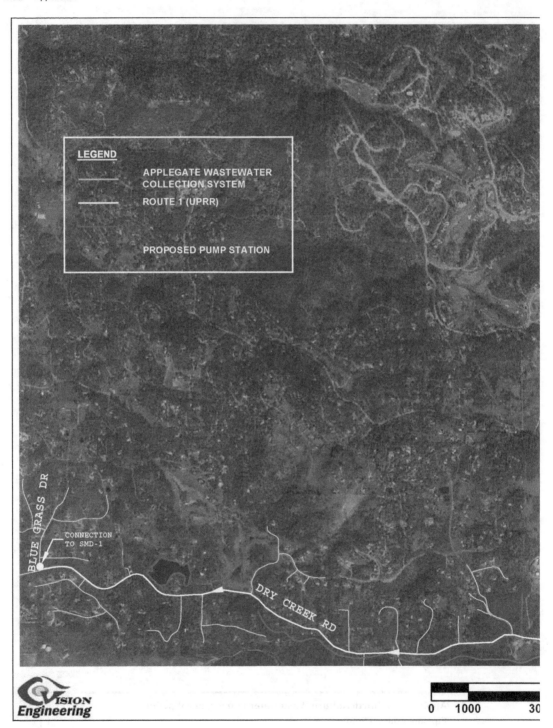

Figure 4.2
Route 2 (Applegate Rd)

- Existing utilities

    o Water mains, gas lines and possibly other underground utilities along Applegate Road;
    o Overhead power and telephone lines along Applegate Road may be too low to provide the necessary clearance for construction vehicles and equipment.

- Existing infrastructure

    o Drainage ditches and culverts;
    o Creek and canal crossings.

- Stakeholders' concerns

    o Air pollution;
    o Construction noise;
    o Road closures;
    o Inconvenience to property owners along the county roads.

4.3.3 Cost Estimate

This route follows Applegate Road for a longer distance and Lake Arthur Road for a shorter distance than Route 3. There will be little difference in cost between Routes 2 and Route 3. There is a high possibility of additional cost for more instances of relocation of utilities along Route 2, due to its path through the community of Applegate. This alternative will pose a greater inconvenience to Applegate's residents since private residence density is higher on Applegate Road than that along Lake Arthur Road. Table 4.2 presents a summary of the preliminary cost estimate for this alternative. A detailed preliminary cost estimate can be found in Appendix H.

Table 4.2- Summary of Preliminary Cost Estimate for Route 2

SUBTOTALS		
FORCE MAIN	$	2,249,820
PUMP STATION	$	445,000
SITE WORK	$	2,667,390
LAND ACQUISITION	$	35,000
ANNUAL MAINTENANCE	$	14,374
TOTAL	$	5,411,584

4.3.4 Alternative Evaluation

This route was developed to stay on County roads to avoid disturbing the environment on undeveloped lands and traversing private property. The route goes through the Applegate community, along the main thoroughfare. Among the three proposed alternatives, Route 2 has

the highest probability of conflicts with existing utilities, particularly from Fairbridge Drive to Spring Creek Road (Appendix B, L2-5). Careful planning, staging and traffic control would need to be performed by contractors to minimize the impact of construction on residents. Local ordinances would need to be observed to lessen the disturbance of construction during specific hours and to avoid generating excessive traffic delays. Staging and observance of ordinances may cause delays during construction.

The jack-and-bore location to cross the I-80 corridor was chosen to keep the pipeline away from Lake Arthur and to avoid additional creek crossings further south on Applegate Road. One of the challenges along this alternative is a creek crossing at Cheryl Lane, which is the Boardman Canal drinking water supply line. The pipeline would need to be incased in a steel pipe and go underneath the channel with a 10 foot minimum clearance. This location also has a difficult terrain and very limited construction space.

Route 2 has the following benefits:

- Less than significant impact on the environment because the pipeline follows a paved road for its entirety;
- Ease of access and maintenance of the new pipeline during the time of construction and operation.

This alternative was pursued in an attempt to mitigate the following:

- Conflict with UPRR;
- Conflict with Kinder Morgan Inc.;
- Minimize the environmental impacts.

4.4   Route 3 (Lake Arthur Road)

The pipeline for Route 3 (Figure 4.3) follows Applegate Road, Lake Arthur Road and Dry Creek Road. It will pick up flows from the existing Applegate sewer pipeline at the same location as Route 2. A lift station will pump the wastewater south, away from the community, along Applegate Road for about 1.4 miles to a jack-and-bore crossing of I-80 between Spring Creek Road and East View Drive (Figure 4.3). The force main will continue to convey flows across I-80 to Lake Arthur Road (Appendix B, X-2). Once across, the force main will continue south along Lake Arthur Road for 3.0 miles and west along Dry Creek Road for 2.5 miles to the connection point with SMD-1. The total length of force main is approximately 6.9 miles.

From the start point at the intersection of Applegate Road and Merry Lane the pipeline will stay within the Placer County right-of-way. A sewer easement would not be necessary because Placer County is to become the pipeline's owner. Crossing underneath I-80 would require either a jack-and-bore or a directional drilling method. A sewer easement would need to be obtained from Caltrans for the jack-and-bore as well as both of the receiving pits. The receiving pits must be outside the Caltrans right-of-way. No surface work is to be done in the Caltrans right-of-way.

4.4.1   Design Details

The design drawings for this alternative can be found in Appendix B. This alternative provides the following unique elements to meet the scope of the project:

- Jack-and-bore crossing of I-80 between Applegate Road and Lake Arthur Road is at the location between Spring Creek Road and East View Drive;
- This alternative requires a single pump at the point of flow interception located at the intersection of Applegate Road and Merry Lane. The total elevation head that needs to be overcame by the pump is roughly 68 feet. The total dynamic head is 118 feet at a pumping rate of 110 gallons per minute, which identifies the pump's operational point. The pump was sized utilizing a published catalog from an industry standard pump manufacturer – Goulds Pumps. Total head losses, system curve and pump selection can be found in Appendix E-4 and Appendix F.
- Route 3 requires the same land acquisition for the pump station site as Route 2. The proposed land acquisition is the above mentioned 350 square feet (Appendix B, PM-4) from parcel 073-120-013-000 owned by 'Daffern Dayle Lorraine.'

4.4.2   Constraints

Table G.3 in Appendix G show the locations along Route 3 where certain physical, environmental or regulatory constraints may be an issue. Permits that would have to be obtained and their areas of concern along the route are also included. This table was used as an aid for comparing the constraints for each of the alternatives. The following is a summary of the constraints for Route 3:

- Existing utilities
    - o Water mains, gas lines and possibly other underground utilities along Applegate Road and Lake Arthur Road;
    - o Overhead power and telephone lines along Lake Arthur Road.

- Existing infrastructure
    - o Drainage ditches and culverts;
    - o Creek and canal crossings.
- Stakeholders' concerns
    - o Air pollution;
    - o Construction noise;
    - o Road closures;
    - o Inconvenience to property owners along Applegate Road and Lake Arthur Road.

4.4.3 Cost Estimate

This route follows Applegate Road for a shorter distance and Lake Arthur Road for a longer distance than Route 3. There will be a little difference in cost between Route 2 and Route 3. However, this route should have less costs associated with utility conflicts since it avoids most of the Applegate community, by crossing I-80 much sooner than Route 2. Table 4.3 presents a summary of the preliminary cost estimate for this alternative. A detailed preliminary cost estimate can be found in Appendix H.

Table 4.3- Summary of Preliminary Cost Estimate for Route 3

SUBTOTALS	
FORCE MAIN	$ 2,234,580
PUMP STATION	$ 445,000
SITE WORK	$ 2,562,310
LAND ACQUISITION	$ 35,000
ANNUAL MAINTENANCE	$ 14,311
TOTAL	$ 5,291,201

4.4.4 Alternative Evaluation

This route was developed to stay on County roads to avoid disturbing the environment on undeveloped lands and traversing across private property. The route follows Applegate Road for a much shorter distance than Route 2 does. Careful planning, staging and traffic control should be performed by contractors to minimize the impact of construction on residents. The route crosses I-80 before going into the populated area along Applegate Road. This delineation will avoid the potential additional utility and private residence conflicts, which are present along Route 2. The location of the jack-and-bore crossing of the I-80 corridor allows for an easier crossing of the Boardman Canal, which makes Route 3 more preferable and easier to construct than Route 2. The crossing of the Boardman Canal drinking water supply line would still require a 10 foot minimum clearance.

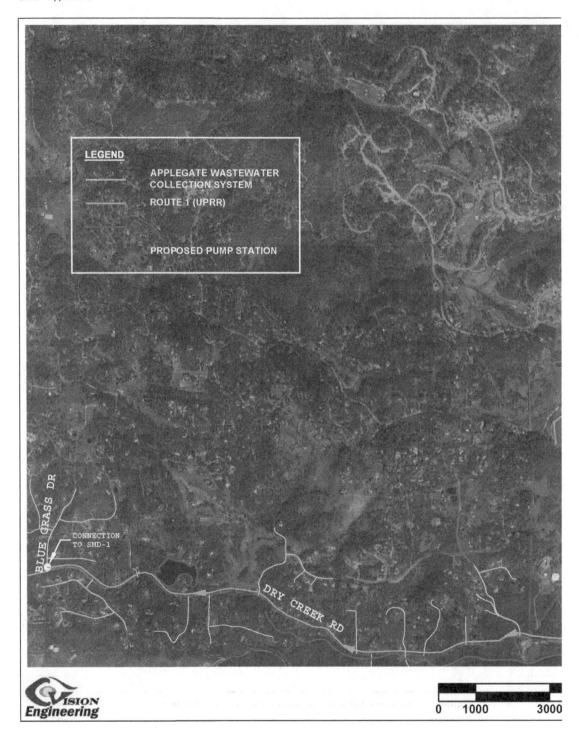

Figure 4.3
Route 3 (Lake Arthur Rd)

Route 3 is proposed with the following benefits:

- Less than significant impact on the environment due to following public right-of-way;
- Significantly less impact on Applegate's residents than Route 2;
- Significantly less conflicts with utilities than Route 2.

This alternative was pursued in an attempt to mitigate the following:

- A difficult creek crossing on Applegate Road;
- Major utility conflicts on Applegate Road;
- Conflict with UPRR;
- Conflict with Kinder Morgan Inc.

5.0 EVALUATION OF ALTERNATIVES

After brainstorming possible alternatives, three feasible alternatives were determined from the results of the gathered research. These three alternatives were then analyzed individually based on the following ranking criteria.

5.1 Ranking Criteria

The evaluation of each alternative was determined by assigning points to each area of constraints which represent a weighted average with respect to each other (Table 5.1). There are a maximum of 100 points that represent 100 percent of the weighted constraint averages. Each area has its own maximum points possible. The higher the points given, the better the alternative is. The alternative with the most points was determined to be the recommended alternative. The following are the nine criteria used for the evaluation of each alternative:

- Physical Constraints
- Permits
- Existing Utility Conflicts
- Environmental Impacts
- Construction Costs
- Impacts on Businesses / Residents
- Safety
- Land Acquisition / Easements
- Maintenance Costs / Accessibility

The weighted averages were chosen for each of the constraints based on their importance to the project and to each other. Since the 6-inch sewer pipeline would follow exiting roads for most of its route, the project would have minimal if any environmental impacts. Therefore, a lower weight was given to the environmental impacts criterion. Maintenance costs, accessibility, construction costs, and impacts to businesses and residents would be of a greater importance to the project. At the same time, maintenance costs and accessibility are long-term conditions so they were rated higher than the construction costs. Conflict with existing utilities is also a major factor for the evaluation of the alternatives, because they may pose significant schedule and budget implications. Since the proposed land acquisitions are minimal, this criterion was assigned lower possible maximum points. Safety should always be a priority, so this area was given a higher importance. Although the working conditions may be difficult along some stretches along the proposed routes, the density and speed of the adjacent traffic is low. This is why the safety criterion was given a lesser importance than some of the other ranking criteria.

December 12, 2008

Table 5.1- Ranking Criteria with Alternatives' Scores

RANKING CRITERIA	POSSIBLE POINTS	ROUTE 1	ROUTE 2	ROUTE 3
Physical Constraints	10	5	7	9
Permits	8	4	6	7
Existing Utility Conflicts	12	10	6	8
Environmental Impacts	8	3	6	7
Construction Costs	15	12	14	15
Impacts on Businesses / Residents	14	11	9	9
Safety	10	7	8	9
Land Acquisition / Easements	5	5	4	4
Maintenance Costs / Accessibility	18	10	18	18
TOTAL POINTS	100	67	78	86

5.2 Alternative Rankings

Each alternative faces a specific set of constraints. The points given to each alternative are meant to reflect the complexity and the difficulty in constructing that route. Although the value of the points assigned to each area is not supported by precise statistical data, they do serve as a reasonable tool in the comparison of the proposed alternatives to one another and in the selection of the recommended alternative.

5.2.1 Route 1 (UPRR)

The following describes the reasons for which a specific number of points was assigned to Route 1 for each of the ranking criteria (Table 5.1):

- Physical Constraints (Points: 5)
    - The sewer line must overcome significant elevation changes due to the hilly terrain along the UPRR right-of-way.
    - Rocky soil conditions (Appendix C) may make trenching for the sewer pipeline difficult and time consuming.
    - Numerous canal and creek crossings along Applegate Road and Dry Creek Road, including a crossing of the Boardman Canal at Clipper Gap Road.
    - Lack of paved roads for construction equipment to access areas along UPRR.
    - The route encroaches on the UPRR and the Caltrans right-of-way.
    - The route traverses across private property at the intersection of Clipper Gap and UPRR.

- Permits (Points: 4)
    - o UPRR has stringent requirements for utilities encroaching on its right-of-way.
    - o UPRR has a long permitting process.
- Existing Utility Conflicts (Points: 10)
    - o Most of the existing utilities, which would be encountered by the route are along Lake Arthur Road and Dry Creek Road, which is typical for all of the proposed alternatives.
- Environmental Impacts (Points: 3)
    - o There is a moderate possibility of removing Blue Oak Trees along the route since it crosses undeveloped areas along the UPRR right-of-way.
    - o There is a slight chance of harming other plant and animal species mentioned in Section 2, since a long stretch of the route is along an undeveloped part of the UPRR right-of-way.
    - o The route passes near Lake Theodore, the Boardman Canal and other creeks.
- Construction Costs (Points: 12)
    - o Highest total cost ($5,828,084).
    - o The route would not have as high construction cost associated with excavation, and re-paving as the other two routes.
- Impacts on Business and Residents (Points: 11)
    - o Since the majority of the construction would take place in the UPRR right-of-way the impact on residents should be minimal
- Safety (Points: 7)
    - o A significant portion of the construction activities for this route would take place in the UPRR right-of-way, but no less than 35 feet away from the railroad tracks. Still, special safety measures should be taken for construction personnel and equipment.
    - o Working near live traffic would be done along the County roads, which is typical for all of the routes.
- Land Acquisition / Easements (Points: 5)
    - o Easements from UPRR are more stringent and time consuming than those from other regulatory agencies.
    - o A limited temporary easement would be required at the intersection of Clipper Gap Road and the railroad tracks.
- Maintenance Costs / Accessibility (Points: 10)
    - o The cost of maintenance costs may be greater for this alternative as it requires a larger and more expensive pump. The accessibility would be difficult for the stretch of the route in the UPRR right-of-way.

5.2.2   Route 2 (Applegate Road)

The following describes the reasons for which a specific number of points was assigned to Route 2 for each of the ranking criteria (Table 5.1):

- Physical Constraints (Points: 7)
    - o   Numerous canal and creek crossings along Applegate Road and Dry Creek Road.
    - o   The route encroaches on the UPRR and the Caltrans right-of-way.
    - o   The proposed pump station encroaches on residential property.
- Permits (Points: 6)
    - o   Placer County does not have permits that are as stringent or time consuming as the UPRR permits.
    - o   Caltrans required permits for transverse encroachments are not as stringent as those required by UPRR.
    - o   Route 2 requires a total number of permits which is higher than that for both Route 1 and Route 3.
- Existing Utility Conflicts (Points: 6)
    - o   Many overhead power lines along Applegate Road and Lake Arthur Road may be a concern for construction equipment.
    - o   Probable underground utility lines across Applegate Road, Lake Arthur Road and Dry Creek Road.
- Environmental Impacts (Points: 6)
    - o   There is a slight chance of harming the plant and animal species mentioned in Section 2 since most of the route is on County roads.
    - o   There are several creek and canal crossings along the route.
- Construction Costs (Points: 14)
    - o   Total cost of $5,411,584 is lower than the one for Route 1 and higher than the one for Route 3.
    - o   Compared to Route 1, this route has higher construction costs associated with excavation and re-paving.
- Impacts on Business and Residents (Points: 9)
    - o   Since most of the construction would take place along County roads passing through densely populated areas, this alternative would cause more inconveniences to residents and businesses in the area than Route 1 and 3.
- Safety (Points: 8)
    - o   Most of the construction activities for this route would take place on narrow County roads accommodating traffic and serving as the only access to private residences. Safety measures should be taken to prevent traffic accidents and protect construction workers and the general public.
- Land Acquisition / Easements (Points: 4)
    - o   Easier process of obtaining easements from Caltrans than from UPRR.
    - o   Minor land acquisition is needed for the footprint of the proposed pump station.
- Maintenance Costs / Accessibility (Points: 18)
    - o   Lower maintenance cost than Route 1.

- o Better accessibility to the project site than that for Route 1 both during construction and operation.

5.2.3 Route 3 (Lake Arthur Road)

The following describes the reasons for which a specific number of points was assigned to Route 3 for each of the ranking criteria (Table 5.1):

- Physical Constraints (Points: 9)
  - o Numerous canal and creek crossings along Applegate Road and Dry Creek Road.
  - o The route encroaches on the UPRR and the Caltrans right-of-way.
  - o The location of the proposed pump station encroaches on private property.
- Permits (Points: 7)
  - o Placer County does not have permits that are as stringent or time consuming as the UPRR permits.
  - o Caltrans required permits for transverse encroachments are not as stringent as those required by UPRR.
  - o Route 3 requires a total number of permits which is higher than that for Route 1 but less than that for Route 2.
- Existing Utility Conflicts (Points: 8)
  - o Probable underground utility lines across Applegate Road, Lake Arthur Road and Dry Creek Road.
  - o Lower probability of utility conflicts than Route 2.
- Environmental Impacts (Points: 7)
  - o There is a slight chance of harming the plant and animal species mentioned in Section 2 since most of the route is on County roads.
  - o There are several creek and canal crossings along the route.
- Construction Costs (Points: 15)
  - o Lowest total cost of $5,291,201.
  - o Higher construction costs associated with excavation and re-paving than Route 1 and Route 2.
- Impacts on Business and Residents (Points: 9)
  - o Less impact on residents and businesses in the area than Route 2.
- Safety (Points: 4)
  - o Like Route 2, most of the construction activities for this route will take place on the County roads. Safety measures should be taken to prevent traffic accidents and protect construction workers and the general public.
- Land Acquisition / Easements (Points: 5)
  - o Same as Route 2.
- Maintenance Costs / Accessibility (Points: 18)
  - o Same as Route 2.

6.0 RECOMMENDATION

This study recommends Route 3, based on the ranking criteria discussed in Section 5 and summarized in Table 5.1. The other two alternatives, though feasible, did not have as many benefits and received a lower score based on the criteria established in this study.

6.1 *Reason for Recommendation*

Reasons for denying Route 1 (UPRR):

- Permits and timing associated with strict UPRR regulations may go beyond the desired deadline for implementation of the design.

- Environmental concerns due to construction on undeveloped lands are lower as compared to the other alternatives.

- Physical constraints, accessibility along undeveloped land and limited space around the railroad infrastructure are more significant constraints for this alternative than for the other two routes.

- Safety concerns for construction along live train traffic and additional bond insurance required per UPRR standards.

Reasons for denying Route 2 (Applegate Road):

- Complex crossing of the Boardman Canal increases the potential for more delays and costs than Route 3.

- Significant inconveniences to residents along Applegate Road.

- Complicated construction staging to minimize impact on residents.

- Construction costs increased due to delays and staging along a more travelled road.

Route 3 follows a route along the same county roads as Route 2, but would impact the community less by crossing I-80 before going through the populated area on Applegate Road. Since Route 3 stays mostly on county roads, construction equipment would not have to go off road to the extent of Route 1. Finally, potential environmental impacts may be the same as Route 2, however, construction cost is lower. This results in minimizing the following:

- Construction delays

- Conflicts with residents

- Conflict with existing utilities

Although Route 3 did not surpass the other alternatives in each criterion, it established itself as the highest overall scoring alternative. Thus, the study identified Route 3 as the recommended route for the implementation of the new Applegate wastewater conveyance pipeline connecting to North Auburn's SMD-1 at Dry Creek Road and Blue Grass Drive.

6.2   Benefits

Based on the established by the study ranking criteria Route 3 has the following benefits:

- It has the potential to be constructed in the least amount of time.
- It is the least likely to be opposed by stakeholders, residents and environmental groups.
- It is the cheapest of the three proposed alternatives.

6.3   Cost

The preliminary cost for Route 3 is estimated to be $5,291,201.

REFERENCES

1. Landis and Associates. (January 1, 1975). *Placer County Service Area No.24. Improvements to Community Sewer System: Applegate.* (Record Drawings).

2. Central Valley Regional Water Quality Board (CVRWQB). (June 23, 2006). *Administrative Civil Liability Complaint No. R5-2006-0510 Against Placer County Service Area and No. 28, Zone 24 Applegate Wastewater Treatment Facility.*

3. Department of Public Works, Placer County, State of California. (August 2005). *General Specifications.* Section 71: Sewers.

4. State of California, Department of Transportation. (January 19, 2004). *Encroachment Permits Manual.* Chapter 6, Utilities Permits. Available: http://www.dot.ca.gov/hq/traffops/developserv/permits/pdf/manual/Chapter_6.pdf. Retrieved: October, 2008.

5. Union Pacific Railroad. (2008). *Pipeline Installation Engineering Specifications: Specifications for Pipelines with Maximum Casing Diameter Of 48 Inches and Encased Gas Transmission Lines Crossing Under Railroad Tracks.* Available: http://www.uprr.com/reus/pipeline/pipespec.shtml. Retrieved: October, 2008.

6. California Department of Public Health. (May 2006). *Pipeline Separation Design and Installation Reference Guide.*

7. Placer County, California. (October 1, 2008). *Placer County Code.* Available: http://qcode.us/codes/placercounty/. Retrieved: September, 2008.

8. Placer County, California. (2005). *General Plan. Placer County General Plan. Auburn, CA.* Available: <http://www.placer.ca.gov/Departments/CommunityDevelopment/Planning/Documents/ComPlans/GenPlanPC.aspx>;. Retrieved: October, 2008.

9. Union Pacific Railroad. (2008). *Procedures for Encroachments.* Available: http://www.uprr.com/reus/encroach/procedur.shtml. Retrieved: October, 2008.

10. Central Valley Regional Water Quality Board (CVRWQB). (2008). *Laws and Regulations.* Available: http://www.swrcb.ca.gov/centralvalley/laws_regulations/. Retrieved: October, 2008.

11. State Water Resources Control Board. (January 2003). *Water Quality Order 99-08-DWQ. National Pollutant Discharge Elimination System (NPDES). Storm Water Discharges associated with Construction Activity (General Permit).*

12. Placer County Air Pollution Control District. (2008). *Air Quality Standards.* Available: http://www.placer.ca.gov/Departments/Air/airquality.aspx. Retrieved: September 2008.

December 12, 2008

13. California Air Resources Board. 2008. *Air Districts (APCD or AQMD). Authority to Construct.* Available: http://www.arb.ca.gov/permits/airdisac.htm. Retrieved: October 2008.

14. U.S. Army Corps of Engineers. (2008). *Regulatory Program Overview.* Available: http://www.usace.army.mil/cw/cecwo/reg/oceover.htm. Retrieved: October 2008.

15. California Department of Parks and Recreation. (February 14, 2002). *National Historic Preservation Act.* Available: http://ceres.ca.gov/wetlands/permitting/nhpa.html. Retrieved: October, 2008.

16. State of California, Department of Transportation. (January 19, 2004). *Project Plans for Construction on State Highway in Placer County, from Route 80/193 Separation to Auburn Ravine Undercrossing and from 0.8km West of Auburn Ravine Road Overcrossing to Route 174/80 Separation.* (Record Drawings).

17. Union Pacific Railroad. (July 26, 2006). *Interim Guidelines for Horizontal Directional Drilling (HDD) Under Union Pacific Right-of-Way.*

18. Kleinfelder. (July 18, 2008). *Geotechnical Data Report. Applegate Wastewater Connection to SMD-1 Collection System.*

19. ICF Jones & Stokes. (September, 2008). *Initial Study/Environmental Assessment for Applegate Wastewater Treatment Plant Closure and Pipeline Project.* Available: http://www.placer.ca.gov/Departments/CommunityDevelopment/EnvCoordSvcs/EIR/ApplegateWastewater.aspx. Retrieved: September 2008.

20. Placer County, California. (February 22, 2005). *Placer County Conservation Plan.* Available: http://www.placer.ca.gov/Departments/CommunityDevelopment/Planning/PCCP.aspx. Retrieved: September 2008.

21. Cooper, Rodolf and Associates. (May 27, 1981). *Improvement plans for Low-Pressure Sewer System, Saddleback, California.* (Record Drawings). Sheet 3/10 (Blue Grass Drive).

22. Placer County, California. (1980). *Southern Pacific Transportation Company Right of Way and Track Map.* (Record Drawings).

December 12, 2008

APPENDICES (*not included*)

A. *American River Watershed Map*

B. *Design Drawings - Route 3*
   - Title Sheet (T-1)
   - Route 1 Layouts (L1-1 to L1-6)
   - Route 2 Layouts (L2-1 to L2-6)
   - Route 3 Layouts (L3-1 to L3-6)
   - Cross-Sections (X1 to X2)
   - Details (C1)
   - Parcel Maps (PM-1 to PM-4)

C. *Site Geologic and Exploration Location Map*

D. *Naturally Occurring Asbestos Map*

E. *Pump Calculations*
   - E-1. Applegate's Existing Flows
   - E-2. Route 1 (UPRR)
   - E-3. Route 2 (Applegate Road)
   - E-4. Route 3 (Lake Arthur Road)

F. *Pump Selections*
   - Route 1 – Goulds Model 3771(12 hour)
   - Route 2 and Route 3 – Goulds Model 3655 (12 hour)

G. *Constraints*
   - Table G.1- Constraints along Route 1 (UPRR)
   - Table G.2- Constraints along Route 1 (Applegate Rd)
   - Table G.3- Constraints along Route 1 (Lake Arthur Rd)

H. *Preliminary Cost Estimates*

Introduction
Ethics
Professional Engagement
History
What Engineers Deliver
Emerging Technology
1 2 3 4 5 6
Engineer's Role in Project
Development
Permitting
Having
a Life
17
A B C D E F
7 8 9 10 11 12 13 14 15 16
Leadership Managing
Executing a Professional
Commission
Sustainability
Client Relationship
Globalization
Legal Aspects Communicating

A p p e n d i x **D**

Example Short Technical Report:

The Benefits of Green Roofs

ABSTRACT

Green roofs are becoming a popular choice for many cities and corporations. The purpose of this report is to discuss the benefits of green roofs. A typical green roof section includes structural support, a roofing membrane, insulation, a root barrier, drainage medium, filter fabric, a growing medium, and vegetation. The two types of green roofs are intensive and extensive. Intensive roofs are characterized by a shallow growing medium and shorter plants, while an extensive roof contains a deeper growing medium and can support a wide range of plant life. The advantages of green roofs include: improved stormwater runoff management, improved air quality, reduced heat island effect, thermal and acoustical insulation, reduced HVAC costs, extended roof life, potential food production, and community social, health, and emotional benefits. Issues that must be kept in mind when considering green roofs are initial costs, maintenance costs, and drainage and irrigation. Green roofs have the potential to address a myriad of urban issues. Given the benefits, all cities should promote the use of green roofs. Green roofs are particularly suited to public buildings and corporate entities with a long-term commitment to their projects.

INTRODUCTION

Green roofs are becoming a popular choice for many cities and corporations. They have many benefits. The purpose of this report is to discuss the benefits of green roofs. The background section contains information on the basic components of a green roof system and explains the two main types of green roofs. The discussion section outlines the benefits of green roofs and briefly explains a few considerations that must be addressed. The conclusion reiterates the benefits of green roofs and explains who realizes each of those benefits. The recommendations include those who should strongly consider the use of green roofs.

BACKGROUND

Most roof assemblies contain structural support, a waterproof roofing membrane, and thermal insulation. However, a green roof assembly requires additional components not used in conventional roof assemblies. Like a conventional roof, a typical green roof section includes structural support, a waterproof roof membrane, and thermal insulation. In addition, a green roof requires a root barrier to protect the roof membrane, a drainage medium, a filter fabric, a growing medium, and vegetation. There are several products available that can be used to fulfill these functions. Figure 1, Green Roof Assembly, from American Wick Drains (American Wick Drains)

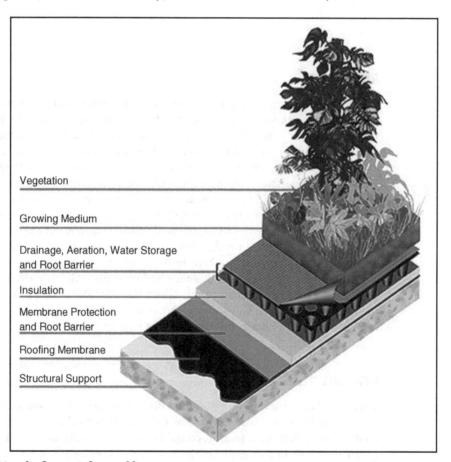

Vegetation

Growing Medium

Drainage, Aeration, Water Storage
and Root Barrier

Insulation

Membrane Protection
and Root Barrier

Roofing Membrane

Structural Support

Figure 1   Green roof assembly

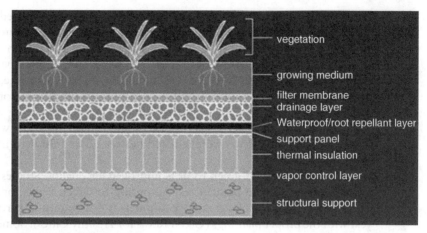

vegetation

growing medium

filter membrane
drainage layer
Waterproof/root repellant layer

support panel

thermal insulation

vapor control layer

structural support

Figure 2 Green roof technologies

and Figure 2, Principle Green Roof Technologies, from National Research Council, Institute for Research in Construction (Green Roofs for Healthy Cities), show the basic components of a green roof assembly.

There are two primary types of green roofs, intensive and extensive. Extensive roofs are characterized by shallow soil depth and low-growing plants. Soil depths are usually less than 8 inches with plantings reaching heights up to 36 inches (Garden the Planet). They are lighter and lower maintenance. They provide habitat for flora and fauna. Intensive roofs can resemble parklike spaces. They are characterized by soil depths up to 4 feet and can support a wide range of plant life (Garden the Planet).

DISCUSSION

Green roofs have many advantages over conventional roofing systems. These advantages include:

- Improved stormwater runoff management
- Improved air quality
- Reduced heat island effect
- Thermal and sound insulation
- Reduced HVAC costs
- Extended roof life
- Potential food production
- Community social, health, and emotional benefits

One of the most touted benefits of green roofs is the ability to substantially reduce stormwater runoff from the buildings they cover. The planting medium absorbs much of the rainwater reducing the load on municipal storm drainage systems. The flow water that is not absorbed is delayed. This delay assists in reducing peak flows and preventing sewage overflows and flooding that often occur during peak flows. A study by North Carolina State University demonstrated reductions in runoff up to 63 percent for a 3-inch-deep green roof. This same study showed peak flow reduction of up to 87 percent (Tokarz) for 0.6 inches of rainfall. Milwaukee, Wisconsin is another city using green roofs to assist in managing stormwater runoff. After the 2003 installation of seven green roofs, Milwaukee conducted a modeling study that showed the volume of stormwater runoff sent to sewer treatment plants was reduced 31 to 37 percent and peak flows were reduced between 5 and 36 percent (Environmental Protection Agency).

Another benefit of green roofs is improved air quality. Studies have shown that green roofs can absorb particulates. According to Green Roofs for Healthy Cities, a 10-squarefoot grass roof can remove as much as 4.4 pounds of particulate matter per year with the proper plantings (Green Roofs for Healthy Cities). Some of the pollutants that can be removed from the air include "nitrogen oxides, sulfur dioxide, carbon monoxide, and ground-level ozone (Massachusetts Department of Environmental Protection). The plant life associated with green roofs also removes carbon dioxide and produces oxygen. Green Roofs for Healthy Cities states 10 square feet of uncut grass on a green roof can generate enough oxygen for one person for a year (Green Roofs for Healthy Cities). This also serves to improve air quality, which has become an important issue for many cities facing increased air pollution and its associated health issues.

Green roofs also mitigate the effects of urban heat islands. Green spaces tend to be cooler than typical urban hardscape areas. This is also true of roofing areas. During the summer months, temperatures of conventional roofs can be substantially higher than ambient temperature reaching 130° F (Holladay) or more. Green roofs have been shown to lower the temperature of the roof areas. For instance, the City of Chicago found that the green roof of City Hall measured a range of 91 to 119° F, while the adjacent conventional roof measured 169° F on a day with a 90° F temperature during the month of August. Figure 3 illustrates the temperature differences between a conventional and a green roof (Environmental Protection Agency).

Figure 3 Temperature difference between a green and conventional roof

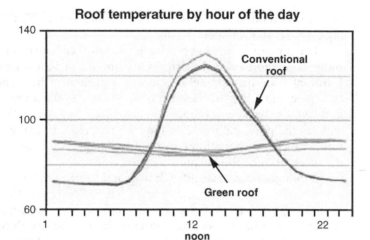

Figure 4   Roof temperature by hour

Figure 4 was developed by the American Society of Heating, Refrigeration, and Air Conditioning Engineers, Inc., and shows how the temperature of a green roof compares to a conventional roof throughout typical a day. The green roof temperature is more consistent and lower overall, reducing the heat radiated into the surrounding area.

Green roofs also act as additional thermal and sound insulation. They stop heat from moving through the roof. In her report titled "Energy Efficiency in Green Roofs," Karen Lui states, "The growing medium and the plants enhanced the thermal performance of the rooftop garden by providing shading, insulation and evaporative cooling. It acted as a thermal mass, which effectively damped the thermal fluctuations going through the roofing system." In addition to thermal insulation, green roofs provide sound attenuation. According to Green Roofs for Healthy Cities, a 4.7-inch substrate has the ability to reduce sound levels 40 decibels, and a 7.9-inch substrate may reduce sound by up to 50 decibels.

The potential for reduced HVAC costs is another benefit of green roofs. By lowering the heat gain and heat loss associated with heat transmission through the roof, the demand for air conditioning can be reduced. In a study performed by Lui, a 6-inch extensive green roof demonstrated a 95 percent reduction in heat gain and a 26 percent reduction in heat loss when compared to a reference roof (Green Roofs for Healthy Cities). The decreased heat load may allow for a reduction in mechanical equipment for the building. It also decreases the energy consumption used to heat and cool the building. Chicago's City Hall realized savings of up to 30 percent in their energy costs after the installation of their green roof (Garden the Planet).

The list of benefits continues with prolonged roof life. The plants and planting medium provide protection from UV rays and thermal extremes that deteriorate conventional roofing membranes. It also provides physical protection from wind, hail, fireworks, and vandalism (International Green Roof Association). It's estimated that this protection doubles the life of the roof membrane. This, in turn, reduces reroofing expenses and landfill materials.

The potential for food production is another benefit of green roofs. Intensive roofs can support a broad spectrum of plant life, including herbs and vegetables. The Fairmont Waterfront Hotel in Vancouver planted an herb and vegetable garden on their green roof. The roof garden produces $30,000 per year of produce that is used in the hotel's restaurant (Green Roofs for Healthy Cities). The idea can be expanded and has the potential to create food sources close to living centers. This could, in turn, reduce transportation environmental impacts as well by reducing the shipping distances for some food items.

Although much more difficult to quantify, green roofs provide communities with social, emotional, and health benefits. The EPA states, "An increasing number of studies suggest that vegetation and green space—two key components of green infrastructure—can have a positive impact on human health. Recent research has linked the presence of trees, plants, and green space to reduced levels of inner-city crime and violence, a stronger sense of community, improved academic performance, and even reductions in the symptoms associated with attention deficit and hyperactivity disorders. One such study discusses the association between neighborhood greenness and the body mass of children. (Environmental Protection Agency).

The benefits of green roofs continue. However, there are some issues that must be accounted for when considering a green roof. These include:

- Initial costs
- Maintenance costs
- Drainage and irrigation

Green roofs require more materials to build and therefore are more expensive than conventional roofs. In addition, the structural capacity of the roof must be sufficient to support the additional weight of the plants, planting medium, and human activities. According to Green Roofs for Healthy Cities and the EPA, green roofs cost $10 to $24 per square foot. While more than conventional roofs, the long-term savings in energy and maintenance costs should be considered.

Maintenance is an important consideration for green roofs. The EPA estimates maintenance costs to be about $0.75 to $1.50 per square foot. Care must be taken to keep woody plants from overgrowing and potentially damaging the roof membrane.

Although the planting medium absorbs much of the rainwater, drainage must still be provided and maintained. Gutters and downspouts must be kept clean and free of debris to avoid roof damage. Provisions for providing irrigation to plant life must be considered as well.

CONCLUSION

The benefits of green roofs cannot be ignored. Green roofs have many long-term benefits. They have the potential to address a myriad of urban issues. Many municipalities are turning to green roofs to address issues of stormwater runoff management, urban air quality, and the effects of urban heat islands. Building owners benefit from the reduced energy costs and extended roof life. Community members gain social, health, and emotional benefits. Green roofs are a long-term investment in community well-being.

RECOMMENDATIONS

Given the benefits of green roofs, all cities should promote the use of green roofs. With the initial costs associated with green roofs, they may not be an appropriate choice for speculative developers who are concerned with keeping initial costs low. However, green roofs are an excellent choice for public buildings and corporate entities with long-term commitment to their projects and who will be able to reap the long-term benefits.

REFERENCES

American Wick Drains. *Green Roof.* April 14, 2009
www.americanwick.com/applications/detail.cfm?app_id=18.

Environmental Protection Agency. *Managing Wet Weather with Green Infrastructure.* December 15, 2008 and April 14, 2009
http://cfpub.epa.gov/npdes/home.cfm?program_id=298.

———. *National Pollution Discharge Elimination System: Green Infrastructure Case Studies,* Milwaukee, Wisconsin. December 9, 2008 and April 14, 2009
http://cfpub.epa.gov/npdes/greeninfrastructure/gicasestudies_specific.cfm?case_id=61.

———. "Reducing Urban Heat Islands Compendium of Strategies Green Roofs." February 9, 2009. *Green Roofs.* April 14, 2009
www.epa.gov/heatisland/resources/pdf/GreenRoofsCompendium.pdf.

Garden the Planet. *Green Roofs.* February 14, 2009 and April 14, 2009
www.gardentheplanet.com/gr_components.htm.

Green Roofs for Healthy Cities. *About Green Roofs 2000–2005.* April 14, 2009
www.greenroofs.org/index.php?option=com_content&task=view&id=26&Itemid=40.

Holladay, April. *Green Roofs Swing Temperatures in Urban Jungles.* April 24, 2006 and April 14, 2009
www.usatoday.com/tech/columnist/aprilholladay/2006-04-24-green-roofs_x.htm.

International Green Roof Association. *Private Benefits, 2009.* April 14, 2009
www.igra-world.com/benefits/private_benefits.php.

Lui, Karen. "Energy Efficiency and Environmental Benefits of Rooftop Gardens." March 2002. *ELT Easy Green*. April 14, 2009 www.eltgreenroofs.com/pdf/ELT-NRCC45345.pdf.

Massachusetts Department of Environmental Protection. *Green Roofs and Storm Water Management*. April 14, 2009 www.mass.gov/dep/water/wastewater/grnroof.htm.

Tokarz, Erika. "CEER Green Roof Project." May 2006. Villanova University, April 14, 2009 http://egrfaculty.villanova.edu/public/Civil_Environmental/WREE/VUSP_Web_Folder/GR_web_folder/GR_paper.html.

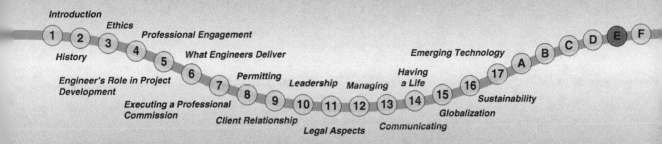

Appendix **E**

Example Specification: Cast-in-Place Concrete

SECTION 03 30 00

CAST-IN-PLACE CONCRETE

PART 1 - GENERAL

1.01   SECTION INCLUDES

A.   Conveying and placing concrete.

B.   Placement under water.

C.   Consolidation.

D.   Construction joints.

E.   Expansion and contraction joints.

F.   Curing and protection.

1.02   RELATED SECTIONS

A.   Portland cement concrete specified in Section 03 05 15 - Portland Cement Concrete.

B.   Finishing and curing of formed and unformed concrete surfaces, including repair and patching of surface defects, are specified in Section 03 35 00 - Concrete Finishing.

C.   Shotcrete is specified in Section 03 37 13 - Shotcrete.

D.   Vapor barrier under slabs on grade is specified in Section 07 26 00 - Vapor Retarders.

1.03   MEASUREMENT AND PAYMENT

A.   General: Measurement and payment for cast-in-place concrete will be either by the lump-sum method or by the unit-price method as determined by the listing of the bid item for cast-in-place concrete indicated in the Bid Schedule of the Bid Form.

B.   Lump Sum: If the Bid Schedule indicates a lump-sum for cast-in-place concrete, the lump-sum method of measurement and payment will be in accordance with Section 01 20 00 - Price and Payment Procedures, Article 1.03.

C.   Unit Prices: If the Bid Schedule indicates a unit price for cast-in-place concrete, the unit-price method of measurement and payment will be as follows:

1.   Measurement:

a.   Except as specified otherwise in other Sections of these Specifications or the Contract Specifications, each class of concrete and type of placement of cast-in-place concrete will be measured for payment by the cubic yard, and quantities will be computed, based on the neat lines or pay lines, section profiles, and dimensions indicated on the Contract Drawings, without deduction for chamfers, reinforcing steel and embedded items, and openings and recesses having an area of less than

RELEASE – R2.0
Issued: 09/30/2008

SECTION 03 30 00
PAGE 1 OF 9

BART FACILITIES STANDARDS
STANDARD SPECIFICATIONS

593

CAST-IN-PLACE CONCRETE

two square feet.

    b.   Additional concrete used to replace overcut or for overbreak, or to repair or replace defective work, will not be measured separately for payment.

    2.   Payment: Cast-in-place concrete will be paid for at the indicated Contract unit prices for the computed quantities as determined by the measurement method specified in Article 1.03.C.1.

1.04   **DEFINITIONS**

A.   The words and terms used in these Specifications conform with the definitions given in ACI 116R.

1.05   **REFERENCES**

A.   American Concrete Institute (ACI):

1.	ACI 116R	Cement and Concrete Terminology
2.	ACI 117	Standard Specification for Tolerances for Concrete Construction and Materials
3.	ACI 301	Standard Specifications for Structural Concrete
4.	ACI 302.1R	Guide for Concrete Floor and Slab Construction
5.	ACI 303.1	Standard Specification for Cast-In-Place Architectural Concrete
6.	ACI 304R	Guide for Measuring, Mixing, Transporting, and Placing Concrete
7.	ACI 304.2R	Placing Concrete by Pumping Methods
8.	ACI 305R	Hot Weather Concreting
9.	ACI 306.1	Standard Specification for Cold Weather Concreting
10.	ACI 308	Standard Practice for Curing Concrete
11.	ACI 309R	Guide for Consolidation of Concrete
12.	ACI 318	Building Code Requirements for Structural Concrete
13.	ACI 503.2	Standard Specification for Bonding Plastic Concrete to Hardened Concrete with a Multi-Component Epoxy Adhesive

B.   American Society for Testing and Materials (ASTM):

1.	ASTM C31	Standard Practice of Making and Curing Concrete Test Specimens in the Field

RELEASE – R2.0
Issued: 09/30/2008

SECTION 03 30 00
PAGE 2 OF 9

BART FACILITIES STANDARDS
STANDARD SPECIFICATIONS

CAST-IN-PLACE CONCRETE

2.   ASTM C94   Specification for Ready-Mixed Concrete

3.   ASTM C881   Specification for Epoxy-Resin-Base Bonding Systems for Concrete

1.06   **SUBMITTALS**

A.   General: Refer to Section 01 33 00 - Submittal Procedures, and Section 01 33 23 - Shop Drawings, Product Data, and Samples, for submittal requirements and procedures.

B.   Shop Drawings:

1.   Submit drawings that indicate the locations of all joints in concrete, including construction joints, expansion joints, isolation joints, and contraction joints. Coordinate with the requirements specified in Section 03 11 00 - Concrete Forming.

2.   Submit drawings that indicate concrete placement schedule, method, sequence, location, and boundaries. Include each type and class of concrete, and quantity in cubic yards.

C.   Product Data: Submit manufacturer's product data for epoxy adhesive.

D.   Records and Reports: Report the location in the finished work of each mix design, and the start and completion times of placement of each batch of concrete placed for each date concrete is placed.

1.07   **QUALITY ASSURANCE**

A.   Tolerances:

1.   Concrete Tolerances: Comply with the requirements of ACI 117 as applicable. Coordinate with the requirements specified in Section 03 11 00 - Concrete Forming.

2.   Tolerances for Slabs and Flatwork: Comply with the requirements specified in Section 03 35 00 - Concrete Finishing.

B.   Architectural Concrete: Where concrete is indicated as architectural concrete exposed to public view, such concrete shall be produced in accordance with applicable requirements of ACI 301 and ACI 303.1.

C.   Site Mock-Ups:

1.   Refer to Section 01 43 38 - Field Samples and Mockups, for mock-up requirements and procedures.

2.   Construct site mock-ups for all architectural concrete work and formed concrete that will be exposed to the public in the finished work, not less than 4 feet by 6 feet in surface area, for review and acceptance by the Engineer, before starting the placement of concrete.

RELEASE – R2.0
Issued: 09/30/2008

SECTION 03 30 00
PAGE 3 OF 9

BART FACILITIES STANDARDS
STANDARD SPECIFICATIONS

CAST-IN-PLACE CONCRETE

3. Approved site mock-ups shall set the standard for the various architectural concrete features, formed finishes, and colors of the concrete. Provide as many mock-ups as required to show all the different features and formed surfaces of the concrete.

D. Cold Joints: Cold joints in concrete will not be permitted unless planned and treated properly as construction joints.

E. Monitoring of Formwork: Provide monitoring of forms and embedded items to detect movement, or forms and embedded items out-of-alignment, from pressure of concrete placement.

1.08 ENVIRONMENTAL REQUIREMENTS

A. Delivering and placing of concrete in hot weather and cold weather shall conform with applicable requirements of ACI 305R and ACI 306.1 and Section 03 05 15 - Portland Cement Concrete.

B. Do not place concrete when the rate of evaporation of surface moisture from concrete exceeds 0.2 pounds per square foot per hour as indicated in Figure 2.1.5 of ACI 305R.

C. Do not place concrete in, or adjacent to, any structure where piles are required until all piles in the structure have been driven or installed.

PART 2 - PRODUCTS

2.01 MATERIALS

A. Formwork: Refer to Section 03 11 00 - Concrete Forming, for requirements.

B. Joint Fillers and Sealers: Refer to Section 03 15 00 - Concrete Accessories, for requirements.

C. Waterstops: Refer to Section 03 15 13 - Waterstops, for requirements.

D. Reinforcing Steel: Refer to Section 03 20 00 - Concrete Reinforcing, for requirements.

E. Portland Cement Concrete: Refer to Section 03 05 15 - Portland Cement Concrete, for mix designs and other requirements.

F. Concrete Curing Materials: Refer to Section 03 35 00 - Concrete Finishing, for requirements.

G. Vapor Barrier Materials: Refer to Section 07 26 00 - Vapor Retarders, for requirements.

H. Epoxy Adhesive: ASTM C881, Type II for non-load-bearing concrete and Type V for load-bearing concrete, Grade and Class as determined by project conditions and requirements.

PART 3 - EXECUTION

3.01 EXAMINATION

A. Inspect forms, earth bearing surfaces, reinforcement, and embedded items, and obtain the

RELEASE – R2.0
Issued: 09/30/2008

SECTION 03 30 00
PAGE 4 OF 9

BART FACILITIES STANDARDS
STANDARD SPECIFICATIONS

CAST-IN-PLACE CONCRETE

Engineer's written approval before placing concrete. Complete and sign a pour card on the form supplied by the Engineer. The Engineer shall countersign the card prior to commencing the pour.

3.02   **PREPARATION**

A.   Place concrete under the observation of the Engineer and with the Contractor's Quality Control Representative present to document requirements and results of the placement.

B.   Whenever possible, place concrete during normal working hours. When concrete- placement schedules require concrete placement at times other than normal working hours, ensure that the Engineer is notified and is present at the time of placement.

C.   Do not place concrete until conditions and facilities for the storage, handling, and transportation of concrete test specimens are in compliance with the requirements of ASTM C31 and Section 03 05 15 - Portland Cement Concrete, and are approved by the Engineer.

D.   Prior to placement of concrete, the subgrade shall be in a firm, well-drained condition, and of adequate and uniform load-bearing nature to support construction personnel, construction materials, construction equipment, and steel reinforcing mats without tracking, rutting, heaving, or settlement. All weak, soft, saturated, or otherwise unsuitable material shall be removed and replaced with structural backfill or lean concrete.

E.   All structure foundations, including those for Stations and for subway box, shall be inspected and approved, in writing, by a qualified, independent geotechnical engineer prior to placement of footings and base slabs, to confirm the adequacy of the supporting soil for concrete placement.

F.   Earth bottoms or bearing surfaces for footings and slabs shall be dampened but not saturated or muddied just before placing concrete.

3.03   **TRANSPORTING**

A.   Concrete shall be central-mixed concrete from a central batch plant, transported to the jobsite in a truck mixer, in accordance with the requirements specified in Section 03 05 15 - Portland Cement Concrete, and ASTM C94.

B.   Transport concrete to the jobsite in a manner that will assure efficient delivery of concrete to the point of placement without adversely altering specified properties with regard to water-cement ratio, slump, air entrainment, and homogeneity.

3.04   **CONVEYING AND PLACING**

A.   Placement Standards: Conveying and placing of concrete shall conform with applicable requirements of ACI 301, ACI 302.1R, ACI 304R, and ACI 318.

RELEASE – R2.0
Issued: 09/30/2008

SECTION 03 30 00
PAGE 5 OF 9

BART FACILITIES STANDARDS
STANDARD SPECIFICATIONS

CAST-IN-PLACE CONCRETE

B.   Handling and Depositing:

1.   Concrete placing equipment shall have sufficient capacity to provide a placement rate that will preclude cold joints and that shall deposit the concrete without segregation or loss of ingredients.

2.   Concrete placement, once started, shall be carried on as a continuous operation until the section of approved size and shape is completed.

3.   Concrete shall be handled as rapidly as practicable from the mixer to the place of final deposit by methods that prevent the separation or loss of ingredients. Concrete shall be deposited, as nearly as practicable, in its final horizontal position to avoid redistribution or flowing.

4.   Concrete shall not be dropped freely where reinforcing will cause segregation, nor shall it be dropped freely more than 5 feet. Concrete shall be deposited to maintain a plastic surface approximately horizontal.

5.   In placing walls, columns, or thin sections (6 inches or less in thickness) of heights greater than 10 feet, concrete placement rate, lift thickness, and time intervals between lifts shall be as indicated on approved Shop Drawings. Openings in the form, elephant trunk tremies, or other approved devices, shall be used that will permit the concrete to be placed without segregation or accumulation of hardened concrete on the forms or metal reinforcement above the level of the fresh concrete.

6.   Concrete that has partially hardened shall not be deposited in the work. The discharge of concrete shall be started not later than 60 minutes after the introduction of mixing water. Placing of concrete shall be completed within 90 minutes after the first introduction of water into the mix.

C.   Pumping:

1.   Concrete may be placed by pumping if the maximum slump can be maintained and if accepted in writing by the Engineer for the location proposed.

2.   Placing concrete by pumping methods shall conform with applicable requirements of ACI 304R and ACI 304.2R.

3.   Equipment for pumping shall be of such size and design as to ensure a continuous flow of concrete at the delivery end without separation of materials. Concrete from end of hose shall have a free fall of less than 5 feet. Pump hoses shall be supported on horses or similar devices so that reinforcement or post-tensioning ducts or tendons are not moved from their original position.

4.   The concrete mix shall be designed to the same requirements as specified in Section 03 05 15 - Portland Cement Concrete, and may be altered for placement purposes with the prior approval of the Engineer.

RELEASE – R2.0
Issued: 09/30/2008

SECTION 03 30 00
PAGE 6 OF 9

BART FACILITIES STANDARDS
STANDARD SPECIFICATIONS

CAST-IN-PLACE CONCRETE

3.05 PLACEMENT UNDER WATER

A. Placement Standards: Placing of concrete in or under water shall conform with requirements of ACI 304R. All concrete to be placed under water shall be placed by the tremie method or by direct pumping.

B. Placement Requirements: Deposit concrete in water only when indicated or approved in writing by the Engineer, and only under the observation of the Engineer. Use only tremie method and direct pumping with equipment that has been accepted by the Engineer.

3.06 CONSOLIDATION

A. Concrete shall be thoroughly consolidated and compacted by mechanical vibration during placement in accordance with the requirements of ACI 309R.

B. The Engineer will inspect concrete placement to confirm that proper placing methods are being employed, and that special techniques are being used in congested areas and around obstructions such as pipes and other embedded items. Check installation of embedded items for correct location and orientation during concrete placement.

C. Conduct vibration in a systematic manner by competent, skilled, and experienced workers, with regularly maintained vibrators, and with sufficient back-up units at the jobsite. Use the largest and most powerful vibrator that can be effectively operated in the given work, with a minimum frequency of 8,000 vibrations or impulses per minute, and of sufficient amplitude to effectively consolidate the concrete.

D. Insert and withdraw the vibrator vertically at uniform spacing over the entire area of the placement. Space the distance between insertions such that "spheres of influence" of each insertion overlap.

E. Conduct vibration so as to produce concrete that is of uniform texture and appearance, free of honeycombing, air and rock pockets, streaking, cold joints, and visible lift lines.

F. On vertical surfaces and on all architectural concrete where an as-cast finish is required, use additional vibration and spading as required to bring a full surface of mortar against the forms, so as to eliminate objectionable air voids, bug holes, and other surface defects. Additional procedures for vibrating concrete shall consist of the following:

1. Reduce the distance between internal vibration insertions and increase the time for each insertion.

2. Insert the vibrator as close to the face of the form as possible, without contacting the form.

3. Use spading as a supplement to vibration at forms to provide fully filled out form surfaces without air holes and rock pockets.

4. Provide vibration of forms only if approved by the Engineer for the location.

RELEASE – R2.0
Issued: 09/30/2008

SECTION 03 30 00
PAGE 7 OF 9

BART FACILITIES STANDARDS
STANDARD SPECIFICATIONS

CAST-IN-PLACE CONCRETE

3.07   CONSTRUCTION JOINTS

A.   Construction joints will be permitted only where indicated or approved by the Engineer.

B.   Provide and prepare construction joints and install waterstops in accordance with the applicable requirements of ACI 301 and ACI 304R, and as specified in Section 03 11 00 - Concrete Forming.

C.   Make construction joints straight and as inconspicuous as possible, and in exact vertical and horizontal alignment with the structure, as the case may be.

D.   Use approved key, at least 1-1/2 inches in depth, at joints unless otherwise indicated or approved by the Engineer.

E.   Thoroughly clean the surface of the concrete at construction joints and remove laitance, loose or defective concrete, coatings, sand, sealing compound and other foreign material. Prepare surfaces of joints by sandblasting or other approved methods to remove laitance and expose aggregate uniformly.

F.   Immediately before new concrete is placed, wet the joint surfaces and remove standing water. To allow for shrinkage, do not place new concrete against the hardened concrete side of a construction joint for a minimum of 72 hours.

G.   Locate joints that are not indicated so that the strength of the structure is not impaired. Joint types and their locations are subject to prior approval of the Engineer.

H.   Ensure that reinforcement is continuous across construction joints.

I.   Place waterstops in construction joints where indicated.

J.   Where bonding of the joint is required, provide epoxy adhesive hereinbefore specified and apply in accordance with ACI 503.2.

K.   Retighten forms and dampen concrete surfaces before concrete placing is continued.

L.   Allow at least 72 hours to elapse before continuing concrete placement at a construction joint. Approval for accelerating the minimum time elapsing between adjacent placements will be based on tests and methods that confirm that a minimum moisture loss at a relatively constant temperature will be maintained for the period as necessary to control the heat of hydration and hardening of concrete, and to prevent shrinkage and thermal cracking.

3.08   EXPANSION AND CONTRACTION JOINTS

A.   Refer to Section 03 11 00 - Concrete Forming, for slab screeds and for formwork where expansion and contraction joints are indicated as architectural features, such as reveals or rustications.

B.   Refer to Section 03 15 00 - Concrete Accessories, for expansion joint filler material and joint sealing compound.

RELEASE – R2.0
Issued: 09/30/2008

SECTION 03 30 00
PAGE 8 OF 9

BART FACILITIES STANDARDS
STANDARD SPECIFICATIONS

CAST-IN-PLACE CONCRETE

C.   Refer to Section 03 35 00 - Concrete Finishing, for finishing of edges of expansion joints in slabs with curved edging tool.

3.09   CURING AND PROTECTION

A.   Curing of concrete shall conform with applicable requirements of ACI 301 and ACI 308, except that the curing duration shall be a minimum period of ten days. HVFAC shall be cured a minimum of 28 days including an initial 10 days of moist curing. Curing with earth, sand, sawdust, straw, and hay will not be permitted.

B.   Keep concrete in a moist condition from the time it is placed until it has cured for at least ten days. Keep forms damp and cool until removal of forms.

C.   Immediately upon removal of forms, exposed concrete surfaces shall be kept moist by applying an approved curing compound or by covering with damp curing materials as specified in Section 03 35 00 - Concrete Finishing.

D.   Concrete shall not be permitted to dry during the curing period because of finishing operations.

E.   Protect fresh concrete from hot sun, drying winds, rain, damage, or soiling. Fog spray freshly placed slabs after bleed water dissipates and after finishing operations commence. Allow no slabs to become dry at any time until finishing operations are complete.

F.   Finishing and curing of slabs are specified in Section 03 35 00 - Concrete Finishing.

G.   Protect concrete from injurious action of the elements and defacement of any kind. Protect exposed concrete corners from traffic or use that will damage them in any way.

H.   Protect concrete during the curing period from mechanical and physical stresses that may be caused by heavy equipment movement, subjecting the concrete to load stress, load shock, or excessive vibration.

3.10   REPAIR OF SURFACE DEFECTS

A.   Refer to Section 03 35 00 - Concrete Finishing, for requirements.

END OF SECTION 03 30 00

RELEASE – R2.0
Issued: 09/30/2008

SECTION 03 30 00
PAGE 9 OF 9

BART FACILITIES STANDARDS
STANDARD SPECIFICATIONS

Introduction
Ethics
Professional Engagement
History
What Engineers Deliver
Emerging Technology
Engineer's Role in Project
Development
Permitting
Leadership
Managing
Having
a Life
Sustainability
Executing a Professional
Commission
Client Relationship
Legal Aspects
Communicating
Globalization

1 2 3 4 5 6 7 8 9 10 11 12 13 14 15 16 17 A B C D E F

Appendix F

Contracts

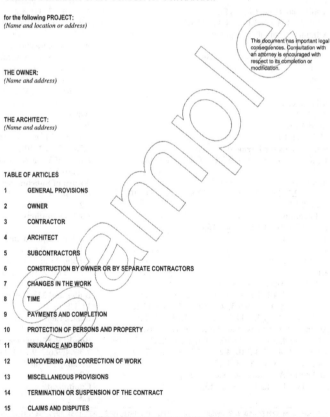

AIA Document A201™ – 2007

General Conditions of the Contract for Construction

for the following PROJECT:
(Name and location or address)

THE OWNER:
(Name and address)

THE ARCHITECT:
(Name and address)

This document has important legal consequences. Consultation with an attorney is encouraged with respect to its completion or modification.

TABLE OF ARTICLES

1   GENERAL PROVISIONS

2   OWNER

3   CONTRACTOR

4   ARCHITECT

5   SUBCONTRACTORS

6   CONSTRUCTION BY OWNER OR BY SEPARATE CONTRACTORS

7   CHANGES IN THE WORK

8   TIME

9   PAYMENTS AND COMPLETION

10   PROTECTION OF PERSONS AND PROPERTY

11   INSURANCE AND BONDS

12   UNCOVERING AND CORRECTION OF WORK

13   MISCELLANEOUS PROVISIONS

14   TERMINATION OR SUSPENSION OF THE CONTRACT

15   CLAIMS AND DISPUTES

INDEX
(Numbers and Topics in Bold are Section Headings)

ARTICLE 1 GENERAL PROVISIONS
§ 1.1 BASIC DEFINITIONS
§ 1.1.1 THE CONTRACT DOCUMENTS

The Contract Documents are enumerated in the Agreement between the Owner and Contractor (hereinafter the Agreement) and consist of the Agreement, Conditions of the Contract (General, Supplementary and other Conditions), Drawings, Specifications, Addenda issued prior to execution of the Contract, other documents listed in the Agreement and Modifications issued after execution of the Contract. A Modification is (1) a written amendment to the Contract signed by both parties, (2) a Change Order, (3) a Construction Change Directive or (4) a written order for a minor change in the Work issued by the Architect. Unless specifically enumerated in the Agreement, the Contract Documents do not include the advertisement or invitation to bid, Instructions to Bidders, sample forms, other information furnished by the Owner in anticipation of receiving bids or proposals, the Contractor's bid or proposal, or portions of Addenda relating to bidding requirements.

§ 1.1.2 THE CONTRACT

The Contract Documents form the Contract for Construction. The Contract represents the entire and integrated agreement between the parties hereto and supersedes prior negotiations, representations or agreements, either written or oral. The Contract may be amended or modified only by a Modification. The Contract Documents shall not be construed to create a contractual relationship of any kind (1) between the Contractor and the Architect or the Architect's consultants, (2) between the Owner and a Subcontractor or a Sub-subcontractor, (3) between the Owner and the Architect or the Architect's consultants or (4) between any persons or entities other than the Owner and the Contractor. The Architect shall, however, be entitled to performance and enforcement of obligations under the Contract intended to facilitate performance of the Architect's duties.

§ 1.1.3 THE WORK

The term "Work" means the construction and services required by the Contract Documents, whether completed or partially completed, and includes all other labor, materials, equipment and services provided or to be provided by the Contractor to fulfill the Contractor's obligations. The Work may constitute the whole or a part of the Project.

§ 1.1.4 THE PROJECT

The Project is the total construction of which the Work performed under the Contract Documents may be the whole or a part and which may include construction by the Owner and by separate contractors.

§ 1.1.5 THE DRAWINGS

The Drawings are the graphic and pictorial portions of the Contract Documents showing the design, location and dimensions of the Work, generally including plans, elevations, sections, details, schedules and diagrams.

§ 1.1.6 THE SPECIFICATIONS

The Specifications are that portion of the Contract Documents consisting of the written requirements for materials, equipment, systems, standards and workmanship for the Work, and performance of related services.

§ 1.1.7 INSTRUMENTS OF SERVICE

Instruments of Service are representations, in any medium of expression now known or later developed, of the tangible and intangible creative work performed by the Architect and the Architect's consultants under their respective professional services agreements. Instruments of Service may include, without limitation, studies, surveys, models, sketches, drawings, specifications, and other similar materials.

§ 1.1.8 INITIAL DECISION MAKER

The Initial Decision Maker is the person identified in the Agreement to render initial decisions on Claims in accordance with Section 15.2 and certify termination of the Agreement under Section 14.2.2.

§ 1.2 CORRELATION AND INTENT OF THE CONTRACT DOCUMENTS

§ 1.2.1 The intent of the Contract Documents is to include all items necessary for the proper execution and completion of the Work by the Contractor. The Contract Documents are complementary, and what is required by one shall be as binding as if required by all; performance by the Contractor shall be required only to the extent consistent with the Contract Documents and reasonably inferable from them as being necessary to produce the indicated results.

§ 1.2.2 Organization of the Specifications into divisions, sections and articles, and arrangement of Drawings shall not control the Contractor in dividing the Work among Subcontractors or in establishing the extent of Work to be performed by any trade.

§ 1.2.3 Unless otherwise stated in the Contract Documents, words that have well-known technical or construction industry meanings are used in the Contract Documents in accordance with such recognized meanings.

§ 1.3 CAPITALIZATION
Terms capitalized in these General Conditions include those that are (1) specifically defined, (2) the titles of numbered articles or (3) the titles of other documents published by the American Institute of Architects.

§ 1.4 INTERPRETATION
In the interest of brevity the Contract Documents frequently omit modifying words such as "all" and "any" and articles such as "the" and "an," but the fact that a modifier or an article is absent from one statement and appears in another is not intended to affect the interpretation of either statement.

§ 1.5 OWNERSHIP AND USE OF DRAWINGS, SPECIFICATIONS AND OTHER INSTRUMENTS OF SERVICE
§ 1.5.1 The Architect and the Architect's consultants shall be deemed the authors and owners of their respective Instruments of Service, including the Drawings and Specifications, and will retain all common law, statutory and other reserved rights, including copyrights. The Contractor, Subcontractors, Sub-subcontractors, and material or equipment suppliers shall not own or claim a copyright in the Instruments of Service. Submittal or distribution to meet official regulatory requirements or for other purposes in connection with this Project is not to be construed as publication in derogation of the Architect's or Architect's consultants' reserved rights.

§ 1.5.2 The Contractor, Subcontractors, Sub-subcontractors and material or equipment suppliers are authorized to use and reproduce the Instruments of Service provided to them solely and exclusively for execution of the Work. All copies made under this authorization shall bear the copyright notice, if any, shown on the Instruments of Service. The Contractor, Subcontractors, Sub-subcontractor, and material or equipment suppliers may not use the Instruments of Service on other projects or for additions to this Project outside the scope of the Work without the specific written consent of the Owner, Architect and the Architect's consultants.

§ 1.6 TRANSMISSION OF DATA IN DIGITAL FORM
If the parties intend to transmit Instruments of Service or any other information or documentation in digital form, they shall endeavor to establish necessary protocols governing such transmissions, unless otherwise already provided in the Agreement or the Contract Documents.

ARTICLE 2 OWNER
§ 2.1 GENERAL
§ 2.1.1 The Owner is the person or entity identified as such in the Agreement and is referred to throughout the Contract Documents as if singular in number. The Owner shall designate in writing a representative who shall have express authority to bind the Owner with respect to all matters requiring the Owner's approval or authorization. Except as otherwise provided in Section 4.2.1, the Architect does not have such authority. The term "Owner" means the Owner or the Owner's authorized representative.

§ 2.1.2 The Owner shall furnish to the Contractor within fifteen days after receipt of a written request, information necessary and relevant for the Contractor to evaluate, give notice of or enforce mechanic's lien rights. Such information shall include a correct statement of the record legal title to the property on which the Project is located, usually referred to as the site, and the Owner's interest therein.

§ 2.2 INFORMATION AND SERVICES REQUIRED OF THE OWNER
§ 2.2.1 Prior to commencement of the Work, the Contractor may request in writing that the Owner provide reasonable evidence that the Owner has made financial arrangements to fulfill the Owner's obligations under the Contract. Thereafter, the Contractor may only request such evidence if (1) the Owner fails to make payments to the Contractor as the Contract Documents require; (2) a change in the Work materially changes the Contract Sum; or (3) the Contractor identifies in writing a reasonable concern regarding the Owner's ability to make payment when due. The Owner shall furnish such evidence as a condition precedent to commencement or continuation of the Work or the portion of the Work affected by a material change. After the Owner furnishes the evidence, the Owner shall not materially vary such financial arrangements without prior notice to the Contractor.

§ 2.2.2 Except for permits and fees that are the responsibility of the Contractor under the Contract Documents, including those required under Section 3.7.1, the Owner shall secure and pay for necessary approvals, easements, assessments and charges required for construction, use or occupancy of permanent structures or for permanent changes in existing facilities.

§ 2.2.3 The Owner shall furnish surveys describing physical characteristics, legal limitations and utility locations for the site of the Project, and a legal description of the site. The Contractor shall be entitled to rely on the accuracy of information furnished by the Owner but shall exercise proper precautions relating to the safe performance of the Work.

§ 2.2.4 The Owner shall furnish information or services required of the Owner by the Contract Documents with reasonable promptness. The Owner shall also furnish any other information or services under the Owner's control and relevant to the Contractor's performance of the Work with reasonable promptness after receiving the Contractor's written request for such information or services.

§ 2.2.5 Unless otherwise provided in the Contract Documents, the Owner shall furnish to the Contractor one copy of the Contract Documents for purposes of making reproductions pursuant to Section 1.5.2.

§ 2.3 OWNER'S RIGHT TO STOP THE WORK
If the Contractor fails to correct Work that is not in accordance with the requirements of the Contract Documents as required by Section 12.2 or repeatedly fails to carry out Work in accordance with the Contract Documents, the Owner may issue a written order to the Contractor to stop the Work, or any portion thereof, until the cause for such order has been eliminated; however, the right of the Owner to stop the Work shall not give rise to a duty on the part of the Owner to exercise this right for the benefit of the Contractor or any other person or entity, except to the extent required by Section 6.1.3.

§ 2.4 OWNER'S RIGHT TO CARRY OUT THE WORK
If the Contractor defaults or neglects to carry out the Work in accordance with the Contract Documents and fails within a ten-day period after receipt of written notice from the Owner to commence and continue correction of such default or neglect with diligence and promptness, the Owner may, without prejudice to other remedies the Owner may have, correct such deficiencies. In such case an appropriate Change Order shall be issued deducting from payments then or thereafter due the Contractor the reasonable cost of correcting such deficiencies, including Owner's expenses and compensation for the Architect's additional services made necessary by such default, neglect or failure. Such action by the Owner and amounts charged to the Contractor are both subject to prior approval of the Architect. If payments then or thereafter due the Contractor are not sufficient to cover such amounts, the Contractor shall pay the difference to the Owner.

ARTICLE 3 CONTRACTOR
§ 3.1 GENERAL
§ 3.1.1 The Contractor is the person or entity identified as such in the Agreement and is referred to throughout the Contract Documents as if singular in number. The Contractor shall be lawfully licensed, if required in the jurisdiction where the Project is located. The Contractor shall designate in writing a representative who shall have express authority to bind the Contractor with respect to all matters under this Contract. The term "Contractor" means the Contractor or the Contractor's authorized representative.

§ 3.1.2 The Contractor shall perform the Work in accordance with the Contract Documents.

§ 3.1.3 The Contractor shall not be relieved of obligations to perform the Work in accordance with the Contract Documents either by activities or duties of the Architect in the Architect's administration of the Contract, or by tests, inspections or approvals required or performed by persons or entities other than the Contractor.

§ 3.2 REVIEW OF CONTRACT DOCUMENTS AND FIELD CONDITIONS BY CONTRACTOR
§ 3.2.1 Execution of the Contract by the Contractor is a representation that the Contractor has visited the site, become generally familiar with local conditions under which the Work is to be performed and correlated personal observations with requirements of the Contract Documents.

§ 3.2.2 Because the Contract Documents are complementary, the Contractor shall, before starting each portion of the Work, carefully study and compare the various Contract Documents relative to that portion of the Work, as well as the information furnished by the Owner pursuant to Section 2.2.3, shall take field measurements of any existing conditions related to that portion of the Work, and shall observe any conditions at the site affecting it. These obligations are for the purpose of facilitating coordination and construction by the Contractor and are not for the purpose of discovering errors, omissions, or inconsistencies in the Contract Documents; however, the Contractor shall promptly report to the Architect any errors, inconsistencies or omissions discovered by or made known to the Contractor as a request for information in such form as the Architect may require. It is recognized that the Contractor's review is made in the Contractor's capacity as a contractor and not as a licensed design professional, unless otherwise specifically provided in the Contract Documents.

§ 3.2.3 The Contractor is not required to ascertain that the Contract Documents are in accordance with applicable laws, statutes, ordinances, codes, rules and regulations, or lawful orders of public authorities, but the Contractor shall promptly report to the Architect any nonconformity discovered by or made known to the Contractor as a request for information in such form as the Architect may require.

§ 3.2.4 If the Contractor believes that additional cost or time is involved because of clarifications or instructions the Architect issues in response to the Contractor's notices or requests for information pursuant to Sections 3.2.2 or 3.2.3, the Contractor shall make Claims as provided in Article 15. If the Contractor fails to perform the obligations of Sections 3.2.2 or 3.2.3, the Contractor shall pay such costs and damages to the Owner as would have been avoided if the Contractor had performed such obligations. If the Contractor performs those obligations, the Contractor shall not be liable to the Owner or Architect for damages resulting from errors, inconsistencies or omissions in the Contract Documents, for differences between field measurements or conditions and the Contract Documents, or for nonconformities of the Contract Documents to applicable laws, statutes, ordinances, codes, rules and regulations, and lawful orders of public authorities.

§ 3.3 SUPERVISION AND CONSTRUCTION PROCEDURES
§ 3.3.1 The Contractor shall supervise and direct the Work, using the Contractor's best skill and attention. The Contractor shall be solely responsible for, and have control over, construction means, methods, techniques, sequences and procedures and for coordinating all portions of the Work under the Contract, unless the Contract Documents give other specific instructions concerning these matters. If the Contract Documents give specific instructions concerning construction means, methods, techniques, sequences or procedures, the Contractor shall evaluate the jobsite safety thereof and, except as stated below, shall be fully and solely responsible for the jobsite safety of such means, methods, techniques, sequences or procedures. If the Contractor determines that such means, methods, techniques, sequences or procedures may not be safe, the Contractor shall give timely written notice to the Owner and Architect and shall not proceed with that portion of the Work without further written instructions from the Architect. If the Contractor is then instructed to proceed with the required means, methods, techniques, sequences or procedures without acceptance of changes proposed by the Contractor, the Owner shall be solely responsible for any loss or damage arising solely from those Owner-required means, methods, techniques, sequences or procedures.

§ 3.3.2 The Contractor shall be responsible to the Owner for acts and omissions of the Contractor's employees, Subcontractors and their agents and employees, and other persons or entities performing portions of the Work for, or on behalf of, the Contractor or any of its Subcontractors.

§ 3.3.3 The Contractor shall be responsible for inspection of portions of Work already performed to determine that such portions are in proper condition to receive subsequent Work.

§ 3.4 LABOR AND MATERIALS
§ 3.4.1 Unless otherwise provided in the Contract Documents, the Contractor shall provide and pay for labor, materials, equipment, tools, construction equipment and machinery, water, heat, utilities, transportation, and other facilities and services necessary for proper execution and completion of the Work, whether temporary or permanent and whether or not incorporated or to be incorporated in the Work.

§ 3.4.2 Except in the case of minor changes in the Work authorized by the Architect in accordance with Sections 3.12.8 or 7.4, the Contractor may make substitutions only with the consent of the Owner, after evaluation by the Architect and in accordance with a Change Order or Construction Change Directive.

§ 3.4.3 The Contractor shall enforce strict discipline and good order among the Contractor's employees and other persons carrying out the Work. The Contractor shall not permit employment of unfit persons or persons not properly skilled in tasks assigned to them.

§ 3.5 WARRANTY

The Contractor warrants to the Owner and Architect that materials and equipment furnished under the Contract will be of good quality and new unless the Contract Documents require or permit otherwise. The Contractor further warrants that the Work will conform to the requirements of the Contract Documents and will be free from defects, except for those inherent in the quality of the Work the Contract Documents require or permit. Work, materials, or equipment not conforming to these requirements may be considered defective. The Contractor's warranty excludes remedy for damage or defect caused by abuse, alterations to the Work not executed by the Contractor, improper or insufficient maintenance, improper operation, or normal wear and tear and normal usage. If required by the Architect, the Contractor shall furnish satisfactory evidence as to the kind and quality of materials and equipment.

§ 3.6 TAXES

The Contractor shall pay sales, consumer, use and similar taxes for the Work provided by the Contractor that are legally enacted when bids are received or negotiations concluded, whether or not yet effective or merely scheduled to go into effect.

§ 3.7 PERMITS, FEES, NOTICES, AND COMPLIANCE WITH LAWS

§ 3.7.1 Unless otherwise provided in the Contract Documents, the Contractor shall secure and pay for the building permit as well as for other permits, fees, licenses, and inspections by government agencies necessary for proper execution and completion of the Work that are customarily secured after execution of the Contract and legally required at the time bids are received or negotiations concluded.

§ 3.7.2 The Contractor shall comply with and give notices required by applicable laws, statutes, ordinances, codes, rules and regulations, and lawful orders of public authorities applicable to performance of the Work.

§ 3.7.3 If the Contractor performs Work knowing it to be contrary to applicable laws, statutes, ordinances, codes, rules and regulations, or lawful orders of public authorities, the Contractor shall assume appropriate responsibility for such Work and shall bear the costs attributable to correction.

§ 3.7.4 Concealed or Unknown Conditions. If the Contractor encounters conditions at the site that are (1) subsurface or otherwise concealed physical conditions that differ materially from those indicated in the Contract Documents or (2) unknown physical conditions of an unusual nature that differ materially from those ordinarily found to exist and generally recognized as inherent in construction activities of the character provided for in the Contract Documents, the Contractor shall promptly provide notice to the Owner and the Architect before conditions are disturbed and in no event later than 21 days after first observance of the conditions. The Architect will promptly investigate such conditions and, if the Architect determines that they differ materially and cause an increase in decrease in the Contractor's cost of, or time required for, performance of any part of the Work, will recommend an equitable adjustment in the Contract Sum or Contract Time, or both. If the Architect determines that the conditions at the site are not materially different from those indicated in the Contract Documents and that no change in the terms of the Contract is justified, the Architect shall promptly notify the Owner and Contractor in writing, stating the reasons. If either party disputes the Architect's determination or recommendation, that party may proceed as provided in Article 15.

§ 3.7.5 If, in the course of the Work, the Contractor encounters human remains or recognizes the existence of burial markers, archaeological sites or wetlands not indicated in the Contract Documents, the Contractor shall immediately suspend any operations that would affect them and shall notify the Owner and Architect. Upon receipt of such notice, the Owner shall promptly take any action necessary to obtain governmental authorization required to resume the operations. The Contractor shall continue to suspend such operations until otherwise instructed by the Owner but shall continue with all other operations that do not affect those remains or features. Requests for adjustments in the Contract Sum and Contract Time arising from the existence of such remains or features may be made as provided in Article 15.

§ 3.8 ALLOWANCES

§ 3.8.1 The Contractor shall include in the Contract Sum all allowances stated in the Contract Documents. Items covered by allowances shall be supplied for such amounts and by such persons or entities as the Owner may direct, but the Contractor shall not be required to employ persons or entities to whom the Contractor has reasonable objection.

§ 3.8.2 Unless otherwise provided in the Contract Documents,

    .1   allowances shall cover the cost to the Contractor of materials and equipment delivered at the site and all required taxes, less applicable trade discounts;

    .2   Contractor's costs for unloading and handling at the site, labor, installation costs, overhead, profit and other expenses contemplated for stated allowance amounts shall be included in the Contract Sum but not in the allowances; and

    .3   whenever costs are more than or less than allowances, the Contract Sum shall be adjusted accordingly by Change Order. The amount of the Change Order shall reflect (1) the difference between actual costs and the allowances under Section 3.8.2.1 and (2) changes in Contractor's costs under Section 3.8.2.2.

§ 3.8.3 Materials and equipment under an allowance shall be selected by the Owner with reasonable promptness.

§ 3.9 SUPERINTENDENT

§ 3.9.1 The Contractor shall employ a competent superintendent and necessary assistants who shall be in attendance at the Project site during performance of the Work. The superintendent shall represent the Contractor, and communications given to the superintendent shall be as binding as if given to the Contractor.

§ 3.9.2 The Contractor, as soon as practicable after award of the Contract, shall furnish in writing to the Owner through the Architect the name and qualifications of a proposed superintendent. The Architect may reply within 14 days to the Contractor in writing stating (1) whether the Owner or the Architect has reasonable objection to the proposed superintendent or (2) that the Architect requires additional time to review. Failure of the Architect to reply within the 14 day period shall constitute notice of no reasonable objection.

§ 3.9.3 The Contractor shall not employ a proposed superintendent to whom the Owner or Architect has made reasonable and timely objection. The Contractor shall not change the superintendent without the Owner's consent, which shall not unreasonably be withheld or delayed.

§ 3.10 CONTRACTOR'S CONSTRUCTION SCHEDULES

§ 3.10.1 The Contractor, promptly after being awarded the Contract, shall prepare and submit for the Owner's and Architect's information a Contractor's construction schedule for the Work. The schedule shall not exceed time limits current under the Contract Documents, shall be revised at appropriate intervals as required by the conditions of the Work and Project, shall be related to the entire Project to the extent required by the Contract Documents, and shall provide for expeditious and practicable execution of the Work.

§ 3.10.2 The Contractor shall prepare a submittal schedule, promptly after being awarded the Contract and thereafter as necessary to maintain a current submittal schedule, and shall submit the schedule(s) for the Architect's approval. The Architect's approval shall not unreasonably be delayed or withheld. The submittal schedule shall (1) be coordinated with the Contractor's construction schedule, and (2) allow the Architect reasonable time to review submittals. If the Contractor fails to submit a submittal schedule, the Contractor shall not be entitled to any increase in Contract Sum or extension of Contract Time based on the time required for review of submittals.

§ 3.10.3 The Contractor shall perform the Work in general accordance with the most recent schedules submitted to the Owner and Architect.

§ 3.11 DOCUMENTS AND SAMPLES AT THE SITE

The Contractor shall maintain at the site for the Owner one copy of the Drawings, Specifications, Addenda, Change Orders and other Modifications, in good order and marked currently to indicate field changes and selections made during construction, and one copy of approved Shop Drawings, Product Data, Samples and similar required submittals. These shall be available to the Architect and shall be delivered to the Architect for submittal to the Owner upon completion of the Work as a record of the Work as constructed.

§ 3.12 SHOP DRAWINGS, PRODUCT DATA AND SAMPLES

§ 3.12.1 Shop Drawings are drawings, diagrams, schedules and other data specially prepared for the Work by the Contractor or a Subcontractor, Sub-subcontractor, manufacturer, supplier or distributor to illustrate some portion of the Work.

§ 3.12.2 Product Data are illustrations, standard schedules, performance charts, instructions, brochures, diagrams and other information furnished by the Contractor to illustrate materials or equipment for some portion of the Work.

§ 3.12.3 Samples are physical examples that illustrate materials, equipment or workmanship and establish standards by which the Work will be judged.

§ 3.12.4 Shop Drawings, Product Data, Samples and similar submittals are not Contract Documents. Their purpose is to demonstrate the way by which the Contractor proposes to conform to the information given and the design concept expressed in the Contract Documents for those portions of the Work for which the Contract Documents require submittals. Review by the Architect is subject to the limitations of Section 4.2.7. Informational submittals upon which the Architect is not expected to take responsive action may be so identified in the Contract Documents. Submittals that are not required by the Contract Documents may be returned by the Architect without action.

§ 3.12.5 The Contractor shall review for compliance with the Contract Documents, approve and submit to the Architect Shop Drawings, Product Data, Samples and similar submittals required by the Contract Documents in accordance with the submittal schedule approved by the Architect or, in the absence of an approved submittal schedule, with reasonable promptness and in such sequence as to cause no delay in the Work or in the activities of the Owner or of separate contractors.

§ 3.12.6 By submitting Shop Drawings, Product Data, Samples and similar submittals, the Contractor represents to the Owner and Architect that the Contractor has (1) reviewed and approved them, (2) determined and verified materials, field measurements and field construction criteria related thereto, or will do so and (3) checked and coordinated the information contained within such submittals with the requirements of the Work and of the Contract Documents.

§ 3.12.7 The Contractor shall perform no portion of the Work for which the Contract Documents require submittal and review of Shop Drawings, Product Data, Samples or similar submittals until the respective submittal has been approved by the Architect.

§ 3.12.8 The Work shall be in accordance with approved submittals except that the Contractor shall not be relieved of responsibility for deviations from requirements of the Contract Documents by the Architect's approval of Shop Drawings, Product Data, Samples or similar submittals unless the Contractor has specifically informed the Architect in writing of such deviation at the time of submittal and (1) the Architect has given written approval to the specific deviation as a minor change in the Work, or (2) a Change Order or Construction Change Directive has been issued authorizing the deviation. The Contractor shall not be relieved of responsibility for errors or omissions in Shop Drawings, Product Data, Samples or similar submittals by the Architect's approval thereof.

§ 3.12.9 The Contractor shall direct specific attention, in writing or on resubmitted Shop Drawings, Product Data, Samples or similar submittals, to revisions other than those requested by the Architect on previous submittals. In the absence of such written notice, the Architect's approval of a resubmission shall not apply to such revisions.

§ 3.12.10 The Contractor shall not be required to provide professional services that constitute the practice of architecture or engineering unless such services are specifically required by the Contract Documents for a portion of the Work or unless the Contractor needs to provide such services in order to carry out the Contractor's responsibilities for construction means, methods, techniques, sequences and procedures. The Contractor shall not be required to provide professional services in violation of applicable law. If professional design services or certifications by a design professional related to systems, materials or equipment are specifically required of the Contractor by the Contract Documents, the Owner and the Architect will specify all performance and design criteria that such services must satisfy. The Contractor shall cause such services or certifications to be provided by a properly licensed design professional, whose signature and seal shall appear on all drawings, calculations, specifications, certifications, Shop Drawings and other submittals prepared by such professional. Shop Drawings and other submittals related to the Work designed or certified by such professional, if prepared by others, shall bear such professional's written approval when submitted to the Architect. The Owner and the Architect shall be entitled to rely upon the adequacy, accuracy and completeness of the services, certifications and approvals performed or provided by such design professionals, provided the Owner and Architect have specified to the Contractor all performance and design criteria that such services must satisfy. Pursuant to this Section 3.12.10, the Architect will review, approve or take other appropriate action on submittals only for the limited purpose of checking for conformance with information given and the design concept expressed in the Contract Documents. The Contractor shall not be responsible for the adequacy of the performance and design criteria specified in the Contract Documents.

§ 3.13 USE OF SITE
The Contractor shall confine operations at the site to areas permitted by applicable laws, statutes, ordinances, codes, rules and regulations, and lawful orders of public authorities and the Contract Documents and shall not unreasonably encumber the site with materials or equipment.

§ 3.14 CUTTING AND PATCHING
§ 3.14.1 The Contractor shall be responsible for cutting, fitting or patching required to complete the Work or to make its parts fit together properly. All areas requiring cutting, fitting and patching shall be restored to the condition existing prior to the cutting, fitting and patching, unless otherwise required by the Contract Documents.

§ 3.14.2 The Contractor shall not damage or endanger a portion of the Work or fully or partially completed construction of the Owner or separate contractors by cutting, patching or otherwise altering such construction, or by excavation. The Contractor shall not cut or otherwise alter such construction by the Owner or a separate contractor except with written consent of the Owner and of such separate contractor; such consent shall not be unreasonably withheld. The Contractor shall not unreasonably withhold from the Owner or a separate contractor the Contractor's consent to cutting or otherwise altering the Work.

§ 3.15 CLEANING UP
§ 3.15.1 The Contractor shall keep the premises and surrounding area free from accumulation of waste materials or rubbish caused by operations under the Contract. At completion of the Work, the Contractor shall remove waste materials, rubbish, the Contractor's tools, construction equipment, machinery and surplus materials from and about the Project.

§ 3.15.2 If the Contractor fails to clean up as provided in the Contract Documents, the Owner may do so and Owner shall be entitled to reimbursement from the Contractor.

§ 3.16 ACCESS TO WORK
The Contractor shall provide the Owner and Architect access to the Work in preparation and progress wherever located.

§ 3.17 ROYALTIES, PATENTS AND COPYRIGHTS
The Contractor shall pay all royalties and license fees. The Contractor shall defend suits or claims for infringement of copyrights and patent rights and shall hold the Owner and Architect harmless from loss on account thereof, but shall not be responsible for such defense or loss when a particular design, process or product of a particular manufacturer or manufacturers is required by the Contract Documents, or where the copyright violations are contained in Drawings, Specifications or other documents prepared by the Owner or Architect. However, if the Contractor has reason to believe that the required design, process or product is an infringement of a copyright or a patent, the Contractor shall be responsible for such loss unless such information is promptly furnished to the Architect.

§ 3.18 INDEMNIFICATION
§ 3.18.1 To the fullest extent permitted by law the Contractor shall indemnify and hold harmless the Owner, Architect, Architect's consultants, and agents and employees of any of them from and against claims, damages, losses and expenses, including but not limited to attorneys' fees, arising out of or resulting from performance of the Work, provided that such claim, damage, loss or expense is attributable to bodily injury, sickness, disease or death, or to injury to or destruction of tangible property (other than the Work itself), but only to the extent caused by the negligent acts or omissions of the Contractor, a Subcontractor, anyone directly or indirectly employed by them or anyone for whose acts they may be liable, regardless of whether or not such claim, damage, loss or expense is caused in part by a party indemnified hereunder. Such obligation shall not be construed to negate, abridge, or reduce

other rights or obligations of indemnity that would otherwise exist as to a party or person described in this Section 3.18.

§ **3.18.2** In claims against any person or entity indemnified under this Section 3.18 by an employee of the Contractor, a Subcontractor, anyone directly or indirectly employed by them or anyone for whose acts they may be liable, the indemnification obligation under Section 3.18.1 shall not be limited by a limitation on amount or type of damages, compensation or benefits payable by or for the Contractor or a Subcontractor under workers' compensation acts, disability benefit acts or other employee benefit acts.

ARTICLE 4 ARCHITECT
§ 4.1 GENERAL
§ **4.1.1** The Owner shall retain an architect lawfully licensed to practice architecture or an entity lawfully practicing architecture in the jurisdiction where the Project is located. That person or entity is identified as the Architect in the Agreement and is referred to throughout the Contract Documents as if singular in number.

§ **4.1.2** Duties, responsibilities and limitations of authority of the Architect as set forth in the Contract Documents shall not be restricted, modified or extended without written consent of the Owner, Contractor and Architect. Consent shall not be unreasonably withheld.

§ **4.1.3** If the employment of the Architect is terminated, the Owner shall employ a successor architect as to whom the Contractor has no reasonable objection and whose status under the Contract Documents shall be that of the Architect.

§ 4.2 ADMINISTRATION OF THE CONTRACT
§ **4.2.1** The Architect will provide administration of the Contract as described in the Contract Documents and will be an Owner's representative during construction until the date the Architect issues the final Certificate For Payment. The Architect will have authority to act on behalf of the Owner only to the extent provided in the Contract Documents.

§ **4.2.2** The Architect will visit the site at intervals appropriate to the stage of construction, or as otherwise agreed with the Owner, to become generally familiar with the progress and quality of the portion of the Work completed, and to determine in general if the Work observed is being performed in a manner indicating that the Work, when fully completed, will be in accordance with the Contract Documents. However, the Architect will not be required to make exhaustive or continuous on-site inspections to check the quality or quantity of the Work. The Architect will not have control over, charge of, or responsibility for, the construction means, methods, techniques, sequences or procedures, or for the safety precautions and programs in connection with the Work, since these are solely the Contractor's rights and responsibilities under the Contract Documents, except as provided in Section 3.3.1.

§ **4.2.3** On the basis of the site visits, the Architect will keep the Owner reasonably informed about the progress and quality of the portion of the Work completed, and report to the Owner (1) known deviations from the Contract Documents and from the most recent construction schedule submitted by the Contractor, and (2) defects and deficiencies observed in the Work. The Architect will not be responsible for the Contractor's failure to perform the Work in accordance with the requirements of the Contract Documents. The Architect will not have control over or charge of and will not be responsible for acts or omissions of the Contractor, Subcontractors, or their agents or employees, or any other persons or entities performing portions of the Work.

§ 4.2.4 COMMUNICATIONS FACILITATING CONTRACT ADMINISTRATION
Except as otherwise provided in the Contract Documents or when direct communications have been specially authorized, the Owner and Contractor shall endeavor to communicate with each other through the Architect about matters arising out of or relating to the Contract. Communications by and with the Architect's consultants shall be through the Architect. Communications by and with Subcontractors and material suppliers shall be through the Contractor. Communications by and with separate contractors shall be through the Owner.

§ **4.2.5** Based on the Architect's evaluations of the Contractor's Applications for Payment, the Architect will review and certify the amounts due the Contractor and will issue Certificates for Payment in such amounts.

§ **4.2.6** The Architect has authority to reject Work that does not conform to the Contract Documents. Whenever the Architect considers it necessary or advisable, the Architect will have authority to require inspection or testing of the

Work in accordance with Sections 13.5.2 and 13.5.3, whether or not such Work is fabricated, installed or completed. However, neither this authority of the Architect nor a decision made in good faith either to exercise or not to exercise such authority shall give rise to a duty or responsibility of the Architect to the Contractor, Subcontractors, material and equipment suppliers, their agents or employees, or other persons or entities performing portions of the Work.

§ 4.2.7 The Architect will review and approve, or take other appropriate action upon, the Contractor's submittals such as Shop Drawings, Product Data and Samples, but only for the limited purpose of checking for conformance with information given and the design concept expressed in the Contract Documents. The Architect's action will be taken in accordance with the submittal schedule approved by the Architect or, in the absence of an approved submittal schedule, with reasonable promptness while allowing sufficient time in the Architect's professional judgment to permit adequate review. Review of such submittals is not conducted for the purpose of determining the accuracy and completeness of other details such as dimensions and quantities, or for substantiating instructions for installation or performance of equipment or systems, all of which remain the responsibility of the Contractor as required by the Contract Documents. The Architect's review of the Contractor's submittals shall not relieve the Contractor of the obligations under Sections 3.3, 3.5 and 3.12. The Architect's review shall not constitute approval of safety precautions or, unless otherwise specifically stated by the Architect, of any construction means, methods, techniques, sequences or procedures. The Architect's approval of a specific item shall not indicate approval of an assembly of which the item is a component.

§ 4.2.8 The Architect will prepare Change Orders and Construction Change Directives, and may authorize minor changes in the Work as provided in Section 7.4. The Architect will investigate and make determinations and recommendations regarding concealed and unknown conditions as provided in Section 3.7.4.

§ 4.2.9 The Architect will conduct inspections to determine the date or dates of Substantial Completion and the date of final completion; issue Certificates of Substantial Completion pursuant to Section 9.8; receive and forward to the Owner, for the Owner's review and records, written warranties and related documents required by the Contract and assembled by the Contractor pursuant to Section 9.10; and issue a final Certificate for Payment pursuant to Section 9.10.

§ 4.2.10 If the Owner and Architect agree, the Architect will provide one or more project representatives to assist in carrying out the Architect's responsibilities at the site. The duties, responsibilities and limitations of authority of such project representatives shall be as set forth in an exhibit to be incorporated in the Contract Documents.

§ 4.2.11 The Architect will interpret and decide matters concerning performance under, and requirements of, the Contract Documents on written request of either the Owner or Contractor. The Architect's response to such requests will be made in writing within any time limits agreed upon or otherwise with reasonable promptness.

§ 4.2.12 Interpretations and decisions of the Architect will be consistent with the intent of, and reasonably inferable from, the Contract Documents and will be in writing or in the form of drawings. When making such interpretations and decisions, the Architect will endeavor to secure faithful performance by both Owner and Contractor, will not show partiality to either and will not be liable for results of interpretations or decisions rendered in good faith.

§ 4.2.13 The Architect's decisions on matters relating to aesthetic effect will be final if consistent with the intent expressed in the Contract Documents.

§ 4.2.14 The Architect will review and respond to requests for information about the Contract Documents. The Architect's response to such requests will be made in writing within any time limits agreed upon or otherwise with reasonable promptness. If appropriate, the Architect will prepare and issue supplemental Drawings and Specifications in response to the requests for information.

ARTICLE 5   SUBCONTRACTORS
§ 5.1 DEFINITIONS
§ 5.1.1 A Subcontractor is a person or entity who has a direct contract with the Contractor to perform a portion of the Work at the site. The term "Subcontractor" is referred to throughout the Contract Documents as if singular in number and means a Subcontractor or an authorized representative of the Subcontractor. The term "Subcontractor" does not include a separate contractor or subcontractors of a separate contractor.

§ 5.1.2 A Sub-subcontractor is a person or entity who has a direct or indirect contract with a Subcontractor to perform a portion of the Work at the site. The term "Sub-subcontractor" is referred to throughout the Contract Documents as if singular in number and means a Sub-subcontractor or an authorized representative of the Sub-subcontractor.

§ 5.2 AWARD OF SUBCONTRACTS AND OTHER CONTRACTS FOR PORTIONS OF THE WORK

§ 5.2.1 Unless otherwise stated in the Contract Documents or the bidding requirements, the Contractor, as soon as practicable after award of the Contract, shall furnish in writing to the Owner through the Architect the names of persons or entities (including those who are to furnish materials or equipment fabricated to a special design) proposed for each principal portion of the Work. The Architect may reply within 14 days to the Contractor in writing stating (1) whether the Owner or the Architect has reasonable objection to any such proposed person or entity or (2) that the Architect requires additional time for review. Failure of the Owner or Architect to reply within the 14-day period shall constitute notice of no reasonable objection.

§ 5.2.2 The Contractor shall not contract with a proposed person or entity to whom the Owner or Architect has made reasonable and timely objection. The Contractor shall not be required to contract with anyone to whom the Contractor has made reasonable objection.

§ 5.2.3 If the Owner or Architect has reasonable objection to a person or entity proposed by the Contractor, the Contractor shall propose another to whom the Owner or Architect has no reasonable objection. If the proposed but rejected Subcontractor was reasonably capable of performing the Work, the Contract Sum and Contract Time shall be increased or decreased by the difference, if any, occasioned by such change, and an appropriate Change Order shall be issued before commencement of the substitute Subcontractor's Work. However, no increase in the Contract Sum or Contract Time shall be allowed for such change unless the Contractor has acted promptly and responsively in submitting names as required.

§ 5.2.4 The Contractor shall not substitute a Subcontractor, person or entity previously selected if the Owner or Architect makes reasonable objection to such substitution.

§ 5.3 SUBCONTRACTUAL RELATIONS

By appropriate agreement, written where legally required for validity, the Contractor shall require each Subcontractor, to the extent of the Work to be performed by the Subcontractor, to be bound to the Contractor by terms of the Contract Documents, and to assume toward the Contractor all the obligations and responsibilities, including the responsibility for safety of the Subcontractor's Work, which the Contractor, by these Documents, assumes toward the Owner and Architect. Each subcontract agreement shall preserve and protect the rights of the Owner and Architect under the Contract Documents with respect to the Work to be performed by the Subcontractor so that subcontracting thereof will not prejudice such rights, and shall allow to the Subcontractor, unless specifically provided otherwise in the subcontract agreement, the benefit of all rights, remedies and redress against the Contractor that the Contractor, by the Contract Documents, has against the Owner. Where appropriate, the Contractor shall require each Subcontractor to enter into similar agreements with Sub-subcontractors. The Contractor shall make available to each proposed Subcontractor, prior to the execution of the subcontract agreement, copies of the Contract Documents to which the Subcontractor will be bound, and, upon written request of the Subcontractor, identify to the Subcontractor terms and conditions of the proposed subcontract agreement that may be at variance with the Contract Documents. Subcontractors will similarly make copies of applicable portions of such documents available to their respective proposed Sub-subcontractors.

§ 5.4 CONTINGENT ASSIGNMENT OF SUBCONTRACTS

§ 5.4.1 Each subcontract agreement for a portion of the Work is assigned by the Contractor to the Owner, provided that

    .1 assignment is effective only after termination of the Contract by the Owner for cause pursuant to Section 14.2 and only for those subcontract agreements that the Owner accepts by notifying the Subcontractor and Contractor in writing; and

    .2 assignment is subject to the prior rights of the surety, if any, obligated under bond relating to the Contract.

When the Owner accepts the assignment of a subcontract agreement, the Owner assumes the Contractor's rights and obligations under the subcontract.

§ 5.4.2 Upon such assignment, if the Work has been suspended for more than 30 days, the Subcontractor's compensation shall be equitably adjusted for increases in cost resulting from the suspension.

§ 5.4.3 Upon such assignment to the Owner under this Section 5.4, the Owner may further assign the subcontract to a successor contractor or other entity. If the Owner assigns the subcontract to a successor contractor or other entity, the Owner shall nevertheless remain legally responsible for all of the successor contractor's obligations under the subcontract.

ARTICLE 6 CONSTRUCTION BY OWNER OR BY SEPARATE CONTRACTORS
§ 6.1 OWNER'S RIGHT TO PERFORM CONSTRUCTION AND TO AWARD SEPARATE CONTRACTS
§ 6.1.1 The Owner reserves the right to perform construction or operations related to the Project with the Owner's own forces, and to award separate contracts in connection with other portions of the Project or other construction or operations on the site under Conditions of the Contract identical or substantially similar to these including those portions related to insurance and waiver of subrogation. If the Contractor claims that delay or additional cost is involved because of such action by the Owner, the Contractor shall make such Claim as provided in Article 15.

§ 6.1.2 When separate contracts are awarded for different portions of the Project or other construction or operations on the site, the term "Contractor" in the Contract Documents in each case shall mean the Contractor who executes each separate Owner-Contractor Agreement.

§ 6.1.3 The Owner shall provide for coordination of the activities of the Owner's own forces and of each separate contractor with the Work of the Contractor, who shall cooperate with them. The Contractor shall participate with other separate contractors and the Owner in reviewing their construction schedules. The Contractor shall make any revisions to the construction schedule deemed necessary after a joint review and mutual agreement. The construction schedules shall then constitute the schedules to be used by the Contractor, separate contractors and the Owner until subsequently revised.

§ 6.1.4 Unless otherwise provided in the Contract Documents, when the Owner performs construction or operations related to the Project with the Owner's own forces, the Owner shall be deemed to be subject to the same obligations and to have the same rights that apply to the Contractor under the Conditions of the Contract, including, without excluding others, those stated in Article 3, this Article 6 and Articles 10, 11 and 12.

§ 6.2 MUTUAL RESPONSIBILITY
§ 6.2.1 The Contractor shall afford the Owner and separate contractors reasonable opportunity for introduction and storage of their materials and equipment and performance of their activities, and shall connect and coordinate the Contractor's construction and operations with theirs as required by the Contract Documents.

§ 6.2.2 If part of the Contractor's Work depends for proper execution or results upon construction or operations by the Owner or a separate contractor, the Contractor shall, prior to proceeding with that portion of the Work, promptly report to the Architect apparent discrepancies or defects in such other construction that would render it unsuitable for such proper execution and results. Failure of the Contractor so to report shall constitute an acknowledgment that the Owner's or separate contractor's completed or partially completed construction is fit and proper to receive the Contractor's Work, except as to defects not then reasonably discoverable.

§ 6.2.3 The Contractor shall reimburse the Owner for costs the Owner incurs that are payable to a separate contractor because of the Contractor's delays, improperly timed activities or defective construction. The Owner shall be responsible to the Contractor for costs the Contractor incurs because of a separate contractor's delays, improperly timed activities, damage to the Work or defective construction.

§ 6.2.4 The Contractor shall promptly remedy damage the Contractor wrongfully causes to completed or partially completed construction or to property of the Owner, separate contractors as provided in Section 10.2.5.

§ 6.2.5 The Owner and each separate contractor shall have the same responsibilities for cutting and patching as are described for the Contractor in Section 3.14.

§ 6.3 OWNER'S RIGHT TO CLEAN UP

If a dispute arises among the Contractor, separate contractors and the Owner as to the responsibility under their respective contracts for maintaining the premises and surrounding area free from waste materials and rubbish, the Owner may clean up and the Architect will allocate the cost among those responsible.

ARTICLE 7   CHANGES IN THE WORK
§ 7.1 GENERAL

§ 7.1.1 Changes in the Work may be accomplished after execution of the Contract, and without invalidating the Contract, by Change Order, Construction Change Directive or order for a minor change in the Work, subject to the limitations stated in this Article 7 and elsewhere in the Contract Documents.

§ 7.1.2 A Change Order shall be based upon agreement among the Owner, Contractor and Architect; a Construction Change Directive requires agreement by the Owner and Architect and may or may not be agreed to by the Contractor; an order for a minor change in the Work may be issued by the Architect alone.

§ 7.1.3 Changes in the Work shall be performed under applicable provisions of the Contract Documents, and the Contractor shall proceed promptly, unless otherwise provided in the Change Order, Construction Change Directive or order for a minor change in the Work.

§ 7.2 CHANGE ORDERS

§ 7.2.1 A Change Order is a written instrument prepared by the Architect and signed by the Owner, Contractor and Architect stating their agreement upon all of the following:

    .1   The change in the Work;
    .2   The amount of the adjustment, if any, in the Contract Sum; and
    .3   The extent of the adjustment, if any, in the Contract Time.

§ 7.3 CONSTRUCTION CHANGE DIRECTIVES

§ 7.3.1 A Construction Change Directive is a written order prepared by the Architect and signed by the Owner and Architect, directing a change in the Work prior to agreement on adjustment, if any, in the Contract Sum or Contract Time, or both. The Owner may by Construction Change Directive, without invalidating the Contract, order changes in the Work within the general scope of the Contract consisting of additions, deletions or other revisions, the Contract Sum and Contract Time being adjusted accordingly.

§ 7.3.2 A Construction Change Directive shall be used in the absence of total agreement on the terms of a Change Order.

§ 7.3.3 If the Construction Change Directive provides for an adjustment to the Contract Sum, the adjustment shall be based on one of the following methods:

    .1   Mutual acceptance of a lump sum properly itemized and supported by sufficient substantiating data to permit evaluation;
    .2   Unit prices stated in the Contract Documents or subsequently agreed upon;
    .3   Cost to be determined in a manner agreed upon by the parties and a mutually acceptable fixed or percentage fee; or
    .4   As provided in Section 7.3.7.

§ 7.3.4 If unit prices are stated in the Contract Documents or subsequently agreed upon, and if quantities originally contemplated are materially changed in a proposed Change Order or Construction Change Directive so that application of such unit prices to quantities of Work proposed will cause substantial inequity to the Owner or Contractor, the applicable unit prices shall be equitably adjusted.

§ 7.3.5 Upon receipt of a Construction Change Directive, the Contractor shall promptly proceed with the change in the Work involved and advise the Architect of the Contractor's agreement or disagreement with the method, if any, provided in the Construction Change Directive for determining the proposed adjustment in the Contract Sum or Contract Time.

§ 7.3.6 A Construction Change Directive signed by the Contractor indicates the Contractor's agreement therewith, including adjustment in Contract Sum and Contract Time or the method for determining them. Such agreement shall be effective immediately and shall be recorded as a Change Order.

§ 7.3.7 If the Contractor does not respond promptly or disagrees with the method for adjustment in the Contract Sum, the Architect shall determine the method and the adjustment on the basis of reasonable expenditures and savings of those performing the Work attributable to the change, including, in case of an increase in the Contract Sum, an amount for overhead and profit as set forth in the Agreement, or if no such amount is set forth in the Agreement, a reasonable amount. In such case, and also under Section 7.3.3.3, the Contractor shall keep and present, in such form as the Architect may prescribe, an itemized accounting together with appropriate supporting data. Unless otherwise provided in the Contract Documents, costs for the purposes of this Section 7.3.7 shall be limited to the following:

    .1   Costs of labor, including social security, old age and unemployment insurance, fringe benefits required by agreement or custom, and workers' compensation insurance;

    .2   Costs of materials, supplies and equipment, including cost of transportation, whether incorporated or consumed;

    .3   Rental costs of machinery and equipment, exclusive of hand tools, whether rented from the Contractor or others;

    .4   Costs of premiums for all bonds and insurance, permit fees, and sales, use or similar taxes related to the Work; and

    .5   Additional costs of supervision and field office personnel directly attributable to the change.

§ 7.3.8 The amount of credit to be allowed by the Contractor to the Owner for a deletion or change that results in a net decrease in the Contract Sum shall be actual net cost as confirmed by the Architect. When both additions and credits covering related Work or substitutions are involved in a change, the allowance for overhead and profit shall be figured on the basis of net increase, if any, with respect to that change.

§ 7.3.9 Pending final determination of the total cost of a Construction Change Directive to the Owner, the Contractor may request payment for Work completed under the Construction Change Directive in Applications for Payment. The Architect will make an interim determination for purposes of monthly certification for payment for those costs and certify for payment the amount that the Architect determines, in the Architect's professional judgment, to be reasonably justified. The Architect's interim determination of cost shall adjust the Contract Sum on the same basis as a Change Order, subject to the right of either party to disagree and assert a Claim in accordance with Article 15.

§ 7.3.10 When the Owner and Contractor agree with a determination made by the Architect concerning the adjustments in the Contract Sum and Contract Time, or otherwise reach agreement upon the adjustments, such agreement shall be effective immediately and the Architect will prepare a Change Order. Change Orders may be issued for all or any part of a Construction Change Directive.

§ 7.4 MINOR CHANGES IN THE WORK
The Architect has authority to order minor changes in the Work not involving adjustment in the Contract Sum or extension of the Contract Time and not inconsistent with the intent of the Contract Documents. Such changes will be effected by written order signed by the Architect and shall be binding on the Owner and Contractor.

ARTICLE 8   TIME
§ 8.1 DEFINITIONS
§ 8.1.1 Unless otherwise provided, Contract Time is the period of time, including authorized adjustments, allotted in the Contract Documents for Substantial Completion of the Work.

§ 8.1.2 The date of commencement of the Work is the date established in the Agreement.

§ 8.1.3 The date of Substantial Completion is the date certified by the Architect in accordance with Section 9.8.

§ 8.1.4 The term "day" as used in the Contract Documents shall mean calendar day unless otherwise specifically defined.

§ 8.2 PROGRESS AND COMPLETION
§ 8.2.1 Time limits stated in the Contract Documents are of the essence of the Contract. By executing the Agreement the Contractor confirms that the Contract Time is a reasonable period for performing the Work.

§ 8.2.2 The Contractor shall not knowingly, except by agreement or instruction of the Owner in writing, prematurely commence operations on the site or elsewhere prior to the effective date of insurance required by Article 11 to be furnished by the Contractor and Owner. The date of commencement of the Work shall not be changed by the effective date of such insurance.

§ 8.2.3 The Contractor shall proceed expeditiously with adequate forces and shall achieve Substantial Completion within the Contract Time.

§ 8.3 DELAYS AND EXTENSIONS OF TIME

§ 8.3.1 If the Contractor is delayed at any time in the commencement or progress of the Work by an act or neglect of the Owner or Architect, or of an employee of either, or of a separate contractor employed by the Owner; or by changes ordered in the Work; or by labor disputes, fire, unusual delay in deliveries, unavoidable casualties or other causes beyond the Contractor's control; or by delay authorized by the Owner pending mediation and arbitration; or by other causes that the Architect determines may justify delay, then the Contract Time shall be extended by Change Order for such reasonable time as the Architect may determine.

§ 8.3.2 Claims relating to time shall be made in accordance with applicable provisions of Article 15.

§ 8.3.3 This Section 8.3 does not preclude recovery of damages for delay by either party under other provisions of the Contract Documents.

ARTICLE 9   PAYMENTS AND COMPLETION
§ 9.1 CONTRACT SUM
The Contract Sum is stated in the Agreement and, including authorized adjustments, is the total amount payable by the Owner to the Contractor for performance of the Work under the Contract Documents.

§ 9.2 SCHEDULE OF VALUES
Where the Contract is based on a stipulated sum or Guaranteed Maximum Price, the Contractor shall submit to the Architect, before the first Application for Payment, a schedule of values allocating the entire Contract Sum to the various portions of the Work and prepared in such form and supported by such data to substantiate its accuracy as the Architect may require. This schedule, unless objected to by the Architect, shall be used as a basis for reviewing the Contractor's Applications for Payment.

§ 9.3 APPLICATIONS FOR PAYMENT
§ 9.3.1 At least ten days before the date established for each progress payment, the Contractor shall submit to the Architect an itemized Application for Payment prepared in accordance with the schedule of values, if required under Section 9.2., for completed portions of the Work. Such application shall be notarized, if required, and supported by such data substantiating the Contractor's right to payment as the Owner or Architect may require, such as copies of requisitions from Subcontractors and material suppliers, and shall reflect retainage if provided for in the Contract Documents.

§ 9.3.1.1 As provided in Section 7.3.9, such applications may include requests for payment on account of changes in the Work that have been properly authorized by Construction Change Directives, or by interim determinations of the Architect, but not yet included in Change Orders.

§ 9.3.1.2 Applications for Payment shall not include requests for payment for portions of the Work for which the Contractor does not intend to pay a Subcontractor or material supplier, unless such Work has been performed by others whom the Contractor intends to pay.

§ 9.3.2 Unless otherwise provided in the Contract Documents, payments shall be made on account of materials and equipment delivered and suitably stored at the site for subsequent incorporation in the Work. If approved in advance by the Owner, payment may similarly be made for materials and equipment suitably stored off the site at a location agreed upon in writing. Payment for materials and equipment stored on or off the site shall be conditioned upon compliance by the Contractor with procedures satisfactory to the Owner to establish the Owner's title to such materials and equipment or otherwise protect the Owner's interest, and shall include the costs of applicable insurance, storage and transportation to the site for such materials and equipment stored off the site.

§ 9.3.3 The Contractor warrants that title to all Work covered by an Application for Payment will pass to the Owner no later than the time of payment. The Contractor further warrants that upon submittal of an Application for Payment all Work for which Certificates for Payment have been previously issued and payments received from the

Owner shall, to the best of the Contractor's knowledge, information and belief, be free and clear of liens, claims, security interests or encumbrances in favor of the Contractor, Subcontractors, material suppliers, or other persons or entities making a claim by reason of having provided labor, materials and equipment relating to the Work.

§ 9.4 CERTIFICATES FOR PAYMENT

§ 9.4.1 The Architect will, within seven days after receipt of the Contractor's Application for Payment, either issue to the Owner a Certificate for Payment, with a copy to the Contractor, for such amount as the Architect determines is properly due, or notify the Contractor and Owner in writing of the Architect's reasons for withholding certification in whole or in part as provided in Section 9.5.1.

§ 9.4.2 The issuance of a Certificate for Payment will constitute a representation by the Architect to the Owner, based on the Architect's evaluation of the Work and the data comprising the Application for Payment, that, to the best of the Architect's knowledge, information and belief, the Work has progressed to the point indicated and that the quality of the Work is in accordance with the Contract Documents. The foregoing representations are subject to an evaluation of the Work for conformance with the Contract Documents upon Substantial Completion, to results of subsequent tests and inspections, to correction of minor deviations from the Contract Documents prior to completion and to specific qualifications expressed by the Architect. The issuance of a Certificate for Payment will further constitute a representation that the Contractor is entitled to payment in the amount certified. However, the issuance of a Certificate for Payment will not be a representation that the Architect has (1) made exhaustive or continuous on-site inspections to check the quality or quantity of the Work, (2) reviewed construction means, methods, techniques, sequences or procedures, (3) reviewed copies of requisitions received from Subcontractors and material suppliers and other data requested by the Owner to substantiate the Contractor's right to payment, or (4) made examination to ascertain how or for what purpose the Contractor has used money previously paid on account of the Contract Sum.

§ 9.5 DECISIONS TO WITHHOLD CERTIFICATION

§ 9.5.1 The Architect may withhold a Certificate for Payment in whole or in part, to the extent reasonably necessary to protect the Owner, if in the Architect's opinion the representations to the Owner required by Section 9.4.2 cannot be made. If the Architect is unable to certify payment in the amount of the Application, the Architect will notify the Contractor and Owner as provided in Section 9.4.1. If the Contractor and Architect cannot agree on a revised amount, the Architect will promptly issue a Certificate for Payment for the amount for which the Architect is able to make such representations to the Owner. The Architect may also withhold a Certificate for Payment or, because of subsequently discovered evidence, may nullify the whole or a part of a Certificate for Payment previously issued, to such extent as may be necessary in the Architect's opinion to protect the Owner from loss for which the Contractor is responsible, including loss resulting from acts and omissions described in Section 3.3.2, because of

.1   defective Work not remedied;

.2   third party claims filed or reasonable evidence indicating probable filing of such claims unless security acceptable to the Owner is provided by the Contractor;

.3   failure of the Contractor to make payments properly to Subcontractors or for labor, materials or equipment;

.4   reasonable evidence that the Work cannot be completed for the unpaid balance of the Contract Sum;

.5   damage to the Owner or a separate contractor;

.6   reasonable evidence that the Work will not be completed within the Contract Time, and that the unpaid balance would not be adequate to cover actual or liquidated damages for the anticipated delay; or

.7   repeated failure to carry out the Work in accordance with the Contract Documents.

§ 9.5.2 When the above reasons for withholding certification are removed, certification will be made for amounts previously withheld.

§ 9.5.3 If the Architect withholds certification for payment under Section 9.5.1.3, the Owner may, at its sole option, issue joint checks to the Contractor and to any Subcontractor or material or equipment suppliers to whom the Contractor failed to make payment for Work properly performed or material or equipment suitably delivered. If the Owner makes payments by joint check, the Owner shall notify the Architect and the Architect will reflect such payment on the next Certificate for Payment.

§ 9.6 PROGRESS PAYMENTS

§ 9.6.1 After the Architect has issued a Certificate for Payment, the Owner shall make payment in the manner and within the time provided in the Contract Documents, and shall so notify the Architect.

§ 9.6.2 The Contractor shall pay each Subcontractor no later than seven days after receipt of payment from the Owner the amount to which the Subcontractor is entitled, reflecting percentages actually retained from payments to the Contractor on account of the Subcontractor's portion of the Work. The Contractor shall, by appropriate agreement with each Subcontractor, require each Subcontractor to make payments to Sub-subcontractors in a similar manner.

§ 9.6.3 The Architect will, on request, furnish to a Subcontractor, if practicable, information regarding percentages of completion or amounts applied for by the Contractor and action taken thereon by the Architect and Owner on account of portions of the Work done by such Subcontractor.

§ 9.6.4 The Owner has the right to request written evidence from the Contractor that the Contractor has properly paid Subcontractors and material and equipment suppliers amounts paid by the Owner to the Contractor for subcontracted Work. If the Contractor fails to furnish such evidence within seven days, the Owner shall have the right to contact Subcontractors to ascertain whether they have been properly paid. Neither the Owner nor Architect shall have an obligation to pay or to see to the payment of money to a Subcontractor, except as may otherwise be required by law.

§ 9.6.5 Contractor payments to material and equipment suppliers shall be treated in a manner similar to that provided in Sections 9.6.2, 9.6.3 and 9.6.4.

§ 9.6.6 A Certificate for Payment, a progress payment, or partial or entire use or occupancy of the Project by the Owner shall not constitute acceptance of Work not in accordance with the Contract Documents.

§ 9.6.7 Unless the Contractor provides the Owner with a payment bond in the full penal sum of the Contract Sum, payments received by the Contractor for Work properly performed by Subcontractors and suppliers shall be held by the Contractor for those Subcontractors or suppliers who performed Work or furnished materials, or both, under contract with the Contractor for which payment was made by the Owner. Nothing contained herein shall require money to be placed in a separate account and not commingled with money of the Contractor, shall create any fiduciary liability or tort liability on the part of the Contractor for breach of trust or shall entitle any person or entity to an award of punitive damages against the Contractor for breach of the requirements of this provision.

§ 9.7 FAILURE OF PAYMENT
If the Architect does not issue a Certificate for Payment, through no fault of the Contractor, within seven days after receipt of the Contractor's Application for Payment, or if the Owner does not pay the Contractor within seven days after the date established in the Contract Documents the amount certified by the Architect or awarded by binding dispute resolution, then the Contractor may, upon seven additional days' written notice to the Owner and Architect, stop the Work until payment of the amount owing has been received. The Contract Time shall be extended appropriately and the Contract Sum shall be increased by the amount of the Contractor's reasonable costs of shut-down, delay and start-up, plus interest as provided for in the Contract Documents.

§ 9.8 SUBSTANTIAL COMPLETION
§ 9.8.1 Substantial Completion is the stage in the progress of the Work when the Work or designated portion thereof is sufficiently complete in accordance with the Contract Documents so that the Owner can occupy or utilize the Work for its intended use.

§ 9.8.2 When the Contractor considers that the Work, or a portion thereof which the Owner agrees to accept separately, is substantially complete, the Contractor shall prepare and submit to the Architect a comprehensive list of items to be completed or corrected prior to final payment. Failure to include an item on such list does not alter the responsibility of the Contractor to complete all Work in accordance with the Contract Documents.

§ 9.8.3 Upon receipt of the Contractor's list, the Architect will make an inspection to determine whether the Work or designated portion thereof is substantially complete. If the Architect's inspection discloses any item, whether or not included on the Contractor's list, which is not sufficiently complete in accordance with the Contract Documents so that the Owner can occupy or utilize the Work or designated portion thereof for its intended use, the Contractor shall, before issuance of the Certificate of Substantial Completion, complete or correct such item upon notification by the Architect. In such case, the Contractor shall then submit a request for another inspection by the Architect to determine Substantial Completion.

§ 9.8.4 When the Work or designated portion thereof is substantially complete, the Architect will prepare a Certificate of Substantial Completion that shall establish the date of Substantial Completion, shall establish responsibilities of the Owner and Contractor for security, maintenance, heat, utilities, damage to the Work and insurance, and shall fix the time within which the Contractor shall finish all items on the list accompanying the Certificate. Warranties required by the Contract Documents shall commence on the date of Substantial Completion of the Work or designated portion thereof unless otherwise provided in the Certificate of Substantial Completion.

§ 9.8.5 The Certificate of Substantial Completion shall be submitted to the Owner and Contractor for their written acceptance of responsibilities assigned to them in such Certificate. Upon such acceptance and consent of surety, if any, the Owner shall make payment of retainage applying to such Work or designated portion thereof. Such payment shall be adjusted for Work that is incomplete or not in accordance with the requirements of the Contract Documents.

§ 9.9 PARTIAL OCCUPANCY OR USE

§ 9.9.1 The Owner may occupy or use any completed or partially completed portion of the Work at any stage when such portion is designated by separate agreement with the Contractor, provided such occupancy or use is consented to by the insurer as required under Section 11.3.1.5 and authorized by public authorities having jurisdiction over the Project. Such partial occupancy or use may commence whether or not the portion is substantially complete, provided the Owner and Contractor have accepted in writing the responsibilities assigned to each of them for payments, retainage, if any, security, maintenance, heat, utilities, damage to the Work and insurance, and have agreed in writing concerning the period for correction of the Work and commencement of warranties required by the Contract Documents. When the Contractor considers a portion substantially complete, the Contractor shall prepare and submit a list to the Architect as provided under Section 9.8.2. Consent of the Contractor to partial occupancy or use shall not be unreasonably withheld. The stage of the progress of the Work shall be determined by written agreement between the Owner and Contractor or, if no agreement is reached, by decision of the Architect.

§ 9.9.2 Immediately prior to such partial occupancy or use, the Owner, Contractor and Architect shall jointly inspect the area to be occupied or portion of the Work to be used in order to determine and record the condition of the Work.

§ 9.9.3 Unless otherwise agreed upon, partial occupancy or use of a portion or portions of the Work shall not constitute acceptance of Work not complying with the requirements of the Contract Documents.

§ 9.10 FINAL COMPLETION AND FINAL PAYMENT

§ 9.10.1 Upon receipt of the Contractor's written notice that the Work is ready for final inspection and acceptance and upon receipt of a final Application for Payment, the Architect will promptly make such inspection and, when the Architect finds the Work acceptable under the Contract Documents and the Contract fully performed, the Architect will promptly issue a final Certificate for Payment stating that to the best of the Architect's knowledge, information and belief, and on the basis of the Architect's on-site visits and inspections, the Work has been completed in accordance with terms and conditions of the Contract Documents and that the entire balance found to be due the Contractor and noted in the final Certificate is due and payable. The Architect's final Certificate for Payment will constitute a further representation that conditions listed in Section 9.10.2 as precedent to the Contractor's being entitled to final payment have been fulfilled.

§ 9.10.2 Neither final payment nor any remaining retained percentage shall become due until the Contractor submits to the Architect (1) an affidavit that payrolls, bills for materials and equipment, and other indebtedness connected with the Work for which the Owner or the Owner's property might be responsible or encumbered (less amounts withheld by Owner) have been paid or otherwise satisfied, (2) a certificate evidencing that insurance required by the Contract Documents to remain in force after final payment is currently in effect and will not be canceled or allowed to expire until at least 30 days' prior written notice has been given to the Owner, (3) a written statement that the Contractor knows of no substantial reason that the insurance will not be renewable to cover the period required by the Contract Documents, (4) consent of surety, if any, to final payment and (5), if required by the Owner, other data establishing payment or satisfaction of obligations, such as receipts, releases and waivers of liens, claims, security interests or encumbrances arising out of the Contract, to the extent and in such form as may be designated by the Owner. If a Subcontractor refuses to furnish a release or waiver required by the Owner, the Contractor may furnish a bond satisfactory to the Owner to indemnify the Owner against such lien. If such lien remains unsatisfied after payments are made, the Contractor shall refund to the Owner all money that the Owner may be compelled to pay in discharging such lien, including all costs and reasonable attorneys' fees.

§ **9.10.3** If, after Substantial Completion of the Work, final completion thereof is materially delayed through no fault of the Contractor or by issuance of Change Orders affecting final completion, and the Architect so confirms, the Owner shall, upon application by the Contractor and certification by the Architect, and without terminating the Contract, make payment of the balance due for that portion of the Work fully completed and accepted. If the remaining balance for Work not fully completed or corrected is less than retainage stipulated in the Contract Documents, and if bonds have been furnished, the written consent of surety to payment of the balance due for that portion of the Work fully completed and accepted shall be submitted by the Contractor to the Architect prior to certification of such payment. Such payment shall be made under terms and conditions governing final payment, except that it shall not constitute a waiver of claims.

§ **9.10.4** The making of final payment shall constitute a waiver of Claims by the Owner except those arising from
 .1 liens, Claims, security interests or encumbrances arising out of the Contract and unsettled;
 .2 failure of the Work to comply with the requirements of the Contract Documents; or
 .3 terms of special warranties required by the Contract Documents.

§ **9.10.5** Acceptance of final payment by the Contractor, a Subcontractor or material supplier shall constitute a waiver of claims by that payee except those previously made in writing and identified by that payee as unsettled at the time of final Application for Payment.

ARTICLE 10 PROTECTION OF PERSONS AND PROPERTY
§ 10.1 SAFETY PRECAUTIONS AND PROGRAMS
The Contractor shall be responsible for initiating, maintaining and supervising all safety precautions and programs in connection with the performance of the Contract.

§ 10.2 SAFETY OF PERSONS AND PROPERTY
§ **10.2.1** The Contractor shall take reasonable precautions for safety of, and shall provide reasonable protection to prevent damage, injury or loss to
 .1 employees on the Work and other persons who may be affected thereby;
 .2 the Work and materials and equipment to be incorporated therein, whether in storage on or off the site, under care, custody or control of the Contractor or the Contractor's Subcontractors or Sub-subcontractors; and
 .3 other property at the site or adjacent thereto, such as trees, shrubs, lawns, walks, pavements, roadways, structures and utilities not designated for removal, relocation or replacement in the course of construction.

§ **10.2.2** The Contractor shall comply with and give notices required by applicable laws, statutes, ordinances, codes, rules and regulations, and lawful orders of public authorities bearing on safety of persons or property or their protection from damage, injury or loss.

§ **10.2.3** The Contractor shall erect and maintain, as required by existing conditions and performance of the Contract, reasonable safeguards for safety and protection, including posting danger signs and other warnings against hazards, promulgating safety regulations and notifying owners and users of adjacent sites and utilities.

§ **10.2.4** When use or storage of explosives or other hazardous materials or equipment or unusual methods are necessary for execution of the Work, the Contractor shall exercise utmost care and carry on such activities under supervision of properly qualified personnel.

§ **10.2.5** The Contractor shall promptly remedy damage and loss (other than damage or loss insured under property insurance required by the Contract Documents) to property referred to in Sections 10.2.1.2 and 10.2.1.3 caused in whole or in part by the Contractor, a Subcontractor, a Sub-subcontractor, or anyone directly or indirectly employed by any of them, or by anyone for whose acts they may be liable and for which the Contractor is responsible under Sections 10.2.1.2 and 10.2.1.3, except damage or loss attributable to acts or omissions of the Owner or Architect or anyone directly or indirectly employed by either of them, or by anyone for whose acts either of them may be liable, and not attributable to the fault or negligence of the Contractor. The foregoing obligations of the Contractor are in addition to the Contractor's obligations under Section 3.18.

§ 10.2.6 The Contractor shall designate a responsible member of the Contractor's organization at the site whose duty shall be the prevention of accidents. This person shall be the Contractor's superintendent unless otherwise designated by the Contractor in writing to the Owner and Architect.

§ 10.2.7 The Contractor shall not permit any part of the construction or site to be loaded so as to cause damage or create an unsafe condition.

§ 10.2.8 INJURY OR DAMAGE TO PERSON OR PROPERTY

If either party suffers injury or damage to person or property because of an act or omission of the other party, or of others for whose acts such party is legally responsible, written notice of such injury or damage, whether or not insured, shall be given to the other party within a reasonable time not exceeding 21 days after discovery. The notice shall provide sufficient detail to enable the other party to investigate the matter.

§ 10.3 HAZARDOUS MATERIALS

§ 10.3.1 The Contractor is responsible for compliance with any requirements included in the Contract Documents regarding hazardous materials. If the Contractor encounters a hazardous material or substance not addressed in the Contract Documents and if reasonable precautions will be inadequate to prevent foreseeable bodily injury or death to persons resulting from a material or substance, including but not limited to asbestos or polychlorinated biphenyl (PCB), encountered on the site by the Contractor, the Contractor shall, upon recognizing the condition, immediately stop Work in the affected area and report the condition to the Owner and Architect in writing.

§ 10.3.2 Upon receipt of the Contractor's written notice, the Owner shall obtain the services of a licensed laboratory to verify the presence or absence of the material or substance reported by the Contractor and, in the event such material or substance is found to be present, to cause it to be rendered harmless. Unless otherwise required by the Contract Documents, the Owner shall furnish in writing to the Contractor and Architect the names and qualifications of persons or entities who are to perform tests verifying the presence or absence of such material or substance or who are to perform the task of removal or safe containment of such material or substance. The Contractor and the Architect will promptly reply to the Owner in writing stating whether or not either has reasonable objection to the persons or entities proposed by the Owner. If either the Contractor or Architect has an objection to a person or entity proposed by the Owner, the Owner shall propose another to whom the Contractor and the Architect have no reasonable objection. When the material or substance has been rendered harmless, Work in the affected area shall resume upon written agreement of the Owner and Contractor. By Change Order, the Contract Time shall be extended appropriately and the Contract Sum shall be increased in the amount of the Contractor's reasonable additional costs of shut-down, delay and start-up.

§ 10.3.3 To the fullest extent permitted by law, the Owner shall indemnify and hold harmless the Contractor, Subcontractors, Architect, Architect's consultants and agents and employees of any of them from and against claims, damages, losses and expenses, including but not limited to attorneys' fees, arising out of or resulting from performance of the Work in the affected area if in fact the material or substance presents the risk of bodily injury or death as described in Section 10.3.1 and has not been rendered harmless, provided that such claim, damage, loss or expense is attributable to bodily injury, sickness, disease or death, or to injury to or destruction of tangible property (other than the Work itself), except to the extent that such damage, loss or expense is due to the fault or negligence of the party seeking indemnity.

§ 10.3.4 The Owner shall not be responsible under this Section 10.3 for materials or substances the Contractor brings to the site unless such materials or substances are required by the Contract Documents. The Owner shall be responsible for materials or substances required by the Contract Documents, except to the extent of the Contractor's fault or negligence in the use and handling of such materials or substances.

§ 10.3.5 The Contractor shall indemnify the Owner for the cost and expense the Owner incurs (1) for remediation of a material or substance the Contractor brings to the site and negligently handles, or (2) where the Contractor fails to perform its obligations under Section 10.3.1, except to the extent that the cost and expense are due to the Owner's fault or negligence.

§ 10.3.6 If, without negligence on the part of the Contractor, the Contractor is held liable by a government agency for the cost of remediation of a hazardous material or substance solely by reason of performing Work as required by the Contract Documents, the Owner shall indemnify the Contractor for all cost and expense thereby incurred.

§ 10.4 EMERGENCIES
In an emergency affecting safety of persons or property, the Contractor shall act, at the Contractor's discretion, to prevent threatened damage, injury or loss. Additional compensation or extension of time claimed by the Contractor on account of an emergency shall be determined as provided in Article 15 and Article 7.

ARTICLE 11 INSURANCE AND BONDS
§ 11.1 CONTRACTOR'S LIABILITY INSURANCE
§ 11.1.1 The Contractor shall purchase from and maintain in a company or companies lawfully authorized to do business in the jurisdiction in which the Project is located such insurance as will protect the Contractor from claims set forth below which may arise out of or result from the Contractor's operations and completed operations under the Contract and for which the Contractor may be legally liable, whether such operations be by the Contractor or by a Subcontractor or by anyone directly or indirectly employed by any of them, or by anyone for whose acts any of them may be liable:

  .1 Claims under workers' compensation, disability benefit and other similar employee benefit acts that are applicable to the Work to be performed;

  .2 Claims for damages because of bodily injury, occupational sickness or disease, or death of the Contractor's employees;

  .3 Claims for damages because of bodily injury, sickness or disease, or death of any person other than the Contractor's employees;

  .4 Claims for damages insured by usual personal injury liability coverage;

  .5 Claims for damages, other than to the Work itself, because of injury to or destruction of tangible property, including loss of use resulting therefrom;

  .6 Claims for damages because of bodily injury, death of a person or property damage arising out of ownership, maintenance or use of a motor vehicle;

  .7 Claims for bodily injury or property damage arising out of completed operations; and

  .8 Claims involving contractual liability insurance applicable to the Contractor's obligations under Section 3.18.

§ 11.1.2 The insurance required by Section 11.1.1 shall be written for not less than limits of liability specified in the Contract Documents or required by law, whichever coverage is greater. Coverages, whether written on an occurrence or claims-made basis, shall be maintained without interruption from the date of commencement of the Work until the date of final payment and termination of any coverage required to be maintained after final payment, and, with respect to the Contractor's completed operations coverage, until the expiration of the period for correction of Work or for such other period for maintenance of completed operations coverage as specified in the Contract Documents.

§ 11.1.3 Certificates of insurance acceptable to the Owner shall be filed with the Owner prior to commencement of the Work and thereafter upon renewal or replacement of each required policy of insurance. These certificates and the insurance policies required by this Section 11.1 shall contain a provision that coverages afforded under the policies will not be canceled or allowed to expire until at least 30 days' prior written notice has been given to the Owner. An additional certificate evidencing continuation of liability coverage, including coverage for completed operations, shall be submitted with the final Application for Payment as required by Section 9.10.2 and thereafter upon renewal or replacement of such coverage until the expiration of the time required by Section 11.1.2. Information concerning reduction of coverage on account of revised limits or claims paid under the General Aggregate, or both, shall be furnished by the Contractor with reasonable promptness.

§ 11.1.4 The Contractor shall cause the commercial liability coverage required by the Contract Documents to include (1) the Owner, the Architect and the Architect's Consultants as additional insureds for claims caused in whole or in part by the Contractor's negligent acts or omissions during the Contractor's operations; and (2) the Owner as an additional insured for claims caused in whole or in part by the Contractor's negligent acts or omissions during the Contractor's completed operations.

§ 11.2 OWNER'S LIABILITY INSURANCE
The Owner shall be responsible for purchasing and maintaining the Owner's usual liability insurance.

§ 11.3 PROPERTY INSURANCE
§ 11.3.1 Unless otherwise provided, the Owner shall purchase and maintain, in a company or companies lawfully authorized to do business in the jurisdiction in which the Project is located, property insurance written on a builder's

29

risk "all-risk" or equivalent policy form in the amount of the initial Contract Sum, plus value of subsequent Contract Modifications and cost of materials supplied or installed by others, comprising total value for the entire Project at the site on a replacement cost basis without optional deductibles. Such property insurance shall be maintained, unless otherwise provided in the Contract Documents or otherwise agreed in writing by all persons and entities who are beneficiaries of such insurance, until final payment has been made as provided in Section 9.10 or until no person or entity other than the Owner has an insurable interest in the property required by this Section 11.3 to be covered, whichever is later. This insurance shall include interests of the Owner, the Contractor, Subcontractors and Sub-subcontractors in the Project.

§ 11.3.1.1 Property insurance shall be on an "all-risk" or equivalent policy form and shall include, without limitation, insurance against the perils of fire (with extended coverage) and physical loss or damage including, without duplication of coverage, theft, vandalism, malicious mischief, collapse, earthquake, flood, windstorm, falsework, testing and startup, temporary buildings and debris removal including demolition occasioned by enforcement of any applicable legal requirements, and shall cover reasonable compensation for Architect's and Contractor's services and expenses required as a result of such insured loss.

§ 11.3.1.2 If the Owner does not intend to purchase such property insurance required by the Contract and with all of the coverages in the amount described above, the Owner shall so inform the Contractor in writing prior to commencement of the Work. The Contractor may then effect insurance that will protect the interests of the Contractor, Subcontractors and Sub-subcontractors in the Work, and by appropriate Change Order the cost thereof shall be charged to the Owner. If the Contractor is damaged by the failure or neglect of the Owner to purchase or maintain insurance as described above, without so notifying the Contractor in writing, then the Owner shall bear all reasonable costs properly attributable thereto.

§ 11.3.1.3 If the property insurance requires deductibles, the Owner shall pay costs not covered because of such deductibles.

§ 11.3.1.4 This property insurance shall cover portions of the Work stored off the site, and also portions of the Work in transit.

§ 11.3.1.5 Partial occupancy or use in accordance with Section 9.9 shall not commence until the insurance company or companies providing property insurance have consented to such partial occupancy or use by endorsement or otherwise. The Owner and the Contractor shall take reasonable steps to obtain consent of the insurance company or companies and shall, without mutual written consent, take no action with respect to partial occupancy or use that would cause cancellation, lapse or reduction of insurance.

§ 11.3.2 BOILER AND MACHINERY INSURANCE
The Owner shall purchase and maintain boiler and machinery insurance required by the Contract Documents or by law, which shall specifically cover such insured objects during installation and until final acceptance by the Owner; this insurance shall include interests of the Owner, Contractor, Subcontractors and Sub-subcontractors in the Work, and the Owner and Contractor shall be named insureds.

§ 11.3.3 LOSS OF USE INSURANCE
The Owner, at the Owner's option, may purchase and maintain such insurance as will insure the Owner against loss of use of the Owner's property due to fire or other hazards, however caused. The Owner waives all rights of action against the Contractor for loss of use of the Owner's property, including consequential losses due to fire or other hazards however caused.

§ 11.3.4 If the Contractor requests in writing that insurance for risks other than those described herein or other special causes of loss be included in the property insurance policy, the Owner shall, if possible, include such insurance, and the cost thereof shall be charged to the Contractor by appropriate Change Order.

§ 11.3.5 If during the Project construction period the Owner insures properties, real or personal or both, at or adjacent to the site by property insurance under policies separate from those insuring the Project, or if after final payment property insurance is to be provided on the completed Project through a policy or policies other than those insuring the Project during the construction period, the Owner shall waive all rights in accordance with the terms of Section 11.3.7 for damages caused by fire or other causes of loss covered by this separate property insurance. All separate policies shall provide this waiver of subrogation by endorsement or otherwise.

§ 11.3.6 Before an exposure to loss may occur, the Owner shall file with the Contractor a copy of each policy that includes insurance coverages required by this Section 11.3. Each policy shall contain all generally applicable conditions, definitions, exclusions and endorsements related to this Project. Each policy shall contain a provision that the policy will not be canceled or allowed to expire, and that its limits will not be reduced, until at least 30 days' prior written notice has been given to the Contractor.

§ 11.3.7 WAIVERS OF SUBROGATION

The Owner and Contractor waive all rights against (1) each other and any of their subcontractors, sub-subcontractors, agents and employees, each of the other, and (2) the Architect, Architect's consultants, separate contractors described in Article 6, if any, and any of their subcontractors, sub-subcontractors, agents and employees, for damages caused by fire or other causes of loss to the extent covered by property insurance obtained pursuant to this Section 11.3 or other property insurance applicable to the Work, except such rights as they have to proceeds of such insurance held by the Owner as fiduciary. The Owner or Contractor, as appropriate, shall require of the Architect, Architect's consultants, separate contractors described in Article 6, if any, and the subcontractors, sub-subcontractors, agents and employees of any of them, by appropriate agreements, written where legally required for validity, similar waivers each in favor of other parties enumerated herein. The policies shall provide such waivers of subrogation by endorsement or otherwise. A waiver of subrogation shall be effective as to a person or entity even though that person or entity would otherwise have a duty of indemnification, contractual or otherwise, did not pay the insurance premium directly or indirectly, and whether or not the person or entity had an insurable interest in the property damaged.

§ 11.3.8 A loss insured under the Owner's property insurance shall be adjusted by the Owner as fiduciary and made payable to the Owner as fiduciary for the insureds, as their interests may appear, subject to requirements of any applicable mortgagee clause and of Section 11.3.10. The Contractor shall pay Subcontractors their just shares of insurance proceeds received by the Contractor, and by appropriate agreements, written where legally required for validity, shall require Subcontractors to make payments to their Sub-subcontractors in similar manner.

§ 11.3.9 If required in writing by a party in interest, the Owner as fiduciary shall, upon occurrence of an insured loss, give bond for proper performance of the Owner's duties. The cost of required bonds shall be charged against proceeds received as fiduciary. The Owner shall deposit in a separate account proceeds so received, which the Owner shall distribute in accordance with such agreement as the parties in interest may reach, or as determined in accordance with the method of binding dispute resolution selected in the Agreement between the Owner and Contractor. If after such loss no other special agreement is made and unless the Owner terminates the Contract for convenience, replacement of damaged property shall be performed by the Contractor after notification of a Change in the Work in accordance with Article 7.

§ 11.3.10 The Owner as fiduciary shall have power to adjust and settle a loss with insurers unless one of the parties in interest shall object in writing within five days after occurrence of loss to the Owner's exercise of this power; if such objection is made, the dispute shall be resolved in the manner selected by the Owner and Contractor as the method of binding dispute resolution in the Agreement. If the Owner and Contractor have selected arbitration as the method of binding dispute resolution, the Owner as fiduciary shall make settlement with insurers or, in the case of a dispute over distribution of insurance proceeds, in accordance with the directions of the arbitrators.

§ 11.4 PERFORMANCE BOND AND PAYMENT BOND

§ 11.4.1 The Owner shall have the right to require the Contractor to furnish bonds covering faithful performance of the Contract and payment of obligations arising thereunder as stipulated in bidding requirements or specifically required in the Contract Documents on the date of execution of the Contract.

§ 11.4.2 Upon the request of any person or entity appearing to be a potential beneficiary of bonds covering payment of obligations arising under the Contract, the Contractor shall promptly furnish a copy of the bonds or shall authorize a copy to be furnished.

ARTICLE 12 UNCOVERING AND CORRECTION OF WORK
§ 12.1 UNCOVERING OF WORK

§ 12.1.1 If a portion of the Work is covered contrary to the Architect's request or to requirements specifically expressed in the Contract Documents, it must, if requested in writing by the Architect, be uncovered for the Architect's examination and be replaced at the Contractor's expense without change in the Contract Time.

§ **12.1.2** If a portion of the Work has been covered that the Architect has not specifically requested to examine prior to its being covered, the Architect may request to see such Work and it shall be uncovered by the Contractor. If such Work is in accordance with the Contract Documents, costs of uncovering and replacement shall, by appropriate Change Order, be at the Owner's expense. If such Work is not in accordance with the Contract Documents, such costs and the cost of correction shall be at the Contractor's expense unless the condition was caused by the Owner or a separate contractor in which event the Owner shall be responsible for payment of such costs.

§ 12.2 CORRECTION OF WORK
§ 12.2.1 BEFORE OR AFTER SUBSTANTIAL COMPLETION
The Contractor shall promptly correct Work rejected by the Architect or failing to conform to the requirements of the Contract Documents, whether discovered before or after Substantial Completion and whether or not fabricated, installed or completed. Costs of correcting such rejected Work, including additional testing and inspections, the cost of uncovering and replacement, and compensation for the Architect's services and expenses made necessary thereby, shall be at the Contractor's expense.

§ 12.2.2 AFTER SUBSTANTIAL COMPLETION
§ **12.2.2.1** In addition to the Contractor's obligations under Section 3.5, if, within one year after the date of Substantial Completion of the Work or designated portion thereof or after the date for commencement of warranties established under Section 9.9.1, or by terms of an applicable special warranty required by the Contract Documents, any of the Work is found to be not in accordance with the requirements of the Contract Documents, the Contractor shall correct it promptly after receipt of written notice from the Owner to do so unless the Owner has previously given the Contractor a written acceptance of such condition. The Owner shall give such notice promptly after discovery of the condition. During the one-year period for correction of Work, if the Owner fails to notify the Contractor and give the Contractor an opportunity to make the correction, the Owner waives the rights to require correction by the Contractor and to make a claim for breach of warranty. If the Contractor fails to correct nonconforming Work within a reasonable time during that period after receipt of notice from the Owner or Architect, the Owner may correct it in accordance with Section 2.4.

§ **12.2.2.2** The one-year period for correction of Work shall be extended with respect to portions of Work first performed after Substantial Completion by the period of time between Substantial Completion and the actual completion of that portion of the Work.

§ **12.2.2.3** The one-year period for correction of Work shall not be extended by corrective Work performed by the Contractor pursuant to this Section 12.2.

§ **12.2.3** The Contractor shall remove from the site portions of the Work that are not in accordance with the requirements of the Contract Documents and are neither corrected by the Contractor nor accepted by the Owner.

§ **12.2.4** The Contractor shall bear the cost of correcting destroyed or damaged construction, whether completed or partially completed, of the Owner or separate contractors caused by the Contractor's correction or removal of Work that is not in accordance with the requirements of the Contract Documents.

§ **12.2.5** Nothing contained in this Section 12.2 shall be construed to establish a period of limitation with respect to other obligations the Contractor has under the Contract Documents. Establishment of the one-year period for correction of Work as described in Section 12.2.2 relates only to the specific obligation of the Contractor to correct the Work, and has no relationship to the time within which the obligation to comply with the Contract Documents may be sought to be enforced, nor to the time within which proceedings may be commenced to establish the Contractor's liability with respect to the Contractor's obligations other than specifically to correct the Work.

§ 12.3 ACCEPTANCE OF NONCONFORMING WORK
If the Owner prefers to accept Work that is not in accordance with the requirements of the Contract Documents, the Owner may do so instead of requiring its removal and correction, in which case the Contract Sum will be reduced as appropriate and equitable. Such adjustment shall be effected whether or not final payment has been made.

ARTICLE 13 MISCELLANEOUS PROVISIONS
§ 13.1 GOVERNING LAW
The Contract shall be governed by the law of the place where the Project is located except that, if the parties have selected arbitration as the method of binding dispute resolution, the Federal Arbitration Act shall govern Section 15.4.

§ 13.2 SUCCESSORS AND ASSIGNS
§ 13.2.1 The Owner and Contractor respectively bind themselves, their partners, successors, assigns and legal representatives to covenants, agreements and obligations contained in the Contract Documents. Except as provided in Section 13.2.2, neither party to the Contract shall assign the Contract as a whole without written consent of the other. If either party attempts to make such an assignment without such consent, that party shall nevertheless remain legally responsible for all obligations under the Contract.

§ 13.2.2 The Owner may, without consent of the Contractor, assign the Contract to a lender providing construction financing for the Project, if the lender assumes the Owner's rights and obligations under the Contract Documents. The Contractor shall execute all consents reasonably required to facilitate such assignment.

§ 13.3 WRITTEN NOTICE
Written notice shall be deemed to have been duly served if delivered in person to the individual, to a member of the firm or entity, or to an officer of the corporation for which it was intended, or if delivered at, or sent by registered or certified mail or by courier service providing proof of delivery to, the last business address known to the party giving notice.

§ 13.4 RIGHTS AND REMEDIES
§ 13.4.1 Duties and obligations imposed by the Contract Documents and rights and remedies available thereunder shall be in addition to and not a limitation of duties, obligations, rights and remedies otherwise imposed or available by law.

§ 13.4.2 No action or failure to act by the Owner, Architect or Contractor shall constitute a waiver of a right or duty afforded them under the Contract, nor shall such action or failure to act constitute approval of or acquiescence in a breach there under, except as may be specifically agreed in writing.

§ 13.5 TESTS AND INSPECTIONS
§ 13.5.1 Tests, inspections and approvals of portions of the Work shall be made as required by the Contract Documents and by applicable laws, statutes, ordinances, codes, rules and regulations or lawful orders of public authorities. Unless otherwise provided, the Contractor shall make arrangements for such tests, inspections and approvals with an independent testing laboratory or entity acceptable to the Owner, or with the appropriate public authority, and shall bear all related costs of tests, inspections and approvals. The Contractor shall give the Architect timely notice of when and where tests and inspections are to be made so that the Architect may be present for such procedures. The Owner shall bear costs of (1) tests, inspections or approvals that do not become requirements until after bids are received or negotiations concluded, and (2) tests, inspections or approvals where building codes or applicable laws or regulations prohibit the Owner from delegating their cost to the Contractor.

§ 13.5.2 If the Architect, Owner or public authorities having jurisdiction determine that portions of the Work require additional testing, inspection or approval not included under Section 13.5.1, the Architect will, upon written authorization from the Owner, instruct the Contractor to make arrangements for such additional testing, inspection or approval by an entity acceptable to the Owner, and the Contractor shall give timely notice to the Architect of when and where tests and inspections are to be made so that the Architect may be present for such procedures. Such costs, except as provided in Section 13.5.3, shall be at the Owner's expense.

§ 13.5.3 If such procedures for testing, inspection or approval under Sections 13.5.1 and 13.5.2 reveal failure of the portions of the Work to comply with requirements established by the Contract Documents, all costs made necessary by such failure including those of repeated procedures and compensation for the Architect's services and expenses shall be at the Contractor's expense.

§ 13.5.4 Required certificates of testing, inspection or approval shall, unless otherwise required by the Contract Documents, be secured by the Contractor and promptly delivered to the Architect.

§ 13.5.5 If the Architect is to observe tests, inspections or approvals required by the Contract Documents, the Architect will do so promptly and, where practicable, at the normal place of testing.

§ 13.5.6 Tests or inspections conducted pursuant to the Contract Documents shall be made promptly to avoid unreasonable delay in the Work.

§ 13.6 INTEREST

Payments due and unpaid under the Contract Documents shall bear interest from the date payment is due at such rate as the parties may agree upon in writing or, in the absence thereof, at the legal rate prevailing from time to time at the place where the Project is located.

§ 13.7 TIME LIMITS ON CLAIMS

The Owner and Contractor shall commence all claims and causes of action, whether in contract, tort, breach of warranty or otherwise, against the other arising out of or related to the Contract in accordance with the requirements of the final dispute resolution method selected in the Agreement within the time period specified by applicable law, but in any case not more than 10 years after the date of Substantial Completion of the Work. The Owner and Contractor waive all claims and causes of action not commenced in accordance with this Section 13.7.

ARTICLE 14 TERMINATION OR SUSPENSION OF THE CONTRACT
§ 14.1 TERMINATION BY THE CONTRACTOR

§ 14.1.1 The Contractor may terminate the Contract if the Work is stopped for a period of 30 consecutive days through no act or fault of the Contractor or a Subcontractor, Sub-subcontractor or their agents or employees or any other persons or entities performing portions of the Work under direct or indirect contract with the Contractor, for any of the following reasons:

    .1   Issuance of an order of a court or other public authority having jurisdiction that requires all Work to be stopped;

    .2   An act of government, such as a declaration of national emergency that requires all Work to be stopped;

    .3   Because the Architect has not issued a Certificate for Payment and has not notified the Contractor of the reason for withholding certification as provided in Section 9.4.1, or because the Owner has not made payment on a Certificate for Payment within the time stated in the Contract Documents; or

    .4   The Owner has failed to furnish to the Contractor promptly, upon the Contractor's request, reasonable evidence as required by Section 2.2.1.

§ 14.1.2 The Contractor may terminate the Contract if, through no act or fault of the Contractor or a Subcontractor, Sub-subcontractor or their agents or employees or any other persons or entities performing portions of the Work under direct or indirect contract with the Contractor, repeated suspensions, delays or interruptions of the entire Work by the Owner as described in Section 14.3 constitute in the aggregate more than 100 percent of the total number of days scheduled for completion, or 120 days in any 365-day period, whichever is less.

§ 14.1.3 If one of the reasons described in Section 14.1.1 or 14.1.2 exists, the Contractor may, upon seven days' written notice to the Owner and Architect, terminate the Contract and recover from the Owner payment for Work executed, including reasonable overhead and profit, costs incurred by reason of such termination, and damages.

§ 14.1.4 If the Work is stopped for a period of 60 consecutive days through no act or fault of the Contractor or a Subcontractor or their agents or employees or any other persons performing portions of the Work under contract with the Contractor because the Owner has repeatedly failed to fulfill the Owner's obligations under the Contract Documents with respect to matters important to the progress of the Work, the Contractor may, upon seven additional days' written notice to the Owner and the Architect, terminate the Contract and recover from the Owner as provided in Section 14.1.3.

§ 14.2 TERMINATION BY THE OWNER FOR CAUSE

§ 14.2.1 The Owner may terminate the Contract if the Contractor

    .1   repeatedly refuses or fails to supply enough properly skilled workers or proper materials;

    .2   fails to make payment to Subcontractors for materials or labor in accordance with the respective agreements between the Contractor and the Subcontractors;

    .3   repeatedly disregards applicable laws, statutes, ordinances, codes, rules and regulations, or lawful orders of a public authority; or

    .4   otherwise is guilty of substantial breach of a provision of the Contract Documents.

§ 14.2.2 When any of the above reasons exist, the Owner, upon certification by the Initial Decision Maker that sufficient cause exists to justify such action, may without prejudice to any other rights or remedies of the Owner and after giving the Contractor and the Contractor's surety, if any, seven days' written notice, terminate employment of the Contractor and may, subject to any prior rights of the surety:

  .1 Exclude the Contractor from the site and take possession of all materials, equipment, tools, and construction equipment and machinery thereon owned by the Contractor;

  .2 Accept assignment of subcontracts pursuant to Section 5.4; and

  .3 Finish the Work by whatever reasonable method the Owner may deem expedient. Upon written request of the Contractor, the Owner shall furnish to the Contractor a detailed accounting of the costs incurred by the Owner in finishing the Work.

§ 14.2.3 When the Owner terminates the Contract for one of the reasons stated in Section 14.2.1, the Contractor shall not be entitled to receive further payment until the Work is finished.

§ 14.2.4 If the unpaid balance of the Contract Sum exceeds costs of finishing the Work, including compensation for the Architect's services and expenses made necessary thereby, and other damages incurred by the Owner and not expressly waived, such excess shall be paid to the Contractor. If such costs and damages exceed the unpaid balance, the Contractor shall pay the difference to the Owner. The amount to be paid to the Contractor or Owner, as the case may be, shall be certified by the Initial Decision Maker, upon application, and this obligation for payment shall survive termination of the Contract.

§ 14.3 SUSPENSION BY THE OWNER FOR CONVENIENCE

§ 14.3.1 The Owner may, without cause, order the Contractor in writing to suspend, delay or interrupt the Work in whole or in part for such period of time as the Owner may determine.

§ 14.3.2 The Contract Sum and Contract Time shall be adjusted for increases in the cost and time caused by suspension, delay or interruption as described in Section 14.3.1. Adjustment of the Contract Sum shall include profit. No adjustment shall be made to the extent

  .1 that performance is, was or would have been so suspended, delayed or interrupted by another cause for which the Contractor is responsible; or

  .2 that an equitable adjustment is made or denied under another provision of the Contract.

§ 14.4 TERMINATION BY THE OWNER FOR CONVENIENCE

§ 14.4.1 The Owner may, at any time, terminate the Contract for the Owner's convenience and without cause.

§ 14.4.2 Upon receipt of written notice from the Owner of such termination for the Owner's convenience, the Contractor shall

  .1 cease operations as directed by the Owner in the notice;

  .2 take actions necessary, or that the Owner may direct, for the protection and preservation of the Work; and

  .3 except for Work directed to be performed prior to the effective date of termination stated in the notice, terminate all existing subcontracts and purchase orders and enter into no further subcontracts and purchase orders.

§ 14.4.3 In case of such termination for the Owner's convenience, the Contractor shall be entitled to receive payment for Work executed, and costs incurred by reason of such termination, along with reasonable overhead and profit on the Work not executed.

ARTICLE 15 CLAIMS AND DISPUTES
§ 15.1 CLAIMS
§ 15.1.1 DEFINITION
A Claim is a demand or assertion by one of the parties seeking, as a matter of right, payment of money, or other relief with respect to the terms of the Contract. The term "Claim" also includes other disputes and matters in question between the Owner and Contractor arising out of or relating to the Contract. The responsibility to substantiate Claims shall rest with the party making the Claim.

§ 15.1.2 NOTICE OF CLAIMS
Claims by either the Owner or Contractor must be initiated by written notice to the other party and to the Initial Decision Maker with a copy sent to the Architect, if the Architect is not serving as the Initial Decision Maker.

Claims by either party must be initiated within 21 days after occurrence of the event giving rise to such Claim or within 21 days after the claimant first recognizes the condition giving rise to the Claim, whichever is later.

§ 15.1.3 CONTINUING CONTRACT PERFORMANCE

Pending final resolution of a Claim, except as otherwise agreed in writing or as provided in Section 9.7 and Article 14, the Contractor shall proceed diligently with performance of the Contract and the Owner shall continue to make payments in accordance with the Contract Documents. The Architect will prepare Change Orders and issue Certificates for Payment in accordance with the decisions of the Initial Decision Maker.

§ 15.1.4 CLAIMS FOR ADDITIONAL COST

If the Contractor wishes to make a Claim for an increase in the Contract Sum, written notice as provided herein shall be given before proceeding to execute the Work. Prior notice is not required for Claims relating to an emergency endangering life or property arising under Section 10.4.

§ 15.1.5 CLAIMS FOR ADDITIONAL TIME

§ 15.1.5.1 If the Contractor wishes to make a Claim for an increase in the Contract Time, written notice as provided herein shall be given. The Contractor's Claim shall include an estimate of cost and of probable effect of delay on progress of the Work. In the case of a continuing delay, only one Claim is necessary.

§ 15.1.5.2 If adverse weather conditions are the basis for a Claim for additional time, such Claim shall be documented by data substantiating that weather conditions were abnormal for the period of time, could not have been reasonably anticipated and had an adverse effect on the scheduled construction.

§ 15.1.6 CLAIMS FOR CONSEQUENTIAL DAMAGES

The Contractor and Owner waive Claims against each other for consequential damages arising out of or relating to this Contract. This mutual waiver includes

- .1 damages incurred by the Owner for rental expenses, for losses of use, income, profit, financing, business and reputation, and for loss of management or employee productivity or of the services of such persons; and
- .2 damages incurred by the Contractor for principal office expenses including the compensation of personnel stationed there, for losses of financing, business and reputation, and for loss of profit except anticipated profit arising directly from the Work.

This mutual waiver is applicable, without limitation, to all consequential damages due to either party's termination in accordance with Article 14. Nothing contained in this Section 15.1.6 shall be deemed to preclude an award of liquidated damages, when applicable, in accordance with the requirements of the Contract Documents.

§ 15.2 INITIAL DECISION

§ 15.2.1 Claims, excluding those arising under Sections 10.3, 10.4, 11.3.9, and 11.3.10, shall be referred to the Initial Decision Maker for initial decision. The Architect will serve as the Initial Decision Maker, unless otherwise indicated in the Agreement. Except for those Claims excluded by this Section 15.2.1, an initial decision shall be required as a condition precedent to mediation of any Claim arising prior to the date final payment is due, unless 30 days have passed after the Claim has been referred to the Initial Decision Maker with no decision having been rendered. Unless the Initial Decision Maker and all affected parties agree, the Initial Decision Maker will not decide disputes between the Contractor and persons or entities other than the Owner.

§ 15.2.2 The Initial Decision Maker will review Claims and within ten days of the receipt of a Claim take one or more of the following actions: (1) request additional supporting data from the claimant or a response with supporting data from the other party, (2) reject the Claim in whole or in part, (3) approve the Claim, (4) suggest a compromise, or (5) advise the parties that the Initial Decision Maker is unable to resolve the Claim if the Initial Decision Maker lacks sufficient information to evaluate the merits of the Claim or if the Initial Decision Maker concludes that, in the Initial Decision Maker's sole discretion, it would be inappropriate for the Initial Decision Maker to resolve the Claim.

§ 15.2.3 In evaluating Claims, the Initial Decision Maker may, but shall not be obligated to, consult with or seek information from either party or from persons with special knowledge or expertise who may assist the Initial Decision Maker in rendering a decision. The Initial Decision Maker may request the Owner to authorize retention of such persons at the Owner's expense.

§ 15.2.4 If the Initial Decision Maker requests a party to provide a response to a Claim or to furnish additional supporting data, such party shall respond, within ten days after receipt of such request, and shall either (1) provide a response on the requested supporting data, (2) advise the Initial Decision Maker when the response or supporting data will be furnished or (3) advise the Initial Decision Maker that no supporting data will be furnished. Upon receipt of the response or supporting data, if any, the Initial Decision Maker will either reject or approve the Claim in whole or in part.

§ 15.2.5 The Initial Decision Maker will render an initial decision approving or rejecting the Claim, or indicating that the Initial Decision Maker is unable to resolve the Claim. This initial decision shall (1) be in writing; (2) state the reasons therefor; and (3) notify the parties and the Architect, if the Architect is not serving as the Initial Decision Maker, of any change in the Contract Sum or Contract Time or both. The initial decision shall be final and binding on the parties but subject to mediation and, if the parties fail to resolve their dispute through mediation, to binding dispute resolution.

§ 15.2.6 Either party may file for mediation of an initial decision at any time, subject to the terms of Section 15.2.6.1.

§ 15.2.6.1 Either party may, within 30 days from the date of an initial decision, demand in writing that the other party file for mediation within 60 days of the initial decision. If such a demand is made and the party receiving the demand fails to file for mediation within the time required, then both parties waive their rights to mediate or pursue binding dispute resolution proceedings with respect to the initial decision.

§ 15.2.7 In the event of a Claim against the Contractor, the Owner may, but is not obligated to, notify the surety, if any, of the nature and amount of the Claim. If the Claim relates to a possibility of a Contractor's default, the Owner may, but is not obligated to, notify the surety and request the surety's assistance in resolving the controversy.

§ 15.2.8 If a Claim relates to or is the subject of a mechanic's lien, the party asserting such Claim may proceed in accordance with applicable law to comply with the lien notice or filing deadlines.

§ 15.3 MEDIATION

§ 15.3.1 Claims, disputes, or other matters in controversy arising out of or related to the Contract except those waived as provided for in Sections 9.10.4, 9.10.5, and 15.1.6 shall be subject to mediation as a condition precedent to binding dispute resolution.

§ 15.3.2 The parties shall endeavor to resolve their Claims by mediation which, unless the parties mutually agree otherwise, shall be administered by the American Arbitration Association in accordance with its Construction Industry Mediation Procedures in effect on the date of the Agreement. A request for mediation shall be made in writing, delivered to the other party to the Contract, and filed with the person or entity administering the mediation. The request may be made concurrently with the filing of binding dispute resolution proceedings but, in such event, mediation shall proceed in advance of binding dispute resolution proceedings, which shall be stayed pending mediation for a period of 60 days from the date of filing, unless stayed for a longer period by agreement of the parties or court order. If an arbitration is stayed pursuant to this Section 15.3.2, the parties may nonetheless proceed to the selection of the arbitrator(s) and agree upon a schedule for later proceedings.

§ 15.3.3 The parties shall share the mediator's fee and any filing fees equally. The mediation shall be held in the place where the Project is located, unless another location is mutually agreed upon. Agreements reached in mediation shall be enforceable as settlement agreements in any court having jurisdiction thereof.

§ 15.4 ARBITRATION

§ 15.4.1 If the parties have selected arbitration as the method for binding dispute resolution in the Agreement, any Claim subject to, but not resolved by, mediation shall be subject to arbitration which, unless the parties mutually agree otherwise, shall be administered by the American Arbitration Association in accordance with its Construction Industry Arbitration Rules in effect on the date of the Agreement. A demand for arbitration shall be made in writing, delivered to the other party to the Contract, and filed with the person or entity administering the arbitration. The party filing a notice of demand for arbitration must assert in the demand all Claims then known to that party on which arbitration is permitted to be demanded.

§ 15.4.1.1 A demand for arbitration shall be made no earlier than concurrently with the filing of a request for mediation, but in no event shall it be made after the date when the institution of legal or equitable proceedings based on the Claim would be barred by the applicable statute of limitations. For statute of limitations purposes, receipt of a written demand for arbitration by the person or entity administering the arbitration shall constitute the institution of legal or equitable proceedings based on the Claim.

§ 15.4.2 The award rendered by the arbitrator or arbitrators shall be final, and judgment may be entered upon it in accordance with applicable law in any court having jurisdiction thereof.

§ 15.4.3 The foregoing agreement to arbitrate and other agreements to arbitrate with an additional person or entity duly consented to by parties to the Agreement shall be specifically enforceable under applicable law in any court having jurisdiction thereof.

§ 15.4.4 CONSOLIDATION OR JOINDER
§ 15.4.4.1 Either party, at its sole discretion, may consolidate an arbitration conducted under this Agreement with any other arbitration to which it is a party provided that (1) the arbitration agreement governing the other arbitration permits consolidation, (2) the arbitrations to be consolidated substantially involve common questions of law or fact, and (3) the arbitrations employ materially similar procedural rules and methods for selecting arbitrator(s).

§ 15.4.4.2 Either party, at its sole discretion, may include by joinder persons or entities substantially involved in a common question of law or fact whose presence is required if complete relief is to be accorded in arbitration, provided that the party sought to be joined consents in writing to such joinder. Consent to arbitration involving an additional person or entity shall not constitute consent to arbitration of any claim, dispute or other matter in question not described in the written consent.

§ 15.4.4.3 The Owner and Contractor grant to any person or entity made a party to an arbitration conducted under this Section 15.4, whether by joinder or consolidation, the same rights of joinder and consolidation as the Owner and Contractor under this Agreement.

▓**AIA**® Document B101™ – 2007

Standard Form of Agreement Between Owner and Architect

AGREEMENT made as of the   day of
in the year of
(In words, indicate day, month and year)

BETWEEN the Architect's client identified as the Owner:
(Name, address and other information)

This document has important legal consequences. Consultation with an attorney is encouraged with respect to its completion or modification.

and the Architect:
(Name, address and other information)

for the following Project:
(Name, location and detailed description)

The Owner and Architect agree as follows.

TABLE OF ARTICLES

EXHIBIT A INITIAL INFORMATION

ARTICLE 1 INITIAL INFORMATION

§ 1.1 This Agreement is based on the Initial Information set forth in this Article 1 and in optional Exhibit A, Initial Information:

(*Complete Exhibit A, Initial Information, and incorporate it into the Agreement at Section 13.2, or state below Initial Information such as details of the Project's site and program, Owner's contractors and consultants, Architect's consultants, Owner's budget for the Cost of the Work, authorized representatives, anticipated procurement method, and other information relevant to the Project.*)

§ 1.2 The Owner's anticipated dates for commencement of construction and Substantial Completion of the Work are set forth below:

    .1   Commencement of construction date:

    .2   Substantial Completion date:

§ 1.3 The Owner and Architect may rely on the Initial Information. Both parties, however, recognize that such information may materially change and, in that event, the Owner and the Architect shall appropriately adjust the schedule, the Architect's services and the Architect's compensation.

ARTICLE 2 ARCHITECT'S RESPONSIBILITIES

§ 2.1 The Architect shall provide the professional services as set forth in this Agreement.

§ 2.2 The Architect shall perform its services consistent with the professional skill and care ordinarily provided by architects practicing in the same or similar locality under the same or similar circumstances. The Architect shall perform its services as expeditiously as is consistent with such professional skill and care and the orderly progress of the Project.

§ 2.3 The Architect shall identify a representative authorized to act on behalf of the Architect with respect to the Project.

§ 2.4 Except with the Owner's knowledge and consent, the Architect shall not engage in any activity, or accept any employment, interest or contribution that would reasonably appear to compromise the Architect's professional judgment with respect to this Project.

§ 2.5 The Architect shall maintain the following insurance for the duration of this Agreement. If any of the requirements set forth below exceed the types and limits the Architect normally maintains, the Owner shall reimburse the Architect for any additional cost:
(Identify types and limits of insurance coverage, and other insurance requirements applicable to the Agreement, if any.)

.1 General Liability

.2 Automobile Liability

.3 Workers' Compensation

.4 Professional Liability

ARTICLE 3 SCOPE OF ARCHITECT'S BASIC SERVICES

§ 3.1 The Architect's Basic Services consist of those described in Article 3 and include usual and customary structural, mechanical, and electrical engineering services. Services not set forth in Article 3 are Additional Services.

§ 3.1.1 The Architect shall manage the Architect's services, consult with the Owner, research applicable design criteria, attend Project meetings, communicate with members of the Project team and report progress to the Owner.

§ 3.1.2 The Architect shall coordinate its services with those services provided by the Owner and the Owner's consultants. The Architect shall be entitled to rely on the accuracy and completeness of services and information furnished by the Owner and the Owner's consultants. The Architect shall provide prompt written notice to the Owner if the Architect becomes aware of any error, omission or inconsistency in such services or information.

§ 3.1.3 As soon as practicable after the date of this Agreement, the Architect shall submit for the Owner's approval a schedule for the performance of the Architect's services. The schedule initially shall include anticipated dates for the commencement of construction and for Substantial Completion of the Work as set forth in the Initial Information. The schedule shall include allowances for periods of time required for the Owner's review, for the performance of the Owner's consultants, and for approval of submissions by authorities having jurisdiction over the Project. Once approved by the Owner, time limits established by the schedule shall not, except for reasonable cause, be exceeded by the Architect or Owner. With the Owner's approval, the Architect shall adjust the schedule, if necessary as the Project proceeds until the commencement of construction.

§ 3.1.4 The Architect shall not be responsible for an Owner's directive or substitution made without the Architect's approval.

§ 3.1.5 The Architect shall, at appropriate times, contact the governmental authorities required to approve the Construction Documents and the entities providing utility services to the Project. In designing the Project, the Architect shall respond to applicable design requirements imposed by such governmental authorities and by such entities providing utility services.

§ 3.1.6 The Architect shall assist the Owner in connection with the Owner's responsibility for filing documents required for the approval of governmental authorities having jurisdiction over the Project.

§ 3.2 SCHEMATIC DESIGN PHASE SERVICES

§ 3.2.1 The Architect shall review the program and other information furnished by the Owner, and shall review laws, codes, and regulations applicable to the Architect's services.

§ 3.2.2 The Architect shall prepare a preliminary evaluation of the Owner's program, schedule, budget for the Cost of the Work, Project site, and the proposed procurement or delivery method and other Initial Information, each in terms of the other, to ascertain the requirements of the Project. The Architect shall notify the Owner of (1) any inconsistencies discovered in the information, and (2) other information or consulting services that may be reasonably needed for the Project.

§ 3.2.3 The Architect shall present its preliminary evaluation to the Owner and shall discuss with the Owner alternative approaches to design and construction of the Project, including the feasibility of incorporating environmentally responsible design approaches. The Architect shall reach an understanding with the Owner regarding the requirements of the Project.

§ 3.2.4 Based on the Project's requirements agreed upon with the Owner, the Architect shall prepare and present for the Owner's approval a preliminary design illustrating the scale and relationship of the Project components.

§ 3.2.5 Based on the Owner's approval of the preliminary design, the Architect shall prepare Schematic Design Documents for the Owner's approval. The Schematic Design Documents shall consist of drawings and other documents including a site plan, if appropriate, and preliminary building plans, sections and elevations; and may include some combination of study models, perspective sketches, or digital modeling. Preliminary selections of major building systems and construction materials shall be noted on the drawings or described in writing.

§ 3.2.5.1 The Architect shall consider environmentally responsible design alternatives, such as material choices and building orientation, together with other considerations based on program and aesthetics, in developing a design that is consistent with the Owner's program, schedule and budget for the Cost of the Work. The Owner may obtain other environmentally responsible design services under Article 4.

§ 3.2.5.2 The Architect shall consider the value of alternative materials, building systems and equipment, together with other considerations based on program and aesthetics in developing a design for the Project that is consistent with the Owner's program, schedule and budget for the Cost of the Work.

§ 3.2.6 The Architect shall submit to the Owner an estimate of the Cost of the Work prepared in accordance with Section 6.3.

§ 3.2.7 The Architect shall submit the Schematic Design Documents to the Owner, and request the Owner's approval.

§ 3.3 DESIGN DEVELOPMENT PHASE SERVICES

§ 3.3.1 Based on the Owner's approval of the Schematic Design Documents, and on the Owner's authorization of any adjustments in the Project requirements and the budget for the Cost of the Work, the Architect shall prepare Design Development Documents for the Owner's approval. The Design Development Documents shall illustrate and describe the development of the approved Schematic Design Documents and shall consist of drawings and other documents including plans, sections, elevations, typical construction details, and diagrammatic layouts of building systems to fix and describe the size and character of the Project as to architectural, structural, mechanical and electrical systems, and such other elements as may be appropriate. The Design Development Documents shall also include outline specifications that identify major materials and systems and establish in general their quality levels.

§ 3.3.2 The Architect shall update the estimate of the Cost of the Work.

§ 3.3.3 The Architect shall submit the Design Development documents to the Owner, advise the Owner of any adjustments to the estimate of the Cost of the Work, and request the Owner's approval.

§ 3.4 CONSTRUCTION DOCUMENTS PHASE SERVICES

§ 3.4.1 Based on the Owner's approval of the Design Development Documents, and on the Owner's authorization of any adjustments in the Project requirements and the budget for the Cost of the Work, the Architect shall prepare Construction Documents for the Owner's approval. The Construction Documents shall illustrate and describe the further development of the approved Design Development Documents and shall consist of Drawings and

Specifications setting forth in detail the quality levels of materials and systems and other requirements for the construction of the Work. The Owner and Architect acknowledge that in order to construct the Work the Contractor will provide additional information, including Shop Drawings, Product Data, Samples and other similar submittals, which the Architect shall review in accordance with Section 3.6.4.

§ 3.4.2 The Architect shall incorporate into the Construction Documents the design requirements of governmental authorities having jurisdiction over the Project.

§ 3.4.3 During the development of the Construction Documents, the Architect shall assist the Owner in the development and preparation of (1) bidding and procurement information that describes the time, place and conditions of bidding, including bidding or proposal forms; (2) the form of agreement between the Owner and Contractor; and (3) the Conditions of the Contract for Construction (General, Supplementary and other Conditions). The Architect shall also compile a project manual that includes the Conditions of the Contract for Construction and Specifications and may include bidding requirements and sample forms.

§ 3.4.4 The Architect shall update the estimate for the Cost of the Work.

§ 3.4.5 The Architect shall submit the Construction Documents to the Owner, advise the Owner of any adjustments to the estimate of the Cost of the Work, take any action required under Section 6.5, and request the Owner's approval.

§ 3.5 BIDDING OR NEGOTIATION PHASE SERVICES
§ 3.5.1 GENERAL
The Architect shall assist the Owner in establishing a list of prospective contractors. Following the Owner's approval of the Construction Documents, the Architect shall assist the Owner in (1) obtaining either competitive bids or negotiated proposals; (2) confirming responsiveness of bids or proposals; (3) determining the successful bid or proposal, if any; and, (4) awarding and preparing contracts for construction.

§ 3.5.2 COMPETITIVE BIDDING
§ 3.5.2.1 Bidding Documents shall consist of bidding requirements and proposed Contract Documents.

§ 3.5.2.2 The Architect shall assist the Owner in bidding the Project by
    .1   procuring the reproduction of Bidding Documents for distribution to prospective bidders;
    .2   distributing the Bidding Documents to prospective bidders, requesting their return upon completion of the bidding process, and maintaining a log of distribution and retrieval and of the amounts of deposits, if any, received from and returned to prospective bidders;
    .3   organizing and conducting a pre-bid conference for prospective bidders;
    .4   preparing responses to questions from prospective bidders and providing clarifications and interpretations of the Bidding Documents to all prospective bidders in the form of addenda; and
    .5   organizing and conducting the opening of the bids, and subsequently documenting and distributing the bidding results, as directed by the Owner.

§ 3.5.2.3 The Architect shall consider requests for substitutions, if the Bidding Documents permit substitutions, and shall prepare and distribute addenda identifying approved substitutions to all prospective bidders.

§ 3.5.3 NEGOTIATED PROPOSALS
§ 3.5.3.1 Proposal Documents shall consist of proposal requirements and proposed Contract Documents.

§ 3.5.3.2 The Architect shall assist the Owner in obtaining proposals by
    .1   procuring the reproduction of Proposal Documents for distribution to prospective contractors, and requesting their return upon completion of the negotiation process;
    .2   organizing and participating in selection interviews with prospective contractors; and
    .3   participating in negotiations with prospective contractors, and subsequently preparing a summary report of the negotiation results, as directed by the Owner.

§ 3.5.3.3 The Architect shall consider requests for substitutions, if the Proposal Documents permit substitutions, and shall prepare and distribute addenda identifying approved substitutions to all prospective contractors.

§ 3.6 CONSTRUCTION PHASE SERVICES
§ 3.6.1 GENERAL

§ 3.6.1.1 The Architect shall provide administration of the Contract between the Owner and the Contractor as set forth below and in AIA Document A201™–2007, General Conditions of the Contract for Construction. If the Owner and Contractor modify AIA Document A201–2007, those modifications shall not affect the Architect's services under this Agreement unless the Owner and the Architect amend this Agreement.

§ 3.6.1.2 The Architect shall advise and consult with the Owner during the Construction Phase Services. The Architect shall have authority to act on behalf of the Owner only to the extent provided in this Agreement. The Architect shall not have control over, charge of, or responsibility for the construction means, methods, techniques, sequences or procedures, or for safety precautions and programs in connection with the Work, nor shall the Architect be responsible for the Contractor's failure to perform the Work in accordance with the requirements of the Contract Documents. The Architect shall be responsible for the Architect's negligent acts or omissions, but shall not have control over or charge of, and shall not be responsible for, acts or omissions of the Contractor or of any other persons or entities performing portions of the Work.

§ 3.6.1.3 Subject to Section 4.3, the Architect's responsibility to provide Construction Phase Services commences with the award of the Contract for Construction and terminates on the date the Architect issues the final Certificate for Payment.

§ 3.6.2 EVALUATIONS OF THE WORK

§ 3.6.2.1 The Architect shall visit the site at intervals appropriate to the stage of construction, or as otherwise required in Section 4.3.3, to become generally familiar with the progress and quality of the portion of the Work completed, and to determine, in general, if the Work observed is being performed in a manner indicating that the Work, when fully completed, will be in accordance with the Contract Documents. However, the Architect shall not be required to make exhaustive or continuous on-site inspections to check the quality or quantity of the Work. On the basis of the site visits, the Architect shall keep the Owner reasonably informed about the progress and quality of the portion of the Work completed, and report to the Owner (1) known deviations from the Contract Documents and from the most recent construction schedule submitted by the Contractor, and (2) defects and deficiencies observed in the Work.

§ 3.6.2.2 The Architect has the authority to reject Work that does not conform to the Contract Documents. Whenever the Architect considers it necessary or advisable, the Architect shall have the authority to require inspection or testing of the Work in accordance with the provisions of the Contract Documents, whether or not such Work is fabricated, installed or completed. However, neither this authority of the Architect nor a decision made in good faith either to exercise or not to exercise such authority shall give rise to a duty or responsibility of the Architect to the Contractor, Subcontractors, material and equipment suppliers, their agents or employees or other persons or entities performing portions of the Work.

§ 3.6.2.3 The Architect shall interpret and decide matters concerning performance under, and requirements of, the Contract Documents on written request of either the Owner or Contractor. The Architect's response to such requests shall be made in writing within any time limits agreed upon or otherwise with reasonable promptness.

§ 3.6.2.4 Interpretations and decisions of the Architect shall be consistent with the intent of and reasonably inferable from the Contract Documents and shall be in writing or in the form of drawings. When making such interpretations and decisions, the Architect shall endeavor to secure faithful performance by both Owner and Contractor, shall not show partiality to either, and shall not be liable for results of interpretations or decisions rendered in good faith. The Architect's decisions on matters relating to aesthetic effect shall be final if consistent with the intent expressed in the Contract Documents.

§ 3.6.2.5 Unless the Owner and Contractor designate another person to serve as an Initial Decision Maker, as that term is defined in AIA Document A201–2007, the Architect shall render initial decisions on Claims between the Owner and Contractor as provided in the Contract Documents.

§ 3.6.3 CERTIFICATES FOR PAYMENT TO CONTRACTOR

§ 3.6.3.1 The Architect shall review and certify the amounts due the Contractor and shall issue certificates in such amounts. The Architect's certification for payment shall constitute a representation to the Owner, based on the Architect's evaluation of the Work as provided in Section 3.6.2 and on the data comprising the Contractor's Application for Payment, that, to the best of the Architect's knowledge, information and belief, the Work has progressed to the point indicated and that the quality of the Work is in accordance with the Contract Documents. The foregoing representations are subject (1) to an evaluation of the Work for conformance with the Contract Documents upon Substantial Completion, (2) to results of subsequent tests and inspections, (3) to correction of minor deviations from the Contract Documents prior to completion, and (4) to specific qualifications expressed by the Architect.

§ 3.6.3.2 The issuance of a Certificate for Payment shall not be a representation that the Architect has (1) made exhaustive or continuous on-site inspections to check the quality or quantity of the Work, (2) reviewed construction means, methods, techniques, sequences or procedures, (3) reviewed copies of requisitions received from Subcontractors and material suppliers and other data requested by the Owner to substantiate the Contractor's right to payment, or (4) ascertained how or for what purpose the Contractor has used money previously paid on account of the Contract Sum.

§ 3.6.3.3 The Architect shall maintain a record of the Applications and Certificates for Payment.

§ 3.6.4 SUBMITTALS

§ 3.6.4.1 The Architect shall review the Contractor's submittal schedule and shall not unreasonably delay or withhold approval. The Architect's action in reviewing submittals shall be taken in accordance with the approved submittal schedule or, in the absence of an approved submittal schedule, with reasonable promptness while allowing sufficient time in the Architect's professional judgment to permit adequate review.

§ 3.6.4.2 In accordance with the Architect-approved submittal schedule, the Architect shall review and approve or take other appropriate action upon the Contractor's submittals such as Shop Drawings, Product Data and Samples, but only for the limited purpose of checking for conformance with information given and the design concept expressed in the Contract Documents. Review of such submittals is not for the purpose of determining the accuracy and completeness of other information such as dimensions, quantities, and installation or performance of equipment or systems, which are the Contractor's responsibility. The Architect's review shall not constitute approval of safety precautions or, unless otherwise specifically stated by the Architect, of any construction means, methods, techniques, sequences or procedures. The Architect's approval of a specific item shall not indicate approval of an assembly of which the item is a component.

§ 3.6.4.3 If the Contract Documents specifically require the Contractor to provide professional design services or certifications by a design professional related to systems, materials or equipment, the Architect shall specify the appropriate performance and design criteria that such services must satisfy. The Architect shall review shop drawings and other submittals related to the Work designed or certified by the design professional retained by the Contractor that bear such professional's seal and signature when submitted to the Architect. The Architect shall be entitled to rely upon the adequacy, accuracy and completeness of the services, certifications and approvals performed or provided by such design professionals.

§ 3.6.4.4 Subject to the provisions of Section 4.3, the Architect shall review and respond to requests for information about the Contract Documents. The Architect shall set forth in the Contract Documents the requirements for requests for information. Requests for information shall include, at a minimum, a detailed written statement that indicates the specific Drawings or Specifications in need of clarification and the nature of the clarification requested. The Architect's response to such requests shall be made in writing within any time limits agreed upon, or otherwise with reasonable promptness. If appropriate, the Architect shall prepare and issue supplemental Drawings and Specifications in response to requests for information.

§ 3.6.4.5 The Architect shall maintain a record of submittals and copies of submittals supplied by the Contractor in accordance with the requirements of the Contract Documents.

§ 3.6.5 CHANGES IN THE WORK

§ 3.6.5.1 The Architect may authorize minor changes in the Work that are consistent with the intent of the Contract Documents and do not involve an adjustment in the Contract Sum or an extension of the Contract Time. Subject to the provisions of Section 4.3, the Architect shall prepare Change Orders and Construction Change Directives for the Owner's approval and execution in accordance with the Contract Documents.

§ 3.6.5.2 The Architect shall maintain records relative to changes in the Work.

§ 3.6.6 PROJECT COMPLETION

§ 3.6.6.1 The Architect shall conduct inspections to determine the date or dates of Substantial Completion and the date of final completion; issue Certificates of Substantial Completion; receive from the Contractor and forward to the Owner, for the Owner's review and records, written warranties and related documents required by the Contract Documents and assembled by the Contractor; and issue a final Certificate for Payment based upon a final inspection indicating the Work complies with the requirements of the Contract Documents.

§ 3.6.6.2 The Architect's inspections shall be conducted with the Owner to check conformance of the Work with the requirements of the Contract Documents and to verify the accuracy and completeness of the list submitted by the Contractor of Work to be completed or corrected.

§ 3.6.6.3 When the Work is found to be substantially complete, the Architect shall inform the Owner about the balance of the Contract Sum remaining to be paid the Contractor, including the amount to be retained from the Contract Sum, if any, for final completion or correction of the Work.

§ 3.6.6.4 The Architect shall forward to the Owner the following information received from the Contractor: (1) consent of surety or sureties, if any, to reduction in or partial release of retainage or the making of final payment; (2) affidavits, receipts, releases and waivers of liens or bonds indemnifying the Owner against liens; and (3) any other documentation required of the Contractor under the Contract Documents.

§ 3.6.6.5 Upon request of the Owner, and prior to the expiration of one year from the date of Substantial Completion, the Architect shall, without additional compensation, conduct a meeting with the Owner to review the facility operations and performance.

ARTICLE 4   ADDITIONAL SERVICES

§ 4.1 Additional Services listed below are not included in Basic Services but may be required for the Project. The Architect shall provide the listed Additional Services only if specifically designated in the table below as the Architect's responsibility, and the Owner shall compensate the Architect as provided in Section 11.2.
(Designate the Additional Services the Architect shall provide in the second column of the table below. In the third column indicate whether the service description is located in Section 4.2 or in an attached exhibit. If in an exhibit, identify the exhibit.)

Additional Services		Responsibility *(Architect, Owner or Not Provided)*	Location of Service Description *(Section 4.2 below or in an exhibit attached to this document and identified below)*
§ 4.1.1	Programming		
§ 4.1.2	Multiple preliminary designs		
§ 4.1.3	Measured drawings		
§ 4.1.4	Existing facilities surveys		
§ 4.1.5	Site Evaluation and Planning (B203™–2007)		
§ 4.1.6	Building information modeling		
§ 4.1.7	Civil engineering		
§ 4.1.8	Landscape design		
§ 4.1.9	Architectural Interior Design (B252™–2007)		
§ 4.1.10	Value Analysis (B204™–2007)		
§ 4.1.11	Detailed cost estimating		
§ 4.1.12	On-site project representation		
§ 4.1.13	Conformed construction documents		
§ 4.1.14	As-designed record drawings		
§ 4.1.15	As-constructed record drawings		
§ 4.1.16	Post occupancy evaluation		
§ 4.1.17	Facility Support Services (B210™–2007)		
§ 4.1.18	Tenant-related services		
§ 4.1.19	Coordination of Owner's consultants		
§ 4.1.20	Telecommunications/data design		

Additional Services	Responsibility *(Architect, Owner or Not Provided)*	Location of Service Description *(Section 4.2 below or in an exhibit attached to this document and identified below)*
§ 4.1.21 Security Evaluation and Planning (B206™–2007)		
§ 4.1.22 Commissioning (B211™–2007)		
§ 4.1.23 Extensive environmentally responsible design		
§ 4.1.24 LEED® Certification (B214™–2007)		
§ 4.1.25 Fast-track design services		
§ 4.1.26 Historic Preservation (B205™–2007)		
§ 4.1.27 Furniture, Finishings, and Equipment Design (B253™–2007)		
§ 4.1.28 Other		

§ 4.2 Insert a description of each Additional Service designated in Section 4.1 as the Architect's responsibility, if not further described in an exhibit attached to this document.

§ 4.3 Additional Services may be provided after execution of this Agreement, without invalidating the Agreement. Except for services required due to the fault of the Architect, any Additional Services provided in accordance with this Section 4.3 shall entitle the Architect to compensation pursuant to Section 11.3 and an appropriate adjustment in the Architect's schedule.

§ 4.3.1 Upon recognizing the need to perform the following Additional Services, the Architect shall notify the Owner with reasonable promptness and explain the facts and circumstances giving rise to the need. The Architect shall not proceed to provide the following services until the Architect receives the Owner's written authorization:

   .1   Services necessitated by a change in the Initial Information, previous instructions or approvals given by the Owner, or a material change in the Project including, but not limited to, size, quality, complexity, the Owner's schedule or budget for Cost of the Work, or procurement or delivery method;
   .2   Services necessitated by the Owner's request for extensive environmentally responsible design alternatives, such as unique system designs, in-depth material research, energy modeling, or LEED® certification;
   .3   Changing or editing previously prepared Instruments of Service necessitated by the enactment or revision of codes, laws or regulations or official interpretations;
   .4   Services necessitated by decisions of the Owner not rendered in a timely manner or any other failure of performance on the part of the Owner or the Owner's consultants or contractors;
   .5   Preparing digital data for transmission to the Owner's consultants and contractors, or to other Owner authorized recipients;
   .6   Preparation of design and documentation for alternate bid or proposal requests proposed by the Owner;
   .7   Preparation for, and attendance at, a public presentation, meeting or hearing;
   .8   Preparation for, and attendance at a dispute resolution proceeding or legal proceeding, except where the Architect is party thereto;
   .9   Evaluation of the qualifications of bidders or persons providing proposals;
   .10 Consultation concerning replacement of Work resulting from fire or other cause during construction; or
   .11 Assistance to the Initial Decision Maker, if other than the Architect.

§ 4.3.2 To avoid delay in the Construction Phase, the Architect shall provide the following Additional Services, notify the Owner with reasonable promptness, and explain the facts and circumstances giving rise to the need. If the Owner subsequently determines that all or parts of those services are not required, the Owner shall give prompt written notice to the Architect, and the Owner shall have no further obligation to compensate the Architect for those services:

.1 Reviewing a Contractor's submittal out of sequence from the submittal schedule agreed to by the Architect;

.2 Responding to the Contractor's requests for information that are not prepared in accordance with the Contract Documents or where such information is available to the Contractor from a careful study and comparison of the Contract Documents, field conditions, other Owner-provided information, Contractor-prepared coordination drawings, or prior Project correspondence or documentation;

.3 Preparing Change Orders and Construction Change Directives that require evaluation of Contractor's proposals and supporting data, or the preparation or revision of Instruments of Service;

.4 Evaluating an extensive number of Claims as the Initial Decision Maker;

.5 Evaluating substitutions proposed by the Owner or Contractor and making subsequent revisions to Instruments of Service resulting therefrom; or

.6 To the extent the Architect's Basic Services are affected, providing Construction Phase Services 60 days after (1) the date of Substantial Completion of the Work or (2) the anticipated date of Substantial Completion identified in Initial Information, whichever is earlier.

§ 4.3.3 The Architect shall provide Construction Phase Services exceeding the limits set forth below as Additional Services. When the limits below are reached, the Architect shall notify the Owner:

.1 ( ) reviews of each Shop Drawing, Product Data item, sample and similar submittal of the Contractor

.2 ( ) visits to the site by the Architect over the duration of the Project during construction

.3 ( ) inspections for any portion of the Work to determine whether such portion of the Work is substantially complete in accordance with the requirements of the Contract Documents

.4 ( ) inspections for any portion of the Work to determine final completion

§ 4.3.4 If the services covered by this Agreement have not been completed within ( ) months of the date of this Agreement, through no fault of the Architect, extension of the Architect's services beyond that time shall be compensated as Additional Services.

ARTICLE 5 OWNER'S RESPONSIBILITIES

§ 5.1 Unless otherwise provided for under this Agreement, the Owner shall provide information in a timely manner regarding requirements for and limitations on the Project, including a written program which shall set forth the Owner's objectives, schedule, constraints and criteria, including space requirements and relationships, flexibility, expandability, special equipment, systems and site requirements. Within 15 days after receipt of a written request from the Architect, the Owner shall furnish the requested information as necessary and relevant for the Architect to evaluate, give notice of or enforce lien rights.

§ 5.2 The Owner shall establish and periodically update the Owner's budget for the Project, including (1) the budget for the Cost of the Work as defined in Section 6.1; (2) the Owner's other costs; and, (3) reasonable contingencies related to all of these costs. If the Owner significantly increases or decreases the Owner's budget for the Cost of the Work, the Owner shall notify the Architect. The Owner and the Architect shall thereafter agree to a corresponding change in the Project's scope and quality.

§ 5.3 The Owner shall identify a representative authorized to act on the Owner's behalf with respect to the Project. The Owner shall render decisions and approve the Architect's submittals in a timely manner in order to avoid unreasonable delay in the orderly and sequential progress of the Architect's services.

§ 5.4 The Owner shall furnish surveys to describe physical characteristics, legal limitations and utility locations for the site of the Project, and a written legal description of the site. The surveys and legal information shall include, as applicable, grades and lines of streets, alleys, pavements and adjoining property and structures; designated wetlands; adjacent drainage; rights-of-way, restrictions, easements, encroachments, zoning, deed restrictions, boundaries and contours of the site; locations, dimensions and necessary data with respect to existing buildings, other improvements and trees; and information concerning available utility services and lines, both public and private, above and below grade, including inverts and depths. All the information on the survey shall be referenced to a Project benchmark.

§ 5.5 The Owner shall furnish services of geotechnical engineers, which may include but are not limited to test borings, test pits, determinations of soil bearing values, percolation tests, evaluations of hazardous materials, seismic evaluation, ground corrosion tests and resistivity tests, including necessary operations for anticipating subsoil conditions, with written reports and appropriate recommendations.

§ 5.6 The Owner shall coordinate the services of its own consultants with those services provided by the Architect. Upon the Architect's request, the Owner shall furnish copies of the scope of services in the contracts between the Owner and the Owner's consultants. The Owner shall furnish the services of consultants other than those designated in this Agreement, or authorize the Architect to furnish them as an Additional Service, when the Architect requests such services and demonstrates that they are reasonably required by the scope of the Project. The Owner shall require that its consultants maintain professional liability insurance as appropriate to the services provided.

§ 5.7 The Owner shall furnish tests, inspections and reports required by law or the Contract Documents, such as structural, mechanical, and chemical tests, tests for air and water pollution, and tests for hazardous materials.

§ 5.8 The Owner shall furnish all legal, insurance and accounting services, including auditing services, that may be reasonably necessary at any time for the Project to meet the Owner's needs and interests.

§ 5.9 The Owner shall provide prompt written notice to the Architect if the Owner becomes aware of any fault or defect in the Project, including errors, omissions or inconsistencies in the Architect's Instruments of Service.

§ 5.10 Except as otherwise provided in this Agreement, or when direct communications have been specially authorized, the Owner shall endeavor to communicate with the Contractor and the Architect's consultants through the Architect about matters arising out of or relating to the Contract Documents. The Owner shall promptly notify the Architect of any direct communications that may affect the Architect's services.

§ 5.11 Before executing the Contract for Construction, the Owner shall coordinate the Architect's duties and responsibilities set forth in the Contract for Construction with the Architect's services set forth in this Agreement. The Owner shall provide the Architect a copy of the executed agreement between the Owner and Contractor, including the General Conditions of the Contract for Construction.

§ 5.12 The Owner shall provide the Architect access to the Project site prior to commencement of the Work and shall obligate the Contractor to provide the Architect access to the Work wherever it is in preparation or progress.

ARTICLE 6 COST OF THE WORK

§ 6.1 For purposes of this Agreement, the Cost of the Work shall be the total cost to the Owner to construct all elements of the Project designed or specified by the Architect and shall include contractors' general conditions costs, overhead and profit. The Cost of the Work does not include the compensation of the Architect, the costs of the land, rights-of-way, financing, contingencies for changes in the Work or other costs that are the responsibility of the Owner.

§ 6.2 The Owner's budget for the Cost of the Work is provided in Initial Information, and may be adjusted throughout the Project as required under Sections 5.2, 6.4 and 6.5. Evaluations of the Owner's budget for the Cost of the Work, the preliminary estimate of the Cost of the Work and updated estimates of the Cost of the Work prepared by the Architect, represent the Architect's judgment as a design professional. It is recognized, however, that neither the Architect nor the Owner has control over the cost of labor, materials or equipment; the Contractor's methods of determining bid prices; or competitive bidding, market or negotiating conditions. Accordingly, the Architect cannot and does not warrant or represent that bids or negotiated prices will not vary from the Owner's budget for the Cost of the Work or from any estimate of the Cost of the Work or evaluation prepared or agreed to by the Architect.

§ 6.3 In preparing estimates of the Cost of Work, the Architect shall be permitted to include contingencies for design, bidding and price escalation; to determine what materials, equipment, component systems and types of construction are to be included in the Contract Documents; to make reasonable adjustments in the program and scope of the Project; and to include in the Contract Documents alternate bids as may be necessary to adjust the estimated Cost of the Work to meet the Owner's budget for the Cost of the Work. The Architect's estimate of the Cost of the Work shall be based on current area, volume or similar conceptual estimating techniques. If the Owner requests detailed cost estimating services, the Architect shall provide such services as an Additional Service under Article 4.

§ **6.4** If the Bidding or Negotiation Phase has not commenced within 90 days after the Architect submits the Construction Documents to the Owner, through no fault of the Architect, the Owner's budget for the Cost of the Work shall be adjusted to reflect changes in the general level of prices in the applicable construction market.

§ **6.5** If at any time the Architect's estimate of the Cost of the Work exceeds the Owner's budget for the Cost of the Work, the Architect shall make appropriate recommendations to the Owner to adjust the Project's size, quality or budget for the Cost of the Work, and the Owner shall cooperate with the Architect in making such adjustments.

§ **6.6** If the Owner's budget for the Cost of the Work at the conclusion of the Construction Documents Phase Services is exceeded by the lowest bona fide bid or negotiated proposal, the Owner shall

.1   give written approval of an increase in the budget for the Cost of the Work;
.2   authorize rebidding or renegotiating of the Project within a reasonable time;
.3   terminate in accordance with Section 9.5;
.4   in consultation with the Architect, revise the Project program, scope, or quality as required to reduce the Cost of the Work; or
.5   implement any other mutually acceptable alternative.

§ **6.7** If the Owner chooses to proceed under Section 6.6.4, the Architect, without additional compensation, shall modify the Construction Documents as necessary to comply with the Owner's budget for the Cost of the Work at the conclusion of the Construction Documents Phase Services, or the budget as adjusted under Section 6.6.1. The Architect's modification of the Construction Documents shall be the limit of the Architect's responsibility under this Article 6.

ARTICLE 7 COPYRIGHTS AND LICENSES

§ **7.1** The Architect and the Owner warrant that in transmitting Instruments of Service, or any other information, the transmitting party is the copyright owner of such information or has permission from the copyright owner to transmit such information for its use on the Project. If the Owner and Architect intend to transmit Instruments of Service or any other information or documentation in digital form, they shall endeavor to establish necessary protocols governing such transmissions.

§ **7.2** The Architect and the Architect's consultants shall be deemed the authors and owners of their respective Instruments of Service, including the Drawings and Specifications, and shall retain all common law, statutory and other reserved rights, including copyrights. Submission or distribution of Instruments of Service to meet official regulatory requirements or for similar purposes in connection with the Project is not to be construed as publication in derogation of the reserved rights of the Architect and the Architect's consultants.

§ **7.3** Upon execution of this Agreement, the Architect grants to the Owner a nonexclusive license to use the Architect's Instruments of Service solely and exclusively for purposes of constructing, using, maintaining, altering and adding to the Project, provided that the Owner substantially performs its obligations, including prompt payment of all sums when due, under this Agreement. The Architect shall obtain similar nonexclusive licenses from the Architect's consultants consistent with this Agreement. The license granted under this section permits the Owner to authorize the Contractor, Subcontractors, Sub-subcontractors, and material or equipment suppliers, as well as the Owner's consultants and separate contractors, to reproduce applicable portions of the Instruments of Service solely and exclusively for use in performing services or construction for the Project. If the Architect rightfully terminates this Agreement for cause as provided in Section 9.4, the license granted in this Section 7.3 shall terminate.

§ **7.3.1** In the event the Owner uses the Instruments of Service without retaining the author of the Instruments of Service, the Owner releases the Architect and Architect's consultant(s) from all claims and causes of action arising from such uses. The Owner, to the extent permitted by law, further agrees to indemnify and hold harmless the Architect and its consultants from all costs and expenses, including the cost of defense, related to claims and causes of action asserted by any third person or entity to the extent such costs and expenses arise from the Owner's use of the Instruments of Service under this Section 7.3.1. The terms of this Section 7.3.1 shall not apply if the Owner rightfully terminates this Agreement for cause under Section 9.4.

§ **7.4** Except for the licenses granted in this Article 7, no other license or right shall be deemed granted or implied under this Agreement. The Owner shall not assign, delegate, sublicense, pledge or otherwise transfer any license granted herein to another party without the prior written agreement of the Architect. Any unauthorized use of the

Instruments of Service shall be at the Owner's sole risk and without liability to the Architect and the Architect's consultants.

ARTICLE 8   CLAIMS AND DISPUTES
§ 8.1 GENERAL
§ 8.1.1 The Owner and Architect shall commence all claims and causes of action, whether in contract, tort, or otherwise, against the other arising out of or related to this Agreement in accordance with the requirements of the method of binding dispute resolution selected in this Agreement within the period specified by applicable law, but in any case not more than 10 years after the date of Substantial Completion of the Work. The Owner and Architect waive all claims and causes of action not commenced in accordance with this Section 8.1.1.

§ 8.1.2 To the extent damages are covered by property insurance, the Owner and Architect waive all rights against each other and against the contractors, consultants, agents and employees of the other for damages, except such rights as they may have to the proceeds of such insurance as set forth in AIA Document A201–2007, General Conditions of the Contract for Construction. The Owner or the Architect, as appropriate, shall require of the contractors, consultants, agents and employees of any of them similar waivers in favor of the other parties enumerated herein.

§ 8.1.3 The Architect and Owner waive consequential damages for claims, disputes or other matters in question arising out of or relating to this Agreement. This mutual waiver is applicable, without limitation, to all consequential damages due to either party's termination of this Agreement, except as specifically provided in Section 9.7.

§ 8.2 MEDIATION
§ 8.2.1 Any claim, dispute or other matter in question arising out of or related to this Agreement shall be subject to mediation as a condition precedent to binding dispute resolution. If such matter relates to or is the subject of a lien arising out of the Architect's services, the Architect may proceed in accordance with applicable law to comply with the lien notice or filing deadlines prior to resolution of the matter by mediation or by binding dispute resolution.

§ 8.2.2 The Owner and Architect shall endeavor to resolve claims, disputes and other matters in question between them by mediation which, unless the parties mutually agree otherwise, shall be administered by the American Arbitration Association in accordance with its Construction Industry Mediation Procedures in effect on the date of the Agreement. A request for mediation shall be made in writing, delivered to the other party to the Agreement, and filed with the person or entity administering the mediation. The request may be made concurrently with the filing of a complaint or other appropriate demand for binding dispute resolution but, in such event, mediation shall proceed in advance of binding dispute resolution proceedings, which shall be stayed pending mediation for a period of 60 days from the date of filing, unless stayed for a longer period by agreement of the parties or court order. If an arbitration proceeding is stayed pursuant to this section, the parties may nonetheless proceed to the selection of the arbitrator(s) and agree upon a schedule for later proceedings.

§ 8.2.3 The parties shall share the mediator's fee and any filing fees equally. The mediation shall be held in the place where the Project is located, unless another location is mutually agreed upon. Agreements reached in mediation shall be enforceable as settlement agreements in any court having jurisdiction thereof.

§ 8.2.4 If the parties do not resolve a dispute through mediation pursuant to this Section 8.2, the method of binding dispute resolution shall be the following:
(Check the appropriate box. If the Owner and Architect do not select a method of binding dispute resolution below, or do not subsequently agree in writing to a binding dispute resolution method other than litigation, the dispute will be resolved in a court of competent jurisdiction.)

    ☐   Arbitration pursuant to Section 8.3 of this Agreement

    ☐   Litigation in a court of competent jurisdiction

    ☐   Other *(Specify)*

§ 8.3 ARBITRATION

§ 8.3.1 If the parties have selected arbitration as the method for binding dispute resolution in this Agreement, any claim, dispute or other matter in question arising out of or related to this Agreement subject to, but not resolved by, mediation shall be subject to arbitration which, unless the parties mutually agree otherwise, shall be administered by the American Arbitration Association in accordance with its Construction Industry Arbitration Rules in effect on the date of this Agreement. A demand for arbitration shall be made in writing, delivered to the other party to this Agreement, and filed with the person or entity administering the arbitration.

§ 8.3.1.1 A demand for arbitration shall be made no earlier than concurrently with the filing of a request for mediation, but in no event shall it be made after the date when the institution of legal or equitable proceedings based on the claim, dispute or other matter in question would be barred by the applicable statute of limitations. For statute of limitations purposes, receipt of a written demand for arbitration by the person or entity administering the arbitration shall constitute the institution of legal or equitable proceedings based on the claim, dispute or other matter in question.

§ 8.3.2 The foregoing agreement to arbitrate and other agreements to arbitrate with an additional person or entity duly consented to by parties to this Agreement shall be specifically enforceable in accordance with applicable law in any court having jurisdiction thereof.

§ 8.3.3 The award rendered by the arbitrator(s) shall be final, and judgment may be entered upon it in accordance with applicable law in any court having jurisdiction thereof.

§ 8.3.4 CONSOLIDATION OR JOINDER

§ 8.3.4.1 Either party, at its sole discretion, may consolidate an arbitration conducted under this Agreement with any other arbitration to which it is a party provided that (1) the arbitration agreement governing the other arbitration permits consolidation; (2) the arbitrations to be consolidated substantially involve common questions of law or fact; and (3) the arbitrations employ materially similar procedural rules and methods for selecting arbitrator(s).

§ 8.3.4.2 Either party, at its sole discretion, may include by joinder persons or entities substantially involved in a common question of law or fact whose presence is required if complete relief is to be accorded in arbitration, provided that the party sought to be joined consents in writing to such joinder. Consent to arbitration involving an additional person or entity shall not constitute consent to arbitration of any claim, dispute or other matter in question not described in the written consent.

§ 8.3.4.3 The Owner and Architect grant to any person or entity made a party to an arbitration conducted under this Section 8.3, whether by joinder or consolidation, the same rights of joinder and consolidation as the Owner and Architect under this Agreement.

ARTICLE 9 TERMINATION OR SUSPENSION

§ 9.1 If the Owner fails to make payments to the Architect in accordance with this Agreement, such failure shall be considered substantial nonperformance and cause for termination or, at the Architect's option, cause for suspension of performance of services under this Agreement. If the Architect elects to suspend services, the Architect shall give seven days' written notice to the Owner before suspending services. In the event of a suspension of services, the Architect shall have no liability to the Owner for delay or damage caused the Owner because of such suspension of services. Before resuming services, the Architect shall be paid all sums due prior to suspension and any expenses incurred in the interruption and resumption of the Architect's services. The Architect's fees for the remaining services and the time schedules shall be equitably adjusted.

§ 9.2 If the Owner suspends the Project, the Architect shall be compensated for services performed prior to notice of such suspension. When the Project is resumed, the Architect shall be compensated for expenses incurred in the interruption and resumption of the Architect's services. The Architect's fees for the remaining services and the time schedules shall be equitably adjusted.

§ 9.3 If the Owner suspends the Project for more than 90 cumulative days for reasons other than the fault of the Architect, the Architect may terminate this Agreement by giving not less than seven days' written notice.

§ 9.4 Either party may terminate this Agreement upon not less than seven days' written notice should the other party fail substantially to perform in accordance with the terms of this Agreement through no fault of the party initiating the termination.

§ 9.5 The Owner may terminate this Agreement upon not less than seven days' written notice to the Architect for the Owner's convenience and without cause.

§ 9.6 In the event of termination not the fault of the Architect, the Architect shall be compensated for services performed prior to termination, together with Reimbursable Expenses then due and all Termination Expenses as defined in Section 9.7.

§ 9.7 Termination Expenses are in addition to compensation for the Architect's services and include expenses directly attributable to termination for which the Architect is not otherwise compensated, plus an amount for the Architect's anticipated profit on the value of the services not performed by the Architect.

§ 9.8 The Owner's rights to use the Architect's Instruments of Service in the event of a termination of this Agreement are set forth in Article 7 and Section 11.9.

ARTICLE 10 MISCELLANEOUS PROVISIONS
§ 10.1 This Agreement shall be governed by the law of the place where the Project is located, except that if the parties have selected arbitration as the method of binding dispute resolution, the Federal Arbitration Act shall govern Section 8.3.

§ 10.2 Terms in this Agreement shall have the same meaning as those in AIA Document A201–2007, General Conditions of the Contract for Construction.

§ 10.3 The Owner and Architect, respectively, bind themselves, their agents, successors, assigns and legal representatives to this Agreement. Neither the Owner nor the Architect shall assign this Agreement without the written consent of the other, except that the Owner may assign this Agreement to a lender providing financing for the Project if the lender agrees to assume the Owner's rights and obligations under this Agreement.

§ 10.4 If the Owner requests the Architect to execute certificates, the proposed language of such certificates shall be submitted to the Architect for review at least 14 days prior to the requested dates of execution. If the Owner requests the Architect to execute consents reasonably required to facilitate assignment to a lender, the Architect shall execute all such consents that are consistent with this Agreement, provided the proposed consent is submitted to the Architect for review at least 14 days prior to execution. The Architect shall not be required to execute certificates or consents that would require knowledge, services or responsibilities beyond the scope of this Agreement.

§ 10.5 Nothing contained in this Agreement shall create a contractual relationship with or a cause of action in favor of a third party against either the Owner or Architect.

§ 10.6 Unless otherwise required in this Agreement, the Architect shall have no responsibility for the discovery, presence, handling, removal or disposal of, or exposure of persons to, hazardous materials or toxic substances in any form at the Project site.

§ 10.7 The Architect shall have the right to include photographic or artistic representations of the design of the Project among the Architect's promotional and professional materials. The Architect shall be given reasonable access to the completed Project to make such representations. However, the Architect's materials shall not include the Owner's confidential or proprietary information if the Owner has previously advised the Architect in writing of the specific information considered by the Owner to be confidential or proprietary. The Owner shall provide professional credit for the Architect in the Owner's promotional materials for the Project.

§ 10.8 If the Architect or Owner receives information specifically designated by the other party as "confidential" or "business proprietary," the receiving party shall keep such information strictly confidential and shall not disclose it to any other person except to (1) its employees, (2) those who need to know the content of such information in order to perform services or construction solely and exclusively for the Project, or (3) its consultants and contractors whose contracts include similar restrictions on the use of confidential information.

ARTICLE 11 COMPENSATION

§ 11.1 For the Architect's Basic Services described under Article 3, the Owner shall compensate the Architect as follows:
(Insert amount of, or basis for, compensation.)

§ 11.2 For Additional Services designated in Section 4.1, the Owner shall compensate the Architect as follows:
(Insert amount of, or basis for, compensation. If necessary, list specific services to which particular methods of compensation apply.)

§ 11.3 For Additional Services that may arise during the course of the Project, including those under Section 4.3, the Owner shall compensate the Architect as follows:
(Insert amount of, or basis for, compensation.)

§ 11.4 Compensation for Additional Services of the Architect's consultants when not included in Section 11.2 or 11.3, shall be the amount invoiced to the Architect plus   percent (%), or as otherwise stated below:

§ 11.5 Where compensation for Basic Services is based on a stipulated sum or percentage of the Cost of the Work, the compensation for each phase of services shall be as follows:

Schematic Design Phase:	percent (	%)
Design Development Phase:	percent (	%)
Construction Documents Phase:	percent (	%)
Bidding or Negotiation Phase:	percent (	%)
Construction Phase:	percent (	%)
Total Basic Compensation	one hundred percent (	100.00%)

§ 11.6 When compensation is based on a percentage of the Cost of the Work and any portions of the Project are deleted or otherwise not constructed, compensation for those portions of the Project shall be payable to the extent services are performed on those portions, in accordance with the schedule set forth in Section 11.5 based on (1) the lowest bona fide bid or negotiated proposal, or (2) if no such bid or proposal is received, the most recent estimate of the Cost of the Work for such portions of the Project. The Architect shall be entitled to compensation in accordance with this Agreement for all services performed whether or not the Construction Phase is commenced.

§ 11.7 The hourly billing rates for services of the Architect and the Architect's consultants, if any, are set forth below. The rates shall be adjusted in accordance with the Architect's and Architect's consultants' normal review practices.
(If applicable, attach an exhibit of hourly billing rates or insert them below.)

§ 11.8 COMPENSATION FOR REIMBURSABLE EXPENSES

§ 11.8.1 Reimbursable Expenses are in addition to compensation for Basic and Additional Services and include expenses incurred by the Architect and the Architect's consultants directly related to the Project, as follows:

.1   Transportation and authorized out-of-town travel and subsistence;

.2   Long distance services, dedicated data and communication services, teleconferences, Project Web sites, and extranets;

.3   Fees paid for securing approval of authorities having jurisdiction over the Project;

.4   Printing, reproductions, plots, standard form documents;

.5   Postage, handling and delivery;

.6   Expense of overtime work requiring higher than regular rates, if authorized in advance by the Owner;

.7   Renderings, models, mock-ups, professional photography, and presentation materials requested by the Owner;

.8   Architect's Consultant's expense of professional liability insurance dedicated exclusively to this Project, or the expense of additional insurance coverage or limits if the Owner requests such insurance in excess of that normally carried by the Architect's consultants;

.9   All taxes levied on professional services and on reimbursable expenses;

.10   Site office expenses; and

.11   Other similar Project-related expenditures.

§ 11.8.2 For Reimbursable Expenses the compensation shall be the expenses incurred by the Architect and the Architect's consultants plus   percent (%) of the expenses incurred.

§ 11.9 COMPENSATION FOR USE OF ARCHITECT'S INSTRUMENTS OF SERVICE

If the Owner terminates the Architect for its convenience under Section 9.5, or the Architect terminates this Agreement under Section 9.3, the Owner shall pay a licensing fee as compensation for the Owner's continued use of the Architect's Instruments of Service solely for purposes of completing, using and maintaining the Project as follows:

§ 11.10 PAYMENTS TO THE ARCHITECT

§ 11.10.1 An initial payment of   Dollars
($) shall be made upon execution of this Agreement and is the minimum payment under this Agreement. It shall be credited to the Owner's account in the final invoice.

§ 11.10.2 Unless otherwise agreed, payments for services shall be made monthly in proportion to services performed. Payments are due and payable upon presentation of the Architect's invoice. Amounts unpaid
(   ) days after the invoice date shall bear interest at the rate entered below, or in the absence thereof at the legal rate prevailing from time to time at the principal place of business of the Architect.
(Insert rate of monthly or annual interest agreed upon.)

§ 11.10.3 The Owner shall not withhold amounts from the Architect's compensation to impose a penalty or liquidated damages on the Architect, or to offset sums requested by or paid to contractors for the cost of changes in the Work unless the Architect agrees or has been found liable for the amounts in a binding dispute resolution proceeding.

§ 11.10.4 Records of Reimbursable Expenses, expenses pertaining to Additional Services, and services performed on the basis of hourly rates shall be available to the Owner at mutually convenient times.

ARTICLE 12   SPECIAL TERMS AND CONDITIONS

Special terms and conditions that modify this Agreement are as follows:

ARTICLE 13 SCOPE OF THE AGREEMENT

§ 13.1 This Agreement represents the entire and integrated agreement between the Owner and the Architect and supersedes all prior negotiations, representations or agreements, either written or oral. This Agreement may be amended only by written instrument signed by both Owner and Architect.

§ 13.2 This Agreement is comprised of the following documents listed below:

.1 AIA Document B101™–2007, Standard Form Agreement Between Owner and Architect

.2 AIA Document E201™–2007, Digital Data Protocol Exhibit, if completed, or the following:

.3 Other documents:
(List other documents, if any, including Exhibit A, Initial Information, and additional scopes of service, if any, forming part of the Agreement.)

This Agreement entered into as of the day and year first written above.

_____   _____
OWNER *(Signature)*   **ARCHITECT** *(Signature)*

_____   _____
(Printed name and title)   *(Printed name and title)*

CAUTION: You should sign an original AIA Contract Document, on which this text appears in RED. An original assures that changes will not be obscured.

▧AIA® Document B101™ – 2007 Exhibit A

Initial Information

for the following PROJECT:
(Name and location or address)

THE OWNER:
(Name and address)

THE ARCHITECT:
(Name and address)

This Agreement is based on the following information.
(Note the disposition for the following items by inserting the requested information or a statement such as "not applicable," "unknown at time of execution" or "to be determined later by mutual agreement.")

ARTICLE A.1 PROJECT INFORMATION
§ A.1.1 The Owner's program for the Project:
(Identify documentation or state the manner in which the program will be developed.)

§ A.1.2 The Project's physical characteristics:
(Identify or describe, if appropriate, size, location, dimensions, or other pertinent information, such as geotechnical reports; site, boundary and topographic surveys; traffic and utility studies; availability of public and private utilities and services; legal description of the site; etc.)

§ A.1.3 The Owner's budget for the Cost of the Work, as defined in Section 6.1:
(Provide total, and if known, a line item break down.)

§ **A.1.4** The Owner's other anticipated scheduling information, if any, not provided in Section 1.2:

§ **A.1.5** The Owner intends the following procurement or delivery method for the Project:
(Identify method such as competitive bid, negotiated contract, or construction management.)

§ **A.1.6** Other Project information:
(Identify special characteristics or needs of the Project not provided elsewhere, such as environmentally responsible design or historic preservation requirements.)

ARTICLE A.2 PROJECT TEAM
§ **A.2.1** The Owner identifies the following representative in accordance with Section 5.3:
(List name, address and other information.)

§ **A.2.2** The persons or entities, in addition to the Owner's representative, who are required to review the Architect's submittals to the Owner are as follows:
(List name, address and other information.)

§ **A.2.3** The Owner will retain the following consultants and contractors:
(List discipline and, if known, identify them by name and address.)

§ **A.2.4** The Architect identifies the following representative in accordance with Section 2.3:
(List name, address and other information.)

§ A.2.5 The Architect will retain the consultants identified in Sections A.2.5.1 and A.2.5.2.
(List discipline and, if known, identify them by name and address.)

§ A.2.5.1 Consultants retained under Basic Services:
  .1   Structural Engineer

  .2   Mechanical Engineer

  .3   Electrical Engineer

§ A.2.5.2 Consultants retained under Additional Services:

§ A.2.6 Other Initial Information on which the Agreement is based:
(Provide other Initial Information.)

Engineers Joint Documents Committee
Design and Construction Related Documents
Instructions and License Agreement

Instructions

Before you use any EJCDC document:
1. Read the License Agreement. You agree to it and are bound by its terms when you use the EJCDC document.

2. Make sure that you have the correct version for your word processing software.

How to Use:
1. While EJCDC has expended considerable effort to make the software translations exact, it can be that a few document controls (e.g., bold, underline) did not carry over.

2. Similarly, your software may change the font specification if the font is not available in your system. It will choose a font that is close in appearance. In this event, the pagination may not match the control set.

3. If you modify the document, you must follow the instructions in the License Agreement about notification.

4. Also note the instruction in the License Agreement about the EJCDC copyright.

License Agreement

You should carefully read the following terms and conditions before using this document. Commencement of use of this document indicates your acceptance of these terms and conditions. If you do not agree to them, you should promptly return the materials to the vendor, and your money will be refunded.

The Engineers Joint Contract Documents Committee ("EJCDC") provides **EJCDC Design and Construction Related Documents** and licenses their use worldwide. You assume sole responsibility for the selection of specific documents or portions thereof to achieve your intended results, and for the installation, use, and results obtained from **EJCDC Design and Construction Related Documents**.

You acknowledge that you understand that the text of the contract documents of **EJCDC Design and Construction Related Documents** has important legal consequences and that consultation with an attorney is recommended with respect to use or modification of the text. You further acknowledge that EJCDC documents are protected by the copyright laws of the United States.

License:
You have a limited nonexclusive license to:

1. Use **EJCDC Design and Construction Related Documents** on any number of machines owned, leased or rented by your company or organization.

2. Use **EJCDC Design and Construction Related Documents** in printed form for bona fide contract documents.

3. Copy **EJCDC Design and Construction Related Documents** into any machine readable or printed form for backup or modification purposes in support of your use of **EJCDC Design and Construction Related Documents**.

You agree that you will:
1. Reproduce and include EJCDC's copyright notice on any printed or machine-readable copy, modification, or portion merged into another document or program. All proprietary rights in **EJCDC Design and Construction Related Documents** are and shall remain the property of EJCDC.

2. Not represent that any of the contract documents you generate from **EJCDC Design and Construction Related Documents** are EJCDC documents unless (i) the document text is used without alteration or (ii) all additions and changes to, and deletions from, the text are clearly shown.

You may not use, copy, modify, or transfer EJCDC Design and Construction Related Documents, or any copy, modification or merged portion, in whole or in part, except as expressly provided for in this license. Reproduction of EJCDC Design and Construction Related Documents in printed or machine-readable format for resale or educational purposes is expressly prohibited.

If you transfer possession of any copy, modification or merged portion of EJCDC Design and Construction Related Documents to another party, your license is automatically terminated.

Term:
The license is effective until terminated. You may terminate it at any time by destroying **EJCDC Design and Construction Related Documents** altogether with all copies, modifications and merged portions in any form. It will also terminate upon conditions set forth elsewhere in this Agreement or if you fail to comply with any term or condition of this Agreement. You agree upon such termination to destroy **EJCDC Design and Construction**

Related Documents along with all copies, modifications and merged portions in any form.

Limited Warranty:

EJCDC warrants the CDs and diskettes on which **EJCDC Design and Construction Related Documents** is furnished to be free from defects in materials and workmanship under normal use for a period of ninety (90) days from the date of delivery to you as evidenced by a copy of your receipt.

There is no other warranty of any kind, either expressed or implied, including, but not limited to the implied warranties of merchantability and fitness for a particular purpose. Some states do not allow the exclusion of implied warranties, so the above exclusion may not apply to you. This warranty gives you specific legal rights and you may also have other rights which vary from state to state.

EJCDC does not warrant that the functions contained in **EJCDC Design and Construction Related Documents** will meet your requirements or that the operation of **EJCDC Design and Construction Related Documents** will be uninterrupted or error free.

Limitations of Remedies:

EJCDC's entire liability and your exclusive remedy shall be:

1. the replacement of any document not meeting EJCDC's "Limited Warranty" which is returned to EJCDC's selling agent with a copy of your receipt, or

2. if EJCDC's selling agent is unable to deliver a replacement CD or diskette which is free of defects in materials and workmanship, you may terminate this Agreement by returning EJCDC Document and your money will be refunded.

In no event will EJCDC be liable to you for any damages, including any lost profits, lost savings or other incidental or consequential damages arising out of the use or inability to use **EJCDC Design and Construction Related Documents** even if EJCDC has been advised of the possibility of such damages, or for any claim by any other party.

Some states do not allow the limitation or exclusion of liability for incidental or consequential damages, so the above limitation or exclusion may not apply to you.

General:

You may not sublicense, assign, or transfer this license except as expressly provided in this Agreement. Any attempt otherwise to sublicense, assign, or transfer any of the rights, duties, or obligations hereunder is void.

This Agreement shall be governed by the laws of the State of Virginia. Should you have any questions concerning this Agreement, you may contact EJCDC by writing to:

> Arthur Schwartz, Esq.
> General Counsel
> National Society of Professional Engineers
> 1420 King Street
> Alexandria, VA 22314
>
> Phone: (703) 684-2845
> Fax: (703) 836-4875
> e-mail: aschwartz@nspe.org

You acknowledge that you have read this agreement, understand it and agree to be bound by its terms and conditions. You further agree that it is the complete and exclusive statement of the agreement between us which supersedes any proposal or prior agreement, oral or written, and any other communications between us relating to the subject matter of this agreement.

This document has important legal consequences; consultation with an attorney is encouraged with respect to its use or modification. This document should be adapted to the particular circumstances of the contemplated Project and the controlling Laws and Regulations.

AGREEMENT
BETWEEN OWNER AND ENGINEER
FOR
STUDY AND REPORT
PROFESSIONAL SERVICES

Prepared by

EJCDC

ENGINEERS JOINT CONTRACT
DOCUMENTS COMMITTEE

and

Issued and Published Jointly by

ACEC
AMERICAN COUNCIL OF ENGINEERING COMPANIES

AGC of America
THE ASSOCIATED GENERAL CONTRACTORS OF AMERICA
Quality People. Quality Projects.

ASCE American Society of Civil Engineers

National Society of Professional Engineers
Professional Engineers in Private Practice

AMERICAN COUNCIL OF ENGINEERING COMPANIES

ASSOCIATED GENERAL CONTRACTORS OF AMERICA

AMERICAN SOCIETY OF CIVIL ENGINEERS

PROFESSIONAL ENGINEERS IN PRIVATE PRACTICE
A Practice Division of the
NATIONAL SOCIETY OF PROFESSIONAL ENGINEERS

Copyright © 2009:

National Society of Professional Engineers
1420 King Street, Alexandria, VA 22314-2794
(703) 684-2882
www.nspe.org

American Council of Engineering Companies
1015 15th Street N.W., Washington, DC 20005
(202) 347-7474
www.acec.org

American Society of Civil Engineers
1801 Alexander Bell Drive, Reston, VA 20191-4400
(800) 548-2723
www.asce.org

Associated General Contractors of America
2300 Wilson Boulevard, Suite 400, Arlington, VA   22201-3308
(703) 548-3118
www.agc.org

The copyright for EJCDC E-525 is owned jointly by the four EJCDC sponsoring organizations listed above. The National Society of Professional Engineers (NSPE) is the Copyright Administrator for the EJCDC documents; please direct all inquiries and requests regarding EJCDC copyrights to NSPE.

NOTE: EJCDC publications may be ordered directly from any of the four sponsoring organizations above. For more information on EJCDC please visit www.ejcdc.org.

TABLE OF CONTENTS

AGREEMENT
BETWEEN OWNER AND ENGINEER
FOR STUDY AND REPORT
PROFESSIONAL SERVICES

THIS IS AN AGREEMENT effective as of _____ , _____ ("Effective Date") between

_____ ("Owner") and

_____ ("Engineer").

Engineer's services under this Agreement are generally described as follows:

_____

_____

_____

_____ ("Assignment").

If Engineer's services under this Agreement are a part of a more extensive project of the Owner, such project is generally identified as follows:

_____

_____ ("Project").

Owner and Engineer further agree as follows:

ARTICLE 1 – SERVICES OF ENGINEER

1.01 *Scope*

A. Engineer shall provide, or cause to be provided, the services set forth herein and in Exhibit A.

ARTICLE 2 – OWNER'S RESPONSIBILITIES

2.01 *General*

A. Owner shall pay Engineer as set forth in Article 4.

B. Owner shall provide Engineer with all criteria and full information as to Owner's requirements for the Assignment, including design objectives and constraints, space, capacity and performance requirements, flexibility, and expandability, and any anticipated funding sources and budgetary limitations.

C. Owner shall furnish to Engineer all existing studies, reports, and other available data pertinent to the Assignment, obtain or authorize Engineer to obtain or provide additional reports and data as required, and furnish to Engineer such services of others as may be necessary for the performance of Engineer's services.

D. Owner shall arrange for safe access to and make all provisions for Engineer to enter upon public and private property as required for Engineer to perform services under the Agreement.

E. Owner shall be responsible for, and Engineer may rely upon, the accuracy and completeness of all requirements, instructions, reports, data, and other information Owner-furnished by Owner to Engineer pursuant to this Agreement. Engineer may use such requirements, instructions, reports, data, and information in performing or furnishing services under this Agreement.

ARTICLE 3 – SCHEDULE FOR RENDERING SERVICES

3.01   *Commencement*

A. Engineer is authorized to begin rendering services as of the Effective Date.

3.02   *Time for Completion*

A. Engineer shall complete its obligations within a reasonable time. Specific periods of time for rendering services are set forth or specific dates by which services are to be completed are provided in Exhibit A, and are hereby agreed to be reasonable.

B. If, through no fault of Engineer, such periods of time or dates are changed, or the orderly and continuous progress of Engineer's services is impaired, or Engineer's services are delayed or suspended, then the time for completion of Engineer's services, and the rates and amounts of Engineer's compensation, shall be adjusted equitably.

ARTICLE 4 – INVOICES AND PAYMENTS

4.01   *Invoices*

A. *Preparation and Submittal of Invoices*: Engineer shall prepare invoices in accordance with its standard invoicing practices and the terms of this Article. Engineer shall submit its invoices to Owner on a monthly basis. Invoices are due and payable within 30 days of receipt.

4.02   *Payments*

A. *Application to Interest and Principal*: Payment will be credited first to any interest owed to Engineer and then to principal.

B. *Failure to Pay*: If Owner fails to make payments due Engineer for services and expenses within 30 days after receipt of Engineer's invoice, then:

   1. Engineer will be entitled to interest on all amounts due and payable at the rate of 1.0% per month (or the maximum rate of interest permitted by law, if less) from said thirtieth day; and

2. Engineer may, after giving seven days written notice to Owner, suspend services under this Agreement until Owner has paid in full all amounts due for services, expenses, and other related charges. Owner waives any and all claims against Engineer for any such suspension.

NOTE TO USER:
Choose one of the three following compensation options.
Delete the other two options.

4.03 *Payment for Basic Services (Lump Sum Basis)*

A. Using the procedures set forth in Paragraph 4.01, Owner shall pay Engineer for Basic Services as follows:

1. A Lump Sum amount of $_____.

B. The portion of the compensation amount billed monthly for Engineer's services will be based upon Engineer's estimate of the percentage of the Assignment actually completed during the billing period.

[or]

4.03 *Payment for Basic Services (Hourly Rates Plus Reimbursable Expenses)*

A. Using the procedures set forth in Paragraph 4.01, Owner shall pay Engineer for Basic Services as follows:

1. An amount equal to the cumulative hours charged to the Assignment by each class of Engineer's employees times standard hourly rates for each applicable billing class for all services performed on the Assignment, plus reimbursable expenses and Engineer's Consultants' charges, if any.

2. Engineer's standard hourly rates are set forth in Exhibit C.

3. The total compensation for services and reimbursable expenses is estimated to be $____.

[or]

4.03 *Payment for Basic Services (Direct Labor Costs Times Factor, Plus Reimbursables)*

A. Using the procedures set forth in Paragraph 4.01, Owner shall pay Engineer a for Basic Services as follows:

1. An amount equal to Engineer's Direct Labor Costs times a factor of _____ for services of Engineer's employees engaged on the Project, plus reimbursable expenses, and Engineer's Consultants' charges, if any.

2. Direct Labor Costs means salaries and wages paid to employees but does not include payroll related costs or benefits.

3. The total compensation for services and reimbursable expenses is estimated to be $_____.

[End of Compensation Options]

4.04   *Payment for Additional Services*

A. For Additional Services, Owner shall pay Engineer an amount equal to the cumulative hours charged to providing the Additional Services under the Assignment by each class of Engineer's employees, times standard hourly rates for each applicable billing class; plus reimbursable expenses and Engineer's Consultants' charges, if any. Engineer's standard hourly rates and reimbursable expenses schedule are set forth in Exhibit C.

4.05   *Disputed Invoices*

A. If Owner contests an invoice, Owner shall promptly advise Engineer of the specific basis for doing so, may withhold only that portion so contested, and must pay the undisputed portion.

ARTICLE 5 – OPINIONS OF COST

5.01   *Opinions of Probable Construction Cost*

A. Engineer's opinions of probable Construction Cost are to be made on the basis of Engineer's experience and qualifications and represent Engineer's estimate as an experienced and qualified professional generally familiar with the construction industry. However, because of the limited and preliminary nature of the Assignment, and because Engineer has no control over the cost of labor, materials, equipment, or services furnished by others, or over contractors' methods of determining prices, or over competitive bidding or market conditions, Engineer cannot and does not guarantee that proposals, bids, or actual Construction Cost will not vary from opinions of probable Construction Cost prepared by Engineer. If Owner requires greater assurance as to probable Construction Cost, Owner must employ an independent cost estimator.

5.02   *Opinions of Total Project Costs*

A. The services, if any, of Engineer with respect to Total Project Costs shall be limited to assisting the Owner in collating the various cost categories which comprise Total Project Costs. Engineer assumes no responsibility for the accuracy of any opinions of Total Project Costs.

ARTICLE 6 – GENERAL CONSIDERATIONS

6.01   *Standards of Performance*

A. *Standard of Care:* The standard of care for all professional engineering and related services performed or furnished by Engineer under this Agreement will be the care and skill ordinarily used by members of the subject profession practicing under similar circumstances at the same time and in the same locality. Engineer makes no warranties, express or implied, under this Agreement or otherwise, in connection with Engineer's services.

B. *Consultants:* Engineer may employ such Consultants as Engineer deems necessary to assist in the performance or furnishing of the services, subject to reasonable, timely, and substantive objections by Owner.

C. *Reliance on Others:* Subject to the standard of care set forth in Paragraph 6.01.A, Engineer and its Consultants may use or rely upon design elements and information ordinarily or customarily furnished by others, including, but not limited to, specialty contractors, manufacturers, suppliers, and the publishers of technical standards.

D. Engineer shall not be required to sign any documents, no matter by whom requested, that would result in the Engineer having to certify, guarantee, or warrant the existence of conditions whose existence the Engineer cannot ascertain. Owner agrees not to make resolution of any dispute with the Engineer or payment of any amount due to the Engineer in any way contingent upon the Engineer signing any such documents.

E. Engineer shall not have any construction-related duties under this Agreement. Engineer shall not at any time supervise, direct, control, or have authority over any contractor's work, nor shall Engineer have authority over or be responsible for the means, methods, techniques, sequences, or procedures of construction selected or used by any contractor, or the safety precautions and programs incident thereto, for security or safety at the Site, nor for any failure of a contractor to comply with Laws and Regulations applicable to such contractor's furnishing and performing of its work.

6.02   *Use of Documents*

A. All Documents are instruments of service, and Engineer shall retain an ownership and property interest therein (including the copyright and the right of reuse at the discretion of the Engineer) whether or not the Assignment or Project is completed. Owner shall not rely, in any way, on any Document unless it is in printed form, signed or sealed by the Engineer or one of its Consultants.

B. Either party to this Agreement may rely that data or information set forth on paper (also known as hard copies) that the party receives from the other party by mail, hand delivery, or facsimile, are the items that the other party intended to send. Files in electronic media format of text, data, graphics, or other types that are furnished by one party to the other are furnished only for convenience, not reliance, by the receiving party. Any conclusion or information obtained or derived from such electronic files will be at the user's sole risk. If there is a discrepancy between the electronic files and the hard copies, the hard copies govern. If the parties agree to other electronic transmittal procedures, such procedures shall be set forth in an exhibit to this Agreement.

C. Because data stored in electronic media format can deteriorate or be modified inadvertently or otherwise during storage or transmittal, the party receiving electronic files agrees that it will perform acceptance tests or procedures within ten days, after which the receiving party shall be deemed to have accepted the data thus transferred. Any data deficiencies detected within the ten-day acceptance period will be corrected, if possible, by the party delivering the electronic files.

D. When transferring documents in electronic media format, the transferring party makes no representations as to long-term compatibility, usability, or readability of such documents resulting from the use of software application packages, operating systems, or computer hardware differing from those used by the documents' creator.

E. Owner may make and retain copies of Documents solely for Owner's information and reference in connection with the specific subject matter of the Documents, subject to receipt by Engineer of full payment for all services relating to preparation of the Documents, and subject to the following limitations: (1) Owner acknowledges that such Documents are not intended or represented to be suitable for use by Owner unless completed by Engineer; (2) the Documents are instruments of study and report services only, and are not final design or construction documents, (3) no Document shall be altered, modified, or reused by Owner or any third party for any purpose except with Engineer's express written consent; (4) any use, reuse, alteration, or modification of the Documents, except as authorized in this Agreement or by Engineer's written consent, will be at Owner's sole risk and without liability or legal exposure to Engineer or to its officers, directors, members, partners, agents, employees, and Consultants; (5) Owner shall indemnify and hold harmless Engineer and its officers, directors, members, partners, agents, employees, and Consultants from all claims, damages, losses, and expenses, including attorneys' fees, arising out of or resulting from any unauthorized use, reuse, alteration, or modification of the Documents; and (6) nothing in this paragraph shall create any rights in third parties.

6.03   *Insurance*

A. Engineer will maintain insurance coverage for Workers' Compensation, General Liability, Professional Liability, and Automobile Liability and will provide certificates of insurance to Owner upon request.

6.04   *Termination*

A. *Termination for Cause:* The obligation to continue performance under this Agreement may be terminated:

   1. By either party upon 30 days written notice in the event of substantial failure by the other party to perform in accordance with the Agreement's terms through no fault of the terminating party. Failure to pay Engineer for its services is a substantial failure to perform and a basis for termination.

   2. By Engineer:

      a. upon seven days written notice if Engineer believes that Engineer is being required by Owner to furnish or perform services contrary to Engineer's responsibilities as a licensed professional; or

      b. upon seven days written notice if the Engineer's services are delayed for more than 90 days for reasons beyond Engineer's control.

      c. Engineer shall have no liability to Owner on account of a termination by Engineer under Paragraph 6.04.A.2.

3. Notwithstanding the foregoing, this Agreement will not terminate as a result of a substantial failure under Paragraph 6.04.A.1. if the party receiving such notice begins, within seven days of receipt of such notice, to correct its substantial failure to perform and proceeds diligently to cure such failure within no more than 30 days of receipt of notice; provided, however, that if and to the extent such substantial failure cannot be reasonably cured within such 30 day period, and if such party has diligently attempted to cure the same and thereafter continues diligently to cure the same, then the cure period provided for herein shall extend up to, but in no case more than, 60 days after the date of receipt of the notice.

B. *Termination for Convenience:* Owner may terminate the Agreement for Owner's convenience effective upon the Engineer's receipt of written notice from Owner.

C. The terminating party under Paragraphs 6.04.A or 6.04.B may set the effective date of termination at a time up to 30 days later than otherwise provided to allow Engineer to complete tasks whose value would otherwise be lost, to prepare notes as to the status of completed and uncompleted tasks, and to assemble Project materials in orderly files.

D. In the event of any termination under Paragraph 6.04, Engineer will be entitled to invoice Owner and to receive full payment for all services performed or furnished in accordance with this Agreement and all reimbursable expenses incurred through the effective date of termination.

6.05 *Controlling Law*

A. This Agreement is to be governed by the law of the state or jurisdiction in which the subject matter of the Assignment is located.

6.06 *Successors, Assigns, and Beneficiaries*

A. Owner and Engineer are hereby bound and the successors, executors, administrators, and legal representatives of Owner and Engineer (and to the extent permitted by Paragraph 6.06.B the assigns of Owner and Engineer) are hereby bound to the other party to this Agreement and to the successors, executors, administrators and legal representatives (and said assigns) of such other party, in respect of all covenants, agreements, and obligations of this Agreement.

B. Neither Owner nor Engineer may assign, sublet, or transfer any rights under or interest (including, but without limitation, moneys that are due or may become due) in this Agreement without the written consent of the other, except to the extent that any assignment, subletting, or transfer is mandated or restricted by law. Unless specifically stated to the contrary in any written consent to an assignment, no assignment will release or discharge the assignor from any duty or responsibility under this Agreement.

C. Unless expressly provided otherwise in this Agreement:

1. Nothing in this Agreement shall be construed to create, impose, or give rise to any duty owed by Owner or Engineer to any Contractor, Subcontractor, Supplier, or other individual or entity, or to any surety for or employee of any of them.

2. All duties and responsibilities undertaken pursuant to this Agreement will be for the sole and exclusive benefit of Owner and Engineer and not for the benefit of any other party. Any and all Documents prepared by Engineer, including but not limited to the Report to be prepared pursuant to Exhibit A, are prepared solely for the use and benefit of Owner, unless expressly agreed otherwise by Engineer.

6.07   *Dispute Resolution*

A. Owner and Engineer agree to negotiate each dispute between them in good faith during the 30 days after notice of dispute. If negotiations are unsuccessful in resolving the dispute, then the dispute shall be mediated. If mediation is unsuccessful, then the parties may exercise their rights at law.

6.08   *Environmental Condition of Site*

A. Owner has disclosed to Engineer in writing the existence of all known and suspected Asbestos, PCBs, Petroleum, Hazardous Waste, Radioactive Material, hazardous substances, and other Constituents of Concern located at or near the Site, including type, quantity, and location.

B. Owner represents to Engineer that to the best of its knowledge no Constituents of Concern, other than those disclosed in writing to Engineer, exist at the Site.

C. If Engineer encounters or learns of an undisclosed Constituent of Concern at the Site, then Engineer shall notify (1) Owner and (2) appropriate governmental officials if Engineer reasonably concludes that doing so is required by applicable Laws or Regulations.

D. It is acknowledged by both parties that Engineer's scope of services does not include any services related to Constituents of Concern. If Engineer or any other party encounters an undisclosed Constituent of Concern, or if investigative or remedial action, or other professional services, are necessary with respect to disclosed or undisclosed Constituents of Concern, then Engineer may, at its option and without liability for consequential or any other damages, suspend performance of services on the portion of the Project affected thereby until Owner: (1) retains appropriate specialist consultants or contractors to identify and, as appropriate, abate, remediate, or remove the Constituents of Concern; and (2) warrants that the Site is in full compliance with applicable Laws and Regulations.

E. If the presence at the Site of undisclosed Constituents of Concern adversely affects the performance of Engineer's services under this Agreement, then the Engineer shall have the option of (1) accepting an equitable adjustment in its compensation or in the time of completion, or both; or (2) terminating this Agreement for cause on 30 days notice.

F. Owner acknowledges that Engineer is performing professional services for Owner and that Engineer is not and shall not be required to become an "owner" "arranger," "operator," "generator," or "transporter" of hazardous substances, as defined in the Comprehensive Environmental Response, Compensation, and Liability Act (CERCLA), as amended, which are or may be encountered at or near the Site in connection with Engineer's activities under this Agreement.

6.09   *Indemnification and Mutual Waiver*

A. *Indemnification by Engineer:* To the fullest extent permitted by law, Engineer shall indemnify and hold harmless Owner, and Owner's officers, directors, members, partners, agents, consultants, and employees from reasonable claims, costs, losses, and damages arising out of or relating to the Assignment or Project, provided that any such claim, cost, loss, or damage is attributable to bodily injury, sickness, disease, or death, or to injury to or destruction of tangible property, including the loss of use resulting therefrom, but only to the extent caused by any negligent act or omission of Engineer or Engineer's officers, directors, members, partners, agents, employees, or Consultants.

B. *Indemnification by Owner:* Owner shall indemnify and hold harmless Engineer and its officers, directors, members, partners, agents, employees, and Consultants as required by Laws and Regulations.

> ***NOTE TO USER:***
> *Many professional service agreements contain mutual indemnifications. If the parties elect to provide a mutual counterpart to the indemnification of Owner by Engineer in Paragraph 6.09.A, then include the following indemnification of Engineer by Owner by inserting it immediately after the sentence ending in the word "Regulations." in 6.09.B immediately above:*

> > *In addition, to the fullest extent permitted by law, Owner shall indemnify and hold harmless Engineer and its officers, directors, members, partners, agents, Consultants, and employees from reasonable claims, costs, losses, and damages arising out of or relating to the Assignment or Project, provided that any such claim, cost, loss, or damage is attributable to bodily injury, sickness, disease, or death, or to injury to or destruction of tangible property, including the loss of use resulting therefrom, but only to the extent caused by any negligent act or omission of Owner or Owner's officers, directors, members, partners, agents, consultants, employees, or others retained by or under contract to the Owner with respect to this Assignment or to the Project.*

C. *Environmental Indemnification:* To the fullest extent permitted by law, Owner shall indemnify and hold harmless Engineer and its officers, directors, members, partners, agents, employees, and Consultants from and against any and all claims, costs, losses, and damages (including but not limited to all fees and charges of engineers, architects, attorneys and other professionals, and all court, arbitration, or other dispute resolution costs) caused by, arising out of, relating to, or resulting from a Constituent of Concern at, on, or under the Site, provided that (1) any such claim, cost, loss, or damage is attributable to bodily injury, sickness, disease, or death, or to injury to or destruction of tangible property (other than the Work itself), including the loss of use resulting therefrom, and (2) nothing in this paragraph shall obligate Owner to indemnify any individual or entity from and against the consequences of that individual's or entity's own negligence or willful misconduct.

D. *Percentage Share of Negligence:* To the fullest extent permitted by law, a party's total liability to the other party and anyone claiming by, through, or under the other party for any cost, loss, or damages caused in part by the negligence of the party and in part by the negligence of the other party or any other negligent entity or individual, shall not exceed the percentage share that the party's negligence bears to the total negligence of Owner, Engineer, and all other negligent entities and individuals.

E. *Mutual Waiver:* To the fullest extent permitted by law, Owner and Engineer waive against each other, and the other's employees, officers, directors, members, agents, insurers, partners, and consultants, any and all claims for or entitlement to special, incidental, indirect, or consequential damages arising out of, resulting from, or in any way related to the Assignment or Project.

6.10 *Limitation of Engineer's Liability*

A. To the fullest extent permitted by law, the total liability, in the aggregate, of Engineer and Engineer's officers, directors, partners, members, employees, agents, and Consultants, or any of them, to Owner and anyone claiming by, through, or under Owner, for any and all injuries, losses, damages and expenses whatsoever arising out of, resulting from, or in any way related to the Assignment, this Agreement, or the Project from any cause or causes including but not limited to the negligence, professional errors or omissions, strict liability, or breach of contract or warranty, express or implied, of Engineer or Engineer's officers, directors, partners, members, employees, agents, or Consultants, or any of them, shall not exceed the total amount of $50,000 or the total compensation paid to Engineer under this Agreement, whichever is greater.

6.11 *Miscellaneous Provisions*

A. *Notices:* Any notice required under this Agreement will be in writing, addressed to the appropriate party at its address on the signature page and given personally, by facsimile, by registered or certified mail postage prepaid, or by a commercial courier service. All notices shall be effective upon the date of receipt.

B. *Survival:* All express representations, waivers, indemnifications, and limitations of liability included in this Agreement will survive its completion or termination for any reason.

C. *Severability:* Any provision or part of the Agreement held to be void or unenforceable under any Laws or Regulations shall be deemed stricken, and all remaining provisions shall continue to be valid and binding upon Owner and Engineer, which agree that the Agreement shall be reformed to replace such stricken provision or part thereof with a valid and enforceable provision that comes as close as possible to expressing the intention of the stricken provision.

D. *Waiver:* A party's non-enforcement of any provision shall not constitute a waiver of that provision, nor shall it affect the enforceability of that provision or of the remainder of this Agreement.

E. *Accrual of Claims:* To the fullest extent permitted by law, all causes of action arising under this Agreement shall be deemed to have accrued, and all statutory periods of limitation shall commence, no later than the date of completion of the Assignment.

ARTICLE 7 – DEFINITIONS

7.01   *Defined Terms*

A. Wherever used in this Agreement (including the Exhibits hereto) terms (including the singular and plural forms) printed with initial capital letters have the meanings indicated in the text above, in the exhibits, or in the following provisions:

1. *Additional Services* – The services to be performed for or furnished to Owner by Engineer in accordance with Part 2 of Exhibit A of this Agreement.

2. *Agreement* – This written contract for study and report professional services between Owner and Engineer, including all exhibits identified in Paragraph 8.01 and any duly executed amendments.

3. *Asbestos* – Any material that contains more than one percent asbestos and is friable or is releasing asbestos fibers into the air above current action levels established by the United States Occupational Safety and Health Administration.

4. *Basic Services* – The services to be performed for or furnished to Owner by Engineer in accordance with Part 1 of Exhibit A of this Agreement.

5. *Constituent of Concern* – Any substance, product, waste, or other material of any nature whatsoever (including, but not limited to, Asbestos, Petroleum, Radioactive Material, and PCBs) which is or becomes listed, regulated, or addressed pursuant to (a) the Comprehensive Environmental Response, Compensation and Liability Act, 42 U.S.C. §§9601 et seq. ("CERCLA"); (b) the Hazardous Materials Transportation Act, 49 U.S.C. §§1801 et seq.; (c) the Resource Conservation and Recovery Act, 42 U.S.C. §§6901 et seq. ("RCRA"); (d) the Toxic Substances Control Act, 15 U.S.C. §§2601 et seq.; (e) the Clean Water Act, 33 U.S.C. §§1251 et seq.; (f) the Clean Air Act, 42 U.S.C. §§7401 et seq.; and (g) any other federal, state, or local statute, law, rule, regulation, ordinance, resolution, code, order, or decree regulating, relating to, or imposing liability or standards of conduct concerning, any hazardous, toxic, or dangerous waste, substance, or material.

6. *Construction Cost* – The cost to Owner of the construction of a recommended solution presented in the Report furnished by Engineer under Exhibit A, or of a specific portion of the Project for which Engineer has agreed to provide opinions of cost. Construction Cost includes the cost of construction labor, services, materials, equipment, insurance, and bonding, but does not include costs of services of Engineer or other design professionals and consultants; cost of land or rights-of-way, or compensation for damages to properties; Owner's costs for legal, accounting, insurance counseling, or auditing services; interest or financing charges incurred in connection with the Project; or the cost of other services to be provided by others to Owner. Construction Cost is one of the items comprising Total Project Costs.

7. *Consultants* – Individuals or entities having a contract with Engineer to furnish services with respect to this Assignment as Engineer's independent professional associates and consultants, subcontractors, or vendors.

8. *Documents* – Data, studies, reports (including the Report referred to in Exhibit A), and other deliverables, whether in printed or electronic media format, provided or furnished by Engineer to Owner pursuant to this Agreement.

9. *Effective Date* – The date indicated in this Agreement on which it becomes effective, but if no such date is indicated, the date on which this Agreement is signed and delivered by the last of the parties to sign and deliver.

10. *Engineer* – The individual or entity named as such in this Agreement.

11. *Hazardous Waste* – The term Hazardous Waste shall have the meaning provided in Section 1004 of the Solid Waste Disposal Act (42 USC Section 6903) as amended from time to time.

12. *Laws and Regulations; Laws or Regulations* – Any and all applicable laws, rules, regulations, ordinances, codes, and orders of any and all governmental bodies, agencies, authorities, and courts having jurisdiction.

13. *Owner* – The individual or entity with which Engineer has entered into this Agreement and for which Engineer's services are to be performed.

14. *PCBs* – Polychlorinated biphenyls.

15. *Petroleum* – Petroleum, including crude oil or any fraction thereof which is liquid at standard conditions of temperature and pressure (60 degrees Fahrenheit and 14.7 pounds per square inch absolute), such as oil, petroleum, fuel oil, oil sludge, oil refuse, gasoline, kerosene, and oil mixed with other non-hazardous waste and crude oils.

16. *Project* – The total study, design, and construction to be carried out by Owner through its employees, agents, design professionals, consultants, contractors, and others, of which the Assignment is a preliminary part.

17. *Radioactive Material* – Source, special nuclear, or byproduct material as defined by the Atomic Energy Act of 1954 (42 USC Section 2011 et seq.) as amended from time to time.

18. *Site* – Lands or areas where the subject matter of the Assignment or the Project is located.

19. *Total Project Costs* – The total cost of study, design, and construction of the Project, including Construction Cost and all other Project construction labor, services, materials, equipment, insurance, and bonding costs, allowances for contingencies, and the total costs of services of Engineer and other design professionals and consultants, together with such other Project-related costs that Owner furnishes for inclusion, including but not limited to cost of land, rights-of-way, compensation for damages to properties, Owner's costs for legal, accounting, insurance counseling, and auditing services, interest and financing charges incurred in connection with the Project, and the cost of other services to be provided by others to Owner.

ARTICLE 8 – EXHIBITS AND SPECIAL PROVISIONS

8.01   *Exhibits Included*

A. Exhibit A, Engineer's Services.

B. Exhibit B, Reserved. *Not Included.*

C. Exhibit C, Standard Hourly Rates and Reimbursable Expenses Schedule

NOTE TO USER:
If an exhibit is not included as part of a specific Agreement,
indicate "not included" after the listed exhibit item.

8.02   *Total Agreement*

A. This Agreement, (together with the exhibits identified above) constitutes the entire agreement between Owner and Engineer and supersedes all prior written or oral understandings. This Agreement may only be amended, supplemented, modified, or canceled by a duly executed written instrument.

8.03   *Designated Representatives*

A. With the execution of this Agreement, Engineer and Owner shall designate specific individuals to act as Engineer's and Owner's representatives with respect to the Assignment and the responsibilities of Owner under this Agreement. Such an individual shall have authority to transmit instructions, receive information, and render decisions relative to the Assignment on behalf of the respective party whom the individual represents.

8.04   *Engineer's Certifications*

A. Engineer certifies that it has not engaged in corrupt, fraudulent, or coercive practices in competing for or in executing the Agreement. For the purposes of this Paragraph 8.04:

1. "corrupt practice" means the offering, giving, receiving, or soliciting of any thing of value likely to influence the action of a public official in the selection process or in the Agreement execution;

2. "fraudulent practice" means an intentional misrepresentation of facts made (a) to influence the selection process or the execution of the Agreement to the detriment of Owner, or (b) to deprive Owner of the benefits of free and open competition;

3. "coercive practice" means harming or threatening to harm, directly or indirectly, persons or their property to influence their participation in the selection process or affect the execution of the Agreement.

IN WITNESS WHEREOF, the parties hereto have executed this Agreement, the Effective Date of which is indicated on page 1.

Owner:	Engineer:
By:	By:
Title:	Title:
Date	Date
Signed:	Signed:
	Engineer License or Firm's
	Certificate No. (if required):
	State of:
Address for giving notices:	Address for giving notices:
Designated Representative (Paragraph 8.03.A):	Designated Representative (Paragraph 8.03.A):
Name:	Name:
Title:	Title:
Phone Number:	Phone Number:
Facsimile Number:	Facsimile Number:
E-Mail Address:	E-Mail Address:

This is **EXHIBIT A, Engineer's Services,** referred to in and part of the **Agreement between Owner and Engineer for Study and Report Professional Services** dated _____, _____.

Engineer's Services

Article 1 of the Agreement is supplemented to include the following agreement of the parties:

Engineer shall provide Basic and Additional Services as set forth below.

PART 1 – BASIC SERVICES

A1.01 *Study and Report Phase*

A. Engineer shall:

1. Consult with Owner regarding fulfillment of Owner's responsibilities under Article 2.

2. Advise Owner of any need for Owner to provide other data or services.

3. Identify, consult with, and analyze requirements of governmental authorities having jurisdiction to review or approve the Report to be prepared by Engineer, other subject matter of the Assignment, or relevant aspects of the Project.

4. Review environmental assessments and impact statements relating to the Assignment, and report to Owner on the effect of any such environmental documents on the subject matter of the Assignment, including contemplated design and construction;

5. Identify and evaluate *[insert specific number or list here]* potential solutions available to Owner and, after consultation with Owner, recommend to Owner those solutions that in Engineer's judgment meet Owner's requirements.

6. Prepare a report (the "Report") which will, as appropriate, contain schematic layouts, sketches, and conceptual design criteria with appropriate exhibits to indicate the agreed-to requirements, considerations involved, and those potential solutions available to Owner that Engineer recommends. For each recommended solution Engineer will provide the following, which will be separately itemized: opinion of probable Construction Cost; proposed allowances for contingencies; the estimated total costs of design, professional, and related services to be provided by Engineer and other design professionals; and a projection of Total Project Costs, based in part on information furnished by Owner for other items and services included within the definition of Total Project Costs.

7. Perform or provide the following additional Study and Report tasks or deliverables: *[here list any such tasks or deliverables]*

A1.02   *Times for Rendering Services*

A. Furnish _____ review copies of the Report to Owner within _____ calendar days of the Effective Date and review it with Owner.

B. Revise the Report in response to Owner's and other parties' comments, as appropriate, and furnish _____ final copies of the revised Report to the Owner within _____ calendar days after completion of reviewing it with Owner.

C. Engineer's Assignment will be considered complete when all deliverables set forth in Exhibit A are submitted to Owner.

PART 2 – ADDITIONAL SERVICES

A2.01   *Additional Services Requiring Owner's Written Authorization*

A. If authorized in writing by Owner, Engineer shall furnish or obtain from others Additional Services of the types listed below.

  1. Preparation of applications and supporting documents (in addition to those furnished under Basic Services) for private or governmental grants, loans, or advances in connection with the Assignment or Project; preparation of environmental assessments and impact statements; and assistance in obtaining approvals of authorities having jurisdiction over the anticipated environmental impact of the Project.

  2. Services to make measured drawings of or to investigate existing conditions or facilities, or to verify the accuracy of drawings or other information furnished by Owner or others.

  3. Services resulting from significant changes in the scope, extent, or character of the of the Assignment including, but not limited to, changes in size, complexity, Owner's schedule, character of construction, or method of financing; and revising previously accepted studies and reports when such revisions are required by changes in Laws and Regulations enacted subsequent to the Effective Date or are due to any other causes beyond Engineer's control.

  4. Services resulting from Owner's request to evaluate additional potential solutions beyond those identified in Paragraph A1.01.A.5.

  5. Services required as a result of Owner providing incomplete or incorrect information to Engineer.

  6. Providing renderings or models for Owner's use.

  7. Undertaking investigations and studies including, but not limited to, detailed consideration of operations, maintenance, and overhead expenses; the preparation of financial feasibility and cash flow studies, rate schedules, and appraisals; assistance in obtaining financing for the Project; evaluating processes available for licensing, and assisting Owner in obtaining process licensing; detailed quantity surveys of materials, equipment, and labor; and audits or inventories required in connection with construction performed by Owner.

8. Providing assistance in responding to the presence of any Constituent of Concern at the Site, in compliance with current Laws and Regulations.

9. Preparing to serve or serving as a consultant or witness for Owner in any litigation, arbitration, or other dispute resolution process related to the Assignment.

10. Other services performed or furnished by Engineer not otherwise provided for in this Agreement.

[EXHIBIT B: RESERVED].

NOTE TO USER:

Exhibit B is not used as a part of EJCDC E-525.

This is **EXHIBIT C, Standard Hourly Rates and Reimbursable Expenses Schedule,** referred to in and part of the Agreement Between Owner and Engineer for Study and Report Professional Services dated _____, _____.

Part One: Standard Hourly Rates Schedule

A. *Standard Hourly Rates:*

1. Standard Hourly Rates are set forth in this Exhibit C and include salaries and wages paid to personnel in each billing class plus the cost of customary and statutory benefits, general and administrative overhead, non-project operating costs, and operating margin or profit.

2. The Standard Hourly Rates apply only as specified in Article 4, and are subject to annual review and adjustment.

B. *Schedule:*

Hourly rates for services performed on or after the date of the Agreement are:

Billing Class VIII	$ _____/hour
Billing Class VII	_____/hour
Billing Class VI	_____/hour
Billing Class V	_____/hour
Billing Class IV	_____/hour
Billing Class III	_____/hour
Billing Class II	_____/hour
Billing Class I	_____/hour
Support Staff	_____/hour

Part Two: Reimbursable Expenses Factors

1. Engineer shall be entitled to reimbursement of expenses under Parts Two and Three only as specified in Article 4.

2. Owner shall reimburse Engineer for the expenses listed in Part Three below subject to an administrative factor of 1.0, unless another factor is indicated for a specific Reimbursable Expense. *[NOTE TO USER: List any Reimbursable Expenses subject to reimbursement at a factor greater than 1.0 here, as part of this Paragraph 2, or indicate the higher factor on the specific line item in Part Three.]*

3. Owner shall reimburse Engineer for Engineer's Consultant's charges subject to an administrative factor of 1._____. *[NOTE TO USER: Specify the administrative factor (typically greater than 1.0) that applies to Consultant's charges.]*

Part Three: Reimbursable Expenses Schedule

Reimbursable Expenses are subject to annual review and adjustment. As of the date of the Agreement, Reimbursable Expenses for services performed are:

Fax	$_____/page
8"x11" Copies/Impressions	_____/page
Blue Print Copies	_____/sq. ft.
Reproducible Copies (Mylar)	_____/sq. ft.
Reproducible Copies (Paper)	_____/sq. ft.
Mileage (auto)	_____/mile
Field Truck Daily Charge	_____/day
Mileage (Field Truck)	_____/mile
Field Survey Equipment	_____/day
Confined Space Equipment	_____/day plus expenses
Resident Project Representative Equipment	_____/month
Specialized Software	_____/hour
CAD Charge	_____/hour
CAE Terminal Charge	_____/hour
Video Equipment Charge	_____/day, $_____/week, or $_____/month
Electrical Meters Charge	_____/week, or $_____/month
Flow Meter Charge	_____/week, or $_____/month
Rain Gauge	_____/week, or $_____/month
Sampler Charge	_____/week, or $_____/month
Dissolved Oxygen Tester Charge	_____/week
Fluorometer	_____/week
Laboratory Pilot Testing Charge	_____/week, or $_____/month
Soil Gas Kit	_____/day
Submersible Pump	_____/day
Water Level Meter	_____/day, or $_____/month

Soil Sampling	_____/sample
Groundwater Sampling	_____/sample
Health and Safety Level D	_____/day
Health and Safety Level C	_____/day
Electronic Media Charge	_____/hour
Long Distance Phone Calls	at cost
Mobile Phone	_____/day
Meals and Lodging	at cost

NOTE TO USER:
Customize this Reimbursable Expenses Schedule to reflect
anticipated reimbursable expenses for this specific Assignment.

STANDARD FORM OF PRELIMINARY AGREEMENT BETWEEN OWNER AND DESIGN-BUILDER

Document No. 520
Second Edition, 2010
© Design-Build Institute of America
Washington, DC

Design-Build Institute of America - Contract Documents
LICENSE AGREEMENT

By using the DBIA Contract Documents, you agree to and are bound by the terms of this License Agreement.

1. **License.** The Design-Build Institute of America ("DBIA") provides DBIA Contract Documents and licenses their use worldwide. You acknowledge that DBIA Contract Documents are protected by the copyright laws of the United States. You have a limited nonexclusive license to: (a) Use DBIA Contract Documents on any number of machines owned, leased or rented by your company or organization; (b) Use DBIA Contract Documents in printed form for bona fide contract purposes; and (c) Copy DBIA Contract Documents into any machine-readable or printed form for backup or modification purposes in support of your permitted use.

2. **User Responsibility.** You assume sole responsibility for the selection of specific documents or portions thereof to achieve your intended results, and for the installation, use, and results obtained from the DBIA Contract Documents. You acknowledge that you understand that the text of the DBIA Contract Documents has important legal consequences and that consultation with an attorney is recommended with respect to use or modification of the text. You will not represent that any of the contract documents you generate from DBIA Contract Documents are DBIA documents unless (a) the document text is used without alteration or (b) all additions and changes to, and deletions from, the text are clearly shown.

3. **Copies.** You may not use, copy, modify, or transfer DBIA Contract Documents, or any copy, modification or merged portion, in whole or in part, except as expressly provided for in this license. Reproduction of DBIA Contract Documents in printed or machine-readable format for resale or educational purposes is expressly prohibited. You will reproduce and include DBIA's copyright notice on any printed or machine-readable copy, modification, or portion merged into another document or program.

4. **Transfers.** You may not transfer possession of any copy, modification or merged portion of DBIA Contract Documents to another party, except that a party with whom you are contracting may receive and use such transferred material solely for purposes of its contract with you. You may not sublicense, assign, or transfer this license except as expressly provided in this Agreement, and any attempt to do so is void.

5. **Term.** The license is effective for one year from the date of purchase. DBIA may elect to terminate it earlier, by written notice to you, if you fail to comply with any term or condition of this Agreement.

6. **Limited Warranty.** DBIA warrants the electronic files or other media by which DBIA Contract Documents are furnished to be free from defects in materials and workmanship under normal use during the Term. There is no other warranty of any kind, expressed or implied, including, but not limited to, the implied warranties of merchantability and fitness for a particular purpose. Some states do not allow the exclusion of implied warranties, so the above exclusion may not apply to you. This warranty gives you specific legal rights and you may also have other rights which vary from state to state. DBIA does not warrant that the DBIA Contract Documents will meet your requirements or that the operation of DBIA Contract Documents will be uninterrupted or error free.

7. **Limitations of Remedies.** DBIA's entire liability and your exclusive remedy shall be: the replacement of any document not meeting DBIA's "Limited Warranty" which is returned to DBIA with a copy of your receipt, or at DBIA's election, your money will be refunded. In no event will DBIA be liable to you for any damages, including any lost profits, lost savings or other incidental or consequential damages arising out of the use or inability to use DBIA Contract Documents even if DBIA has been advised of the possibility of such damages, or for any claim by any other party. Some states do not allow the limitation or exclusion of liability for incidental or consequential damages, so the above limitation or exclusion may not apply to you.

8. **Acknowledgement.** You acknowledge that you have read this agreement, understand it and agree to be bound by its terms and conditions and that it will be governed by the laws of the District of Columbia. You further agree that it is the complete and exclusive statement of your agreement with DBIA which supersedes any proposal or prior agreement, oral or written, and any other communications between the parties relating to the subject matter of this agreement.

INSTRUCTIONS

For DBIA Document No. 520 Standard Form of Preliminary Agreement Between Owner and Design-Builder (2010 Edition)

Checklist

Use this Checklist to ensure that the Agreement is fully completed and all exhibits are attached.

_____	Page 1	Owner's name, address and form of business
_____	Page 1	Design-Builder's name, address and form of business
_____	Page 1	Project name and address
_____	Section 2.7	Attach exhibit for Additional Services (optional)
_____	Section 4.2.2	Complete blanks for additional sum for use of Work Product
_____	Section 5.1	Complete blanks for calendar days
_____	Section 5.2	Attach exhibit for interim milestone dates (optional)
_____	Section 6.1	Insert the Contract Price
_____	Section 7.1	Insert the payment method
_____	Section 7.2	Complete blanks for interest rate
_____	Section 9.8	Insert any other provisions (optional)
_____	Last Page	Owner's and Design-Builder's execution of the Agreement

General Instructions

No.	Subject	Instruction
1.	Standard Forms	Standard form contracts have long served an important function in the United States and international construction markets. The common purpose of these forms is to provide an economical and convenient way for parties to contract for design and construction services. As standard forms gain acceptance and are used with increased frequency, parties are able to enter into contracts with greater certainty as to their rights and responsibilities.
2.	DBIA Standard Form Contract Documents	Since its formation in 1993, the Design-Build Institute of America ("DBIA") has regularly evaluated the needs of owners, design-builders, and other parties to the design-build process in preparation for developing its own contract forms. Consistent with DBIA's mission of promulgating best design-build practices, DBIA believes that the design-build contract should reflect a balanced approach to risk that considers the legitimate interests of all parties to the design-build process. DBIA's Standard Form Contract Documents reflect a modern risk allocation approach, allocating each risk to the party best equipped to manage and minimize that risk, with the goal of promoting best design-build practices.
3.	Use of Non-DBIA Documents	To avoid inconsistencies among documents used for the same project, DBIA's Standard Form Contract Documents should not be used in conjunction with non-DBIA documents unless the non-DBIA documents are appropriately modified on the advice of legal counsel.
4.	Legal Consequences	DBIA Standard Form Contract Documents are legally binding contracts with important legal consequences. Contracting parties are advised and encouraged to seek legal counsel in completing or modifying these Documents.
5.	Reproduction	DBIA hereby grants to purchasers a limited license to reproduce its documents consistent with the License Agreement accompanying these Documents. At least two original versions of the Agreement should be signed by the parties. Any other reproduction of DBIA Documents is prohibited.

No.	Subject	Instruction
6.	Modifications	Effective contracting is accomplished when the parties give specific thought to their contracting goals and then tailor the contract to meet the unique needs of the project and the design-build team. For that reason, these Documents may require modification for various purposes including, for example, to comply with local codes and laws, or to add special terms. Also, in some instances, these Documents must be modified to indicate the selection of a particular contract term. Any modifications to these Documents should be underlined to distinguish them from original language. Any modifications should be initialed by the parties. To delete provisions, strike through the printed words so that original language remains legible. At no time should a document be re-typed in its entirety. Re-creating the document violates copyright laws and destroys one of the advantages of standard forms - familiarity with the terms.
7.	Execution	It is good practice to execute two original copies of the Agreement. Only persons authorized to sign for the contracting parties may execute the Agreement.

Specific Instructions

Section	Title	Instruction
General	Purpose of This Document	DBIA Document No. 520, *Standard Form of Preliminary Agreement Between Owner and Design-Builder* ("Agreement") is for preliminary services only, not for construction services, and shall be used when Owner decides not to contract for the complete design and construction at one time. Use of this Agreement anticipates a two-stage approach to the Project, whereby Owner retains the Design-Builder to assist in the review and/or development of Owner's Project Criteria and for preliminary Schematic Design Documents. Then, depending upon the Design-Builder's Proposal, Owner has the option of contracting for final design and construction services by executing either DBIA Document No. 525, *Standard Form of Agreement Between Owner and Design-Builder – Lump Sum*, 2010 Edition, or DBIA Document No. 530, *Standard Form of Agreement Between Owner and Design-Builder - Cost Plus Fee with an Option for a Guaranteed Maximum Price*, 2010 Edition. DBIA Document No. 525 and DBIA Document No. 530 can also be used when Owner desires preliminary services as part of a complete design-build contract. Under this Agreement, Design-Builder provides a Schematic Design and a Proposal for the completion of the design and construction. If Owner has not completed its Project Criteria before executing this Agreement, the Agreement allows for Owner to pay Design-Builder to assist in the development of Owner's Project Criteria as an Additional Service. If Owner does not accept the Proposal Design-Builder prepares under this Agreement, Owner may select another design-builder to complete the final design and construction. This Agreement allows Owner a limited license to use the Schematic Design and other Work Product created by Design-Builder under this Agreement to complete the Project, providing that Owner indemnifies Design-Builder for claims arising out of the use of the Work Product, and further agrees to compensate Design-Builder for the use of its Work Product. It is anticipated that Owner and Design-Builder will negotiate the compensation for the use of the Work Product prior to the execution of this Agreement.
General	Purpose of These Instructions	These Instructions are not part of this Agreement, but are provided to aid the parties in their understanding of the Agreement and in completing the Agreement.

Section	Title	Instruction
General	Related Documents	This Agreement includes its own abbreviated general conditions and does not require the use of DBIA Document No. 535, *Standard Form of General Conditions of Contract Between Owner and Design-Builder*, 2010 Edition ("General Conditions of Contract").
		Upon completion of the services under this Agreement, the parties may complete the final design and construction of the Project by executing either DBIA Document No. 525 or DBIA Document No. 530, and the accompanying General Conditions of Contract.
General	Date	On Page 1, enter the date when both parties reach a final understanding. It is possible, due to logistical reasons, that the dates when the parties execute the Agreement may be different. Once both parties execute the Agreement, the effective date of the Agreement will be the date recorded on Page 1. This date does not, however, determine Contract Time, which is measured according to the terms of Article 5.
General	Parties: Owner and Design-Builder	On Page 1, enter the legal name and full address of Owner and Design-Builder, as well as the legal form of each entity, e.g., corporation, partnership, limited partnership, limited liability company, or other.
1.2	Definitions	Although this Agreement is a stand-alone document, terms, words and phrases used in the Agreement shall have the same meanings used in the General Conditions of Contract.
2.1	Design Services	The parties should be aware that in addition to requiring compliance with state licensing laws for design professionals, some states also require that the design professional have a corporate professional license.
2.2	Preliminary Services	If Owner's Project Criteria are provided, Design-Builder's review and written evaluation of the Project Criteria will promote a clear understanding of Owner's program prior to Design-Builder's preparation of Schematic Design Documents.
		This Agreement acknowledges that Owner may not have developed its Project Criteria prior to the execution of this Agreement, and provides that Owner may pay Design-Builder an additional fee to assist in this effort pursuant to Section 2.7, Additional Services.
2.4	Proposal	Upon completion of the Schematic Design Documents, Design-Builder shall prepare its Proposal, which shall contain the information described in Sections 2.4.1, 2.4.2, 2.4.3, and 2.4.4. If the parties agree to additional or other requirements, state these requirements in Section 9.8, Other Provisions, or modify Section 2.4 appropriately.
2.4.2	Schedule	Given that expedited delivery is one of the primary factors driving many owners to select the design-build method, DBIA strongly believes that the parties should discuss and understand what each party must do to support the Project schedule. The entire Work, both design and construction, should be scheduled. The schedule should indicate the dates for the start and completion of the various stages of the Work, including the date when Owner information and approvals are required and any Owner created constraints.
2.4.3	Other information	Other information may be required to enter into a subsequent agreement for final design and construction. For example, if a Guaranteed Maximum Price ("GMP") is proposed, Design-Builder will need to provide all documents used as the basis for the GMP and identify them in a GMP Exhibit. For a Lump Sum proposal, Design-Builder may need to create a Design-Builder's Deviation List to identify any deviations from Owner's Project Criteria.
		To identify other information that may be required, Design-Builder should familiarize itself with the terms of DBIA Document No. 525 or DBIA Document No. 530, and the accompanying General Conditions of Contract.

Section	Title	Instruction
2.6	Completion of the Agreement	If Design-Builder and Owner are unable to reach agreement on mutually acceptable revisions to the Proposal, and Owner does not accept the Proposal, Design-Builder will have no further involvement in the Project. Design-Builder's ownership of the Work Product prepared under this Agreement, and Owner's limited license to its use are described in Article 4, Ownership of Work Product.
2.7	Additional Services	Attach as a separate exhibit to this Agreement the scope of work for any Additional Services to be performed by Design-Builder, such as the development of Owner's Project Criteria pursuant to Section 2.2.2.
Article 4	Ownership of Work Product	This Agreement provides that unless the parties select the optional provisions set forth in Article 4, Design-Builder shall retain ownership of the Work Product it produces, but obligates Design-Builder to grant a limited license to Owner to use the Work Product conditioned on the terms of Sections 4.2.1 and 4.2.2. DBIA recognizes that the critical decisions affecting the success of the Project and the greatest intellectual effort are typically developed during the preliminary phase. The purpose of Article 4 is to balance the interests of Owner, whose schedule will be adversely affected if it cannot use the Work Product created under this Agreement, and Design-Builder, who may not have been compensated for the full market value of its preliminary work, and who must be protected from liability for design that it does not complete or construct.
4.2.2	Additional Compensation	To minimize disputes, the parties should negotiate prior to execution of the Agreement the amount of additional compensation Owner will pay Design-Builder for the right to use the Work Product. Enter the amount of this additional compensation.
5.1	Commencement Date	Design-Builder will commence its services within five (5) days of its receipt of Owner's Notice to Proceed, and complete its services no later than the calendar day duration of time negotiated between the parties. Enter the calendar days duration of this negotiated Contract Time.
5.2	Interim Dates	Attach an exhibit for interim dates, if any.
6.1	Contract Price	Insert the Contract Price, or the basis for its calculation as agreed to by the parties.
7.1	Payment	Insert the method agreed upon by Owner and Design-Builder for partial and final payment.
7.2	Interest	Enter the rate at which interest will accrue on Design-Builder's payments, if unpaid five (5) days after due.
9.1	Dispute Resolution	DBIA endorses the use of partnering, negotiation, mediation and arbitration for the prevention and resolution of disputes. This Agreement provides for mandatory, non-binding mediation followed by binding arbitration for any dispute not resolved by mediation. The parties are encouraged to attempt to negotiate a mutually satisfactory resolution of any claim, dispute, or controversy prior to resorting to mediation.
9.8	Other Provisions	Insert any other provisions.

TABLE OF CONTENTS

Standard Form of Preliminary Agreement Between Owner and Design-Builder

This document has important legal consequences. Consultation with
an attorney is recommended with respect to its completion or modification.

This **AGREEMENT** is made as of the _____ day of _____ in the year of
20 _____, by and between the following parties, for services in connection with the Project identified
below.

OWNER:
(Name and address)

DESIGN-BUILDER:
(Name and address)

PROJECT:
(Include Project name and location as it will appear in the Contract Documents)

In consideration of the mutual covenants and obligations contained herein, Owner and Design-Builder agree as
set forth herein.

Article 1

General

1.1   **Duty to Cooperate**. Owner and Design-Builder commit at all times to cooperate fully with each other, and proceed on the basis of trust and good faith to permit each party to realize the benefits afforded under this Agreement.

1.2   **Definitions.** Terms, words and phrases used in this Agreement shall have the meanings given them in DBIA Document No. 535, *Standard Form of General Conditions of Contract Between Owner and Design-Builder* (2010 Edition) ("General Conditions of Contract").

Article 2

Design-Builder's Services and Responsibilities

2.1   **Design Services.** Design-Builder shall, consistent with applicable state licensing laws, provide design services, including architectural, engineering and other design professional services, required by this Agreement. Such design services shall be provided through qualified, licensed design professionals who are either (i) employed by Design-Builder, or (ii) procured by Design-Builder from independent sources. Nothing in this Agreement is intended to create any legal or contractual relationship between Owner and any independent design professional.

2.2   **Preliminary Services.**

   2.2.1   Owner shall provide Design-Builder with Owner's Project Criteria describing Owner's program requirements and objectives for the Project. Owner's Project Criteria shall include Owner's use, space, price, time, site, performance and expandability requirements. Owner's Project Criteria may include conceptual documents, design specifications, design performance specifications and other technical materials and requirements prepared by or for Owner.

   2.2.2   If Owner's Project Criteria have not been developed prior to the execution of this Agreement, Design-Builder will assist Owner in developing Owner's Project Criteria, with such service deemed to be an Additional Service pursuant to Section 2.7 hereof. If Owner has developed Owner's Project Criteria prior to executing this Agreement, Design-Builder shall review and prepare a written evaluation of such criteria, including recommendations to Owner for different and innovative approaches to the design and construction of the Project. The parties shall meet to discuss Design-Builder's written evaluation of Owner's Project Criteria and agree upon what revisions, if any, should be made to such criteria.

2.3   **Schematic Design Documents.** Design-Builder shall prepare Schematic Design Documents based on Owner's Project Criteria, as may be revised in accordance with Section 2.2.2 hereof. The Schematic Design Documents shall include design criteria, drawings, diagrams and specifications setting forth the requirements of the Project. The parties shall meet to discuss the Schematic Design Documents and agree upon what revisions, if any, should be made. Design-Builder shall perform such agreed-upon revisions.

2.4   **Proposal.** Based on Owner's Project Criteria, the Schematic Design Documents, as each may be revised pursuant to Sections 2.2.2 and 2.3 above, and any other Basis of Design Documents upon which the parties may agree, Design-Builder shall submit a proposal to Owner (the "Proposal"), which shall include the following unless the parties mutually agree otherwise:

   2.4.1   a proposed contract price for the design and construction of the Project, which price shall be in the form of a lump sum or the cost of the work plus a fee with an option for a Guaranteed Maximum Price ("GMP");

2.4.2   a schedule and date of Substantial Completion of the Project upon which the Contract Price for the Project is based;

2.4.3   all other information necessary for the parties to enter into DBIA Document No. 525, *Standard Form of Agreement Between Owner and Design-Builder - Lump Sum* (2010 Edition) or DBIA Document No. 530, *Standard Form of Agreement Between Owner and Design-Builder - Cost Plus Fee with an Option for a Guaranteed Maximum Price* (2010 Edition), with the accompanying General Conditions of Contract, DBIA Document 535; and

2.4.4   the time limit for acceptance of the Proposal.

2.5   **Review of Proposal.** Design-Builder and Owner shall meet to discuss and review the Proposal. If Owner has any comments regarding the Proposal, or finds any inconsistencies or inaccuracies in the information presented, it shall promptly give written notice to Design-Builder of such comments or findings. If Design-Builder finds the revisions acceptable, Design-Builder shall, upon receipt of Owner's notice, adjust the Proposal.

2.6   **Completion of This Agreement.** Design-Builder's services under this Agreement shall be deemed completed upon meeting with Owner to discuss the Proposal and making those revisions to the Proposal, if any, Design-Builder finds acceptable.

2.7   **Additional Services.** Design-Builder shall perform the Additional Services set forth in a separate exhibit to this Agreement. The cost for such services shall be as mutually agreed upon by Owner and Design-Builder, with the Contract Price for this Agreement, as set forth in Section 6.1 hereof, being adjusted accordingly.

Article 3

Owner's Services and Responsibilities

3.1   **Timely Performance.** Owner shall throughout the performance of this Agreement cooperate with Design-Builder. Owner shall perform its responsibilities, obligations and services, including its reviews and approvals of Design-Builder's submissions, in a timely manner so as not to delay or interfere with Design-Builder's performance of its obligations under this Agreement.

3.2   **Owner's Project Criteria.** Owner shall provide Design-Builder with Owner's Project Criteria. If Owner desires that Design-Builder assist Owner in developing such criteria as an Additional Service under Section 2.7 hereof, Owner shall provide Design-Builder with its objectives, limitations and other relevant information regarding the Project.

3.3   **Owner Provided Information.** Owner shall provide, at its own cost and expense, for Design-Builder's information and use, the following, all of which Design-Builder is entitled to rely upon in performing its obligations hereunder:

3.3.1   Surveys describing the property, boundaries, topography and reference points for use during construction, including existing service and utility lines;

3.3.2   Geotechnical studies describing subsurface conditions, and other surveys describing other latent or concealed physical conditions at the Site;

3.3.3   Temporary and permanent easements, zoning and other requirements and encumbrances affecting land use or necessary to permit the proper design and construction of the Project;

3.3.4   A legal description of the Site;

3.3.5   To the extent available, as-built and record drawings of any existing structures at the Site; and

3.3.6   To the extent available, environmental studies, reports and impact statements describing the environmental conditions, including, but not limited to, Hazardous Conditions, in existence at the Site.

Article 4

Ownership of Work Product

4.1   **Work Product.** All drawings, specifications and other documents and electronic data furnished by Design-Builder to Owner under this Agreement ("Work Product") are deemed to be instruments of service and Design-Builder shall retain the ownership and property interests therein, including but not limited to any intellectual property rights, copyrights and/or patents, subject to the provisions set forth below.

4.2   **Owner's Limited License.** If Owner fails to enter into a contract on this Project with Design-Builder to complete the design and construction of the Project and Owner proceeds to design and construct the Project through its employees, agents or third parties, Design-Builder, upon payment in full of the amounts due Design-Builder under this Agreement, shall grant Owner a limited license to use the Work Product to complete the Project, conditioned on the following:

   4.2.1   Use of the Work Product is at Owner's sole risk without liability or legal exposure to Design-Builder or anyone working by or through Design-Builder, including Design Consultants of any tier (collectively the "Indemnified Parties"). Owner shall defend, indemnify and hold harmless the Indemnified Parties from and against any and all claims, damages, liabilities, losses and expenses, including attorneys' fees, arising out of or resulting from the use of the Work Product; and

   4.2.2   Owner agrees to pay Design-Builder the additional sum of _____ Dollars ($ _____) as compensation for the right to use the Work Product in accordance with this Article 4.

[At the parties' option, one of the following may be used in lieu of Section 4.2]:

☐   If Owner fails to enter into a contract on this Project with Design-Builder to complete the design and construction of the Project and Owner proceeds to design and construct the Project through its employees, agents or third parties, Design-Builder, upon payment in full of the amounts due Design-Builder under this Agreement: (a) grants Owner a limited license to use the Work Product in connection with the Owner's completion of the Project; and (b) transfers all ownership and property interests, including but not limited to any intellectual property rights, copyrights and/or patents, in those portions of the Work Product that consist of architectural and other design elements and specifications that are unique to the Project. The parties shall specifically designate those portions of the Work Product for which ownership in the Work Product shall be transferred. Such grant and transfer are conditioned on the following:

   4.2.1   Use of the Work Product is at Owner's sole risk without liability or legal exposure to Design-Builder or anyone working by or through Design-Builder, including Design Consultants of any tier (collectively the "Indemnified Parties"). Owner shall defend, indemnify and hold harmless the Indemnified Parties from and against any and all claims, damages, liabilities, losses and expenses, including attorneys' fees, arising out of or resulting from the use of the Work Product; and

   4.2.2   Owner agrees to pay Design-Builder the additional sum of _____ Dollars ($_____) as compensation for the right to use the Work Product in accordance with this Article 4.

or

☐ If Owner fails to enter into a contract on this Project with Design-Builder to complete the design and construction of the Project and Owner proceeds to design and construct the Project through its employees, agents or third parties, Design-Builder, upon payment in full of the amounts due Design-Builder under this Agreement, transfers to Owner all ownership and property interests, including but not limited to any intellectual property rights, copyrights and/or patents, in the Work Product. Such transfer is conditioned on the following:

4.2.1 Use of the Work Product is at Owner's sole risk without liability or legal exposure to Design-Builder or anyone working by or through Design-Builder, including Design Consultants of any tier (collectively the "Indemnified Parties"). Owner shall defend, indemnify and hold harmless the Indemnified Parties from and against any and all claims, damages, liabilities, losses and expenses, including attorneys' fees, arising out of or resulting from the use of the Work Product; and

4.2.2 Owner agrees to pay Design-Builder the additional sum of _____ Dollars ($ _____) as compensation for the right to use the Work Product in accordance with this Article 4.

Article 5

Contract Time

5.1 **Commencement Date.** Design-Builder shall commence performance of the services set forth in this Agreement within five (5) days of Design-Builder's receipt of Owner's Notice to Proceed ("Date of Commencement") unless the parties mutually agree otherwise in writing. Design-Builder shall complete such services no later than _____ (_____) calendar days after the Date of Commencement.

5.2 **Interim Dates.** Interim milestone dates, if any, of identified portions of the services set forth in this Agreement shall be achieved as described in a separate exhibit to this Agreement.

Article 6

Contract Price

6.1 **Contract Price.** The Contract Price for this Agreement is as set forth below: *(Provide for a fixed lump sum amount, cost of the work plus a fee with a GMP, hourly rates, or some other basis of compensation)*

6.2 **Scope of Contract Price.** The Contract Price shall be the full compensation due Design-Builder for the performance of all services set forth in this Agreement, and shall be deemed to include all the sales, use, consumer and other taxes mandated by applicable Legal Requirements. The Contract Price shall be adjusted to reflect any Additional Services agreed upon by the parties after execution of this Agreement.

Article 7

Procedure for Payment

7.1 **Payment.** Design-Builder and Owner agree upon the following method for partial and final payment to Design-Builder for the services hereunder: *(Insert terms)*

7.2   **Interest.** Payments due and unpaid by Owner to Design-Builder shall bear interest commencing five (5) days after payment is due at the rate of _____percent (_____%).

Article 8

Electronic Data

8.1   **Electronic Data.**

8.1.1   The parties recognize that Contract Documents, including drawings, specifications and three-dimensional modeling (such as Building Information Models) and other Work Product may be transmitted among Owner, Design-Builder and others in electronic media as an alternative to paper hard copies (collectively "Electronic Data").

8.2   **Transmission of Electronic Data.**

8.2.1   Owner and Design-Builder shall agree upon the software and the format for the transmission of Electronic Data. Each party shall be responsible for securing the legal rights to access the agreed-upon format, including, if necessary, obtaining appropriately licensed copies of the applicable software or electronic program to display, interpret and/or generate the Electronic Data.

8.2.2   Neither party makes any representations or warranties to the other with respect to the functionality of the software or computer program associated with the electronic transmission of Work Product. Unless specifically set forth in the Agreement, ownership of the Electronic Data does not include ownership of the software or computer program with which it is associated, transmitted, generated or interpreted.

8.2.3   By transmitting Work Product in electronic form, the transmitting party does not transfer or assign its rights in the Work Product. The rights in the Electronic Data shall be as set forth in Article 4 of the Agreement. Under no circumstances shall the transfer of ownership of Electronic Data be deemed to be a sale by the transmitting party of tangible goods.

8.3   **Electronic Data Protocol.**

8.3.1   The parties acknowledge that Electronic Data may be altered or corrupted, intentionally or otherwise, due to occurrences beyond their reasonable control or knowledge, including but not limited to compatibility issues with user software, manipulation by the recipient, errors in transcription or transmission, machine error, environmental factors, and operator error. Consequently, the parties understand that there is some level of increased risk in the use of Electronic Data for the communication of design and construction information and, in consideration of this, agree, and shall require their independent contractors, Subcontractors and Design Consultants to agree, to the following protocols, terms and conditions set forth in this Section 8.3.

8.3.2   Electronic Data will be transmitted in the format agreed upon in Section 8.2.1 above, including file conventions and document properties, unless prior arrangements are made in advance in writing.

8.3.3   The Electronic Data represents the information at a particular point in time and is subject to change. Therefore, the parties shall agree upon protocols for notification by the author to the recipient of any changes which may thereafter be made to the Electronic Data, which protocol shall also address the duty, if any, to update such information if such information changes prior to Final Completion.

8.3.4   The transmitting party specifically disclaims all warranties, expressed or implied, including, but not limited to, implied warranties of merchantability and fitness for a particular purpose, with respect to the media transmitting the Electronic Data. However, transmission of the Electronic Data via electronic means shall not invalidate or negate any duties pursuant to the applicable standard of care with respect to the creation of the Electronic Data, unless such data is materially changed or altered after it is transmitted to the receiving party, and the transmitting party did not participate in such change or alteration.

Article 9

Other Provisions

9.1   Initial Dispute Resolution. The parties agree that any claim, dispute or controversy arising out of or relating to this Agreement or the breach thereof that cannot be resolved through discussions by the parties shall be submitted to non-binding mediation administered by a mutually agreeable impartial mediator, or if the parties cannot so agree, a mediator designated by the American Arbitration Association ("AAA") pursuant to the Construction Industry Mediation Rules then in effect. Any claim, dispute, or controversy arising out of or relating to this Agreement or the breach thereof which has not been resolved by mediation shall be submitted to binding arbitration administered by the AAA pursuant to the Construction Industry Arbitration Rules then in effect.

9.2   Confidentiality. Confidential Information is defined as information which is determined by the transmitting party to be of a confidential or proprietary nature and: (i) the transmitting party identifies it as either confidential or proprietary; (ii) the transmitting party takes steps to maintain the confidential or proprietary nature of the information; and (iii) the document is not otherwise available in or considered to be in the public domain. The receiving party agrees to maintain the confidentiality of the Confidential Information and agrees to use the Confidential Information solely in connection with the services set forth in this Agreement.

9.3   Assignment. Neither Design-Builder nor Owner shall without the written consent of the other party assign, transfer, or sublet any portion or part of its obligations under this Agreement.

9.4   Governing Law. This Agreement shall be governed by the laws of the place of the Project, without giving effect to its conflict of law principles.

9.5   Severability. If any provision or any part of a provision of this Agreement shall be finally determined to be superseded, invalid, illegal, or otherwise unenforceable pursuant to applicable laws by any authority having jurisdiction, such determination shall not impair or otherwise affect the validity, legality, or enforceability of the remaining provisions or parts of the provision of this Agreement, which shall remain in full force and effect as if the unenforceable provision or part was deleted.

9.6   Amendments. This Agreement may not be changed, altered, or amended in any way except in writing signed by a duly authorized representative of both parties.

9.7   Entire Agreement. This Agreement forms the entire agreement between Owner and Design-Builder. No oral representations or other agreements have been made by the parties except as specifically stated in this Agreement.

9.8   Other Provisions. Other provisions, if any, are as follows:

In executing this Agreement, Owner and Design-Builder each individually represents that it has the necessary financial resources to fulfill its obligations under this Agreement, and each has the necessary corporate approvals to execute this Agreement, and perform the services described herein

OWNER:	DESIGN-BUILDER:
_____	_____
(Name of Owner)	*(Name of Design-Builder)*
_____	_____
(Signature)	*(Signature)*
_____	_____
(Printed Name)	*(Printed Name)*
_____	_____
(Title)	*(Title)*
Date: _____	Date: _____

Caution: You should sign an original DBIA document which has this caution printed in blue. An original assures that changes will not be obscured as may occur when documents are reproduced.

Index